Modern Vibrational Spectroscopy and Micro-Spectroscopy

Modern Vibrational Spectroscopy and Micro-Spectroscopy

Theory, Instrumentation and Biomedical Applications

MAX DIEM
Northeastern University, USA

WILEY

This edition first published 2015

Registered office
John Wiley & Sons Ltd, The Atrium, Southern Gate, Chichester, West Sussex, PO19 8SQ, United Kingdom

For details of our global editorial offices, for customer services and for information about how to apply for permission to reuse the copyright material in this book please see our website at www.wiley.com.

Library of Congress Cataloging-in-Publication Data

Diem, Max, 1947-
Modern vibrational spectroscopy and micro-spectroscopy : theory, instrumentation, and biomedical applications / Max Diem.
pages cm
Includes bibliographical references and index.
ISBN 978-1-118-82486-3 (pbk.)
1. Spectrum analysis–Diagnostic use 2. Vibrational spectra. I. Title.
RC78.7.S65D54 2015
616.07′54–dc23

2015019354

A catalogue record for this book is available from the British Library.

Set in 10/12pt TimesLTStd by SPi Global, Chennai, India

1 2015

I would like to dedicate this book to my wife, Mary Jo, who put up with my hiding away in my home office for many an evening and my absentmindedness for 16 months during which this book was written.

I also would like to acknowledge the excellent and enthusiastic crew of post-docs, graduate and undergraduate students in my research laboratory at North-eastern University who was responsible for the majority of the work presented here (listed chronologically).

Miloš, Christian, Susie, Melissa, Brian, Tatjana, Jen (I), Ben, Ellen, Kostas, Antonella, Evgenia, Erin, Jen (II), Christina, Kathleen, and Doug.

Max Diem

Contents

Preface

Although this book was conceived as a second edition of *Introduction to Modern Vibrational Spectroscopy* published by the author in 1993, it really is not a second edition, but a completely rewritten monograph on a subject that has changed so much in 20 years that the old edition appears seriously antiquated. In fact, few other areas of spectroscopy have undergone such radical changes in the past two decades as vibrational spectroscopy has: subjects that then were cutting edge technology have become so common that they have been part of undergraduate physical chemistry core laboratories for quite some time, and areas that were not even thought about 20 years ago are on the verge of commercialization. This enormous progress was spawned by a fortuitous co-incidence of many factors, such as instrumental advances that allow the collection of 10 000 infrared spectra within a few minutes, or the collection of Raman spectra of samples in the picogram regime in a few hundred milliseconds. The ready availability of pico- and femtosecond tunable laser sources has made routine acquisition possible of several nonlinear spectroscopic effects, based on excitation of short-lived vibrational states. Last not least, the enormous increase in computational power over the past 20 years has opened entirely new avenues for data processing and statistical analyses of the plethora of data collected in short times. In fact, the increased computational power is certainly one major enabling factor of an area of vibrational spectroscopic imaging, in which between 10 000 and 100 000 individual pixel spectra are collected through specialized microscopes and converted to pseudo-color images based on the vibrational spectroscopic features. This increase in computational power can easily be felt considering that in 1993, a top-end personal computer (PC) incorporated less than 1 MB of RAM, and was still based on Intel 386/387 processors and much of the scientific programming was carried out using FORTRAN compilers. Twenty years later, workstations with between 16 and 48 GB of RAM and multicore Pentium type processors are routine and available for a similar dollar amount required for the purchase of top-end PCs in 1993. The accessibility of thousands of routines for data processing and analysis, and being contained in the MATLAB environment, have offered sophisticated statistical and analysis routines even to the non-expert. Finally, theoretical developments have solved some puzzling aspects of vibrational spectroscopy; the prime example here is the understanding of the theoretical foundations of surface-enhanced Raman spectroscopy (SERS). In 1993, an overall agreement on the theoretical foundations of this effect had not been reached. Furthermore, the field seemed to be plagued by low reproducibility, which disappeared once the theory of surface enhancement was properly developed.

The need for inclusion of new techniques in vibrational spectroscopy, as well as the shifting research interests of the author, contributed to the very new format of this book, and the material contained herein. Furthermore, the strategies for teaching concepts of vibrational spectroscopy have changed both at the graduate and the undergraduate levels. Finally, the intended readership of this book has changed since the first edition addressed an audience of first year graduate students in chemistry while the present edition addresses, in addition, a readership interested primarily in the medical imaging and diagnostics fields. For these readers, the theory sections have been streamlined, and imaging aspects have been added. The first edition still contained a detailed discussion on how to perform empirical calculation of molecular force fields and vibrational frequencies. This subject has gone the path of dinosaurs, since such calculations are now all performed at the *ab initio* molecular orbital level. In addition, methods of data handling and analysis have been added that are necessary for both students and researchers in modern vibrational spectroscopy.

All these changes have forced a total rewriting of the book. Part I of the reworked monograph contains the theory of vibrational spectroscopy, presented both from the view of classical mechanics as well as from quantum mechanics. Although the later parts of the book deal mostly with large biological molecules, a short review of the group theoretical foundations of vibrational spectra of small molecules is included, as well as basic instrumental aspects. New techniques – surface and nanostructure-based spectroscopies, nonlinear effects, and time-resolved methods – are introduced as well. Thus, Part I represents a short course into "Modern Vibrational Spectroscopy" and is somewhat comparable to the first edition.

Part II deals with biophysical, medical, and diagnostic applications of vibrational spectroscopy. It starts with a review of the biophysical applications of macroscopic vibrational spectroscopy, a field that has produced ten thousands of papers on biomolecular structure, dynamics, and interactions. Subsequently, vibrational microspectroscopy (also referred to as vibrational microscopy) will be introduced. Although infrared and Raman microspectroscopy were certainly known and applied in 1993, their relevance was relatively low, and they were not included in the first edition. In contrast, both techniques are so prevalent now that they contribute to about 30% of sales in Raman and infrared spectral instrumentation. Both these techniques present their own challenges in instrumentation, data manipulation, and analysis, and are discussed in detail.

Vibrational microspectroscopy allows the detection and analysis of individual bacterial cells. In 1993, a few brave souls had embarked into this field and found that infrared spectra proved phenomenally sensitive in distinguishing different bacterial species [1]. The analysis of cells and tissue has undergone an explosive expansion during the past decade, and is now on the verge of becoming a major diagnostic and prognostic tool. Specialized journals, such as *Biospectroscopy* (now part of Biopolymers) and the *Journal of Biophotonics* are devoted to the application of (mostly) vibrational spectroscopy toward biological sciences and medicine. Many journals that used to concentrate on classical analytical chemistry have devoted entire issues to the emerging field of spectral diagnostics (see, for example, *Analyst*, Volume 135 and *J. Biophotonics*, Volumes 3 and 6).

Thus, the author hopes that this book will provide a detailed background, from the quantum mechanical foundation to the specific applications, for researchers in, or entering, the exciting and re-emerging field of vibrational spectroscopy.

In the years since the first edition was published, several new books have appeared discussing in detail several aspects of vibrational spectroscopy, such as D.A. Long's book on Raman theory [2], L.A. Nafie's book on Vibrational Optical Activity [3], volumes on Quantum Mechanics or Group Theory [4], monographs on microspectroscopy, and many more. The present book can impossibly compete with these specialized books on the details presented, since it still was conceived as an "Introduction" to modern vibrational spectroscopy. Therefore, the subjects discussed here are more appropriate for a researcher entering this field, or for advanced undergraduate students in Chemistry or the Life Sciences.

Finally, the author herewith apologizes categorically for one aspect in this book that is presented inconsistently throughout the chapters: the presentation of spectra from left to right and from right to left. Historically, Raman spectra have been presented mostly from "left to right," that is, from low to high wavenumber. Infrared spectra were originally presented from high wavenumber to low wavenumber, or from "right to left." Of course, this was due to the fact that the wavelength increases from "left to right" in this presentation. Although there are some recommendations by IUPAC on the representation of spectra – Raman spectra from "left to right" and infrared spectra from "right to left" – too many researchers have not abided by this rule; consequently, spectra are displayed in the literature both ways. Since many figures in this book are taken from the work of many researchers, these figures could not be reversed to a standard theme. Thus, the reader is reminded to pay particular attention to the direction of the wavenumber scale in the figures.

Boston, July 2014

References

1. Naumann, D., Fijala, V., Labischinski, H. and Giesbrecht, P. (1988) The rapid differentiation of pathogenic bacteria using Fourier transform infrared spectroscopic and multivariate statistical analysis. *J. Mol. Struct.*, **174**, 165–170.
2. Long, D.A. (2002) *The Raman Effect: A Unified Treatment of the Theory of Raman Scattering by Molecules*, John Wiley & Sons, Ltd, Chichester.
3. Nafie, L.A. (2011) *Vibrational Optical Activity: Principles and Applications*, John Wiley & Sons, Ltd, Chichester.
4. Cotton, F.A. (1990) *Chemical Applications of Group Theory*, 3rd edn, John Wiley & Sons, Inc., New York.

Preface to *Introduction to Modern Vibrational Spectroscopy* (1994)

The aim of this book is to provide a text for a course in modern vibrational spectroscopy. The course is intended for advanced undergraduate students, who have had an introductory course in Quantum Chemistry and have been exposed to group theoretical concepts in an inorganic chemistry course, or for graduate students who have passed a graduate level course in Quantum Chemistry.

There are probably a dozen or so recent books in vibrational spectroscopy, and a few classical texts over three decades old. This large number seems to discourage any efforts to produce yet another book on the subject, unless one is willing to pursue a novel approach in presenting the material. This is, of course, exactly what was attempted with the present book, and the approach taken will be outlined in the following paragraphs.

There are two classical and comprehensive texts on vibrational spectroscopy, *Molecular Spectra and Molecular Structure. II. Infrared and Raman Spectra of Polyatomic Molecules* by Herzberg [1] and *Molecular Vibrations* by Wilson *et al.* [2]. Both of these books are absolutely essential for an in-depth understanding of vibrational spectroscopy, and devote hundreds of pages to theoretical derivations. However, due to the rapid progress in instrumental techniques and computational methods and due to the fact that thousands of molecules have been studied since these two books were written, the practical aspects of these books are certainly limited. However, the value of these classic books for the serious vibrational spectroscopist is immeasurable, since they provide many of the fine points needed for a detailed understanding of the subject.

Among the more recent books, the reader will find either very specialized works dealing with one or a few specific topics of vibrational spectroscopy, or books that are more a compilation of data than a comprehensive text. The more practically oriented books often emphasize correlations of observed spectra with molecular structural features, and may contain large compilations of spectra and group frequencies, and only cursory treatment of theoretical principles. These books are essential for researchers who wish to employ vibrational spectroscopy as a qualitative structural tool.

However, neither of these could be used as a text book in a course, nor could they be used by a researcher who wants to gain insight into modern aspects of vibrational spectroscopy. Thus, the author was faced with the challenging task of writing a text that incorporates some theoretical background material which is necessary for the understanding of the principles of vibrational spectroscopy, in addition to computational methods, instrumental aspects, novel developments in vibrational spectroscopy, and a number of relatively detailed examples for the interpretations of vibrational spectra. Since the scope of this book is much broader than any of the aforementioned specialized texts, some of the theoretical material needed to be adjusted accordingly. Thus, the quantum mechanics of molecular vibrations, time-dependent perturbations, transition moments, and many other topics are only summarized in this text, and detailed derivations are omitted. For details, the reader is referred to specialized books, such as any one of the many available text books on Quantum Chemistry (see, for example, Levine, *Quantum Chemistry*, Volumes (I) and (II) [3]). For a detailed theoretical background on symmetry aspects, the classical book (Cotton, *Chemical Applications of Group Theory* [4]) is recommended, and the aforementioned books on vibrational spectroscopy for a more thorough treatment of theoretical aspects of molecular vibrations.

Thus, the present book does not supersede any of the classical texts, but is a further extension of them, and intends to bring the reader to a more practical and up-to-date level of understanding in the field of vibrational spectroscopy. Subjects such as Raman spectroscopy, which has become a major area of research in vibrational spectroscopy, is not treated as an afterthought, but experimental and theoretical aspects are discussed in detail. Items of historical significance, such as the Toronto arc for excitation of Raman spectra (which was actually mentioned as a viable source for Raman spectroscopy in a recent text), or the manual solution of the vibrational secular equation, have been banished from this book. Instead, modern experimental aspects, such as multi-channel Raman instrumentation, time-resolved and resonance Raman techniques, nonlinear Raman effects, and Fourier transform infrared and Raman techniques are introduced. In addition, computational methods for the calculation of normal modes of vibration are treated in detail in this book.

One chapter is devoted to the biological applications of vibrational spectroscopy. This is a rapidly developing field, and perhaps the most fascinating, for the molecules are often very large and difficult to study due to low solubility and solvent interference. It is in this area that the enormous progress of modern vibrational spectroscopy can best be gauged, since this field is not even discussed in the books of 30 or 40 years ago. The final chapter is devoted to a new branch of vibrational spectroscopy carried out with circularly polarized light. The new techniques introduced here combine principles of vibrational spectroscopy with those of optical activity measurements of chiral molecules. Applications of these techniques to biological molecules, and to simple chiral systems, are presented.

The author would like to thank his colleague, Prof. John Lombardi from the Department of Chemistry, City University of New York, City College, for his encouragement about this book, and for correcting a large number of errors in the original manuscript. The Graduate Spectroscopy class (U761) at the City University of New York in Spring, 1992, also was instrumental in pointing out inconsistencies in the manuscript, when an early version was used for the first time. The author is grateful for the input he received from these students.

New York, January 1993

References

1. Herzberg, G. (1945) *Molecular Spectra and Molecular Structure*, Van Nostrand Reinhold Company, New York.
2. Wilson, E.B., Decius, J.C. and Cross, P.C. (1955) *Molecular Vibrations*, McGraw-Hill.
3. Levine, I. (1970) *Quantum Chemistry*, vol. **1 and 2**, Allyn & Bacon, Boston.
4. Cotton, F.A. (1990) *Chemical Applications of Group Theory*, 3rd edn, John Wiley & Sons, Inc., New York.

Part I

Modern Vibrational Spectroscopy and Micro-spectroscopy: Theory, Instrumentation and Biomedical Applications

Introduction

I.1 Historical Perspective of Vibrational Spectroscopy

The subject of this monograph, vibrational spectroscopy, derives its name from the fact that atoms in molecules undergo continuous vibrational motion about their equilibrium position, and that these vibrational motions can be probed *via* one of two major techniques: infrared (IR) absorption spectroscopy and Raman scattering, and several variants of these two major categories. IR spectroscopy experienced a boon in the years of World War II, when the US military was involved in an effort to produce and characterize synthetic rubber. Vibrational spectroscopy, for which industrial application guides were published as early as 1944 [1], turned out to be a fast and accurate way to identify different synthetic products. The rapid growth of the field of vibrational spectroscopy can be gauged by the fact that by the mid-1940s, the field was firmly established as a scientific endeavor, and the volume "Molecular Spectra and Molecular Structure" by Herzberg (one volume of a trilogy of incredibly advanced treatises on molecular spectroscopy [2]) reported infrared and Raman results on hundreds of small molecules. Similarly, the earliest efforts to use infrared spectroscopy as a means to distinguish normal and diseased tissue – the subject of Part II of this book – were reported by Blout and Mellors [3] and by Woernley [4] by the late 1940s and early 1950s. In the 1960s, the petroleum industry added another industrial use for these spectroscopic techniques when it was realized that hydrocarbons of different chain lengths and degrees of saturation produced distinct infrared spectral patterns, and for the first time, computational methods to understand, reproduce, and predict spectral patterns were introduced [5]. By then, commercial scanning infrared spectrometers were commercially available.

The 1970 produced another boon when commercial Fourier transform (FT) infrared spectrometers became commercially available, and gas lasers replaced the mercury arc lamp as excitation source for Raman spectroscopy. Before lasers, Raman spectroscopy was a somewhat esoteric technique, since large sample volumes and a lot of time were required to collect Raman data with the prevailing Hg arc excitation sources. Yet, after the introduction of laser sources, Raman spectra could be acquired rapidly and as easily as infrared spectra. After the introduction and wide acceptance of interferometry, infrared spectroscopy became the method of choice for many routine and quality control applications. During the ensuing decade, the field of vibrational spectroscopy blossomed at a phenomenal rate, and it is safe to state that no other spectroscopy grew at such a pace than vibrational spectroscopy, perhaps with the exception of nuclear magnetic resonance techniques. Other spectroscopic methods, such as ultraviolet/visible (UV/vis), microwave, or electron paramagnetic resonance (EPR) spectroscopy certainly profited from theoretical and technical advances; however, in vibrational spectroscopy, the sensitivity of the measurements increased by orders of magnitude while the time requirements for data acquisition dropped similarly.

After the introduction of tunable, high power, and pulsed lasers, not only were faster and more sensitive techniques developed (for example, resonance Raman and time-resolved techniques), but also, entirely new spectroscopic methods were discovered, among them a number of "non-linear" Raman techniques (such as Hyper-Raman and Coherent Anti-Stokes Raman Scattering, CARS) in which the effect depends non-linearly on the laser field strength. Dramatic progress was also achieved in IR spectroscopy, due to the advent of infrared lasers and further refinements of interferometric methods.

Aside from small molecule applications of vibrational spectroscopy, the past three decades have seen an ever increasing use of vibrational spectroscopy in biophysical, biochemical, and biomedical studies. This field has enormously enhanced the ability to determine solution conformations of biological molecules, their interaction and even reaction pathways. All major classes of biological molecules, proteins, nucleic acids, and lipids, exhibit vibrational spectra that are enormously sensitive to structure and structural changes, and thousands of research papers have been published demonstrating the value of these methods for the understanding of biochemical processes.

With the advent of vibrational micro-spectroscopic instruments, that is, infrared or Raman spectrometers coupled to optical microscopes, cell biological and medical application of vibrational spectroscopy was practical since the spatial resolution in Raman and infrared absorption microscopy is sufficient to distinguish, by spectral features, parts of cells and tissue. These methods are poised to enter the medical diagnostic field as inherently reproducible and objective tools. This monograph will provide an introduction to many of these fields mentioned above.

I.2 Vibrational Spectroscopy within Molecular Spectroscopy

Molecular spectroscopy is a branch of science in which the interactions of electromagnetic radiation and matter are studied. While the theory of these interactions itself is the subject of ongoing research, the aim and goal of the discussions here is the elucidation of information on molecular structure and dynamics, the environment of the sample molecules and their state of association, interactions with solvent, and many other topics.

Molecular (or atomic) spectroscopy is usually classified by the wavelength ranges (or energies) of the electromagnetic radiation (e.g., microwave or infrared spectroscopies) interacting with the molecular systems. These spectral ranges are summarized in Table I.1.

In Table I.1, NMR and EPR stand for nuclear magnetic and electron paramagnetic resonance spectroscopy, respectively. In both these spectroscopic techniques, the transition energy of a proton or electron spin depends on the applied magnetic field strength. All techniques listed in Table I.1 can be described by absorption processes (see below) although other descriptions, such as bulk magnetization in NMR, are possible as well.

The interaction of the radiation with molecules or atoms that was referred to above as an "absorption process" requires that the energy difference between two (molecular or atomic) "stationary states" exactly matches the energy of the photon:

$$\Delta E_{\text{molecule}} = (h\nu)_{\text{photon}}$$

Table I.1 *Table of photon energies and spectroscopic ranges*

	ν_{photon}	λ_{photon}	E_{photon} (J)	E_{photon} (kJ mol^{-1})	E_{photon} (cm^{-1})	Transition
Radio	750 MHz	0.4 m	5×10^{-25}	3×10^{-4}	0.025	NMR
Microwave	3 GHz	10 cm	2×10^{-24}	0.001	0.1	EPR
Microwave	30 GHz	1 cm	2×10^{-23}	0.012	1	Rotational
Infrared	3×10^{13} Hz	10 µm	2×10^{-20}	12	1 000	Vibrational
UV-visible	10^{15}	300 nm	6×10^{-19}	360	30 000	Electronic
X-ray	10^{18}	0.3 nm	6×10^{-16}	3.6×10^{5}	3×10^{7}	X-ray absorption

where h denotes Planck's constant ($h = 6.6\times10^{-34}$Js) and ν the frequency of the photon, in s^{-1}. As seen from Table I.1, these photon energies are between 10^{-16} and 10^{-25} J/molecule or about 10^{-4} to 10^{5} kJ/(mol photons). In an absorption process, one photon interacts with one atom or molecule to promote it into a state of higher excitation, and the photon is annihilated. The reverse process also occurs where an atomic or molecular system undergoes a transition from a more highly excited to a less highly excited state; in this process, a photon is created. However, this view of the interaction between light and matter is somewhat restrictive, since radiation interacts with matter even if its wavelength is far different than the specific wavelength at which a transition occurs. Thus, a classification of spectroscopy, which is more general than that given by the wavelength range alone, would be a resonance/off-resonance distinction. Many of the effects described and discussed in spectroscopy books are observed as resonance interactions where the incident light, indeed, possesses the exact energy of the molecular transition in question. IR and UV/vis absorption spectroscopy, microwave spectroscopy, or EPR are examples of such resonance interactions. However, interaction of light and matter occurs, in a more subtle way, even if the wavelength of light is different from that of a molecular transition. These off-resonance interactions between electromagnetic radiation and matter give rise to well-known phenomena such as the refractive index of dielectric materials, and the anomalous dispersion of the refractive index with wavelength. The normal (non-resonant) Raman effect also is a phenomenon that is best described in terms of off-resonance models. A discussion of non-resonance effects ties together many well-known aspects of classical optics and spectroscopy.

This interplay between classical optics and spectroscopy will be emphasized throughout the book, and the study of vibrational or other fields of spectroscopy exposes students to a more unified picture of physical phenomena than individual courses in chemistry or physics provide. As such, vibrational spectroscopy has enormous pedagogical values, because it provides the link between different scientific fields: data can be collected easily by students in the laboratory, often on compounds synthesized or prepared by them. The spectral results are tangible, qualitatively interpretable and can be used to identify compounds. A deeper exposure to the material can be used to explain eigenvector/eigenvalue problems and demonstrate quantum mechanical principles such as allowed and forbidden transitions, breakdown of first order approximations and many more. Symmetry and group theory can be introduced logically when discussing vibrational spectroscopy, since the symmetry of atomic displacements during a normal mode of vibrations can be visualized easily and provide an intuitive approach for teaching the concepts of symmetry. Furthermore, vibrational spectroscopy is a useful probe for the structure of small molecules, because vibrational spectra can be predicted from symmetry considerations (group theory) and group frequencies. In fact, microwave (rotational) and vibrational spectroscopies were instrumental in determining the structure and symmetries of many small molecules. Rules for predicting the structures and shapes of small molecules, such as the VSEPR (Valence Shell Electron Pair Repulsion) model taught routinely in introductory chemistry classes are partially based on vibrational spectroscopic results of the 1950s and 1960s. This aspect of the importance of vibrational spectroscopy becomes apparent when one reads the chapter on "Individual Molecules" in Molecular Spectra and Molecular Structure [2] because in the 1940s, many structures were not known for certain. In fact, for many molecules that are now known to exhibit tetrahedral shapes, alternative structures were offered then, and the implications of other structures on the observed vibrational spectra were discussed.

Thus, vibrational spectroscopy cannot be regarded as a static field which has outlived its usefulness. Quite contrary, it is a very dynamic and innovative field, and there is no reason to believe that the progress in this field is slowing down. Thus, this book emphasizes many of these new developments. As mentioned in the foreword, this volume is not intended to replace the previous books on vibrational spectroscopy, but to combine in one volume the necessary background to understand vibrational spectroscopy, and to introduce the reader to the many new and fascinating techniques and results.

References

1. Barnes, R.B., Gore, R.C., Liddel, U., *et al.* (1944) *Infrared Spectroscopy. Industrial Applications and Biography*, Reinhold Publishing Co, New York.
2. Herzberg, G. (1945) *Molecular Spectra and Molecular Structure*, Van Nostrand Reinhold Co, New York.
3. Blout, E.R. and Mellors, R.C. (1949) Infrared spectra of tissues. *Science*, **110**, 137–138.
4. Woernley, D.L. (1952) Infrared absorption curves for normal and neoplastic tissues and related biological substances. *Cancer Res.*, **12**, 516–523.
5. Schachtschneider, J.H. (1965) *Vibrational Analysis of Polyatomic Molecules: FORTRAN IV Programs for solving the vibrational secular equation and least-square refinement of force constants*, The Shell Development Company, Emeryville, USA.

1

Molecular Vibrational Motion

The atoms in matter – be it in gaseous, liquid, or condensed phases – are in constant motion. The amplitude of this motion increases with increasing temperature; however, even at absolute zero temperature, it never approaches zero or perfect stillness. Furthermore, the amplitude of the atomic motion is a measure of the thermodynamic heat content as measured by the product of the specific heat times the absolute temperature. If one could observe the motion in real time – which is not possible because the motions occur at a timescale of about 10^{13} Hz – one would find that it is completely random and that the atoms are most likely to be found in ellipsoidal regions in space, such as the ones depicted in X-ray crystallographic structures. Yet, the random motion can be decomposed into distinct "normal modes of vibration." These normal modes can be derived from classical physical principles (see Section 1.2) and are defined as follows: in a normal mode, all atoms vibrate, or oscillate, at the same frequency and phase, but with different amplitudes, to produce motions that are referred as symmetric and antisymmetric stretching, deformation, twisting modes, and so on. In general, a molecule with N atoms will have $3N - 6$ normal modes of vibrational normal modes.

At this point, a discrepancy arises between the classical (Newtonian) description of the motion of atoms in a molecule and the quantum mechanical description. While in the classical description the amplitude of the motion, and thereby the kinetic energy of the moving atoms, can increase in arbitrarily small increments, the quantum mechanical description predicts that the increase in energy is quantized, and that infrared (IR) photons can be absorbed by a vibrating molecular system to increase the energy along one of the normal modes of vibration.

In the discussion to follow, the concepts of normal modes of vibration will be introduced for a system of spring-coupled masses, as shown in Figure 1.1, a typical mechanical model for a molecular system. Through a series of mathematical steps, the principle of normal modes will be derived from Newtonian laws of motion. Once this set of "normal modes" is defined, it is relatively trivial to extend these coordinates to a quantum mechanical description that results in the basic formalism for stationary vibrational states in molecules, and the transitions between these stationary states that are observed in IR and Raman spectroscopies.

Sections 1.1–1.7 are aimed at presenting a summary of the physical principles required for understanding the principles of vibrational spectroscopy. They do not present the subject with the mathematical rigor presented in earlier treatments, for example, in Wilson *et al.* [1] for the classical description of normal modes nor the quantum mechanical detail found in typical texts such as those by Kauzman [2] or Levine [3]. However, sufficient detail is provided to expose the reader to the necessary physical principles such as normal modes

Modern Vibrational Spectroscopy and Micro-Spectroscopy: Theory, Instrumentation and Biomedical Applications, First Edition.
Max Diem.

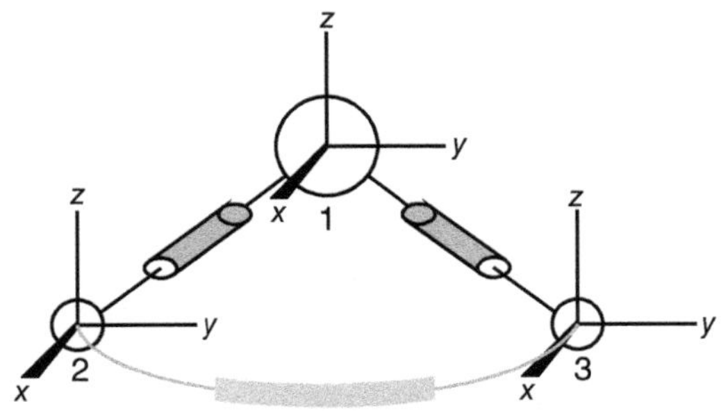

Figure 1.1 "Mass-and-spring" model of Cartesian displacement vectors for a triatomic molecule. The gray cylinders represent springs obeying Hook's law

of vibration and normal coordinates, the basic quantum mechanics of vibrating systems, and the transition moment, but in addition to the aforementioned texts, it will introduce the reader to many practical aspects of vibrational spectroscopy, as well as branches of vibrational spectroscopy that were not included in the earlier treatments.

For the remainder of this book, a standard convention for expressing vibrational energies in wavenumber units will be followed. Although energies should be expressed in units of Joule ($1\,J = kg\,m^2\,s^{-2}$), these numbers are unyielding, and wavenumber units are used throughout. The following energy unit conversions apply:

$$E = h\nu = \frac{hc}{\lambda} = h\,c\tilde{\nu}$$

Here, h has the value of 6.6×10^{-34} J s. Using $h\nu = hc\tilde{\nu}$, one finds that $1\,cm^{-1} \approx 30\,GHz = 2 \times 10^{-23}$ J (see also the table in the introduction and comments after Eqs. 1.31 and 1.63). Herewith, the author categorically apologizes for referring to transitions, expressed in wavenumber units, as "energies."

1.1 The concept of normal modes of vibration

Consider a set of masses connected by springs that obey Hook's law, as shown in Figure 1.1. Two of these springs act as restoring force when the "bonds" between atoms 1 and 2 and between atoms 1 and 3 are elongated or compressed, whereas one spring acts to restore the bond angle between atoms 2–1–3. Furthermore, we assume that the force required to move one atom along a coordinate depends on the momentary position of all other atoms: the force required to extend bond 1–2 may decrease if bond 1–3 is elongated. In a mechanical system, this situation is referred to as a coupled spring ensemble, where the stiffness of a spring depends on all coordinates. In a molecular system, this corresponds to electron rearrangement when the molecular shape changes, with a concomitant change in bond strength. In order to describe the atomic motions in a vibrating system, one attaches a system of Cartesian displacement coordinates to every atom, as shown in Figure 1.1. A normal mode of vibration then can be described as a combination of properly scaled displacement vector components.

1.2 The separation of vibrational, translational, and rotational coordinates

Based on Figure 1.1, one may expect a system of N atoms to exhibit $3N$ degrees of vibrational freedom: a degree of freedom for all three Cartesian displacement coordinates of each atom, or any linear combinations of them. However, vibrational spectroscopy depends on a restoring force to bring the atoms of a molecule back to their equilibrium position. If, for example, all atoms in a molecule move simultaneously in the x-direction, by the same amount, no bonds are being compressed or elongated. Thus, this motion is not that of an internal vibrational coordinate but that of a translation. There are three translational degrees of freedom, corresponding to a motion of all atoms along the x, y, or z axes by the same amount. Another view of the same fact is that these modes have zero frequencies as there is no restoring force acting during the atomic displacements. This view will be favored in the derivation of the concepts of normal modes presented in Section 1.3, which is carried out in (mass-weighted) Cartesian displacement coordinates. Later on, a different and simpler coordinate system will be introduced as well.

Similarly, one can argue that certain combinations of Cartesian displacements correspond to a rotation of the entire molecule, where there is no change in the intermolecular potential of the atoms. There are three degrees of rotational freedom of a molecule, corresponding to rotations about the three axes of inertia. Subtracting these from the remaining number of degrees of freedom, one arrives at $3N-6$ degrees of vibrational freedom for a polyatomic, nonlinear molecule. Linear molecules have one more ($3N-5$) degree of vibrational freedom, because they have only two moments of inertia. This is because one assumes a zero moment of inertia for a rotation about the longitudinal axis.

Mathematically, the separation of rotation and translation from the vibration of a molecule proceeds as follows: in order for the translational energy of the molecule to be zero at all times, one defines a coordinate system that translates with the molecule. In this coordinate system, the translational energy is zero by definition. The translating coordinate system is defined such that the center of mass of the molecule is at the origin of the coordinate system at all times. This leads to the condition

$$\sum\nolimits_{\alpha=1}^{N} m_\alpha \xi_\alpha = 0 \tag{1.1}$$

where ξ denotes the X, Y, and Z coordinates of the α'th atom of a molecule with N atoms. Similarly, the rotational energy can be reduced to zero by defining a coordinate system that rotates with the molecule. This requires that the angular moments of the molecule in the rotating coordinate frame are zero, which leads to three more equations of zero frequencies. These six equations are needed to define a coordinate system in which both translational and rotational energies are zero. Details of this derivation can be found in Wilson *et al.* [1, Chapters 2 and 11].

1.3 Classical vibrations in mass-weighted Cartesian displacement coordinates

The concept of normal modes of vibration, necessary for understanding the quantum mechanical description of vibrational spectroscopy and obtaining a pictorial description of the atomic motions, can be introduced by the previously described classical model of a molecule consisting of a number of point masses, held in their equilibrium positions by springs. This kind of discussion is treated in complete detail in the classic books on vibrational spectroscopy, for example, in Chapter 2 of Wilson *et al.* [1].

The treatment starts with Lagrange's equation of motion:

$$\frac{d}{dt}\frac{\partial T}{\partial \dot{x}_i} + \frac{\partial V}{\partial x_i} = 0 \tag{1.2}$$

where T and V are the kinetic and potential energies, respectively, the x_i are the Cartesian displacement coordinates, and the dot denotes the derivative with respect to time. Equation 1.2 is another statement of Newton's equation of motion (Eq. 1.3) expressed in terms of the kinetic and potential energies, rather than terms of force and acceleration:

$$F_i = m_i \frac{d^2 x_i}{dt^2} \tag{1.3}$$

In Eq. 1.3, F represents the force, which is related to the potential energy by

$$F_i = -\frac{dV}{dx_i} = kx_i \tag{1.4}$$

Equation 1.4 is Hook's law, which states that the force F needed to elongate or compress a spring depends on the spring's stiffness, expressed by the force constant k, multiplied by the elongation of the spring, x. The acceleration (d^2x_i/dt^2) in Eq. 1.3 can be related to the kinetic energy as follows:

$$T = \frac{1}{2}\left(\sum_i m_i \dot{x}_i^2\right) \tag{1.5}$$

Thus,

$$\frac{dT}{d\dot{x}_i} = m_i \dot{x}_i \tag{1.6}$$

Substituting Eqs. 1.4 and 1.6 into Eq. 1.3 yields Lagrange's equation of motion in which the expressions for kinetic and potential energies appear separately. This formulation is advantageous for writing the quantum mechanical Hamiltonian, *cf.* the following section.

Next, one rewrites Eq. 1.2 in terms of mass-weighted displacement coordinates, q_i. The reason for this is that the amplitude of a particle's oscillation depends on its mass. When mass-weighted coordinates are used, all amplitudes are properly adjusted for the different masses of the particles. In addition, the use of mass-weighted coordinates simplifies the formalism quite a bit. Let

$$q_i = \sqrt{m_i} x_i \tag{1.7}$$

Then, the kinetic energy can be written as

$$2T = \sum_i^{3N} (\dot{q}_i)^2 \tag{1.8}$$

Note that Eq. 1.8 contains only "diagonal terms"; that is, no cross terms q_{ij} appear in the summation. (The term "diagonal" here refers to a matrix notation to be introduced shortly.) Next, the potential energy of the particles needs to be defined. For masses connected by springs obeying Hook's law, one may assume that the potential energy along each Cartesian displacement coordinate is given by

$$V = \frac{1}{2}\sum_{i=1}^{3N}\sum_{j=1}^{3N}\left(\frac{\partial^2 V}{\partial q_i \partial q_j}\right) dq_i dq_j \tag{1.9}$$

or

$$V = \frac{1}{2}\sum_{i=1}^{3N}\sum_{j=1}^{3N} f_{ij} dq_i dq_j \tag{1.10}$$

with

$$f_{ij} = \left(\frac{\partial^2 V}{\partial q_i \partial q_j}\right) \tag{1.11}$$

The f_{ij} are known as mass-weighted Cartesian force constants and differ from the force constant k defined in Eq. 1.4 by the fact that latter does not explicitly contain the masses of the atoms. The constants f_{ij} express the change in potential energy as an atom or group is moved along the directions given by q_i and q_j. For small displacements about the equilibrium positions, Eq. 1.10 can be written as

$$2V = \sum_{i=1}^{3N}\sum_{j=1}^{3N} f_{ij}\, q_i q_j \tag{1.12}$$

In contrast to the kinetic energy expression in mass-weighted Cartesian coordinates (Eq. 1.8), the potential energy depends on diagonal ($f_{ii}q_i^2$) and off-diagonal ($f_{ij}q_iq_j$) terms as pointed out earlier. Taking the required derivatives and substituting the expressions for $\frac{d}{dt}\frac{\partial T}{\partial \dot{q}_i}$ and $\frac{\partial V}{\partial q_i}$ into Lagrange's equation of motion (Eq. 1.2) yields

$$\ddot{q}_i + \sum_{j=1}^{3N} f_{ij}\, q_j = 0 \tag{1.13}$$

Here, $\ddot{q}$ denotes the second derivative of q with respect to time. Equation 1.13 is a short form for a set of $3N$ simultaneous differential equations, with the index i running from 1 to $3N$. Note that the double summation in Eq. 1.12 disappears when the derivative with respect to one of the displacements, $\partial V/\partial q_i$, is taken. In expanded form, Eq. 1.13 can be presented as:

$$\begin{aligned}
\frac{d^2 q_1}{dt^2} + f_{11}q_1 + f_{12}q_2 + f_{13}q_3 + \cdots + f_{1,3N}q_{3N} = 0 \\
\frac{d^2 q_2}{dt^2} + f_{21}q_1 + f_{22}q_2 + f_{23}q_3 + \cdots + f_{2,3N}q_{3N} = 0 \\
\frac{d^2 q_{3N}}{dt^2} + f_{3N,1}q_1 + f_{3N,2}q_2 + f_{3N,3}q_3 + \cdots + f_{3N,3N}q_{3N} = 0
\end{aligned} \tag{1.14}$$

In each equation, only one term in the summation $\sum f_{ij}q_j$ has the same index as the term containing the time derivative. Thus, these equations can be simplified to read

$$\ddot{q}_i + f_{ii}q_i + C = 0 \tag{1.15}$$

where C is a constant. There are $3N$ solutions to these simultaneous, linear differential equations:

$$q_i = A_i \sin(\sqrt{\lambda} t + \varepsilon) \tag{1.16}$$

where the A_i are amplitude factors, ε are phase angles, and λ is a quantity related to the frequency and determined by the force constants (*cf.* below). Following standard practice in solving linear differential equations, one takes the solution given by Eq. 1.16, differentiates twice with respect to time,

$$\frac{d^2 q_i}{dt^2} = -\lambda q_i \tag{1.17}$$

and substitutes Eq. 1.15 back into Eq. 1.14. One obtains, after canceling the terms $\sin(\sqrt{\lambda} t + \varepsilon)$ from each equation:

$$\begin{aligned}
-A_1\lambda + f_{11}A_1 + f_{12}A_2 + f_{13}A_3 + \cdots + f_{1,3N}A_{3N} = 0 \\
f_{21}A_1 - A_2\lambda + f_{22}A_2 + f_{23}A_3 + \cdots + f_{2,3N}A_{3N} = 0 \\
f_{31}A_1 + f_{32}A_2 - A_3\lambda + f_{33}A_3 + \cdots + f_{3,3N}A_{3N} = 0 \\
\cdots
\end{aligned} \tag{1.18}$$

or

$$\begin{aligned}
(f_{11} - \lambda)A_1 + f_{12}A_2 + f_{13}A_3 + \cdots + f_{1,3N}A_{3N} = 0 \\
f_{21}A_1 + (f_{22} - \lambda)A_2 + f_{23}A_3 + \cdots + f_{2,3N}A_{3N} = 0 \\
f_{31}A_1 + f_{32}A_2 + (f_{33} - \lambda)A_3 + \cdots + f_{3,3N}A_{3N} = 0 \\
\cdots
\end{aligned} \tag{1.19}$$

Thus, $3N$ simultaneous homogeneous linear equations were obtained from $3N$ simultaneous linear differential equations. Homogeneous equations have two kinds of solutions. One of them is the so-called trivial solution in which all coefficient A_i are zero. This condition indeed fulfills Eq. 1.19 but is of no interest here because it implies that all particles are at rest; that is, there is no vibrational motion at all. The other solution for Eq. 1.19 is obtained when the determinant of the coefficients of A is zero in order for the left-hand side of Eq. 1.19 to be zero:

$$|(f_{ij} - \delta_{ij}\lambda)| = 0 \tag{1.20}$$

This is called the nontrivial solution, which requires

$$\begin{aligned} f_{11} - \lambda + f_{12} + f_{13} + \cdots + f_{1,3N} &= 0 \\ f_{21} + f_{22} - \lambda + f_{23} + \cdots + f_{2,3N} &= 0 \\ f_{31} + f_{32} + f_{33} - \lambda + \cdots + f_{3,3N} &= 0 \\ \cdots \end{aligned} \tag{1.21}$$

Equation 1.21 is known as the vibrational secular equation. The solution of this equation gives the eigenvalues λ, which are related to the vibrational frequencies for each of the normal modes. The amplitude factors A_i in Eq. 1.19 are not determined, but the relative magnitude of the displacement vectors can provide a view of the relative amplitudes of all atoms during a normal mode of vibration.

A normal mode of vibration is defined to by one of the $3N$ solutions of Eq. 1.21, where all atoms oscillate with the same frequency and in-phase but with different amplitudes. This definition is one of the most important ones in vibrational spectroscopy. It implies that all atoms are in motion during a normal mode of vibration, which is required to maintain the center of mass of the molecule. If $3N$ mass-weighted Cartesian displacement coordinates are defined, six rotational and translational modes will appear in these calculations as eigenvalues with zero frequencies, as discussed earlier. The displacement vectors will confirm that these motions are, indeed, translations and rotations.

Before continuing the discussion of the normal modes in polyatomic molecules, the simpler case of the vibration of diatomic molecules will be presented. For a diatomic molecule (which, of course, must be linear), Eq. 1.21 described a set of six equations, five of which have zero frequencies (namely the three translational and the two rotational coordinates). That leaves one equation,

$$f_{11} - \lambda = 0 \tag{1.22}$$

or $\lambda = f_{11}$. Next, it is instructive to visualize that Eq. 1.22 actually represents a vibrational frequency. For a diatomic molecule, the vibrational frequency can also be derived, starting with Newton's second law,

$$F = m\frac{d^2 x}{dt^2} \tag{1.3}$$

and assuming a harmonic restoring force obeying Hook's law:

$$F = -kx \tag{1.23}$$

Thus, one can write the equation of motion for a diatomic molecule as

$$\frac{d^2x}{dt^2} + \frac{k}{m}x = 0 \tag{1.24}$$

Here, k is the spring's (bond's) force constant, as discussed earlier (Eq. 1.4) that corresponds to the terms f_{ij} in Eq. 1.19, and m is the reduced mass defined as

$$m = \frac{m_1 m_2}{m_1 + m_2} \tag{1.25}$$

with m_1 and m_2 the individual masses of the two atoms.

One valid solution of the differential equation of motion (Eq. 1.24) is

$$x(t) = A\ \sin(\omega t + \varepsilon) \tag{1.26}$$

where

$$\omega = 2\pi = \sqrt{\frac{k}{m}} \tag{1.27}$$

where ω is the angular frequency and ε is a phase angle. Note that for a classical vibrational problem, the amplitude A is arbitrary, but that the frequency is defined by Eq. 1.25. This implies that for larger amplitudes, the velocity of the motion of the particles increases, but the frequency remains constant.

Rewriting Eq. 1.27 as

$$k = m\omega^2 \tag{1.28}$$

and comparing Eqs. 1.22 and 1.28, one finds that

$$\sqrt{\lambda} = \omega \tag{1.29}$$

when using mass-weighted Cartesian displacement coordinates. This relationship is true for diatomic and polyatomic molecules. At this point, a quick analysis of magnitudes and units is appropriate. The force constant k acting in a diatomic molecule such as gaseous H—Cl typically is about $500\ \mathrm{N\,m^{-1}} = 500\ \mathrm{kg\,s^{-2}}$, corresponding to a relatively stiff spring in classical mechanics. The reduced mass of an H—Cl molecule is, according to Eq. 1.25, approximately

$$m = 1.56 \times 10^{-27}\ \mathrm{kg} \tag{1.30}$$

and thus, the vibrational frequency for the H—Cl molecule is found to be

$$\begin{aligned} \nu &= \frac{1}{2\pi}\sqrt{\frac{k}{m}} = \frac{1}{2\pi}\sqrt{\frac{5.0 \times 10^2}{1.56 \times 10^{-27}}} = \frac{1}{2\pi}\, 5.67 \times 10^{14} \\ &= 9.0 \times 10^{13}((\mathrm{kg\,s^{-2}})/\mathrm{kg})^{1/2} = \mathrm{Hz} \end{aligned} \tag{1.31}$$

Using the frequency/wavenumber conversion $c\tilde{\nu} = \nu$ gives a value close to the observed stretching frequency for gaseous H—Cl of $3 \times 10^3\ \mathrm{cm^{-1}}$. When working in mass-weighted Cartesian displacement coordinates, the reduced mass in Eq. 1.27 disappears, and the frequency of the vibration is given by

$$\omega = \sqrt{\lambda} = \sqrt{f_{11}} \quad \text{or} \quad \nu = \frac{\sqrt{\lambda}}{2\pi} \tag{1.32}$$

Thus, Eq. 1.22, indeed, denotes the vibrational frequency of diatomic or polyatomic molecules.

Returning to the secular equation 1.21 for a triatomic molecule, one finds that one needs 81 force constants to describe the problem. Of these 81 force constants, only 9 are nonzero, because there are $3N-6$, or 3 degrees of vibrational freedom. However, it is clear that even this reduced problem cannot be solved, because there are three equations with nine unknowns. Symmetry arguments reduce this number of unknown force constants even further, but the number of unknowns still exceeds the number of equations. In the past, this problem was alleviated by transferring the diagonal force constants, that is, those with common indexes (f_{ii}) from similar molecules or from isotopic species, assuming that the diagonal force constants should not depend on isotopic substitution. Off-diagonal force constants were fitted to reproduce observed frequencies. More recently, all force constants – diagonal and off-diagonal – are determined computationally *via ab initio* methods (*cf.* Chapter 6). However, the meaning of the off-diagonal force constants needs to be pointed out in more detail. Referring to Figure 1.1, the diagonal force constants f_{11}, f_{22}, and f_{33} refer to the force required to move atom 1 in the x, y, and z directions, respectively. The off-diagonal constant f_{12} describes the force required to move atom 1 in the y direction after displacement in the x-direction.

To obtain vibrational frequencies and a depiction of the normal modes, the system of simultaneous homogeneous linear equations (Eq. 1.21) needs to be solved. As described earlier, values for force constants are substituted into the force constant matrix (also referred to potential energy matrix) that is subsequently diagonalized numerically according to:

$$\mathbf{L}^{\mathbf{T}}\, \boldsymbol{F}\, \mathbf{L} = \Lambda \tag{1.33}$$

Here, the matrix $\boldsymbol{F}$ corresponds to the Cartesian force constant defined in Eq. 1.21, and the eigenvector matrix $\boldsymbol{L}$ that diagonalizes the potential energy matrix is also the matrix that transforms from the mass-weighted Cartesian coordinate system to a new coordinate system Q, known as the "normal coordinates." These normal coordinates are defined such that each of the $3N-6$ normal modes of vibration is associated with one and only one normal coordinate Q. For the discussion of the relationship between mass-weighted displacement coordinates q and the normal coordinates Q, it is advantageous to cast the previously obtained equations into matrix notation. In the following discussion, bold quantities imply matrices or vectors. The two relationships derived earlier, expressing the potential and kinetic energies in terms of mass-weighted Cartesian coordinates, are written in matrix notations as

$$2T = \sum_{i}^{3N} (\dot{q}_i)^2 \quad \text{or} \quad \mathbf{2T} = \dot{\mathbf{q}}^{\mathbf{T}}\dot{\mathbf{q}} \tag{1.8}$$

and

$$2V = \sum_{i=1}^{3N} \sum_{j=1}^{3N} f_{ij}\, q_i q_j \quad \text{or} \quad \mathbf{2V} = \mathbf{q}^{\mathbf{T}}\, \boldsymbol{F}\mathbf{q} \tag{1.12}$$

Here, the superscript $\mathbf{T}$ denotes the transpose of a matrix; thus, the column vector $\dot{\mathbf{q}}$ becomes a row vector upon transposition. The dot implies, as before, the time derivative of the coordinates. $\boldsymbol{F}$ denotes the matrix of mass-weighted Cartesian force constants, as defined in Eq. 1.33.

Normal coordinates are defined such that

$$\mathbf{2T} = \dot{\mathbf{Q}}^{\mathbf{T}}\, \dot{\mathbf{Q}}$$
$$\text{and} \quad \mathbf{2V} = \mathbf{Q}^{\mathbf{T}}\Lambda\mathbf{Q} \tag{1.34}$$

In normal coordinate space, both the kinetic and potential energy matrices are diagonal. As the kinetic energy is diagonal in both q and Q space, the problem simplifies to finding the transformation (eigenvector) matrix $\boldsymbol{L}$ that diagonalizes the potential energy matrix $\boldsymbol{F}$. This matrix also transforms from the mass-weighted Cartesian displacement space into normal coordinate space according to

$$\mathbf{Q} = \mathbf{L}\,\mathbf{q} \tag{1.35}$$

Thus, the diagonalization of the potential energy matrix provides the vibrational frequencies of the system, according to

$$\omega_k = \sqrt{\lambda_k} \tag{1.36}$$

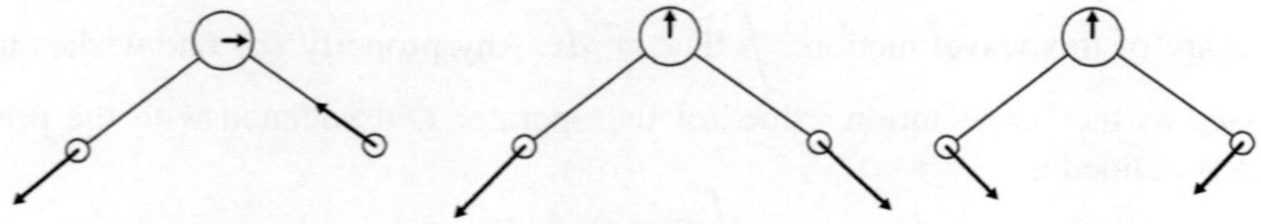

Figure 1.2 *Display of the atomic displacement vectors and the symmetries (see Chapter 2) for the three normal modes of the water molecule. The magnitude of the displacement vectors is not known, but the relative displacements are drawn approximately to scale. The terms A1, and B1 refer to the symmetry species of the coordinates (see Chapter 2).*

as well as the transformation to visualize each normal mode of vibration in terms of a normal coordinate and therewith, in terms of the displacement vectors. The displacement vectors for the normal modes of vibration of the water molecule are shown in Figure 1.2.

1.4 Quantum mechanical description of molecular vibrations

1.4.1 Transition from classical to quantum mechanical description

Next, a connection between the classical normal mode picture and the quantum mechanical description will be presented. The approach here starts with a simple case of a diatomic molecule for which the classical equation of motion was derived earlier (Eqs. 1.23–1.27) and for which the solution of the Schrödinger equation is relatively straightforward. Once the (one-dimensional) situation of the diatomic molecule has been introduced, the transition to polyatomic molecules is fairly simple because the concept of the normal coordinates can be used. As pointed out earlier, the classical description of the vibrations of a diatomic or polyatomic molecule predicts the vibrational coordinates (the normal modes) and their frequency, but not the transitions that are observed in IR absorption or Raman spectroscopy. In order to explain the observed spectra, quantum mechanics has to be invoked.

Quantum mechanics presents an approach to the behavior of microscopic particles very different from that in classical mechanics. While, in classical mechanics, the position and momentum of a moving particle can be established simultaneously, Heisenberg's uncertainty principle prohibits the simultaneous determination of those two quantities. This is manifested by Eq. 1.37 for the one-dimensional case:

$$\Delta p_x \Delta x \geq \frac{\hbar}{2} \tag{1.37}$$

which implies that the uncertainty in the momentum and position always exceeds $\hbar/2$. Mathematically, Eq. 1.37 follows from the fact that the operators responsible for defining position and momentum, $\hat{x}$ and $\hat{p}_x$, do not commutate; that is, $[\hat{x}, \hat{p}_x] \neq 0$. The incorporation of this uncertainty into the picture of the motion of microscopic particles leads to discrepancies between classical and quantum mechanics: classical physics has a deterministic outcome, which implies that if the position and velocity (trajectory) of a moving body are established, and it is possible to predict with certainty where it is going to be found in the future. Quantum mechanical systems obey a probabilistic behavior. As the position and momentum can never be determined at the starting point, the position (or momentum) in the future cannot be precisely predicted, only the probability of either of them. This is manifested in the postulate that all properties, present or future, of a particle are contained in a quantity known as the wavefunction Ψ of a system. This function, in general, depends on spatial coordinates and time; thus, for a one-dimensional motion (to be discussed first), the wavefunction is written as $\Psi(x, t)$. The probability of finding a quantum mechanical system is given by the integral of the square of this wavefunction: $\int \Psi(x,t)^2\,dx$. Any property $\langle o \rangle$ one wishes to observe for the system is expressed as the "expectation value" of the operator $\hat{O}$ associated with the property, where the expectation value is defined as

$$\langle o \rangle = \frac{\int \Psi(x,t)\ \hat{O}\ \Psi(x,t)dx}{\int \Psi(x,t)\ \Psi(x,t)dx} \tag{1.38}$$

As discussed earlier, a diatomic molecule possesses only one degree of vibrational freedom, the periodic elongation and compression of the bond connecting the two atoms that will be designated the x coordinate in the following discussion. Thus, the total energy, in analogy to Eq. 1.2, of an oscillating diatomic molecule can be written as the sum of kinetic energy T and potential energy V:

$$E = T + V \tag{1.39}$$

The kinetic energy is written in terms of classical physics as

$$T = \frac{1}{2}mv^2 \quad \text{or} \quad 2T = \frac{p^2}{2m} \tag{1.40}$$

where the momentum is given by $p = mv$. Here, v is the velocity and m is the reduced mass of the oscillating diatomic molecule, defined earlier (Eq. 1.25). In quantum mechanics, the classical momentum is substituted by the momentum operator $\hat{p}$,

$$\hat{p} = \frac{\hbar}{i}\frac{d}{dx} \tag{1.41}$$

where $\hbar$ is Planck's constant, divided by 2π, and i is the imaginary unit, defined by $\sqrt{-1} = i$. This substitution of the classical momentum by a differential operator is often considered the central postulate of quantum mechanics because it cannot be derived, although it can be visualized from the classical wave equation. Equation 1.41 is a mathematical instruction that requires taking the spatial derivative of the wavefunction $\hat{p}$ is operating on, and multiplying the results by $\hbar/i = -i\hbar$ to obtain the equivalent of the classical momentum. Examples of the use of such an operator are given in the following sections.

1.4.2 Diatomic molecules: harmonic oscillator

The potential energy for a diatomic vibrating system is discussed next. This potential function is shown schematically in Figure 1.3, and can be obtained by detailed quantum mechanical calculations, in which the electronic energy is computed as a function of the internuclear distance. This potential energy can be approximated by the Morse potential, given by

$$V(x) = D_e\,\{1 - e^{-a(x-x_0)}\}^2 \tag{1.42}$$

with

$$a = \sqrt{\frac{k}{2D_e}}\quad [(\mathrm{N\,m^{-1}/J})^{1/2} = \mathrm{m^{-1}}]$$

The function has a minimum at the bond equilibrium distance x_0. When compressing the bond beyond x_0, the potential energy rises sharply because of the repulsion of the two atoms. When the bond is elongated toward large interatomic distances, the potential function eventually levels out, and the bond breaks. One normally defines the potential energy at very large interatomic distances as the zero energy (no bonding interaction takes

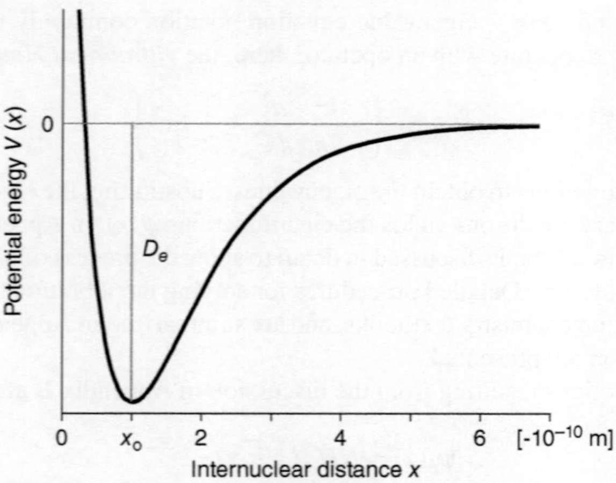

Figure 1.3 *Graph of the potential energy function for a diatomic molecule. Parameters are specified in Eq. 1.58*

place at large distances); thus, the potential energy of the bond is at a negative minimum at the equilibrium distance. The energy difference between zero potential energy and the minimum potential energy at point x_0 is referred as the bond dissociation energy, D_e.

Solving the quantum mechanical equations for the vibrations of a diatomic molecule with the potential function shown in Figure 1.3 would be difficult. Thus, one approximates the shape of the potential function in the vicinity of the potential energy minimum by a more simplistic function by expanding the potential energy $V(x)$ in a power series about the equilibrium distance:

$$V(x) = V(x_0) + (x - x_0)\left(\frac{dV}{dx}\right)_{x=x_0} + \frac{1}{2!}(x - x_0)^2\left(\frac{d^2V}{dx^2}\right)_{x=x_0} + \frac{1}{3!}(x - x_0)^3\left(\frac{d^3V}{dx^3}\right)_{x=x_0} + \cdots \quad (1.43)$$

$V(x_0)$ is an offset along the Y-axis and does not affect the curvature of the potential energy. The term containing the first derivative of the potential energy with respect to x is zero because the equilibrium geometry corresponds to an energy minimum. Terms higher than the quadratic expression in Eq. 1.43 are ignored at this point. Thus, one approximates the potential energy $V(x)$ by

$$V = (x - x_0)^2\left(\frac{d^2V}{dx^2}\right)_{x=x_0} \approx \frac{1}{2}kx^2 \quad (1.44)$$

which also could have been obtained by integrating

$$F = -kx \quad (1.23)$$

for a system obeying Hook's law.

Thus, to a first approximation, one assumes that the chemical bond in a diatomic molecule obeys Hook's law, just as the motion of two spring-coupled masses. Combining Eqs. 1.39, 1.41, and 1.43, the vibrational Schrödinger equation for a two-particle system with one degree of freedom (x) is then:

$$\left\{-\frac{\hbar^2}{2m}\frac{d^2}{dx^2} + \frac{1}{2}kx^2\right\}\psi(x) = E\psi(x) \quad (1.45)$$

This differential equation is known as "Hermite's" differential equation, in which the wavefunctions $\psi(x)$ are the time-independent (stationary-state) vibrational wavefunctions, and E denotes the vibrational eigenvalues.

Equation 1.45 is a typical operator – eigenvalue equation notation commonly used in linear algebra. This formalism is an instruction to operate with an operator, here, the *vibrational Hamiltonian* $\hat{H}_{\text{vib}}$

$$\hat{H}_{\text{vib}} = \left\{ -\frac{\hbar^2}{2m}\frac{d^2}{dx^2} + \frac{1}{2}kx^2 \right\} \tag{1.46}$$

on a set of (yet unknown) functions to obtain the eigenvalues. Substituting the eigenvalues into a trial solution and considering the boundary conditions yields the eigenfunctions $\psi(x)$. In Appendix A, a much simpler system, the so-called particle-in-a-box, is discussed in detail to show the process of obtaining the eigenvalues and eigenfunctions in a model system. Detailed procedures for solving the vibrational Schrödinger equation 1.46 can be found in most quantum chemistry textbooks, and are summarized in Appendix B. Here, only the results of the mathematical solution are presented.

The vibrational wavefunctions resulting from the discussion in Appendix B are of the form

$$\psi_n(x) = NH_n\,(\sqrt{\alpha}\,x)\,e^{-\frac{\alpha x^2}{2}} \tag{1.47}$$

where N is a normalization constant, $N = \left(\frac{\alpha}{\pi}\right)^{1/4}$ and $H_n(x)$ are the Hermite polynomials of order n in the variable x, n is an integer, the vibrational quantum number, which may take values from zero to infinite, and

$$\alpha = \frac{2\pi \nu m}{\hbar} \tag{1.48}$$

All other symbols have their usual meaning. Setting $z = (\sqrt{\alpha}\,x)$, the Hermite polynomials in the variable z are

$$\begin{aligned} H_0(z) &= 1 \\ H_1(z) &= 2z \\ H_2(z) &= 4z^2 - 2 \\ H_3(z) &= 8z^3 - 12z \end{aligned} \tag{1.49}$$

The order n of the Hermite polynomial determines the highest power in which the variable z occurs in each polynomial. Thus, more highly excited states (i.e., those with higher quantum number n) correspond to Hermite polynomials with higher power of z. This aspect is important because the power of z determines the shape of the wavefunctions.

The higher members of the Hermite polynomials can be derived from the recursion formula:

$$z\,H_n(z) = nH_{n-1}(z) + \frac{1}{2}\,H_{n+1}(z) \tag{1.50}$$

Thus, the Hermite polynomial of degree n is related to the previous and subsequent polynomials. The first few vibrational wavefunctions thus are:

$$\begin{aligned} \psi_0(x) &= \left(\frac{\alpha}{\pi}\right)^{1/4} e^{-\frac{\alpha x^2}{2}} \\ \psi_1(x) &= \left(\frac{\alpha}{\pi}\right)^{1/4} 2\,(\sqrt{\alpha}x)\,e^{-\frac{\alpha x^2}{2}} \\ \psi_2(x) &= \left(\frac{\alpha}{\pi}\right)^{1/4} (4\alpha x^2 - 2)\,e^{-\frac{\alpha x^2}{2}} \\ &\ldots \end{aligned} \tag{1.51}$$

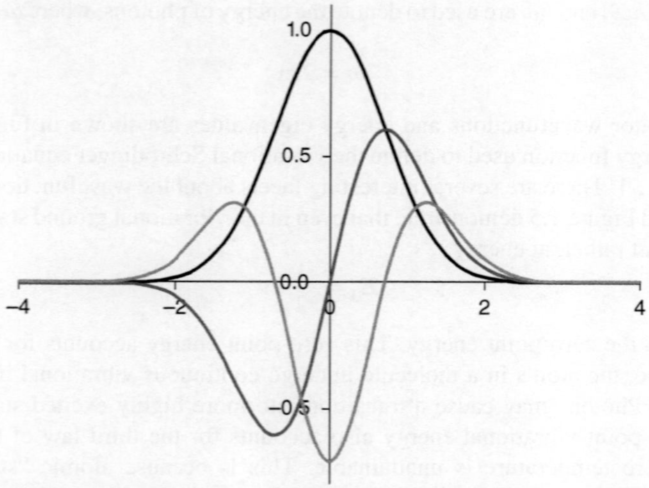

Figure 1.4 *Plot of the first three vibrational wavefunctions, which are products of a polynomial function that becomes very large with increasing x and a Gaussian function that decreases exponentially with x, resulting in the depicted curves for n = 0 (black), n = 1 (dark gray), and n = 2 (light gray)*

These functions, which are shown in Figure 1.4, form an orthonormal vector space. Orthonormality implies that

$$\int_{-\infty}^{\infty} \psi_i(x)\psi_j(x)\, dx = \delta_{ij} = \begin{cases} 1 & \text{if } i = j \\ 0 & \text{if } i \neq j \end{cases} \tag{1.52}$$

Here, δ_{ij} represents the Kronecker delta, which equals to one if $i = j$ and zero otherwise.

For example,

$$\int_{-\infty}^{\infty} \psi_0(x)\psi_0(x)\, dx = \sqrt{\frac{\alpha}{\pi}} \int_{-\infty}^{\infty} e^{-\alpha x^2} dx = \sqrt{\frac{\alpha}{\pi}}\sqrt{\frac{\pi}{\alpha}} = 1 \quad \text{and}$$

$$\int_{-\infty}^{\infty} \psi_0(x)\psi_1(x)\, dx = \sqrt{\frac{\alpha}{\pi}} \int_{-\infty}^{\infty} 2\,(\sqrt{\alpha}x)e^{-2\alpha x^2} dx = 0 \tag{1.53}$$

In Eq. 1.53, the integral relationship $\int_{-\infty}^{\infty} e^{-\alpha x^2} dx = \sqrt{\frac{\pi}{\alpha}}$ was used.

The orthogonality of the vibrational wavefunctions can also be demonstrated by graphical integration, as demonstrated in Section 1.5.

The eigenvalues of the vibrational Schrödinger equation (Eq. 1.46) are given by

$$E_n = \left(n + \frac{1}{2}\right) h\nu \tag{1.54}$$

In Eq. 1.54, the frequency ν is written in units of s^{-1} such that the term $h\nu$ has units of energy (J). Vibrational spectroscopists, however, prefer to use wavenumber as a unit of energy; thus, in the remainder of the book the vibrational energy is expressed as

$$E_n = \left(n + \frac{1}{2}\right) hc\tilde{\nu} \tag{1.55}$$

and the expressions $h\nu$, $hc\tilde{\nu}$, and $\hbar\omega$ are used to denote the energy of photons, where ω is the angular frequency, defined by

$$\omega = 2\pi\nu \tag{1.56}$$

The harmonic oscillator wavefunctions and energy eigenvalues are shown in Figure 1.5, along with the quadratic potential energy function used to define the vibrational Schrödinger equation, and the energy levels corresponding to Eq. 1.54. There are several interesting facets about the wavefunctions and energy level plot. First of all, Eq. 1.54 and Figure 1.5 demonstrate that even in the vibrational ground state with $n = 0$, the system is not at zero energy, but rather, at energy

$$E_0 = \frac{1}{2} h\nu \tag{1.57}$$

which is referred to as the zero-point energy. This zero-point energy accounts for the fact that even in its vibrational ground state, the atoms in a molecule undergo continuous vibrational motion along the normal modes derived earlier. Photons may cause a transition into more highly excited states of these vibrational coordinates. This zero-point vibrational energy also accounts for the third law of thermodynamics, which states that absolute zero temperature is unattainable. This is because atomic "stand-still" is impossible because of the residual vibrational energy. This is also in line with Heisenberg's uncertainty principle (Eq. 1.37), because a vibrational amplitude of zero would define both position and momentum simultaneously. Figure 1.5 also indicates some degree of "tunneling," or a finite probability of the oscillating system to be found outside the potential energy curve. Furthermore, the wavefunctions are symmetric ($n = 0, 2, 4, \ldots$) or antisymmetric ($n = 1, 3, 5, \ldots$) with respect to the X_0 line. This aspect will become particularly important in the discussion of the allowed and forbidden transitions in the harmonic oscillator approximation. Finally, the quadratic potential depicted in Figure 1.5 would not explain bond breakage at sufficiently high energy, because the potential function – the restoring force between the oscillating energy increases steadily in the "harmonic approximation." Therefore, the concept of an anharmonic potential needs to be introduced.

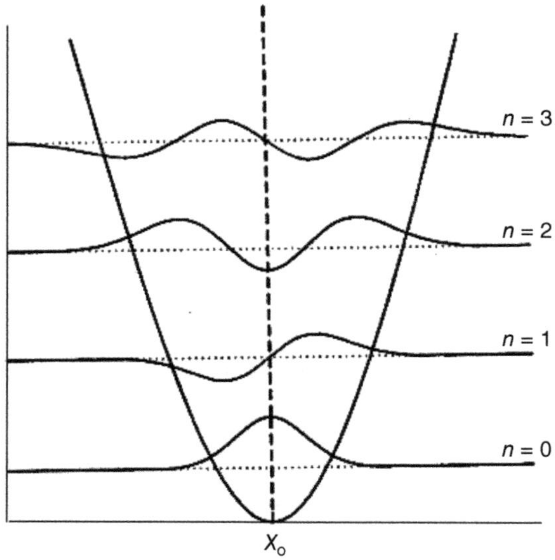

Figure 1.5 *Quadratic potential energy function* $V = 1/2kx^2$ *for a diatomic molecule and the resulting quantum mechanical vibrational wavefunctions*

1.4.3 Diatomic molecules: anharmonicity

The potential energy function used so far, $V = \frac{1}{2}k\,x^2$ (Eq. 1.44), is a gross approximation of the real potential energy function shown in Figure 1.3, and therefore, needs to be modified. The next level of approximation involves the inclusion of higher terms in the original expression (Eq. 1.43) for the potential energy:

$$V(x) = \frac{1}{2!}\,(x - x_0)^2\left(\frac{d^2V}{dx^2}\right)_{x=x_0} + \frac{1}{3!}\,(x - x_0)^3\left(\frac{d^3V}{dx^3}\right)_{x=x_0} \tag{1.58}$$

Using the same simplification of the series expansion of the potential energy and setting the cubic force constant

$$k' = \left(\frac{d^3V}{dx^3}\right)_{x=x_0} \tag{1.59}$$

one can write the anharmonic vibrational Schrödinger equation as

$$\left\{-\frac{\hbar^2}{2m}\frac{d^2}{dx^2} + \frac{1}{2}\,kx^2 + \frac{1}{6}\,k'x^3\right\}\psi(x) = E\psi(x) \tag{1.60}$$

Equation 1.60 is solved by perturbation methods and yields the perturbed energy eigenvalues for a diatomic molecule

$$E_n = \left(n + \frac{1}{2}\right)h\nu - \left(n + \frac{1}{2}\right)^2 h\nu\chi = \left(n + \frac{1}{2}\right)h\nu - \frac{\left(n + \frac{1}{2}\right)^2 h^2\nu^2}{4D_e} \tag{1.61}$$

with

$$\chi = \frac{h\nu}{4D_e} \tag{1.62}$$

As χ always is a positive number, the energy levels of the anharmonic case are always lowered as compared to the harmonic oscillator; this lowering increases with the square of the quantum number n. This is depicted in Figure 1.6, which shows a comparison between harmonic and anharmonic oscillator energy levels.

In addition to lowering the energy values, the wavefunctions will be shifted toward longer internuclear distances; consequently, the symmetry of the wavefunctions changes and they are no longer symmetric or antisymmetric with respect to the x_0 position. This fact changes the selection rules that determine which transitions are allowed for diatomic molecules.

At this point, some calculations and examples are presented in order to transmit a feeling for magnitudes involved. For a common chemical moiety, the bond energy is typically around 1000 kJ mol^{-1}. Expressing this quantity in molecular, rather than molar units (dividing by Avogadro's number), one finds that the bond dissociation energy D_e used in Eq. 1.42 is

$$D_e = 1.5 \times 10^{-18}\ \text{J/molecule} = 75{,}000\ \text{cm}^{-1} \tag{1.63}$$

In Eq. 1.63, the energy conversion $E = hc\tilde{\nu}$ or $1\ \text{J} = 6.6 \times 10^{-34} \times 3 \times 10^{10} \approx 2 \times 10^{-23}\ \text{cm}^{-1}$ was used with the velocity of light c expressed in cm s^{-1}. Vibrational force constants, as defined in Eq. 1.31, are found to be around 500 N m^{-1} for a typical chemical bond. Equation 1.42 was used with

$$a = 1 \times 10^{10}\ \text{m}^{-1} \quad \text{and} \quad x_0 = 100\ \text{pm} = 10^{-10}\ \text{m}$$

to calculate the potential energy curve shown in Figure 1.3. As pointed out earlier, a typical vibrational energy for a diatomic molecules with a force constant of 500 N m^{-1} is about 3000 cm^{-1}; thus, one finds that the spacing between (harmonic) vibrational energy levels is only about 1/25th of the depth of the potential well for a strong bond. Thus, many vibrational quanta of light are necessary to raise the energy of a bond

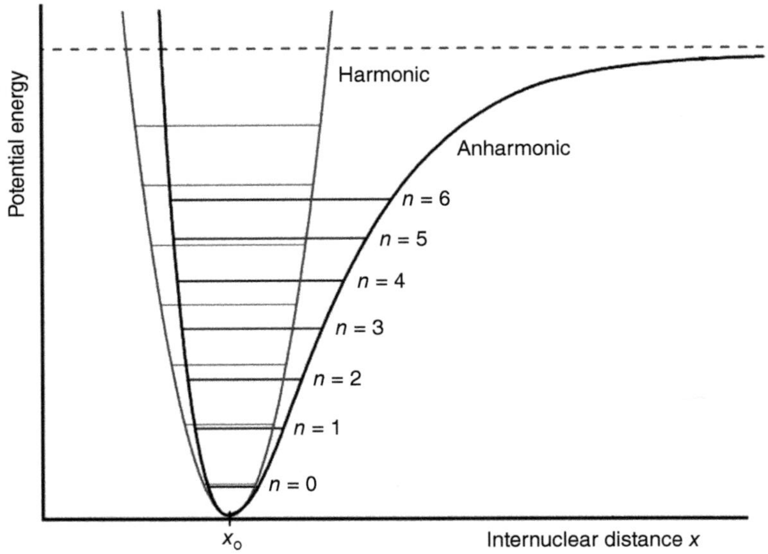

Figure 1.6 *Comparison of energy levels for harmonic and anharmonic oscillators*

stretching vibration such that bond dissociation occurs. However, bond breakage virtually never happens through consecutive absorption of IR photons because of several factors, among them are the short lifetime of vibrational states and the fast (nonradiative) dissipation of the vibrational energy into other vibrational modes. In addition, because the spacing between vibrational levels gets smaller for real molecular systems owing to the anharmonicity of the potential function, photons of different energy would be needed to raise the energy to the higher levels through consecutive absorption processes. On the other hand, the process of raising the vibrational energy from the ground state directly to an excited state near the dissociation limit is low because of the small transition moment. Thus, one concludes that bond breakage occurs mostly through mechanisms different from direct vibrational excitation.

In summary, diatomic molecules were described in both the harmonic and anharmonic approximations. The vibrational wavefunctions defined in Eq. 1.51, and depicted in Figure 1.5, represent the time-independent or stationary-state vibrational wavefunctions for a diatomic molecule. The stationary-state wavefunctions of the particle-in-a-box, discussed in Appendix A, have similar appearance and symmetries, albeit completely different equations. The symmetry of wavefunctions plays an important role determining whether transitions from one to another state are allowed or forbidden. It was pointed out earlier that the unperturbed harmonic oscillator wavefunctions are symmetric with respect to X_0 (*cf.* Figure 1.5) for even quantum numbers n and antisymmetric for odd numbers of n. Therefore, one often refers to "even" and "odd" parity when describing these wavefunctions, and it is easy to see that the particle-in-a-box wavefunctions follow a similar even/odd pattern as do the harmonic oscillator wavefunctions. Furthermore, as the symmetry of the wavefunctions is distorted in the anharmonic approximation, the selection rules will change drastically.

1.4.4 Polyatomic molecules

Section 1.3 demonstrated how the treatment of diatomic molecules could be extended toward polyatomic molecules, and introduced the concepts of normal modes of vibration and normal coordinates for the classical description of the vibrations of polyatomic molecules. In complete analogy, the quantum mechanical treatment can be extended from the case of diatomic molecules toward polyatomic molecules.

For a diatomic molecule, the vibrational Schrödinger equation was found to be

$$\left\{-\frac{\hbar^2}{2m}\frac{d^2}{dx^2}+\frac{1}{2}kx^2\right\}\psi(x)=E\psi(x) \tag{1.45}$$

In analogy, the vibrational Schrödinger equation for a polyatomic molecule, written in terms of the normal coordinates (*cf.* Eq. 1.34), is

$$-\frac{\hbar^2}{2}\frac{d^2\psi_{\text{vib}}}{dQ^2}+\frac{1}{2}\Lambda Q^2\psi_{\text{vib}} = E_{\text{vib}}\psi_{\text{vib}} \tag{1.64}$$

where ψ_{vib} is the total vibrational wavefunction of the molecule. Next, one writes ψ_{vib} as a product of the wavefunctions along each of the $3N-6$ normal coordinates:

$$\psi_{\text{vib}} = \psi_1(Q_1)\cdot\psi_2(Q_2)\cdot\psi_3(Q_3)\cdots \tag{1.65}$$

This definition of the total vibrational wavefunction as products of wavefunctions associated with one and only one normal coordinate succeeds because the expressions for kinetic and potential energies are both diagonal in normal coordinate space (*cf.* Eqs. 1.33 and 1.34). Substitution of Eq. 1.65 into Eq. 1.64 yields the Schrödinger equation in terms of the $3N-6$ normal coordinates:

$$-\frac{\hbar^2}{2}\sum_{k=1}^{3N-6}\frac{d^2\psi_k}{dQ_k^2}+\frac{1}{2}\sum_{k=1}^{3N-6}\Lambda_k Q_k^2\psi_k = E_{\text{vib}}\psi_k \tag{1.66}$$

As the normal coordinates Q_k are orthogonal functions, and because – in normal coordinate space – the kinetic and potential energies are in diagonal form (see Eqs. 1.33 and 1.34), Eq. 1.66 can be separated into $3N-6$ individual differential equations of the form

$$-\frac{\hbar^2}{2}\frac{d^2\psi_k}{dQ_k^2}+\frac{1}{2}\Lambda_k Q_k^2\psi_k = E_k\psi_k \tag{1.67}$$

with the energy eigenvalues (in wavenumber units) given by

$$E_k(n)=\left(n_k+\frac{1}{2}\right)hc\tilde{\nu}_k \tag{1.68}$$

and

$$E_{\text{vib}}=\sum_{k=1}^{3N-6}E_k \tag{1.69}$$

In analogy with the diatomic molecules, one writes the solutions of Eq. 1.67 as the eigenfunctions

$$\psi_n(Q_k)=NH_n(\sqrt{\alpha}\,Q_k)\,e^{-\frac{\alpha Q_k^2}{2}} \tag{1.70}$$

The results of this quantum mechanical discussion are shown in Figure 1.7 for a triatomic molecule, such as water. Here, the vibrational frequencies associated with each normal coordinate are approximately given by

$$\tilde{\nu}_1(Q_1)=3750\ \text{cm}^{-1}\quad \tilde{\nu}_2(Q_2)=3650\ \text{cm}^{-1}\quad \tilde{\nu}_3(Q_3)=1620\ \text{cm}^{-1}$$

The energy level diagram, in the harmonic approximation, consists of three energy ladders (see Figure 1.7), one for each normal coordinate. Note that each ladder starts at

$$E_k(0)=\left(\frac{1}{2}\right)hc\tilde{\nu}_k \tag{1.71}$$

that is, at approximately 1875, 1825, and 810 cm^{-1} for Q_1, Q_2, and Q_3, respectively.

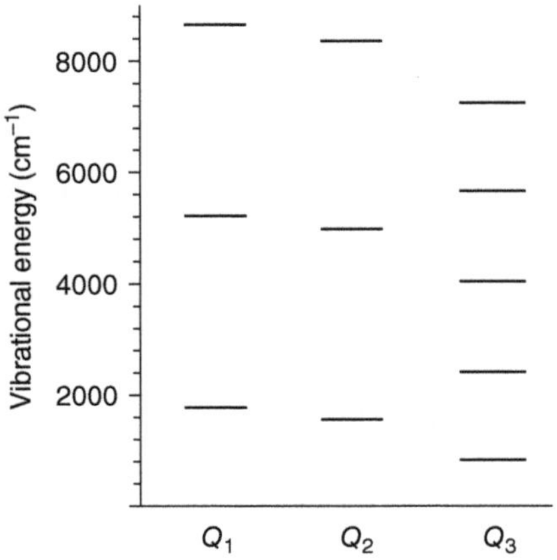

Figure 1.7 *Energy ladder diagram for a triatomic molecule within the harmonic oscillator approximation*

Thus, the total zero-point vibrational energy of water is given by $E_0^T = E_0(Q_1) + E_0(Q_2) + E_0(Q_3) = 1875 + 1825 + 810$ or approximately $4510\,\text{cm}^{-1}$. In any of the energy ladders, a transition to the more highly excited vibrational state may occur if a photon of proper wavenumber (3750, 1650, or $1620\,\text{cm}^{-1}$) interacts with the sample and certain requirements are fulfilled. Whether or not a transition occurs does not solely depend on the presence of photons with the correct energy, but also on some symmetry considerations. In the following section, the basic formalism is introduced that describe the interaction of light with molecular systems in their stationary energy states. This interaction is based on time-dependent processes and requires the use of time-dependent wavefunctions and the time-dependent Hamiltonian.

1.5 Time-dependent description and the transition moment

1.5.1 Time-dependent perturbation of stationary states by electromagnetic radiation

Time-independent quantum mechanics introduced in the previous sections describes the energy expressions and wavefunctions of stationary states, that is, states that do not change with time. Stationary-state wavefunctions and energies were obtained by solving the appropriate Schrödinger equations for the given system. Next, a description is needed that describes how a system can transition from one stationary state to another when a perturbation, typically electromagnetic radiation, is applied. For this, one needs to invoke the time-dependent Schrödinger equation, which for the one-dimensional case is

$$i\hbar \frac{\partial \boldsymbol{\Psi}(x,t)}{t} = \mathbf{H}\boldsymbol{\Psi}(x,t) \tag{1.72}$$

Equation 1.72 is solved by perturbation methods in which a perturbation operator H' is introduced

$$\mathbf{H} = H + H' \tag{1.73}$$

One assumes that there exist exact eigenfunctions for the operator H, and that the perturbation due to H' is small. If the perturbation applied to the system is due to electromagnetic radiation, one may write the perturbation operator as

$$H'(t) = -\mathbf{E}_x^0 \sum\nolimits_i e_i x_i \left\{ \cos\left(2\pi\nu t + \frac{2\pi z}{\lambda}\right)\right\} \tag{1.74}$$

In Eq. 1.74, electromagnetic radiation propagating in the z-direction is assumed, with the electric vector along the x-direction. The electric field

$$\mathbf{E}_x = \mathbf{E}_x^0 \left\{ \cos\left(2\pi\nu t + \frac{2\pi z}{\lambda}\right) \right\} \tag{1.75}$$

will exert a force

$$\mathbf{F} = e\mathbf{E} \tag{1.76}$$

on particles with charge e. If the molecular system consists of i charged particles found at positions x_i, one defines the "electric dipole operator" $\boldsymbol{\mu}$ according to

$$\boldsymbol{\mu} = \sum_i e_i x_i \tag{1.77}$$

and rewrites Eq. 1.75 as

$$H'(t) = -\mathbf{E}_x^0 \,\boldsymbol{\mu} \left\{ \cos\left(2\pi\nu t + \frac{2\pi z}{\lambda}\right) \right\} \tag{1.78}$$

The time-dependent Schrödinger equation

$$i\hbar \frac{\partial \boldsymbol{\Psi}(x,t)}{\partial t} = \left[H - \mathbf{E}_x^0 \boldsymbol{\mu} \left\{ \cos\left(2\pi\nu t + \frac{2\pi z}{\lambda}\right) \right\} \right] \boldsymbol{\Psi}(x,t) \tag{1.79}$$

subsequently is solved with the unperturbed eigenfunctions ψ of the H operator

$$H\psi(x) = E\psi(x) \tag{1.80}$$

which may be, for example, the time-independent (unperturbed) wavefunctions of the harmonic oscillator. The time dependence of each of the wavefunctions $\psi(x)$ is introduced as follows:

$$\Psi(x,t) = \varphi(t)\psi(x) \tag{1.81}$$

where

$$\varphi(t) = e^{i\omega t} \tag{1.82}$$

Equation 1.81 is the general definition of a time-dependent wavefunction that consists of a stationary, time-independent part, $\psi(x)$ and the time evolution of this wavefunction, given by $\varphi(t) = e^{i\omega t}$.

The time-dependent wavefunctions $\boldsymbol{\Psi}(x,t)$ of the system undergoing a transition subsequently are expressed in terms of time-dependent coefficients $c_k(t)$ and the time-dependent wavefunctions $\boldsymbol{\Psi}(x,t)$ according to

$$\boldsymbol{\Psi}(x,t) = \sum_k c_k(t)\Psi(x,t) = \sum_k c_k(t)\varphi(t)\psi(x) \tag{1.83}$$

The coefficients $c_k(t)$ describe the time-dependent response of the quantum mechanical system to the perturbation; in particular, the change in population of the excited state in response to the perturbation. An example may serve to illustrate the procedure invoked so far. Consider a two-state system in the absence of a perturbation, where the splitting between the ground-state energy level $\psi_g(x)$ and the excited-state energy level $\psi_e(x)$ is much larger than thermal energy (see Eqs. 1.101–1.103).

Thus, the system is in the ground state, and can be described by $c_g(t) = 1$ and $c_e(t) = 0$ or

$$\boldsymbol{\Psi}(x,t) = 1\,\varphi_g(t)\,\psi_g(x) + 0\,\varphi_e(t)\,\psi_e(x) \tag{1.84}$$

After a perturbation is applied, the coefficients $c_g(t)$ and $c_e(t)$ change to account for the system undergoing a transition into the excited state that can be described by

$$\boldsymbol{\Psi}(x,t) = 0\,\varphi_g(t)\,\psi_g(x) + 1\,\varphi_e(t)\,\psi_e(x) \tag{1.85}$$

Thus, the overall time-dependent Schrödinger equation that accounts for the response of the system is

$$i\hbar\frac{\partial\left[\sum_k c_k(t)\ \varphi(t)\psi(x)\right]}{\partial t} = \left[H - \mathbf{E}_x^0\boldsymbol{\mu}\ \left\{\cos\left(2\pi\nu t + \frac{2\pi z}{\lambda}\right)\right\}\right]\left[\sum_k c_k(t)\,\varphi(t)\,\psi(x)\right] \tag{1.86}$$

The solution of this equation can be found in many texts on quantum mechanics (*cf.* Ref. [3, vol. II, Chapter 2]), and it proceeds by taking the necessary derivatives and integrating the resulting terms $\frac{dc_k}{dt}$ between time 0 and the duration of the perturbation. This procedure yields an expression for the time evolution of the expansion coefficients $c_m(t)$:

$$c_m(t) = \delta_{nm} + \frac{iE^0}{2\hbar}\langle\psi_n|\ \mu|\psi_m\rangle\left[\frac{e^{i(\omega_{nm}+\omega)t}-1}{(\omega_{nm}+\omega)} + \frac{e^{i(\omega_{nm}-\omega)t}-1}{(\omega_{nm}-\omega)}\right] \tag{1.87}$$

Equation 1.87 is one of the most important equations for understanding spectroscopic processes because it outlines three major features of the response of a molecule when exposed to electromagnetic radiation. First, it implies that the term $\langle\psi_n|\mu|\psi_m\rangle$, known as the transition moment, must be nonzero. The transition moment describes the action of the dipole operator μ defined in Eq. 1.77 on the stationary-state wavefunctions ψ_n and ψ_n between which the transition is induced. The transition moment, in general, determines the selection rules, depending on the exact nature of the wavefunctions and their symmetries. This aspect is discussed in the following section. Second, the term containing the amplitude of the electric field, E^0, indicates that the "light must be on" for a transition to occur. Third, the part in the square brackets in Eq. 1.87 describes how the system responds to electromagnetic radiation of different frequency or wavelength. Thus, it is this term that prescribes that the frequency of the light must match the energy difference between the molecular energy levels. This can be seen from the following discussion. The second term in the square bracket becomes very large at the resonance condition,

$$\omega_{nm} = \omega \tag{1.88}$$

which implies that when the frequency ω of the incident radiation is equal to, or very close to, the energy difference ω_{nm} between states n and m, a transition between these states may occur and a photon with the corresponding energy $\hbar\omega$ may be absorbed, if the transition moment is nonzero.

Similarly, the first term in the square bracket in Eq. 1.87 becomes very large if

$$\omega_{nm} = -\ \omega \tag{1.89}$$

This case corresponds to the situation of stimulated emission, where a photon of the proper energy impinges onto a molecular or atomic system in the excited state and causes this state to emit a photon, thereby returning to the lower energy state. This time-dependent part of Eq. 1.87 also contains explicitly the expressions needed to explain certain off-resonance phenomena, such as molecular polarizability, to be discussed in Section 1.6 and Chapter 4. The magnitude of resonance *vs.* off-resonance effects can be estimated from the expression in square brackets as well.

Equation 1.87 holds for one-photon absorption and emission situations that include standard IR (vibrational), microwave (rotational), and visible/ultraviolet (electronic) absorption spectroscopies. The time-dependent part in the square bracket of Eq. 1.87 can be summarized as

$$\Delta E_{\text{molecule}} = (h\nu)_{\text{photon}} \tag{1.90}$$

This equation was first mentioned in the introduction, and it corresponds exactly to the condition as described in Eqs. 1.88 and 1.89:

$$(\omega_{nm} = \pm\,\omega) \tag{1.91}$$

1.5.2 The vibrational transition moment for absorption: harmonic diatomic molecules

The previous section demonstrated that three conditions are necessary for a transition to occur in an atomic or a molecular system under the influence of a perturbation by electromagnetic radiation. First, radiation must impinge on the molecular system ($E^0 \neq 0$); second, the radiation must possess the proper energy, or frequency, corresponding to the energy difference between the molecular or atomic states. This part is fulfilled by the conditions listed in Eqs. 1.88–1.91. The third condition that must be fulfilled is that the dipole transition moment must be nonzero:

$$\langle \boldsymbol{\mu} \rangle_{nm} = \int \psi_n \mu \psi_m d\tau = \langle \psi_n | \mu | \psi_m \rangle \tag{1.92}$$

In Eq. 1.92, the integration extends over all space, as indicated by the differential $d\tau$. Here, as pointed out earlier, ψ_n and ψ_m are the stationary-state wavefunctions associated with energy levels n and m. For IR absorption spectroscopy, these wavefunctions are the vibrational wavefunctions defined in Eq. 1.51; furthermore, as given in Eqs. 1.102–1.104, the lower energy state and wavefunction normally correspond to the ground state, ψ_0. Thus, one may rewrite Eq. 1.92 as

$$\langle \boldsymbol{\mu} \rangle_{e0} = \langle \psi_e | \mu \; | \psi_0 \rangle \neq 0 \tag{1.93}$$

where ψ_e represents a vibrationally excited state. For a transition to occur, the transition moment $\langle \boldsymbol{\mu} \rangle_{e0}$ must be nonzero. Whether the transition moment for a vibrational transition is zero or not depends on the geometry of the molecule and on the polarity of the atoms, which will be discussed later in detail. However, it is instructive even at this early point in this text to point out that certain vibrational modes have a zero transition moment and cannot be observed in the IR absorption spectrum. A typical example for this is the symmetric stretching mode of CO_2, where both oxygen atoms move away from the central C atom in phase and with equal amplitudes (see Chapter 2). Although the energy of this vibrational mode is well known (from the Raman spectrum), irradiation of CO_2 molecules with IR radiation of this energy does not produce a transition, because the transition moment is zero for symmetry reasons.

Next, the transition moment

$$\langle \boldsymbol{\mu} \rangle_{e0} = \langle \psi_e | \mu | \psi_0 \rangle \neq 0 \tag{1.93}$$

for harmonic oscillator wavefunctions is presented, with particular emphasis of the transition between the ground and first excited states, for which Eq. 1.93 can be written as

$$\begin{aligned}\langle \boldsymbol{\mu} \rangle_{10} &= \langle \psi_1 | \mu | \psi_0 \rangle = \int_{-\infty}^{\infty} \psi_1(x) \mu \psi_0(x)\, dx \\ &= e \int_{-\infty}^{\infty} \psi_1(x)\, x \psi_0(x)\, dx\end{aligned} \tag{1.94}$$

because the dipole operator for a single charged particle is just $\mu = ex$.

First, the orthogonality of the vibrational wavefunctions is demonstrated graphically in Figure 1.8(a) and (b), which demonstrates by graphical integration that the integrals $\int_{-\infty}^{\infty} \psi_1(x)\psi_0(x)\, dx$ and $\int_{-\infty}^{\infty} \psi_2(x)\psi_1(x)\, dx$ are, indeed, zero and that the vibrational wavefunctions are orthogonal. Next, it will be shown by graphical integration that the transition moment between the ground and first excited states is nonzero. When evaluating the transition moment defined by Eq. 1.94, the two wavefunctions are multiplied by the transition operator, shown by the straight line in Figure 1.8(c). This makes the product function $\psi_1(x)\; \mu\psi_0(x)$ positive for both positive and negative values of x; thus, integration along the x-axis yields a nonzero value. This result could also have been obtained by analytical integration. At this point, it is also advantageous to introduce the concept of parity. The wavefunctions shown in Figure 1.5 were previously referred to as "symmetric" or "antisymmetric" with respect to the dashed line corresponding to the equilibrium position of the harmonic oscillator.

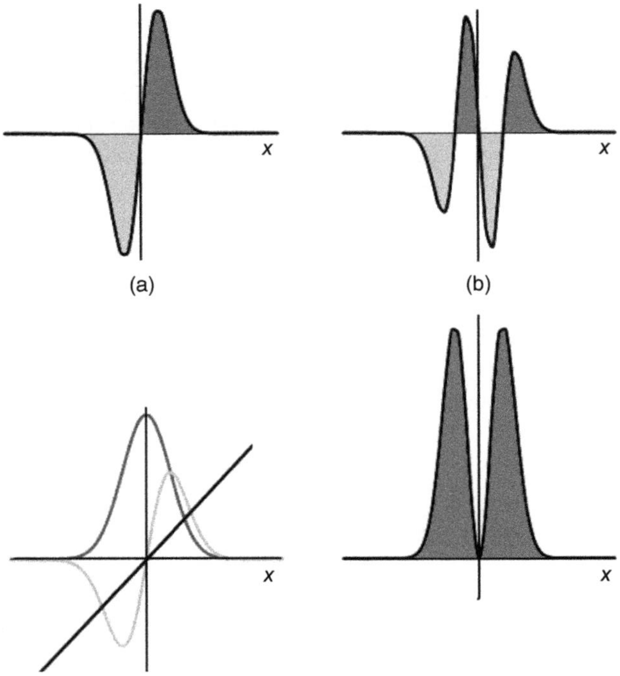

Figure 1.8 *Graphical representation of the orthogonality of vibrational wavefunctions and the vibrational transition moment. (a) Product of $\psi_0 \cdot \psi_1$. The light and dark gray regions under the curves have equal areas; thus, integration along x results in zero net area, and the functions are orthogonal. (b) Product of $\psi_1 \cdot \psi_2$. The same argument demonstrates that the functions are orthogonal. (c) Plot of ψ_0 (gray), ψ_1 (light gray), and the dipole operator $\mu =$ ex (black). (d) Integration of $\langle \psi_1 | \mu | \psi_0 \rangle$ along x-axis yields a nonzero transition moment*

Similar, one can argue that the wavefunctions with quantum numbers $n = 0, 2, 4, \ldots$ have "even" parity, that is, $f(x) = f(-x)$, whereas the wavefunctions with $n = 1, 3, 5, \ldots$ have "odd" parity with $f(x) = -f(-x)$. The multiplication table for parity is

$$\text{even} \times \text{even} = \text{odd} \times \text{odd} = \text{even}$$

$$\text{even} \times \text{odd} = \text{odd} \times \text{even} = \text{odd} \tag{1.95}$$

These relationships can be useful to intuitively evaluate whether or not transition moments are zero. Figure 1.8(c) and (d) demonstrates that the product of two odd functions and one even function produces an even function, as indicated by the fact that the shaded areas in (d) both are above the x-axis.

Next, the proper mathematical derivation of the selection rule for a one-photon emission or absorption for the harmonic oscillator will be presented. This derivation establishes that

$$\int_{-\infty}^{\infty} \psi_n(x)\mu\, \psi_m(x)\, dx \neq 0 \quad \text{if} \quad n = m \pm 1 \tag{1.96}$$

This proof proceeds as follows. As established earlier, the vibrational wavefunctions are of the form

$$\psi_n(x) = \left(\frac{\alpha}{\pi}\right)^{\frac{1}{4}} H_n(\sqrt{\alpha}\, x)\, e^{-\frac{\alpha x^2}{2}} \tag{1.47}$$

The Gaussian function $e^{-\frac{\alpha x^2}{2}}$ has even parity; thus, it does not affect the integral described by Eq. 1.96, and the following discussion can concentrate on the parity of the Hermite polynomials alone. Thus, the transition moment given in Eq. 1.96 can be simplified to

$$\langle \boldsymbol{\mu} \rangle_{nm} = \int H_n(x)\, x\, H_m(x)\, dx \tag{1.97}$$

where the factor $\sqrt{\alpha}$ was set to 1. Recalling the recursion formula for the Hermite polynomials,

$$xH_m(x) = mH_{m-1}(x) + \frac{1}{2}H_{m+1}(x) \tag{1.50}$$

the term $xH_m(x)$ in Eq. 1.97 can be substituted by the right-hand side of Eq. 1.50 to yield

$$\langle \boldsymbol{\mu} \rangle_{nm} = \int H_n(x) \left[mH_{m-1}(x) + \frac{1}{2}H_{m+1}(x) \right] dx \tag{1.98}$$

$$= m\int H_n(x)H_{m-1}(x)dx + \frac{1}{2}\int H_n(x)\, H_{m+1}(x)\, dx \tag{1.99}$$

As the vibrational wavefunctions, as well as the Hermite polynomials, are orthogonal (see Eq. 1.52 and Figure 1.8(a) and (b)), the two integrals in Eq. 1.99 are nonzero if and only if

$$\int H_n(x)\, H_{m-1}(x)dx = \delta_{n,m-1} = \begin{cases} 1 & \text{if } n = m-1 \\ 0 & \text{if } i \neq m-1 \end{cases}$$

$$\int H_n(x)H_{m+1}(x)\, dx = \delta_{n,m+1} = \begin{cases} 1 & \text{if } n = m+1 \\ 0 & \text{if } i \neq m+1 \end{cases} \tag{1.100}$$

Equation 1.100 implies that transitions are allowed only if the vibrational quantum number n changes by one unit, that is,

$$\Delta n = \pm 1 \tag{1.101}$$

This selection rule implies that only transitions between adjacent energy levels are allowed for the harmonic oscillator. This is shown schematically in Figure 1.9, where solid arrows indicate allowed transitions and dashed arrows indicate forbidden transitions.

1.5.3 The vibrational transition moment for absorption: anharmonic diatomic molecules

Inspection of Figure 1.6 suggests that different selection rules may hold for anharmonic diatomic molecules, because the symmetry of the wavefunctions is distorted and they cannot be classified in terms of odd/even functions with respect to the equilibrium position (or the energy minimum). Consequently, transitions that are forbidden in the harmonic oscillator, for example, transitions with $\Delta n = \pm 2, \pm 3, \ldots$ become weakly allowed for the anharmonic oscillator. Furthermore, as the energy levels are no longer equally spaced by $h\nu$ or $hc\tilde{\nu}$, the transition from $n = 1$ to $n = 2$ will have a lower energy (lower wavenumber) than the transition from $n = 0$ to $n = 1$. Thus, the vibrational spectrum of an anharmonic diatomic molecule is characterized by the presence of higher harmonics; that is, the transitions with $\Delta n = \pm 2, \pm 3$. This is shown in Figure 1.10 for gaseous bromine, Br_2. This homonuclear diatomic molecule is nonpolar; therefore, it does not exhibit an IR absorption spectrum; consequently, Figure 1.10 depicts the gas phase Raman spectra. The principles of vibrational energies of higher harmonics hold equally for IR and Raman spectra; however, the selection rules are different, which will be discussed later.

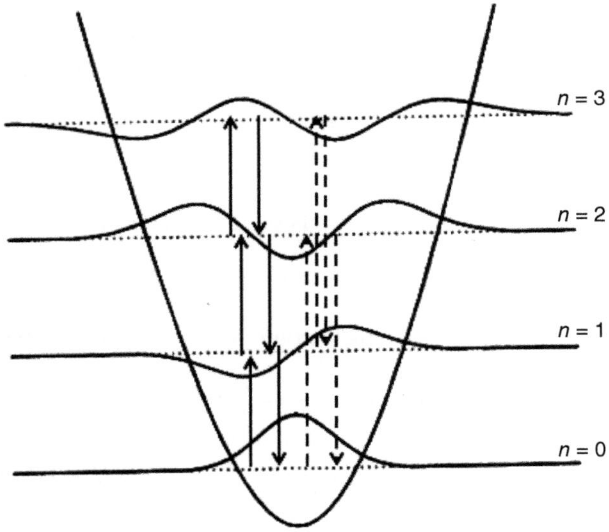

Figure 1.9 *Schematic of allowed (solid arrows) and forbidden (dashed arrows) transitions for the harmonic oscillator*

The spectra shown in Figure 1.10 further demonstrate a number of important features of vibrational spectroscopy. The inset shows an expanded region of the fundamental transition that is split into several peaks. These are, in part, due to the fact that Br_2 is a mixture of isotopic species (isotopic effects are discussed in the following section): $^{79}Br_2$, ^{79}Br–^{81}Br, and $^{81}Br_2$ with an abundance ratio of 1:2:1. Thus, the fundamental transitions exhibit these peaks at 322, 320, and 318 cm^{-1} (features f, e, and d in the inset of Figure 1.10). Furthermore, the spectrum will contain "hot bands," namely the transitions arising from states higher than the vibrational ground state. The name "hot bands" refers to the fact that these transitions are observed predominantly at elevated temperatures, to be discussed next.

So far in the discussion, it was assumed that the molecular systems are in their vibrational ground state at room temperature. This assumption is based on the Boltzmann distribution, which can be stated as

$$\frac{n_2}{n_1} = \frac{g_2}{g_1} e^{-\frac{\Delta E_{21}}{kT}} \tag{1.102}$$

In Eq. 1.102, n_2 and n_1 refer to the populations of a higher and a lower energy states, respectively, and g_2 and g_1 to their degeneracy (see Chapter 2). ΔE_{21} refers to the energy difference between the states, with $E_2 > E_1$, T is the absolute temperature, and k is Boltzmann's constant, defined as the gas constant R divided by Avogadro's number, N:

$$k = \frac{R}{N} \tag{1.103}$$

k has the value of 1.38×10^{-23} J K^{-1} or 0.695 cm^{-1} K^{-1}. The effect of the Boltzmann distribution on the population of ground and first vibrational states is best illustrated by two examples. For HCl, the vibrational transition between the ground and first excited states occurs at about 2990 cm^{-1}; that is, ΔE_{21} in Eq. 1.102 is 2990 cm^{-1}. Thus, the ratio of first excited to ground state populations is given at room temperature (298 K) by

$$\frac{n_2}{n_1} = e^{\frac{-2990}{0.695 \cdot 298}} = e^{-14.44} \approx 5.4 \times 10^{-7} \tag{1.104}$$

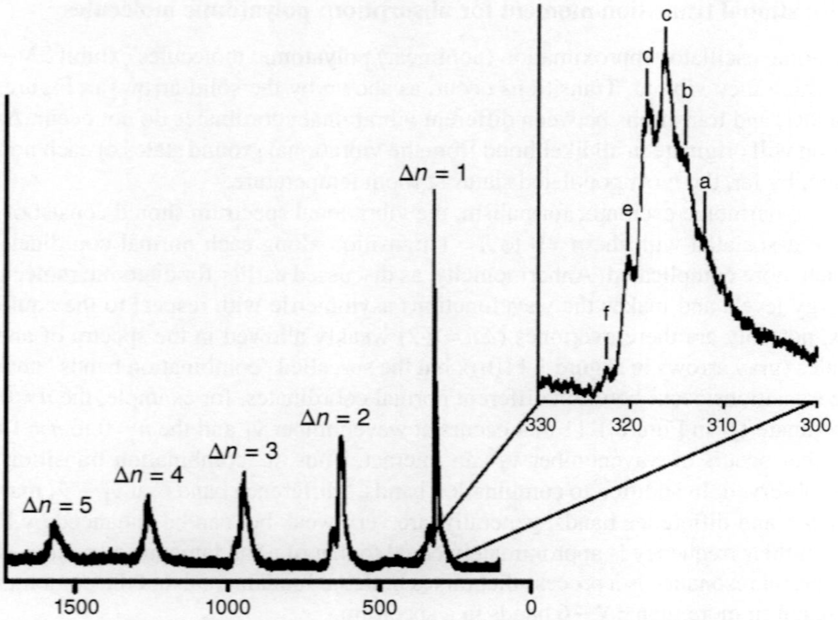

Figure 1.10 *Gas phase Raman spectrum of Br_2 with overtones and hot bands. See text for details. Baierl and Kiefer, 1975, [4]. Reproduced with permission from AIP*

Thus, for HCl at room temperature, one finds that the population of the ground state exceeds that of the excited state by a factor of more than 1 million to 1. The situation is markedly different for a heavy diatomic molecule, such as Br_2, for which the vibrational stretching frequency is much lower, namely 320 cm^{-1}:

$$\frac{n_2}{n_1} = e^{\frac{-320}{0.695 \cdot 298}} = e^{-1.545} \approx 0.21 \tag{1.105}$$

Here, the ratio of populations between the first excited and ground states is on the order of 20%. Raising the temperature of a sample, of course, has a similar effect than lowering the vibrational energy difference between states because the exponent gets smaller, and the ratio of population gets larger. This explains the "hot band" designation of the low-frequency shoulder of the stretching fundamental absorption as the temperature is raised. As indicated earlier, the frequency of the "hot band" is lower than that of the fundamental because the spacing between the $n=1$ and $n=2$ states is smaller than that of the $n=0$ and $n=1$ states. Some of the hot bands observed in the vibrational spectrum of Br_2 are shown in Figure 1.10 as well. These features are marked a, b, and c in the inset. As expected, their transition wavenumbers are lower than those of the $n=0$ to $n=1$ transition; their intensities here are exaggerated by the fact that several hot bands of several isotopomers occur at the same wavenumber [4].

In principle, one could also observe transitions from the $n=1$ to $n=2$ state from species that were not excited thermally (by collisional energy transfer) but rather by prior absorption of a photon that promoted the system into the excited state. Such cases are, however, extremely rare, because the lifetime of vibrational states, in general, is quite short (on the order of 10^{-12}–10^{-14} s); thus, to induce successive transitions, a very high photon flux is required.

1.5.4 The vibrational transition moment for absorption: polyatomic molecules

Within the harmonic oscillator approximation (nonlinear) polyatomic molecules exhibit $3N-6$ normal coordinates along which they vibrate. Transitions occur, as shown by the solid arrows in Figure 1.11(a), within each energy ladder, and transitions between different vibrational coordinates do not occur. As discussed earlier, the transition will originate in all likelihood from the vibrational ground states of each normal coordinate, because they are, by far, the most populated states at room temperature.

Thus, within the harmonic oscillator formalism, the vibrational spectrum should consist of $3N-6$ spectral bands, each one associated with the $n=0$ to $n=1$ transition along each normal coordinate. In reality, the situation is much more complicated. Anharmonicity, as discussed earlier for diatomic molecules, reduces the spacing of energy levels and makes the wavefunctions asymmetric with respect to the equilibrium distance position. Thus, not only are there overtones ($\Delta n=\pm 2$) weakly allowed in the spectra of anharmonic, polyatomic molecules (gray arrows in Figure 1.11(b)), but the so-called "combination bands" are weakly allowed as well. These transitions occur between different normal coordinates, for example, the $n=0$ to $n=1$ transition along coordinate Q_1 in Figure 1.11 that occurs at wavenumber $\tilde{\nu}_1$ and the $n=0$ to $n=1$ transition along coordinate Q_2 that occurs at wavenumber $\tilde{\nu}_2$ can interact. Thus, a "combination transition" that occurs at $\tilde{\nu}_1 + \tilde{\nu}_2$ may be observed. In addition to combination bands, "difference bands" at $\tilde{\nu}_1 - \tilde{\nu}_2$ may be observed as well. Combination and difference bands, generally, are very weak but can be enhanced by Fermi resonance (see Chapter 5) if their frequency is approximately equal to that of a fundamental transition of the same symmetry species. Fermi resonance is a process that mixes both the fundamental and the combination or overtone band and can result in more than $3N-6$ bands in a spectrum.

On the other hand, in highly symmetric molecules, the number of observed bands in a spectrum can be reduced substantially as discussed in Chapter 2. This reduction of observed bands has two causes: first, the transition moment for absorption for highly symmetric moments may be zero, and the transition is not allowed in IR absorption. The symmetric stretching vibration for methane, CH_4, is a typical example: as this mode does not change the dipole transition moment, it is "infrared forbidden," although it is allowed in Raman scattering. This demonstrates that the vibration along the symmetric stretching coordinate certainly does occur, but that this vibration cannot interact with the electromagnetic field to cause absorption. The second reason for reduced number of observed bands in a vibrational spectrum is the possibility of degeneracy. Degenerate vibrations

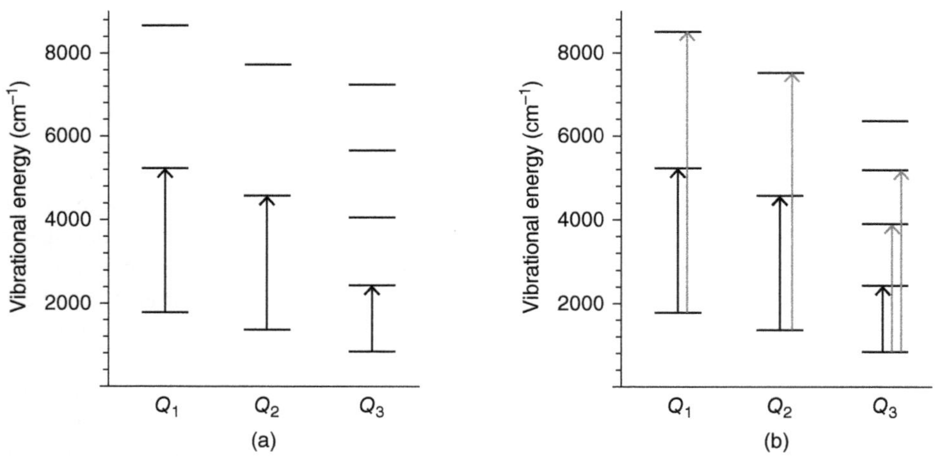

Figure 1.11 *(a) Allowed transitions (solid black arrows) for a polyatomic molecule within the harmonic oscillator approximation. (b) Additional overtone transitions (gray arrows) for an anharmonic polyatomic molecule*

may be visualized as certain vibrational displacements (normal modes) that occur at exactly the same energy. Doubly degenerate (*E* symmetry) and triply degenerate (*T* symmetry) modes, to be discussed in more detail in Chapter 2, will account for two or three degrees of freedom, respectively, and will reduce the number of bands observed in a spectrum.

1.5.5 Isotopic effects: diatomic molecules

The dependence of the vibrational frequency on the reduced mass was discussed earlier for (harmonic) diatomic molecules:

$$\omega = 2\pi\nu = \sqrt{\frac{k}{m}} \tag{1.27}$$

with the reduced mass given by

$$m = \frac{m_1 m_2}{m_1 + m_2} \tag{1.25}$$

When one substitutes an H atom in a diatomic molecule, such as HCl by a deuterium atom with a mass of 2.0 amu, and assumes the same potential energy function (force constant), one finds that the vibrational frequency ω is reduced by a factor of about 1.4 or $\sqrt{2}$. This statement is discussed in detail next. First of all, the assumption that the potential energy curve (see Figure 1.6) is independent of the mass makes sense because the additional neutron in the nucleus of deuterium, as compared to hydrogen, is not expected to influence the potential the electron pair binding the two atoms together.

Second, a comparison of the reduced masses of HCl and DCl yields for the ^{35}Cl isotopic species:

$$m_{\mathrm{HCl}} = \frac{1 \times 35}{1 + 35} = 0.972 \tag{1.106}$$

$$m_{\mathrm{DCl}} = \frac{2 \times 35}{2 + 35} = 1.892 \tag{1.107}$$

which allows an estimate of the vibrational frequency of DCl as follows:

$$\frac{\nu_{\mathrm{DCl}}}{\nu_{\mathrm{HCl}}} = \frac{\sqrt{\frac{k}{m_{\mathrm{DCl}}}}}{\sqrt{\frac{k}{m_{\mathrm{HCl}}}}} = \sqrt{\frac{m_{\mathrm{HCl}}}{m_{\mathrm{DCl}}}} = \sqrt{\frac{0.972}{1.892}} = 0.717 \approx \frac{1}{\sqrt{2}} \tag{1.108}$$

$$\nu_{\mathrm{DCl}} = 2990 \times 0.717 = 2144\ \mathrm{cm}^{-1} \tag{1.109}$$

This frequency agrees reasonably well with the observed frequency of 2091 cm^{-1}. The transition wavenumber for HCl, in Eq. 1.109, was introduced earlier (see Eqs. 1.103 and 1.104).

Third, the lowered vibrational frequency of the isotopic species causes the ground-state vibrational energy, E_0, to be at 1045 cm^{-1} *vs* 1495 cm^{-1} (for HCl). As for anharmonic species, the vibrational potential energy function is slightly asymmetric (see Figure 1.6) and the DCl species is at a lower energy level than HCl, its bond length is slightly lower and its dissociation energy is higher than that of HCl. These findings account for the kinetic isotope effect that states that the rates of chemical reactions that involve the exchange of hydrogen in the transition state may be significantly slower for deuterated species.

The arguments presented here for a diatomic molecule can be applied to polyatomic molecules, such as chloroform and deuterochloroform ($HCCl_3$ and $DCCl_3$, see Section 2.6) by assuming that the —CCl_3 moiety, to a first approximation, acts as a pseudo-atom of mass 117, and that its vibrations do not couple strongly with the C—H stretching vibration. This rather crude assumption actually produces reasonably good results. Similarly, a deuterated methyl group, —CD_3, exhibits stretching vibrations that are shifted by approximately a factor of 0.71 as compared to a —CH_3 group (see Section 2.6).

1.6 Basic infrared and Raman spectroscopies

In this section, the fundamental principles of IR and Raman spectroscopies, the two most common modalities of vibrational spectroscopy, are explored. Both these techniques are commonplace in research and teaching laboratories and in quality control in industrial environments. Furthermore, these techniques can now be found in forensic applications because they both provide highly specific spectral fingerprints for the identification of contraband materials. In this section, a very basic view of the two techniques is presented, emphasizing the differences and commonalities between the methods and introducing some of the instrumental aspects. A more detailed view is presented after the discussion of symmetry and group theoretical aspects (Chapter 2) in Chapters 3 and 4.

1.6.1 Infrared absorption spectroscopy

As pointed out in Section 1.5, IR absorption spectra are observed mostly between the ground and first excited vibrational states; the transitions observed are dipole-mediated, which implies that the normal coordinate along which the transition occurs must change at least one of the Cartesian components of the electric dipole moment (see Chapter 2) for a transition to take place. As the name implies, this modality is observed in "transmission" mode (and converted into absorbance units, see below) where the attenuation (absorption) of IR radiation passing through the sample is observed. According to the condition that the photon energy must match the energy difference between the vibrational states, that is,

$$\Delta E_{\text{molecule}} = E_{\text{photon}} \tag{1.90}$$

only certain "colors" (wavelengths or wavenumbers) of the exciting light will be absorbed, and a spectrum is obtained that is lacking these "colors," or has these colors at reduced intensities. Thus, IR absorption spectroscopy is observed by ratioing the intensity of the incident IR radiation with the intensity of the light transmitted by the sample. This ratio is generally referred as the "transmittance" of the sample. Thus, obtaining an IR absorption spectrum really involves the collection of two separate spectral patterns, the "background" and the "sample" spectra. A typical background spectrum is shown in Figure 1.12(a). Given that the output of an IR source as well as the response of an IR detector generally vary gently with wavelength, the intensity variations in the background spectrum may appear as a surprise. The overall shape of this curve depends on the intensity distribution of the source, typically a black-body (*cf.* Chapter 3) operating at temperatures between 1200 and 2000 °C. The radiance of such a source increases toward the visible part of the electromagnetic spectrum and, therefore, should be higher at 4000 cm^{-1} than at 1000 cm^{-1}. However, the detector used to collect the background spectrum shown in Figure 1.12 was a typical cryogenic HgCdTe detector (see Chapter 3). The peak sensitivity of such a detector occurs at about 1200 cm^{-1} and gradually decreases toward higher wavenumber. Thus, the overall shape of the background curve depends on the source emission profile convoluted with the detector sensitivity profile and the response of the optical components in the beam path, and it is represented in Figure 1.12(a). Frequently, sharp spectral features are observed in the background spectrum between 1300 and 1900 cm^{-1}, and between 3600 and 3900 cm^{-1}; these features are rot-vibrational transitions of atmospheric water vapor, which can present a major problem (see Section 5.4.3). Similarly, a broader feature around 2300 cm^{-1} is due to the rot-vibrational spectrum of carbon dioxide. Both these spectral interferences are discussed later.

In principle, the background spectral features will disappear from the observed IR spectra because of the ratioing between sample and background spectra. However, if there are any changes in the concentration of water and CO_2 between the time the sample and background spectra were collected, their spectral features may be contained in the observed IR spectra of the sample. Thus, it is advantageous to purge the entire instrument

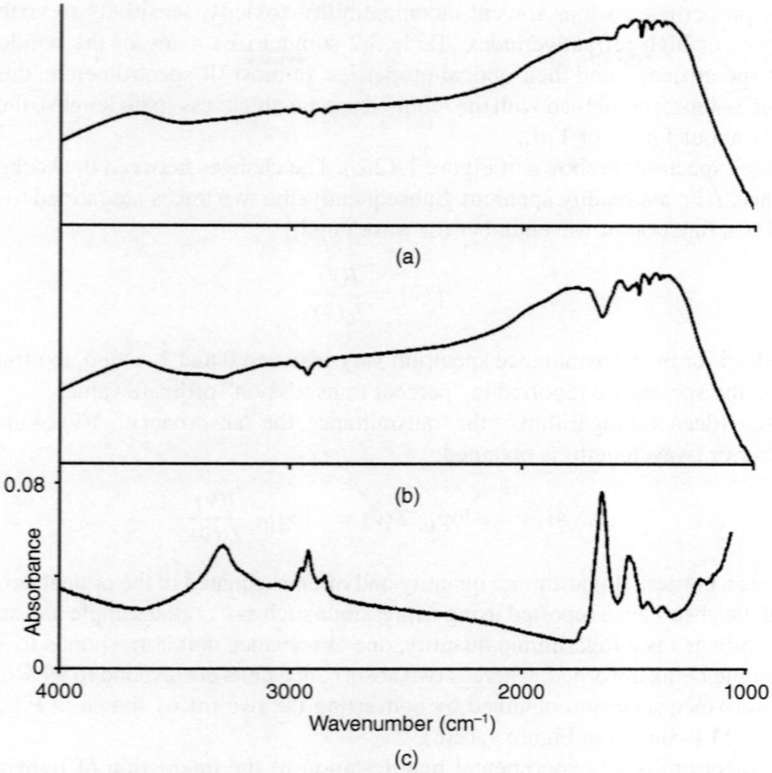

Figure 1.12 *Example of background, sample, and absorption spectra in infrared spectroscopy. (a) Single-beam background spectrum measured in transmission of a single CaF_2 window. The weak features around 2900 and 1500 cm^{-1} are due to a coating on the beam splitter. Note the absence of water vapor and CO_2 features in a well-purged instrument. (b) Single-beam sample spectrum. The intensity scale in (a) and (b) is in arbitrary units. (c) Absorbance spectrum calculated from traces (a) and (b) according to Eq. 1.111. The upward slope of the absorbance spectrum below 1100 cm^{-1} is due to the CaF_2 cutoff*

with dry, CO_2-free air for high-quality data acquisition. Once the background spectrum has been collected, the sample is inserted into the optical beam, and the sample transmission spectrum is collected.

Samples for IR spectroscopy can be gaseous, liquid, or solid. Detailed sampling methodologies are presented in Chapter 3; for the time being, the collection of IR spectra for a neat liquid is discussed. Molar extinction coefficients (see below) for vibrational transitions typically are in the range 10–1000 L (mol cm)$^{-1}$, which requires that neat liquids be measured as thin layers, typically between 10 and 100 μm thick. To this end, the sample is introduced into a liquid cell made of IR transparent windows separated by a spacer of appropriate thickness. The choice of IR transparent window materials is by no means trivial. While ultraviolet–visible (UV–vis) spectroscopy can be carried out in glass or quartz cells, both these materials do not transmit mid-IR radiation and are unsuitable for us in IR spectroscopy. This is because both materials contain Si—O bonds whose vibrations occur, like those of most other molecules, in the mid-IR spectral region. Thus, materials for windows (and, for that matter, for all other optical components such as lenses or substrates) must be found that are devoid of molecular bonds. This leaves a fairly restricted choice based mostly on ionic salts such as NaCl, KCl, KBr, AgCl, AgBr, and CaF_2 and a few more esoteric materials. Many of these window materials exhibit rather undesirable properties, such as solvent incompatibility, toxicity, sensitivity to visible light, restricted optical transmission, or high refractive index. Table 3.2 summarizes many of the window materials commonly used in IR spectroscopy and their optical properties. In most IR spectrometers, the beam diameter at the sample is about 3–4 mm; combined with the required sample thickness (path length), the necessary sample volume is typically about 1 mm^3 or 1 μL.

A typical "sample" spectrum is shown in Figure 1.12(b). The changes between the background trace, $I_0(\tilde{\nu})$, and the sample trace, $I(\tilde{\nu})$, are readily apparent. Subsequently, the two traces are ratioed to yield the transmittance spectrum T as a function of wavenumber (or wavelength):

$$T(\tilde{\nu}) = \frac{I(\tilde{\nu})}{I_0(\tilde{\nu})} \tag{1.110}$$

The ordinate values for the transmittance spectrum vary between 0 and 1. Often, the transmittance is multiplied by 100, and the spectra are reported in "percent transmission" ordinate values.

Taken the negative (decadic) logarithm of the transmittance, the "absorbance" $A(\tilde{\nu})$ of the sample as a function of wavenumber or (wavelength) is obtained:

$$A(\tilde{\nu}) = -\log_{10} T(\tilde{\nu}) = -\log_{10} \frac{I(\tilde{\nu})}{I_0(\tilde{\nu})} \tag{1.111}$$

The absorbance is a unitless (logarithmic) quantity and often is equated to the optical density (OD). Thus, it is common to find the absorbance reported using expressions such as "... the sample had an absorbance of 1.5 OD units." As absorbance is a logarithmic quantity, one absorbance unit corresponds to 90% of all photons incident on the sample being absorbed, whereas two absorbance units correspond to 99% of all photons being absorbed. The absorbance spectrum obtained by converting the two traces shown in Figure 1.12(a) and (b) according to Eq. 1.111 is shown in Figure 1.12(c).

An absorbance spectrum is an experimental manifestation of the interaction of light with the molecular transitions in the sample. It can be related to the theoretical quantities derived in Section 1.5 according to the Lambert–Beer equation:

$$A(\tilde{\nu}) = \epsilon(\tilde{\nu})\, C\, l \tag{1.112}$$

where $\varepsilon(\tilde{\nu})$ is the molar extinction coefficient (with units of (L mol^{-1} cm^{-1})), C the concentration of the sample (mol L^{-1}), and l the length the light beam travels through the sample (the path length or thickness of the sample), expressed in centimeters. As discussed earlier, the sample thickness is typically between 10 and 100 μm or 0.001 and 0.01 cm. The molar extinction coefficient is related directly to the transition moment according to

$$D_{01} \propto \langle \psi_1 |\, \mu | \psi_0 \rangle^2 = \frac{1}{\tilde{\nu}_0} \int_{\text{band}} \varepsilon(\tilde{\nu})\, d\tilde{\nu} \tag{1.113}$$

Here, D_{01} is known as the dipole strength of the transition, $\tilde{\nu}_0$ is the center of the band position, and the integration is over the entire band profile that is assumed to follow Gaussian (Eq. 1.114), Lorentzian (Eq. 1.115), or mixed Gaussian/Lorentzian profiles, shown in Figure 1.13. These band envelopes are given by

$$f(x) = I e^{-\frac{(x-x_0)^2}{2a}} \tag{1.114}$$

for Gaussian and

$$f(x) = I \frac{2a}{(x - x_0)^2 - a^2} \tag{1.115}$$

for Lorentzian bands.

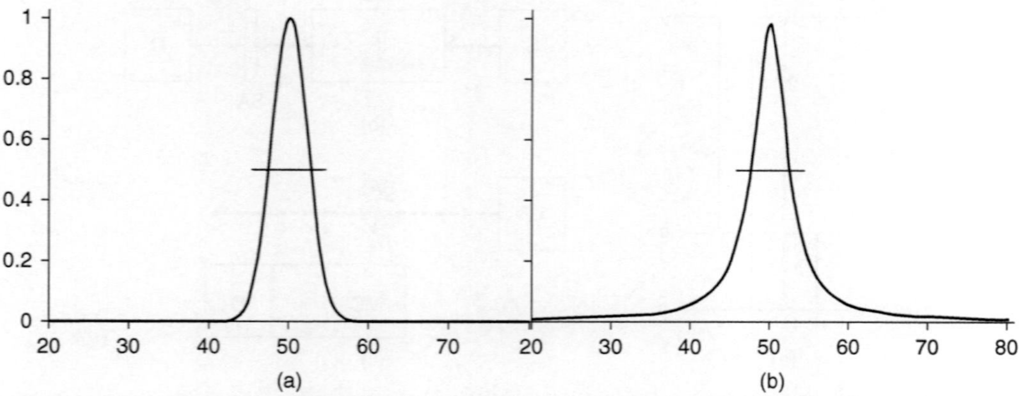

Figure 1.13 *Gaussian (a) and Lorentzian (b) line profiles. Note that the areas under the line profiles are unequal; however, both bands have a full width "a" at half maximum (FWHM) of 5 and an intensity I = 1.0*

Before about 1970, nearly all IR spectroscopy was carried out using dispersive instruments; that is, spectrometers that incorporated a monochromator to separate the broadband radiation from the IR source into narrow wavenumber bands that were tuned in time to expose the sample to the different IR "colors." These monochromators utilized gratings or prisms to disperse the light and slits of appropriate widths to select an appropriate wavenumber band to irradiate the sample. As only one spectral element (band) was sampled at a time, acquisition of spectral data took minutes to hours. These instruments generally operated in "double-beam" mode were a chopper alternatingly directed the beam of light through a reference (background) and sample optical path. Such, the two spectra shown in Figure 1.12(a) and (b) were acquired quasi-simultaneously and ratioed by analog electronics.

Since the advent of laboratory minicomputers, such as the PDP8 by the Digital Equipment Corporation in the mid-1970s, or personal computers in the 1980s, nearly all IR spectroscopy is being carried out by Fourier transform methods, and the techniques is referred to Fourier transform infrared (FTIR) spectroscopy. In FTIR, the light is not encoded sequentially into different colors by a monochromator but simultaneously by an interferometer to yield an interferogram. Fourier transform of the interferogram reveals the IR spectrum. In general, FTIR spectrometers are not operated in dual-beam mode; thus, the sample and background spectra shown in Figure 1.12 were collected consecutively. A more detailed discussion of the principles of the instrumentation, data acquisition, and many other aspects of IR spectroscopy are presented in Chapter 3.

1.6.2 Raman (scattering) spectroscopy

In Raman scattering, a totally different mechanism of excitation from the ground state to the excited state occurs, which is discussed in more detail in Chapter 4. Here, a qualitative approach is taken to introduce the differences and similarities between Raman and IR spectroscopies.

Although the vibrational excitation results in the same final state in both Raman and IR spectroscopies, the excitation process is a result of a two-photon process in the former: a "momentary absorption" into a "virtual state" of a photon with much higher energy than required for the vibrational transition, accompanied by a simultaneous re-emission (scattering) of a photon from the virtual state into either the ground vibrational state (Rayleigh scattering) or the vibrationally excited state (Stokes Raman scattering). These processes are depicted schematically in Figure 1.14(a). Whereas the Rayleigh-scattered photon has the same energy as the incident laser photon, the (inelastically) scattered Raman photon has lost part of its energy to the molecular

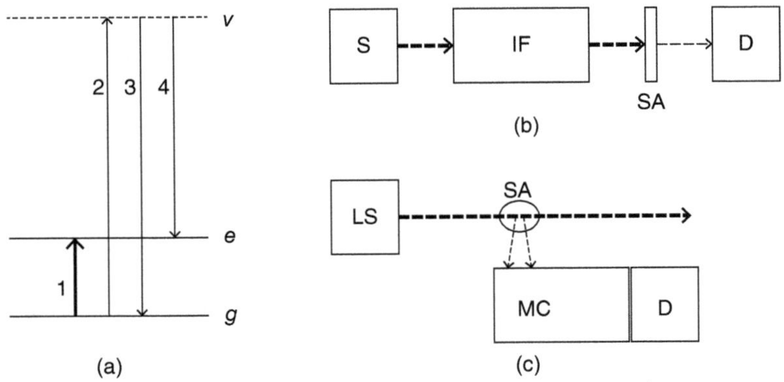

Figure 1.14 *(a) Comparison of infrared absorption and Raman process. g: ground vibrational state, e: excited vibrational state, and v: virtual state. 1: infrared transition, 2: momentary absorption of (laser) photon into virtual state, 3: Rayleigh scattering, and 4: Stokes Raman scattering. (b) Schematic of infrared absorption experiment: S: source, IF: interferometer, SA: sample, and D: detector. The dashed arrows represent the light path. (c) Schematic of Raman scattering experiment. LS: laser source, SA: sample, MC: monochromator, and D: detector*

system, and, therefore, appears at a red-shifted (lower energy) frequency as compared to the incident photon. A comparison of the processes occurring in IR and Raman spectroscopies is shown in Figure 1.14, along with schematic representations of the instruments used for their observation. A third scattering possibility, known as anti-Stokes Raman scattering, is introduced in Chapter 4.

Both Rayleigh and Raman scatterings are extremely weak processes: the Rayleigh scattering process occurs with a likelihood of about 1 in a million; the (nonresonant) spontaneous Stokes Raman process with a likelihood of about $1{:}10^{10}$ to $1{:}10^{12}$, whereas in IR spectroscopy, the chance of a photon incident on the sample to be absorbed can be quite high (in a band with an intensity of one absorbance unit, 90% of the photons are being absorbed). Thus, effective excitation of Stokes Raman spectra is possible only with laser sources that can produce 10^{15}–10^{18} visible photons per second. This fact explains why Raman spectroscopy was a field of very limited applicability before the availability of commercial lasers that occurred in the mid-1970s. Both FTIR and Raman spectroscopies experienced a rebirth at that point in time, and a still ongoing exponential growth ever since.

Raman-scattered light can be collected along the propagation direction of the laser (0° scattering), perpendicular to the propagation direction (90° scattering), or in backscattering (180° scattering). As explained in Chapter 4, 90° scattering was the preferred way to collect Raman spectra until holographic cut-off filters were commercially available to separate the Rayleigh scattering from the Raman scattering; thus, the 90° scattering geometry used prior to this point in time will be used here to explain the collection of Raman spectra data. Similar to fluorescence spectra that are often observed at right angle, the light emitted (scattered) by the sample is collected through a low *f*-number (large aperture) lens and focused onto the entrance slit of a monochromator that separates it into its different color components. For the observation of Raman spectra, the collected light also contains the Rayleigh scattering at the laser wavelength, as well as the much weaker and red-shifted Raman scattering. Within the monochromator, the light is dispersed spatially by a grating, and the intensities of the different, red-shifted Raman components are measured by a solid-state detector such as a charge-coupled device (CCD) camera. Using high-quality detectors and excitation powers between 10 and 100 mW, a good quality Raman spectrum can be collected in about 1 s over the entire spectral range, from about 300 to 3600 cm^{-1} shift from the Rayleigh line.

While IR spectroscopy is caused by direct absorption of a photon to promote the molecule into a vibrationally excited state and requires a change of the dipole moment along a vibrational coordinate, Raman spectra require the polarizability of a molecule to change along a vibrational coordinate. The polarizability of a molecule is its response to the incident radiation far from an electronic transition. This can be visualized as follows. In a clear, colorless liquid, for example, there are no electronic transitions between about 400 and 750 nm, or about 25,000 and 13,000 cm^{-1} (hence the material is colorless). Visible electromagnetic radiation, however, still can interact with the molecule by setting in motion the electron clouds, particularly those in multiple bonds. This phenomenon is referred to as molecular polarization, mediated by the polarizability. Thus, even far from resonance, the light causes an induced dipole moment μ_{ind}, which is proportional to the field strength E of the electromagnetic field:

$$\mu_{ind} = \alpha E \tag{1.116}$$

where the proportionality constant α is referred to as the polarizability. As both the induced dipole moment and the electric field are vectors, the polarizability is actually a tensor. This is an important aspect that is discussed in detail in Chapter 4.

This polarizability varies as the molecule oscillates along its normal coordinates, Q_k, because the polarizability (i.e., the ease with which electrons can be moved around) depends very much on the nuclear coordinates and thereby on the vibrational modes of the molecule. Thus, one may expand the polarizability in a Taylor series about the equilibrium position according to

$$\alpha = \alpha_0 + \left(\frac{\partial \alpha}{\partial Q_k}\right) Q_k \cos(\omega_k t) + \cdots \tag{1.117}$$

where ω_k is the vibrational frequency of normal mode Q_k.

As the exciting electric field of the incident laser radiation can be represented by

$$E = E_0 \cos(\omega_L t) \tag{1.118}$$

and as the molecule modulates the polarizability according to Eq. 1.117, it is obvious that the induced dipole moment emits radiation that contains the product of the two cosine functions that is, according to a well-known trigonometric identity

$$\cos(\omega_L t)\cos(\omega_k t) = \frac{1}{2}[\cos(\omega_L + \omega_k)t + \cos(\omega_L - \omega_k)t] \tag{1.119}$$

The beat frequencies $\omega_L + \omega_k$ and $\omega_L - \omega_k$ are the aforementioned anti-Stokes and Stokes Raman frequencies, respectively. Thus, a simple, classical description of the (off-resonance) interaction of light with a polarizable molecular system can explain some aspects of Raman scattering.

One important aspect of Raman spectroscopy is that the scattered photon generally still is a visible photon. If laser radiation at 488 nm wavelength or 20,491 cm^{-1} (the strong, blue laser emission of an Argon ion laser) is scattered from a molecule that has one vibrational mode, such as CO (with a stretching wavenumber of $\tilde{\nu} = 2145$ cm^{-1}), the Stokes-scattered photon occurs at 20,491 − 2145 = 18,346 cm^{-1} or 545 nm, which is a green, visible photon. If the sample was illuminated by the output of a He—Ne laser (632.8 nm or 15,803 cm^{-1}), the scattered photon would occur at 15,803 − 2145 = 13,658 cm^{-1} or 732 nm, which is at the end of the visible (red) range of the spectrum. Visible light is more easily detected and requires no special optical materials; thus, one of the big advantages of Raman spectroscopy is the use of visible optical detectors, and optical components. This example also shows that the incident photon interacts with the sample regardless of its color: blue and red photons equally create their own virtual states, although the blue photons are scattered with much higher efficiency.

In both IR and Raman spectroscopies, the observed spectrum of CO should consist of one single band at 2145 cm^{-1}; however, in gaseous CO, a more complex spectrum that includes rot-vibrational transitions is observed (see Chapter 5).

1.7 Concluding remarks

In the preceding sections, the classical and quantum mechanical formulation of molecular vibrations was reviewed. This review introduced the concept of normal modes of vibration, which is based strictly on a classical mechanical model. In order to explain the observed vibrational spectra, a transition to quantum mechanical models needs to be made. However, these models also invoke the concept of normal coordinates, because the observed transitions are associated with the normal modes of vibration. This description explains the vibrational (absorption or emission) spectra that consist of spectral "bands" at given frequencies. Each of these absorption bands corresponds to a transition along one normal coordinate.

Thus, at least two pieces of information may be extracted from a spectral band: its position (frequency, wavelength, or wavenumber) and intensity. The intensities of each transition are determined by their transition moments, whereas the frequencies are determined by the energy eigenvalues of the vibrational Hamiltonian (Eq. 1.64) that are based on the energy eigenvalues of the "mass-and-spring" secular determinant (Eq. 1.34).

The square of the transition moment is the actual observable in IR absorption spectroscopy: the integrated intensity of an absorption band is the dipole strength D of a transition, which can be determined experimentally as the area under an IR band, plotted in units of the molar extinction coefficient ε *vs.* wavenumber (see Eq. 1.108) and shown in Figure 1.13(a) and (b). Both the intensity and the energy of a transition can be calculated from quantum mechanical principles if the wavefunctions of the system are known to a high degree of accuracy. In fact, *ab initio* quantum mechanical calculations of vibrational frequencies and intensities now are practical for molecules with up to 50 heavy (not hydrogen) atoms (see Chapter 9).

References

1. Wilson, E.B., Decius, J.C., and Cross, P.C. (1955) *Molecular Vibrations: The Theory of Infrared and Raman vibrational Spectra*, McGraw-Hill co., New York
2. Kauzman, W. (1957) *Quantum Chemistry*, Academic Press, New York.
3. Levine, I. (1970) *Quantum Chemistry*, vol. 1 and 2., Allyn & Bacon, Boston.
4. Baierl, P. and Kiefer, W. (1975) Hot band and isotopic structure in the resonance Raman spectra of bromine vapor. *J. Chem. Phys.*, **62**, 306–308.

2

Symmetry Properties of Molecular Vibrations

This book mostly deals with Raman and infrared (IR) spectroscopy of macromolecules, which – due to their size – generally have low molecular symmetry. However, vibrational spectroscopy is a prime technique to probe molecular symmetry, and the symmetry properties of vibrations, in small molecules. Thus, an introduction to symmetry and group theoretical principles underlying vibrational spectroscopy is presented in this chapter.

It was pointed out in Chapter 1 and Appendix A that inspection of the wavefunction and the transition operator sometimes can lead to a decision on whether the transition moment is zero, that is, whether a transition is allowed or forbidden in a given modality of vibrational spectroscopy. In particular, the concept of parity was introduced in Section 1.5 to allow an assessment of a transition being allowed: it was pointed out that the product of even and even parity wavefunctions and the odd transition operator gives an odd product, indicating a forbidden transition, whereas the product of odd and even wavefunctions and the odd transition operator yields an even function, indicating an allowed transition. However, the argument based on the symmetry (parity) of the wavefunction has to be modified when discussing real molecules. In diatomic molecules, such as H_2, N_2, Cl_2, and so on, the $n = 0$ to $n = 1$ vibrational transition is forbidden in infrared absorption (although seemingly allowed by the selection rules for the harmonic oscillator, Eq. 1.102) because these molecules have no permanent dipole moment and the stretching vibration does not change that dipole moment. However, these molecules exhibit allowed Raman transitions for their stretching vibrations. One can easily extend this argument to molecules such as CO_2. This linear molecule, in which the two carbon–oxygen bonds have equal length, is nonpolar as well. As a linear molecule, it has $3N-5$ normal modes, of which the two deformation modes are perpendicular to each other and degenerate. Therefore, there are three normal modes of vibration that change the dipole moment (i.e., distort the molecule such that is has a dipole moment), the degenerate bending (Figure 2.1(c)) and antisymmetric stretching vibration (Figure 2.1(a)). One vibrational mode, the symmetric stretching vibration, does not change the dipole moment. The vibrational modes that change the dipole moment can be excited by absorption of an infrared photon of appropriate wavenumber, whereas the symmetric stretching mode is inactive, or forbidden in infrared absorption spectroscopy. In addition, for this molecule, a "mutual exclusion" rule holds that states that infrared-allowed transitions cannot

Modern Vibrational Spectroscopy and Micro-Spectroscopy: Theory, Instrumentation and Biomedical Applications, First Edition.
Max Diem.

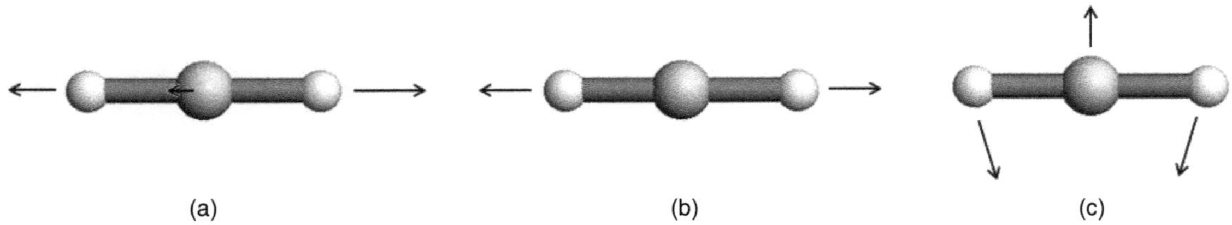

Figure 2.1 Depiction of the vibrational modes of CO_2. (a) Antisymmetric stretching mode; (b) symmetric stretching mode; and (c) deformation mode (degenerate with deformation direction out of the paper plane)

occur in the Raman spectra, and *vice versa*. (In CO_2, however, other effects such as Fermi resonance, to be discussed in Chapter 5, will aggravate the situation somewhat.)

In polyatomic molecules, the arguments on whether a transition is allowed in Raman or infrared spectroscopy become more complicated. This is because a molecule consisting of N atoms has $3N - 6$ degrees of vibrational freedom, and because the total vibrational wavefunctions of ground and excited states are the products of the wavefunctions of each vibrational coordinate, as discussed in Section 1.4. Thus, the symmetry properties of each of the normal modes of vibration and their associated wavefunctions need to be evaluated. One finds that each normal mode of vibration can be described by a symmetry representation and that the transformation properties of these representations will determine whether a given normal mode of vibration is symmetry allowed.

This is, of course, the subject of group theory. This topic is much too broad to be treated in full detail in just one chapter; thus, the reader is referred to specialized textbooks such as the classic "Chemical Applications of Group Theory" [1] for derivations and more examples. However, the basic concepts of group theory, as applied to vibrational spectroscopy, will be presented in this chapter, and a connection between this relatively unfamiliar branch of mathematics to multidimensional vector spaces and linear algebra will be made. The aim of this chapter is to present sufficient information for the discussion of vibrational spectra and the derivation of symmetry-based selection rules.

2.1 Symmetry operations and symmetry groups

Molecules are classified into symmetry-related categories, called "symmetry groups," according to the number and nature of symmetry operations that can be carried out on a molecule. A symmetry operation is a procedure applied to a molecule that leaves it in an arrangement that is indistinguishable from the arrangement before the operation was carried out. An example of such a symmetry operation is shown in Figure 2.2: when the water molecule is rotated by 180° about the axis shown, it appears identical to and indistinguishable

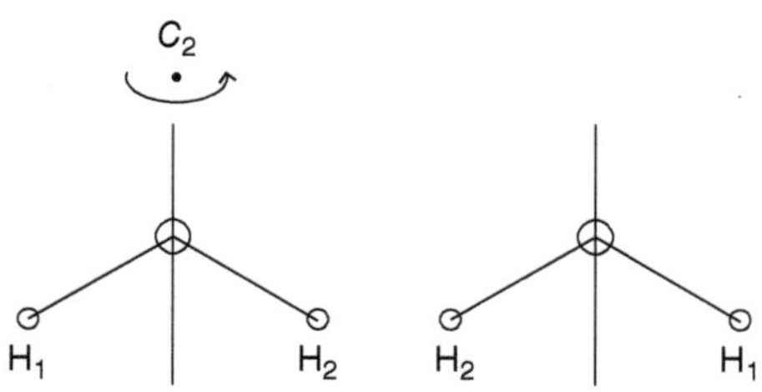

Figure 2.2 Example of a symmetry operation (C_2)

from its original arrangement, since the two H atoms are indistinguishable. This particular symmetry operation is called a proper, twofold rotation about a symmetry axis, and referred to as a C_2 symmetry operation. The symbol "*C*" will be used for all proper rotations, which are discussed below in more detail.

Investigating the symmetry properties of a number of arbitrary molecules or even macroscopic objects such as a snow crystal, a cube, certain letters, and many other items that one intuitively associates as being "symmetric," one finds that there exist only a few distinct and independent symmetry operations.

These can be combined and repeated to create other symmetry operations. For a discussion of the symmetry properties of small molecules for molecular spectroscopy, there are five symmetry operations of interest: the identity element, proper axes of rotation, reflection by a mirror plane, the center of inversion, and rotation–reflection axes, also known as improper axes of rotation. These operations are discussed next.

1. *The identity element*, designated *E*. This operation leaves the molecule unchanged, and the necessity for such an operation initially appears superfluous at first. However, the identity operation is necessary for two reasons. The first of these is purely mathematical and will become clear later during the discussion of symmetry groups. The other reason can be viewed intuitively as follows: if one defines a symmetry operation as a procedure that will return the molecule into a state that is indistinguishable from the state before the operation, a formalism is needed to include the possibility that the molecule actually was, indeed, left unchanged, that is, the molecule was not rotated or operated on at all.
2. *Proper axes of rotation* (C_n). This operation was introduced above for a twofold rotation. In general, one designates C_n operations as rotations about an axis where "*n*" describes how often the operation has to be repeated until a rotation by 360° has occurred. Thus, C_n indicates a rotation about an axis by 360°/*n*.
3. *Reflection by a mirror plane* (σ). Reflection operations can be visualized as a reflection by a mirror plane that can be parallel or perpendicular to a symmetry axis. The former are called σ_v, and the latter as σ_h. Such mirror planes are shown in Figure 2.3.
4. *Center of inversion* (*i*). When pairs of atoms are at positions for which the coordinates are identically the same except that all their signs are reversed, then these atoms are related by a center of inversion. Planar ethene (C_2H_4) is a molecule that incorporates a center of inversion (see Figure 2.4), located halfway between the two C atoms. The hydrogen atoms numbered 1 and 4 in Figure 2.4 have equal and opposite coordinates, as do hydrogen atoms 2 and 3 and the two C atoms.
5. *Rotation–reflection axes or improper axes of rotation* (S_n). This operation consists of a rotation about an axis by 360°/*n*, followed by a reflection by a plane perpendicular to the axis. The hydrogen atoms in ethane, for example, are related by an S_6 operation (Figure 2.5): rotation of the molecule about the direction of the C—C bond by 60° brings the top three hydrogen atoms into a position opposite to the original position of the lower three hydrogen atoms. Reflection of the three upper hydrogen atoms by a plane bisecting the C—C bond brings the upper three hydrogen atoms into the positions originally occupied by the lower three hydrogen atoms.

It is a matter of practice to identify all the symmetry elements of a molecule. Benzene, with a perfect hexagonal structure, has the following symmetry elements: two C_6 axis, two C_3 axes coincident with the C_6 axis, one C_2 axis coincident with the C_6 axis, three C_2 axes that bisect opposing sides, three C_2 axes that contain opposing atoms, two S_6 and two S_3 axes, a horizontal symmetry plane σ_h, three σ_v planes that contain opposing atoms, and three σ_d planes that bisect opposing bonds. When one hydrogen atom is substituted by deuterium, or one carbon by a ^{13}C atom, the symmetry is lowered enormously to just contain one each of the following elements: *E*, C_2, σ_v, and σ'_v.

Certain molecules, such as water, benzene-d_1, and formaldehyde, H_2CO, possess the same symmetry elements, namely, *E*, C_2, σ_v, and σ'_v. (The prime denotes that there are two mutually perpendicular planes both containing the major axis. These symmetry elements are also denoted as $\sigma(yz)$ and $\sigma(xz)$, respectively.) The

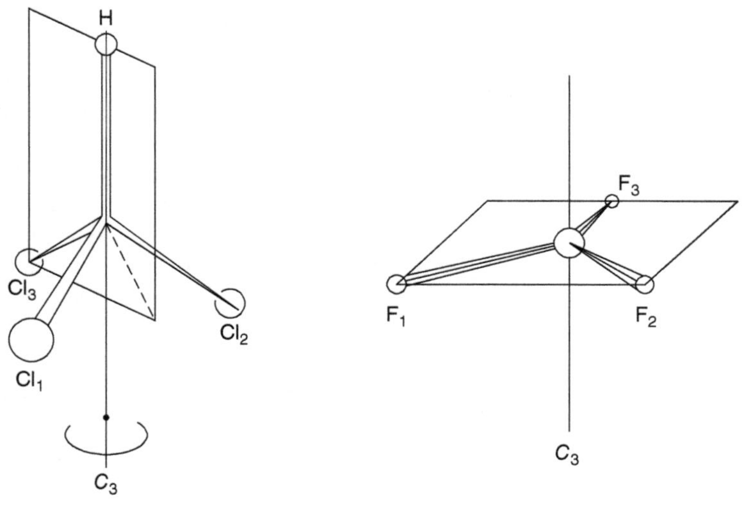

Figure 2.3 *(a) Example of one of three σ_v mirror planes in the molecule chloroform. Notice that the major symmetry axis, marked C_3, is contained in the symmetry plane. (b) Example of a σ_h plane that is perpendicular to the C_3 axis in the molecule boron trifluoride*

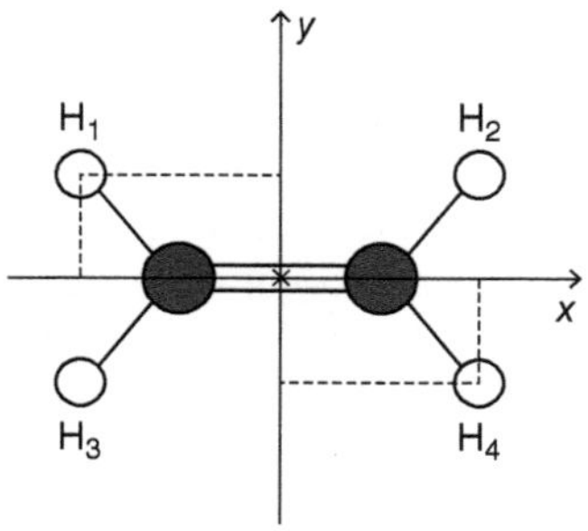

Figure 2.4 *Definition of a center of inversion, located at the coordinate origin, in planar ethene (C_2H_4). The coordinates of H_1 and H_4, for example, are equal in value, but opposite in sign*

occurrence of the same symmetry elements in different molecules suggests that the grouping of the four symmetry elements above is, indeed, special, and it can be shown that the four symmetry elements form a group. A group in the mathematical sense is a collection of elements that are connected according to certain rules. These rules are listed below for the case of symmetry groups but can similarly be written for other mathematical groups:

1. Successive applications of any two or more operations of the group produce another symmetry operation of the group. For example, successive applications of the C_2 operation bring the molecule back to its original orientation. Thus, in the symmetry group of water and formaldehyde,

$$C_2C_2 = E \tag{2.1}$$

with E being another operation of the group. Similarly,

$$\sigma_v \sigma_v' = C_2 \tag{2.2}$$

2. There exists an identity element E such that for any operation A

$$AE = A \tag{2.3}$$

As pointed out above, the E operation leaves the molecule unchanged; thus, any operation A followed by the E operation is equivalent to performing the operation A only.

3. The associative law holds. Let A, B, and C be symmetry operations in a group. Then the order in which they are applied to a molecule does not matter:

$$A(BC) = (AB)C \tag{2.4}$$

4. Every operation must have an inverse operation. The inverse operation A^{-1} for a symmetry operation A is defined such that $AA^{-1} = E$. That is, the inverse operation negates the original operation A and leaves the molecule unchanged (or operated on by the E operator).

Groups are common concepts in mathematics. An example other than symmetry groups is presented here to demonstrate that the concept of groups is common in mathematics. Integer numbers, for example, form a mathematical group for the addition operation. The four conditions above hold for all integers:

1. Addition of integers yields integers only.
2. The number zero is the identity element, since addition of zero leaves any integer unchanged.
3. The order of addition is immaterial.
4. There exists an inverse, namely, the negative of each number, such that zero is obtained by adding a number and its inverse.

For the discussion of molecular symmetry relevant to molecular vibrations, one needs to be concerned with about 40 different symmetry groups, of which only about a dozen or so are common in molecules. The groups are given special designations depending on the symmetry elements they contain, and the particular group

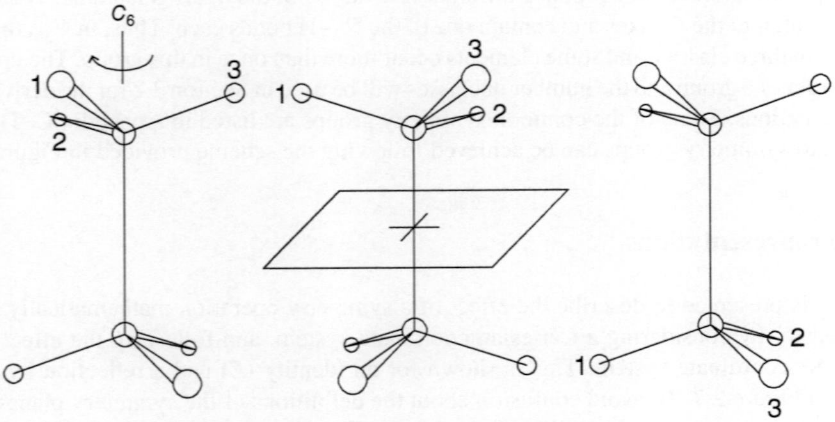

Figure 2.5 *Description of an improper rotation operation* (S_6) *for ethane*

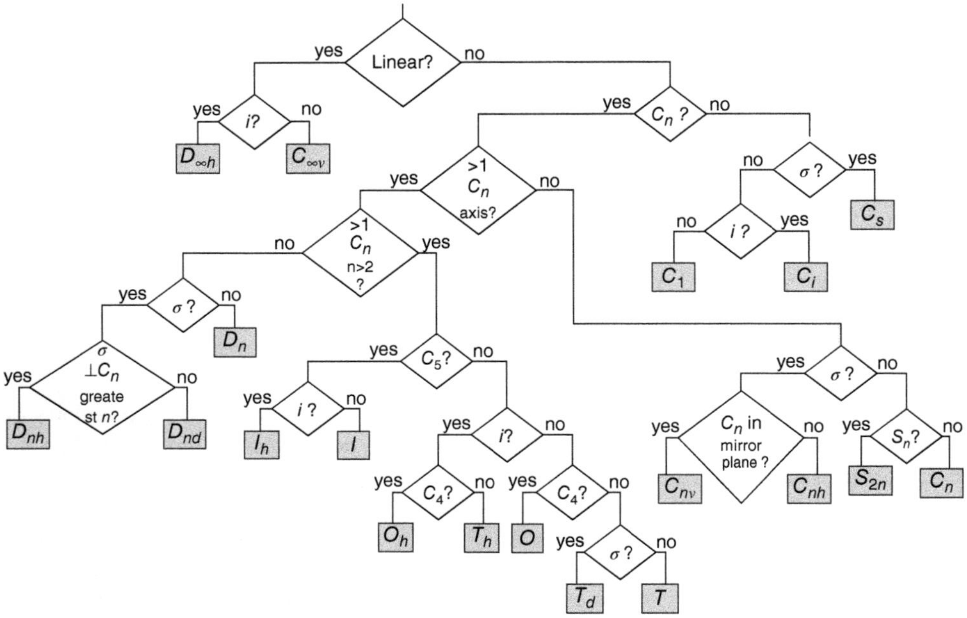

Figure 2.6 *Scheme for classifying molecules into symmetry groups. From Ref. 2. (*See insert for color representation of this figure.*)*

with the symmetry elements discussed above (E, C_2, σ_v, and σ_v') is known as the C_{2v} point group. In this symmetry, or point group, there are four symmetry elements, each occurring only once. Thus, one defines the "order h" of this point group to be 4.

Ammonia, NH_3, belongs to a symmetry group known as C_{3v}, which contains the following elements:

$$C_{3v} : E, 2C_3, 3\sigma_v$$

The two threefold axes are clockwise and counterclockwise rotation by 120° (see Eq. 2.19), which are two different operations since they produce different orientation of the hydrogen atoms. The three symmetry planes are all parallel to the C_3 axes and contain one of the N—H bonds each. Thus, in C_{3v}, there are six symmetry elements in three classes, and some elements occur more than once in this group. The order of the group is $h = 6$. The order of a group and the number of classes will be used in Section 2.2 for the derivation of the irreducible representations. Some of the common symmetry groups are listed in Appendix C. The classification of molecules into symmetry groups can be achieved following the scheme provided in Figure 2.6 [2].

2.2 Group representations

Next, a scheme is presented to describe the effect of a symmetry operation mathematically. This is accomplished most easily by considering a Cartesian coordinate system, and following the effect of a symmetry operation on this coordinate system. This is shown for an identity (E) and a reflection by a mirror plane, (the yz-plane) in Figure 2.7. To avoid confusion about the definitions of the symmetry planes σ_v and σ_v' (see Section 2.1) these symmetry planes will henceforth be referred to as σ_{xz} and σ_{yz}. For the discussion in this chapter, the following conventions will be adopted. When a symmetry operation is performed, the original

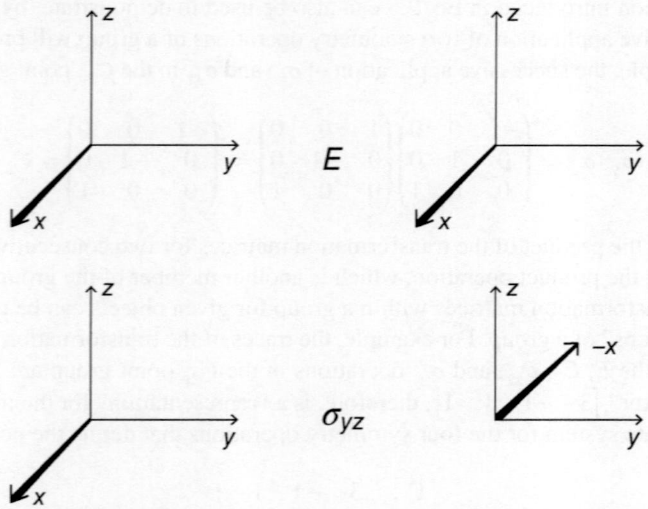

Figure 2.7 *Effects of symmetry operations* E *and* σ_{yz} *on a Cartesian coordinate system*

coordinates are referred to as the *x*, *y*, and *z* coordinates, whereas the coordinates after the transformation are designated as x', y', and z' coordinates. Then, each symmetry operation can be represented by a transformation matrix between the old and the new coordinate system. For example, the *E* operation in Figure 2.7 can then be described by

$$\begin{pmatrix} x' \\ y' \\ z' \end{pmatrix} = \begin{pmatrix} 1 & 0 & 0 \\ 0 & 1 & 0 \\ 0 & 0 & 1 \end{pmatrix} \begin{pmatrix} x \\ y \\ z \end{pmatrix} \tag{2.5}$$

whereas the σ_{yz} reflection is described by

$$\begin{pmatrix} x' \\ y' \\ z' \end{pmatrix} = \begin{pmatrix} -1 & 0 & 0 \\ 0 & 1 & 0 \\ 0 & 0 & 1 \end{pmatrix} \begin{pmatrix} x \\ y \\ z \end{pmatrix} \tag{2.6}$$

These 3×3 transformation matrices can be verified by multiplying out, for example, Eq. 2.6, which yields

$$x' = -x; y' = y; z' = z \tag{2.7}$$

This is what one expects by inspection of Figure 2.7.

Thus, the transformation matrices of a Cartesian coordinate system under the four symmetry operations of the point group C_{2v} can be summarized as:

$$E = \begin{pmatrix} 1 & 0 & 0 \\ 0 & 1 & 0 \\ 0 & 0 & 1 \end{pmatrix} \quad C_2 = \begin{pmatrix} -1 & 0 & 0 \\ 0 & -1 & 0 \\ 0 & 0 & 1 \end{pmatrix}$$
$$\sigma_{yz} = \begin{pmatrix} -1 & 0 & 0 \\ 0 & 1 & 0 \\ 0 & 0 & 1 \end{pmatrix} \quad \sigma_{xz} = \begin{pmatrix} 1 & 0 & 0 \\ 0 & -1 & 0 \\ 0 & 0 & 1 \end{pmatrix} \tag{2.8}$$

The matrix representation introduced in Eq. 2.8 can also be used to demonstrate, by simple matrix multiplication, that the successive application of two symmetry operations of a group will produce another operation of the group. For example, the successive application of σ_{yz} and σ_{xz} in the C_{2v} point group can be represented as follows:

$$\sigma_{yz}\,\sigma_{xz} = \begin{pmatrix} -1 & 0 & 0 \\ 0 & 1 & 0 \\ 0 & 0 & 1 \end{pmatrix} \begin{pmatrix} 1 & 0 & 0 \\ 0 & -1 & 0 \\ 0 & 0 & 1 \end{pmatrix} = \begin{pmatrix} -1 & 0 & 0 \\ 0 & -1 & 0 \\ 0 & 0 & 1 \end{pmatrix} = C_2 \tag{2.9}$$

This result implies that the product of the transformation matrices for two consecutive operations must equal the matrix representing the product operation, which is another member of the group, namely C_2.

The traces of the transformation matrices within a group for given objects can be used to construct what is known as "representations" of a group. For example, the traces of the transformation matrices for a Cartesian coordinate system for the E, C_2, σ_{xz}, and σ_{yz} operations in the C_{2v} point group are 3, −1, 1, 1, respectively (see Eq. 2.8). The "vector" $\begin{bmatrix} 3 & -1 & 1 & 1 \end{bmatrix}$, therefore, is a "representation" for the transformation properties of a Cartesian coordinate system for the four symmetry operations that define the point group C_{2v}:

$$\Gamma: \quad 3 \quad -1 \quad 1 \quad 1 \tag{2.10}$$

where the symbol Γ designates any representation, and the numeric values in the "representation vector" are given the symbols X. Thus, the notation $X(\Gamma_i)$ implies any of the numeric values listed in Eq. 2.10 for a given representation.

It is shown below that this particular representation, $[3, \; -1, \; 1, \; 1]$, can be interpreted as a vector in the four-dimensional space defined by C_{2v} that can be decomposed into the contributions along four one-dimensional unit vectors. These unit vectors are referred to as "irreducible representations" of a group, and are discussed next. These irreducible representations can be considered a way to represent the transformation properties of the very simplest objects, such as a general point. Under C_{2v}, a point lying along the z-axis (the rotation axes) would transform symmetrically (+1) under all symmetry operations of the group and thus have a representation of $[1, \; 1, \; 1, \; 1]$ for the four operations E, C_2, σ_{xz}, and σ_{yz}. A point on the x-axis, however, would transform antisymmetrically (−1) under C_2 and σ_{yz} and thus, would have a representation $[1, \; -1, \; 1, \; -1]$. Similarly, a point on the y-axis will transform antisymmetrically under C_2 and σ_{xz} and has a representation $[1, \; -1, \; -1, \; 1]$. It can be shown that these irreducible representations are the equivalents of unit vectors in a space whose dimension is given by the number of symmetry elements in a group. Thus, C_{2v}, which has four symmetry elements, will have four irreducible representations, three of which have been visualized so far:

$$[1, \; 1, \; 1, \; 1], \; [1, \; -1, \; 1, \; -1] \text{ and } [1, \; -1, \; -1, \; 1]$$

The theoretical derivation of the irreducible representations is quite complicated and follows directly from the *orthogonality theorem of group theory* (see, for example, [1]), but is not discussed here in detail. Some of the consequences of this theorem are listed below:

1. The number of irreducible representations (unit vectors) in a group is equal to the number of symmetry classes. Thus C_{2v} will have four irreducible representations, and the C_{3v} point group, introduced in Section 2.1, has three irreducible representations, although it has six symmetry operations.
2. If one defines a number l_i to be the dimensions of the ith irreducible representation (the largest number $X(\Gamma)$), then

$$\sum_i l_i^2 = h \tag{2.11}$$

In Eq. 2.11, *h* denotes the order of the group, which was defined above (Section 2.1) to be the total number of symmetry operations in a group. Thus, for C_{2v}

$$l_1^2 + l_2^2 + l_3^2 + l_4^2 = 4 \tag{2.12}$$

and

$$l_1 = l_2 = l_3 = l_4 = 1 \tag{2.13}$$

Thus, C_{2v} has four one-dimensional representations, three of which are listed above. The fourth irreducible representation can be derived from the orthogonality theorem, which postulates (among other conditions) that the four representations (being unit vectors) must be orthogonal to each other. Thus, in addition to the three irreducible representations listed above, one writes a fourth, orthogonal vector as

$$[1, \;\; 1, \;\; -1, \;\; -1]$$

which results in a (reorganized) table.

$$\begin{array}{llrrr} \Gamma_1: & 1 & 1 & 1 & 1 \\ \Gamma_2: & 1 & 1 & -1 & -1 \\ \Gamma_3: & 1 & -1 & 1 & -1 \\ \Gamma_4: & 1 & -1 & -1 & 1 \end{array} \tag{2.14}$$

Irreducible representations for a few selected other point groups will be discussed later. Next, the orthogonality of the four representations shown in Eq. 2.14 will be demonstrated. For this, an analogy in three-dimensional space, with unit vectors ***i***, ***j***, and ***k*** will be used. In three-dimensional space, defined by a Cartesian coordinate system, these unit vectors can be represented by the matrix

$$\begin{array}{lccc} & x & y & z \\ \boldsymbol{i}: & 1 & 0 & 0 \\ \boldsymbol{j}: & 0 & 1 & 0 \\ \boldsymbol{k}: & 0 & 0 & 1 \end{array} \tag{2.15}$$

In the matrix defining the *i*, *j*, and *k* vectors, the orthogonality of the unit vectors can be demonstrated by summing the products of the elements corresponding any two unit vectors, for example, *i* and *j*, over the two directions *x* and *y*:

$$\begin{array}{l} 1\cdot 0 + 0\cdot 1 + 0\cdot 0 = 0 \;\; \text{for } i \text{ and } j, \text{ and} \\ 1\cdot 0 + 0\cdot 0 + 0\cdot 1 = 0 \;\; \text{for } i \text{ and } k. \end{array} \tag{2.16}$$

In complete analogy, orthogonality in the four-dimensional space defined by C_{2v} is established by summing over all symmetry operations of a group the products of the *X*'s for any two unit vectors, *i* and *j*:

$$\sum\nolimits_\alpha X_\alpha(\Gamma_i)\, X_\alpha(\Gamma_j) \tag{2.17}$$

For the table of irreducible representations for C_{2v} listed in Eq. 2.12, the vector dot product, designated by the symbol ⊗, of Γ_1 and Γ_2 is

$$\Gamma_1 \otimes \Gamma_2 = 1\cdot 1 + 1\cdot 1 + 1\cdot(-1) + 1\cdot(-1) = 0 \tag{2.18}$$

that is, the two unit vectors Γ_1 and Γ_2 are orthogonal. It is easy to demonstrate that any binary combination of the unit vectors designated $\Gamma_1 \cdots \Gamma_4$ are, indeed, orthogonal.

Next, it is advantageous to discuss some more complicated symmetry point groups, such as C_{3v}, which was introduced earlier (Section 2.1):

$$C_{3v} : E, 2C_3, 3\sigma_v$$

The dimension (order) of this point group is 6 ($h = 6$), since there are six symmetry elements in three classes, and some elements occur more than once in this group. First, the matrix representation of the C_3 operation (a clockwise rotation by an angle of $\theta = 120°$) will be introduced:

$$\begin{pmatrix} x' \\ y' \\ z' \end{pmatrix} = \begin{pmatrix} \cos\theta & -\sin\theta & 0 \\ \sin\theta & \cos\theta & 0 \\ 0 & 0 & 1 \end{pmatrix} \begin{pmatrix} x \\ y \\ z \end{pmatrix} = \begin{pmatrix} -0.5 & -0.866 & 0 \\ 0.866 & -0.5 & 0 \\ 0 & 0 & 1 \end{pmatrix} \begin{pmatrix} x \\ y \\ z \end{pmatrix} \tag{2.19}$$

Equation 2.19 shows that for a rotation by 120°, the trace of the transformation is zero.

According to Eq. 2.11, C_{3v} will have three irreducible representations, but in order to fulfill Eq. 2.12, one of these, Γ_3, must be two-dimensional:

$$\begin{array}{llll} \Gamma_1 : & 1 & 1 & 1 \\ \Gamma_2 : & 1 & 1 & -1 \\ \Gamma_3 : & 2 & -1 & 0 \end{array} \tag{2.20}$$

since Eq. 2.12 requires that

$$l_1^2 + l_2^2 + l_3^2 = 1 + 1 + 4 = 6 \tag{2.21}$$

The dimensionality of a representation was alluded to above in Eq. 2.11. If the largest number in an irreducible representation is two, this representation is referred to as "doubly degenerate," if it is three, the representation is "triply degenerate." The concept of two- and three-dimensional representations needs to be explained in more detail. In three-dimensional space, a two-dimensional representation could be visualized by involving, for example, the x–y plane in a coordinate system that rotates in space about the z-axis. In this case, the x–y plane would be a two-dimensional coordinate. Similarly, one can argue that a representation that has elements larger than 1 represents two one-dimensional directions that cannot be separated.

In groups where the number of symmetry elements per class is larger than 1, such as the two C_3 elements or the three σ_v planes in C_{3v}, the orthogonality condition shown in Eq. 2.17 needs to be modified to include a factor g that indicates the number of times an operation occurs in a group. In C_{3v}, for example, this factor g would be 2 for the C_3 operations, and 3 for the reflection planes. The orthogonality condition then reads:

$$\sum\nolimits_{\alpha} g_\alpha X_\alpha(\Gamma_i)\, X_\alpha(\Gamma_j) \tag{2.22}$$

Equation 2.22 is the more general form of the orthogonality condition that was defined previously (Eq. 2.17).

To demonstrate the orthogonality of the representation listed in Eq. 2.20, one needs to determine the products

$$\begin{aligned} \Gamma_1 \otimes \Gamma_2 &= \mathbf{1} \cdot 1 \cdot 1 + \mathbf{2} \cdot 1 \cdot 1 + \mathbf{3} \cdot 1 \cdot (-1) = 0 \\ \Gamma_1 \otimes \Gamma_3 &= \mathbf{1} \cdot 1 \cdot 2 + \mathbf{2} \cdot 1 \cdot (-1) + \mathbf{3} \cdot 1 \cdot 0 = 0 \\ \Gamma_2 \otimes \Gamma_3 &= \mathbf{1} \cdot 1 \cdot 2 + \mathbf{2} \cdot 1 \cdot (-1) + \mathbf{3} \cdot (-1) \cdot 0 = 0 \end{aligned} \tag{2.23}$$

and finds, indeed, that the three irreducible representations are orthogonal. In Eq. 2.23, the bold numbers indicate the "g" factors, and the following digits the values $X_\alpha(\Gamma_i)$ and $X_\alpha(\Gamma_j)$.

In general, one reserves the designations $\Gamma_1 \cdots \Gamma_N$ for molecular representations (such as the symmetry properties of molecular vibrations, or molecular orbitals), and uses a different nomenclature for the *irreducible* representations of a group. This nomenclature is as follows: the totally symmetric representation of a group, that is, the irreducible representations whose $X(\Gamma)$ values are all +1, is generally given the designation A_x, where the subscript x can be "g," "1," or "1g" (depending on whether a group includes a center of inversion operation). In both C_{2v} and C_{3v}, the totally symmetric representation is designated A_1. The subscripts g and u denote symmetric (gerade) and antisymmetric (ungerade) with respect to a center of inversion.

There is always a totally symmetric representation in a group that is always written as the first row in the character table. The irreducible representation that transforms symmetrically (+1) with respect to the highest axis of symmetry will be given the designation A_2 or A_{2g}. Representations that transform antisymmetrically (−1) with respect to the major axis of symmetry are given the symbol B. Doubly degenerate (two-dimensional) representations are referred to as E, and three-dimensional representations as T. With these designations, the irreducible representations (Eq. 2.14) of the symmetry group C_{2v} can be written as

C_{2v}	E	C_2	σ_v	σ_v'
A_1	1	1	1	1
A_2	1	1	−1	−1
B_1	1	−1	1	−1
B_2	1	−1	−1	1

(2.24)

The construct given by Eq. 2.24 is referred to as a "character table." Each symmetry group has its own characteristic character table, listed in Appendix C. Similarly, the character table for C_{3v} is given by

C_{3v}	E	$2C_3$	$3\sigma_v$
A_1	1	1	1
A_2	1	1	−1
E	2	−1	0

(2.25)

In the discussion of specific molecules, a few more character tables will be introduced and used to derive the symmetry species of these molecules' vibrational modes, and their activity in either infrared or Raman spectroscopy.

Any reducible representation Γ, such as the one given in Eq. 2.10, can be decomposed into irreducible representations just as a vector in three-dimensional space can be decomposed into its Cartesian components along the unit vectors. This is accomplished by projecting the vector Γ onto the directions of the irreducible component unit vectors to determine its component along this direction. The representation derived in Eq. 2.10 for the transformation properties of a Cartesian coordinate system under the operations of the C_{2v} point group will serve as an example of how this decomposition is accomplished. The representation of the transformation properties of a Cartesian coordinate system,

$$\Gamma_{\text{CCS}} : \; 3 \quad -1 \quad 1 \quad 1 \tag{2.10}$$

henceforth will be referred to as Γ_{CCS}, where the subscript CCS stands for "Cartesian coordinate system." The decomposition of Γ_{CCS} into the four unit vectors A_1, A_2, B_1, and B_2 is carried out *via* the reduction (projection) formula:

$$n_i = \left(\frac{1}{h}\right) \sum\nolimits_{\alpha} g_\alpha X_\alpha(\Gamma_i)\, X_\alpha(\Gamma_{\text{CCS}}) \tag{2.26}$$

In Eq. 2.26, n_i indicates the number of times an irreducible representation i occurs in the reducible representation, here Γ_{CCS}. This can be summarized in the following set of equations:

$$n_{A_1} = \left(\frac{1}{4}\right) [\mathbf{1} \cdot 1 \cdot 3 + \mathbf{1} \cdot 1 \cdot (-1) + \mathbf{1} \cdot 1 \cdot 1 + \mathbf{1} \cdot 1 \cdot 1] = \left(\frac{1}{4}\right) [3 - 1 + 1 + 1] = 1$$

$$n_{A_2} = \left(\frac{1}{4}\right) [\mathbf{1} \cdot 1 \cdot 3 + \mathbf{1} \cdot 1 \cdot (-1) + \mathbf{1} \cdot (-1) \cdot 1 + \mathbf{1} \cdot (-1) \cdot 1] = \left(\frac{1}{4}\right) [3 - 1 - 1 - 1] = 0$$

$$n_{B_1} = \left(\frac{1}{4}\right)[\mathbf{1}\cdot 1\cdot 3+\mathbf{1}\cdot(-1)\cdot(-1)+\mathbf{1}\cdot 1\cdot 1+\mathbf{1}\cdot(-1)\cdot 1] = \left(\frac{1}{4}\right)[3+1+1-1] = 1$$

$$n_{B_2} = \left(\frac{1}{4}\right)[\mathbf{1}\cdot 1\cdot 3+\mathbf{1}\cdot(-1)\cdot(-1)+\mathbf{1}\cdot(-1)\cdot 1+\mathbf{1}\cdot 1\cdot 1] = \left(\frac{1}{4}\right)[3+1-1+1] = 1 \quad (2.27)$$

In these equations, the factor "*g*" is indicated in bold typeface. The results presented in Eq. 2.27 indicate that the "reducible representation" Γ_{CCS} can be decomposed into the following unit vectors:

$$\Gamma_{CCS} = 1A_1 + 1B_1 + 1B_2 \quad (2.28)$$

The same procedure will be utilized, in Section 2.3, to derive symmetry species of the individual vibrational modes of polyatomic molecules. At this point, a few more comments about the character tables introduced above are appropriate. In addition to the information given in Eqs. 2.24 and 2.25, character tables of molecular point groups contain two more columns, as shown below:

C_{2v}	E	C_2	σ_v	σ_v'		
A_1	1	1	1	1	T_z	x^2, y^2, z^2
A_2	1	1	−1	−1	R_z	xy
B_1	1	−1	1	−1	T_x, R_y	xz
B_2	1	−1	−1	1	T_y, R_x	yz

(2.29)

The first of these additional columns contains entries of the form T_α and R_α, where α stands for *x*, *y*, or *z* and *T* and *R* for translation or rotation, respectively. Thus, this column indicates how a translation or rotation of the coordinate would transform under the symmetry operations of the group. It can be visualized easily that a translation of a Cartesian coordinate system along the *z*-direction transforms symmetrically under all operations of the group. Often, the translational components T_x, T_y, and T_z are just written as *x*, *y*, and *z* in character tables. The reason for this will be appreciated later (Section 2.4) since the Cartesian components of the dipole operator, μ_x, μ_y, and μ_z, have the same transformation properties as do the translations in the *x*, *y*, and *z* directions; thus, this column in the character table determines whether a given representation is allowed in any spectroscopy that depends on the change in the molecular dipole moment or one of its components. A rotation of the coordinate system about the *z*-axis transforms as a C_2 operation, which transforms as A_2 (see Eq. 2.9).

The last column in Eq. 2.29 indicates the transformation properties of binary combinations of Cartesian coordinates. This column, therefore, determines whether a given symmetry representation is allowed in Raman spectroscopy, since in Raman spectroscopy, one or more of the polarizability tensor components must change during a vibration, and these tensor components transform as the binary combinations of the Cartesian coordinates.

2.3 Symmetry representations of molecular vibrations

As pointed out in the introduction of this chapter, the main goal of the discussion of group theory is to determine which of the vibrational modes of a molecule are allowed in infrared absorption or Raman scattering spectroscopy. This will be accomplished by establishing the symmetry (i.e., the irreducible representations) of given vibrational coordinates, and determining whether these representations transform like *x*, *y*, or *z* (for infrared spectroscopy), or any binary combinations of *x*, *y*, or *z* for Raman spectroscopy.

Thus, one needs to determine how a given normal coordinate transforms under the symmetry operations of a group. In order to accomplish this, one assumes that the normal modes of vibration can be written as linear combinations of *x*, *y*, or *z* Cartesian *displacement* coordinates. Thus, one attaches a *Cartesian displacement*

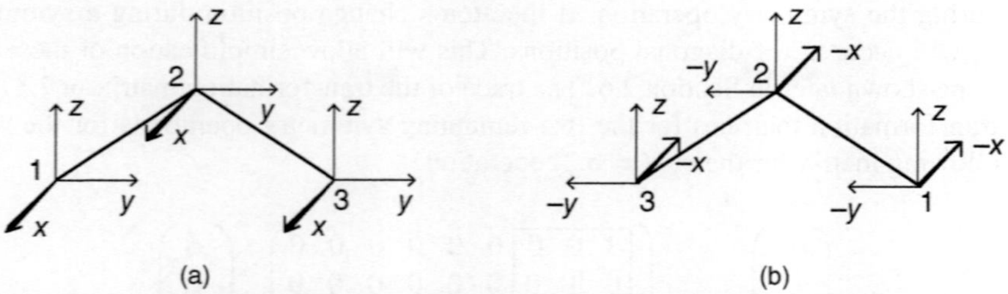

Figure 2.8 *Transformation of Cartesian displacement vectors under* C_2 *operation*

coordinate system to each atom, and follows how each of these coordinate components transforms under the symmetry operations of the group. The previously discussed example of the normal modes of the water molecule will be used to demonstrate these principles (see Figure 1.2). This molecule, along with a set of Cartesian displacement coordinates for each of the atoms, is shown in Figure 2.8(a). As before, when deriving the reducible representation of a coordinate system under the symmetry operations of a group, one needs to establish the traces of the transformation matrices for each of the atomic displacement coordinate systems. Figure 2.8(b) shows the molecule after applying a C_2 operation. As before, one designates the new coordinates (after the symmetry operation) as the primed coordinates and the original coordinates as the unprimed ones, and obtains the transformation matrix for the C_2 operation as shown below:

$$\begin{pmatrix} x_1' \\ y_1' \\ z_1' \\ x_2' \\ y_2' \\ z_2' \\ x_3' \\ y_3' \\ z_3' \end{pmatrix} = \begin{pmatrix} 0 & 0 & 0 & 0 & 0 & 0 & -1 & 0 & 0 \\ 0 & 0 & 0 & 0 & 0 & 0 & 0 & -1 & 0 \\ 0 & 0 & 0 & 0 & 0 & 0 & 0 & 0 & 1 \\ 0 & 0 & 0 & -1 & 0 & 0 & 0 & 0 & 0 \\ 0 & 0 & 0 & 0 & -1 & 0 & 0 & 0 & 0 \\ 0 & 0 & 0 & 0 & 0 & 1 & 0 & 0 & 0 \\ -1 & 0 & 0 & 0 & 0 & 0 & 0 & 0 & 0 \\ 0 & -1 & 0 & 0 & 0 & 0 & 0 & 0 & 0 \\ 0 & 0 & 1 & 0 & 0 & 0 & 0 & 0 & 0 \end{pmatrix} \begin{pmatrix} x_1 \\ y_1 \\ z_1 \\ x_2 \\ y_2 \\ z_2 \\ x_3 \\ y_3 \\ z_3 \end{pmatrix} \tag{2.30}$$

The trace of this transformation matrix is −1. Similarly, the trace of the transformation matrix under the identity operation is 9, since each Cartesian displacement vector component is unchanged under the E operation:

$$\begin{pmatrix} x_1' \\ y_1' \\ z_1' \\ x_2' \\ y_2' \\ z_2' \\ x_3' \\ y_3' \\ z_3' \end{pmatrix} = \begin{pmatrix} 1 & 0 & 0 & 0 & 0 & 0 & 0 & 0 & 0 \\ 0 & 1 & 0 & 0 & 0 & 0 & 0 & 0 & 0 \\ 0 & 0 & 1 & 0 & 0 & 0 & 0 & 0 & 0 \\ 0 & 0 & 0 & 1 & 0 & 0 & 0 & 0 & 0 \\ 0 & 0 & 0 & 0 & 1 & 0 & 0 & 0 & 0 \\ 0 & 0 & 0 & 0 & 0 & 1 & 0 & 0 & 0 \\ 0 & 0 & 0 & 0 & 0 & 0 & 1 & 0 & 0 \\ 0 & 0 & 0 & 0 & 0 & 0 & 0 & 1 & 0 \\ 0 & 0 & 0 & 0 & 0 & 0 & 0 & 0 & 1 \end{pmatrix} \begin{pmatrix} x_1 \\ y_1 \\ z_1 \\ x_2 \\ y_2 \\ z_2 \\ x_3 \\ y_3 \\ z_3 \end{pmatrix} \tag{2.31}$$

From Eqs. 2.30 and 2.31, it becomes obvious that the "molecular" transformation matrix (i.e., the 9 × 9 element matrix 2.30) for the water molecule contains three transformation sub-matrices, one for each individual Cartesian displacement coordinate system, and that these transformation sub-matrices are located at the diagonal of the overall transformation matrix if the particular atom does not change its position with another atom during the symmetry operation. If the atoms change position during a symmetry operation, the sub-matrices will occur at off-diagonal positions. This will allow simplification of these transformation matrices, as will be shown later in Section 2.6. The trace of the transformation matrix in 2.31 is 9.

Writing the transformation matrices for the two remaining symmetry operations for the water molecule, one finds the following matrix for the σ_v (= σ_{yz}) operation

$$\begin{pmatrix} x_1' \\ y_1' \\ z_1' \\ x_2' \\ y_2' \\ z_2' \\ x_3' \\ y_3' \\ z_3' \end{pmatrix} = \begin{pmatrix} -1 & 0 & 0 & 0 & 0 & 0 & 0 & 0 & 0 \\ 0 & 1 & 0 & 0 & 0 & 0 & 0 & 0 & 0 \\ 0 & 0 & 1 & 0 & 0 & 0 & 0 & 0 & 0 \\ 0 & 0 & 0 & -1 & 0 & 0 & 0 & 0 & 0 \\ 0 & 0 & 0 & 0 & 1 & 0 & 0 & 0 & 0 \\ 0 & 0 & 0 & 0 & 0 & 1 & 0 & 0 & 0 \\ 0 & 0 & 0 & 0 & 0 & 0 & -1 & 0 & 0 \\ 0 & 0 & 0 & 0 & 0 & 0 & 0 & 1 & 0 \\ 0 & 0 & 0 & 0 & 0 & 0 & 0 & 0 & 1 \end{pmatrix} \begin{pmatrix} x_1 \\ y_1 \\ z_1 \\ x_2 \\ y_2 \\ z_2 \\ x_3 \\ y_3 \\ z_3 \end{pmatrix} \tag{2.32}$$

where each of the sub-matrices is identical to the one given for the σ_v operation in Eq. 2.6. Reflection of the entire molecule by the yz-plane leaves all atoms in place; therefore, all displacement coordinate sub-matrices appear on the diagonal, and the trace of the overall transformation matrix is 3.

Finally, the $\sigma_v{}'$ (= σ_{xz}) operation exchanges the positions of atoms 1 and 3, and the individual transformation sub-matrices for each atom are given by Eq. 2.8:

$$\begin{pmatrix} x_1' \\ y_1' \\ z_1' \\ x_2' \\ y_2' \\ z_2' \\ x_3' \\ y_3' \\ z_3' \end{pmatrix} = \begin{pmatrix} 0 & 0 & 0 & 0 & 0 & 0 & 1 & 0 & 0 \\ 0 & 0 & 0 & 0 & 0 & 0 & 0 & -1 & 0 \\ 0 & 0 & 0 & 0 & 0 & 0 & 0 & 0 & 1 \\ 0 & 0 & 0 & 1 & 0 & 0 & 0 & 0 & 0 \\ 0 & 0 & 0 & 0 & -1 & 0 & 0 & 0 & 0 \\ 0 & 0 & 0 & 0 & 0 & 1 & 0 & 0 & 0 \\ 1 & 0 & 0 & 0 & 0 & 0 & 0 & 0 & 0 \\ 0 & -1 & 0 & 0 & 0 & 0 & 0 & 0 & 0 \\ 0 & 0 & 1 & 0 & 0 & 0 & 0 & 0 & 0 \end{pmatrix} \begin{pmatrix} x_1 \\ y_1 \\ z_1 \\ x_2 \\ y_2 \\ z_2 \\ x_3 \\ y_3 \\ z_3 \end{pmatrix} \tag{2.33}$$

with a trace of 1. Thus, the (reducible) representation of the nine displacement coordinates, from which the normal modes of vibration can be constructed, is

$$\Gamma_{H_2O} = [9 \quad -1 \quad 3 \quad 1] \tag{2.34}$$

Next, the reduction formula discussed above (Eq. 2.26) will be used to determine the contributions of the four irreducible representations to the reducible representation of the displacement coordinates for water:

$$n_i = \left(\frac{1}{h}\right) \sum\nolimits_\alpha g_\alpha X_\alpha(\Gamma_i)\, X_\alpha(\Gamma_{H_2O}) \tag{2.26}$$

which reveals

$$n(A_1) = \frac{1}{4}[(1)\cdot(1)\cdot(9) + (1)\cdot(1)\cdot(-1) + (1)\cdot(1)\cdot(3) + (1)\cdot(1)\cdot(1)] = \frac{1}{4}[9-1+3+1] = 3$$

$$n(A_2) = \frac{1}{4}[(1)\cdot(1)\cdot(9) + (1)\cdot(1)\cdot(-1) + (1)\cdot(-1)\cdot(3) + (1)\cdot(-1)\cdot(1)] = \frac{1}{4}[9-1-3-1] = 1$$

$$n(B_1) = \frac{1}{4}[(1)\cdot(1)\cdot(9) + (1)\cdot(-1)\cdot(-1) + (1)\cdot(1)\cdot(3) + (1)\cdot(-1)\cdot(1)] = \frac{1}{4}[9+1+3-1] = 3$$

$$n(B_2) = \frac{1}{4}[(1)\cdot(1)\cdot(9) + (1)\cdot(-1)\cdot(-1) + (1)\cdot(-1)\cdot(3) + (1)\cdot(1)\cdot(1)] = \frac{1}{4}[9+1-3+1] = 2 \quad (2.35)$$

Thus, the reducible representation $\Gamma_{H_2O} = [9 \quad -1 \quad 3 \quad 1]$ can be decomposed into, or reduced to,

$$3A_1 + 1A_2 + 3B_1 + 2B_2 \quad (2.36)$$

At this point, one should check that these contributions, indeed, add up to nine degrees of freedom, (3 + 1 + 3 + 2 = 9) since nine displacement coordinates were attached to the water molecule (In case of doubly or triply degenerate representations, the degeneracy needs to be accounted for in this check). Furthermore, one may wish to ascertain that the sum of the four representations, multiplied by their abundance, reproduces the original reducible representation:

$3\ A_1$:	3	3	3	3
$1\ A_2$:	1	1	−1	−1
$3\ B_1$:	3	−3	3	−3
$2\ B_2$:	2	−2	−2	2
	9	−1	3	1

This decomposition presented in Eq. 2.35 revealed nine degrees of freedom, since three Cartesian degrees of freedom were assigned to each atom (*cf.* Figure 2.8). However, since there are only 3 $N-6$ or three degrees of vibrational freedom for the water molecule, the other six degrees are three translational and three rotational degrees of freedom. These translational and rotational modes originate from certain combinations of displacement coordinates: For example, if all atoms are displaced simultaneously in the *z*-direction, the combined motion is a translation of the entire molecule in the *z*-direction. Similarly, if atom 1 moves in the positive *x*-direction and atom 3 moves in the negative *x*-direction, the entire molecule rotates about the *z*-axis. As pointed out above, these translational and rotational degrees of freedom are listed in the character table of the symmetry group C_{2v} (Eq. 2.29) in the column containing the entries T_x, T_y, T_z, R_x, R_y, and R_z. This table reveals that the translation of the water molecule along the positive *z*-axis transforms as the A_1 reducible representation, while the T_x and T_y translations transform as B_1 and B_2, respectively. In order to obtain the symmetry properties of the three vibrational modes of water, one subtracts the translational and rotational representations from the total decomposition of the nine displacement coordinates given by Eq. 2.36 and obtains

$$3A_1 + 1A_2 + 3B_1 + 2B_2 - A_1 - B_1 - B_2 - A_2 - B_1 - B_2 = 2A_1 + B_1$$
$$\qquad T_z \quad T_x \quad T_y \quad R_z \quad R_y \quad R_x \quad (2.37)$$

Thus, one ends up with two A_1 and one B_1 irreducible representations for the symmetry species of the three normal modes of water, which were depicted in Figure 1.2.

This procedure is generally applicable for any molecule. For ammonia with C_{3v} symmetry, for example, the reducible representation of the Cartesian displacement coordinates is

$$\Gamma_{NH_3} = [12 \quad 0 \quad 2] \quad (2.38)$$

This reducible representation can be derived easily using Eq. 2.19 for the rotation matrix applied to each of the Cartesian displacement coordinate sets. The trace of Eq. 2.19, with $\theta = 120°$, is 0, and the trace for each of the σ_v operation is +1. This trace enters into the overall transformation matrix only for the atoms that do not change position during the reflection operation.

Using the reduction formula, Eq. 2.26 discussed above, the reducible representation shown in Eq. 2.38 can be decomposed into

$$3A_1 + 1A_2 + 4E \tag{2.39}$$

Since each of the degenerate E representations accounts for two degrees of freedom, Eq. 2.39, indeed, represents 12 degrees of freedom. Next, the translational and rotational degrees of freedom will be subtracted. These are available from the complete character table for C_{3v}:

C_{3v}	E	$2C_3$	$3\sigma_v$		
A_1	1	1	1	T_z	x^2+y^2, z^2
A_2	1	1	−1	R_z	
E	2	−1	0	(T_x, T_y) (R_x, R_y)	$(x^2-y^2, xy)(xz, yz)$

(2.25)

Notice that the two-dimensional representation E accounts for two degrees of translation (T_x, T_y) and two degrees of rotation (R_x, R_y). The vibrational degrees of freedom, therefore, are composed of $2A_1$ and $2E$ irreducible representations, accounting for six ($3N - 6 = 12 - 6 = 6$) degrees of freedom. At the end of this chapter, one more complete example for obtaining the reducible representation, the decomposition into the vibrational degrees of freedom, and the assignment of observed infrared and Raman transitions to the derived symmetry species will be presented.

2.4 Symmetry-based selection rules for absorption processes

Next, the symmetry rules for infrared absorption, based on the symmetry species for given vibrational transitions is discussed. This discussion follows the parity arguments presented before (Chapter 1 – Eqs. 1.94–1.102 and Figure 1.8), but explicitly includes the symmetry aspects discussed in this section.

As discussed earlier, the vibrational energy of a molecule is determined by the (time independent) vibrational Schrödinger equation, presented earlier (Eq. 1.67). This equation yields the vibrational energy eigenvalues, which are independent of symmetry; that is, a symmetry operation performed on the vibrational wavefunction will not affect the energy. This last statement, in mathematical terms, implies that the vibrational Hamiltonian, $\hat{H}$, and any arbitrary symmetry operator, $\hat{R}$, commutate:

$$[\hat{H}, \hat{R}] = 0 \tag{2.40}$$

that is, the same eigenfunctions ψ_{vib} are solutions both for the vibrational energy and the symmetry operations.

The total vibrational wavefunction for a polyatomic molecule was introduced earlier:

$$\psi_{\text{vib}} = \psi_1(Q_1) \cdot \psi_2(Q_2) \cdot \psi_3(Q_3) \cdots \tag{1.66}$$

which may be written in an abbreviated form as

$$\psi_{\text{vib}} = \prod_{i=1}^{3N-6} \psi_i^0(Q_i) \tag{2.41}$$

Here, ψ_i^0 denotes the ground state vibrational wavefunction of the normal coordinate Q_i. If a transition along coordinate Q_j is excited, the excited state wavefunction can be written as

$$\psi_j^1(Q_j) \prod_{i \neq j}^{3N-7} \psi_i^0(Q_i) \tag{2.42}$$

According to Eq. 2.40, the transition described by

$$\prod_{i=1}^{3N-6} \psi_i^0(Q_i) \rightarrow \psi_j^1(Q_j)\prod_{i\neq j}^{3N-7} \psi_i^0(Q_i) \tag{2.43}$$

occurs such that both sides of Eq. 2.43 are equal. Here, only the symmetry of the one wavefunction, along which the excitation occurs, changes and one may concentrate the future discussion on this coordinate:

$$\psi_j^0(Q_j) \rightarrow \psi_j^1(Q_j) \tag{2.44}$$

For this coordinate, the transition moment

$$\int \psi_j^1(Q_j)\boldsymbol{\mu}\psi_j^0(Q_j)d\tau \tag{2.45}$$

needs to be evaluated, which can be written in terms of the Cartesian components of the transition operator μ:

$$\int \psi_j^1(Q_j)\mu_x\psi_j^0(Q_j)dx + \int \psi_j^1(Q_j)\mu_y\psi_j^0(Q_j)dy + \int \psi_j^1(Q_j)\mu_z\psi_j^0(Q_j)dz \tag{2.46}$$

Equation 2.46 implies that for a transition to occur, the integral over the excited and ground state wavefunctions and at least one component of the dipole operator must be non-zero. The ground state wavefunction for any normal coordinate always transforms as the totally symmetric representation of the symmetry group. This may appear counterintuitive at first since it appears that the normal coordinate should transform as the representation of the excited state. However, the ground state wavefunction of any normal coordinate is given by

$$\psi_j^0(Q_j) = N\ H_n(\sqrt{\alpha}\,Q_j)\,e^{-\frac{\alpha Q_j^2}{2}} \quad \text{with } \alpha = \frac{2\pi\nu}{\hbar} \quad \text{and } H_0 = 1 \tag{2.47}$$

which yields

$$\psi_j^0(Q_j) = N\,e^{-\frac{\alpha\,Q_j^2}{2}} \tag{2.48}$$

Equation 2.48 describes a Gaussian distribution in the coordinate Q_j that is totally symmetric.

The first excited state of the same coordinate, $\psi_j^1(Q_j)$, of course, has the symmetry of the normal coordinate Q_j since the Hermite polynomial for the first excited state,

$$H_1(\sqrt{\alpha}Q_i) = 2\,(\sqrt{\alpha}Q_i) \tag{2.49}$$

and thus, has symmetry of the excited state that is determined by the irreducible representation of the particular normal mode of vibration. Integrals of the form $\int \psi_j^1(Q_j)\mu_z\psi_j^0(Q_j)dz$ will henceforth be abbreviated as $\int f_1f_2f_3\ d\tau$, where f_3 denotes the ground state wavefunction that always transforms as the totally symmetric representation (see Eq. 2.49). f_1 and f_2 represent the excited state wavefunction and transition operator, respectively. Thus, for the total integral $\int f_1f_2f_3\ d\tau$ to be non-zero or even, the product $f_1\,f_2$ must be even as well. This is the case if f_1, the excited state wavefunction, transforms as one of the dipole moment components of the group, since then, its product with f_2 contains the totally symmetric representation of the group. The transformation properties of the components of the dipole operator were discussion before; see Eq. 2.29.

To illustrate the points in the last paragraph, the example of the representations of the vibrations of the water molecule is discussed. Equation 2.37 demonstrates that the vibrations of the water molecule belong to the irreducible representations $A_1\ +\ B_1$: the symmetric stretching mode and the deformation mode transform (see Figure 1.2) as A_1 and the antisymmetric stretching mode as B_1. The question now arises as to which of these modes is allowed in infrared absorption. According to the character table for C_{2v}, reproduced below,

C_{2v}	E	C_2	σ_v	σ_v'		
A_1	1	1	1	1	T_z	x^2, y^2, z^2
A_2	1	1	−1	−1	R_z	xy
B_1	1	−1	1	−1	T_x, R_y	xz
B_2	1	−1	−1	1	T_y, R_x	yz

(2.29)

the A_1 irreducible representation transforms as a translation along the z-direction. Thus, the dipole operator component μ_z also transforms as A_1. For either of the transitions of A_1 symmetry (i.e., the deformation and the symmetric stretching mode), the ground vibrational state is totally symmetric and transforms as A_1. The excited vibrational state for either of these modes also transforms as A_1, as pointed out before (Eq. 2.49). Since the transition moment μ_z also transforms as A_1, the product of the excited state and the dipole operator component (both A_1) certainly contains the totally symmetric representation, and both transitions are allowed in absorption.

For the $0 \rightarrow 1$ transition of the antisymmetric stretching mode of B_1 symmetry one proceeds as follows. Again, the ground state vibrational mode transforms as A_1 (see Eq. 2.48). The excited state transforms as B_1, as does one of the components of the dipole operator, μ_x. Thus, the product $f_1 f_2$ transforms as $B_1 B_1$, and it is easy to see that this product $\begin{bmatrix}1 & 1 & 1 & 1\end{bmatrix}$ contains the totally symmetric representation (or transforms as the totally symmetric representation) of the group, and the antisymmetric stretching vibration of B_1 symmetry is allowed as well.

In summary, the discussion in Section 2.4 demonstrated that any vibrational transition will be allowed in absorption if its irreducible representation contains a component of the electric dipole moment, μ_x, μ_y, or μ_z (or the translational directions T_x, T_y, or T_z).

2.5 Selection rules for Raman scattering

The principles discussed in the previous section can be applied to Raman scattering as well, but the Cartesian components of the dipole transition moment in Eq. 2.46 need to be replaced by the polarizability elements, since Raman spectra arise from the changes in polarizability α during a normal coordinate (see Eq. 1.118). Thus, while infrared transitions require

$$\left(\frac{\partial \mu_\alpha}{\partial Q_k}\right) \neq 0 \tag{2.50}$$

the corresponding condition for a transition in Raman spectroscopy is

$$\left(\frac{\partial \alpha_{\alpha\beta}}{\partial Q_k}\right) \neq 0 \tag{2.51}$$

In Eqs. 2.50 and 2.51, the subscripts α and $\alpha\beta$ denote permutations of the coordinates x, y, and z; $\boldsymbol{\mu}$ is the dipole operator, α the polarizability, and Q_k any normal coordinate. As elaborated upon in Chapter 1, a direct transition from a lower to a higher vibrational state requires absorption of exactly one photon; this absorption process is mediated by the transition dipole operator. Raman spectroscopy, on the other hand, is a process that involves the interaction of two photons. This was briefly introduced in Section 1.6, and will be elaborated upon in more detail in Chapter 4. Thus, the transition moment is determined by expressions of the form

$$\langle \psi_0 | \mu_\alpha | \psi_{\text{int}} \rangle \langle \psi_{\text{int}} | \mu_\beta | \psi_1 \rangle \tag{2.52}$$

where ψ_0 and ψ_1 are the vibrational ground and excited states, respectively, of a given normal coordinate Q_k and ψ_{int} is a virtual (vibronic) intermediate state (see Figure 1.13). Since two transition moments are formally involved in the transition process, different selection rules apply for Raman spectroscopy. The irreducible representations that support Raman transitions, therefore, have binary combinations of the Cartesian coordinates listed, for example, $x^2, y^2, z^2, xy, xz, yz$, or others. The character table for the symmetry group C_{2v}, shown in Eq. 2.29, has these binary combinations listed in a column to the right of the column containing the translational and rotational components. In complete analogy to the selection rules for absorption, a vibration transforming as a given irreducible representation will be allowed in Raman scattering if at least one of the binary combinations of the Cartesian coordinates is listed.

Inspection of the character tables in Appendix C reveals that symmetry groups that contain the inversion symmetry element, *i*, never have irreducible representations that simultaneously contain elements of the dipole operator (i.e., *x*, *y*, and *z*) and any binary combinations of these coordinates. This observation leads to what is known as the "mutual exclusion rule" that states that in a point group that contains a center of inversion, *i*, vibrational transitions cannot be simultaneously active in Raman scattering and infrared absorption. Thus, in molecules such as CO_2 (see Figure 2.1) or octahedral SF_6, vibrations that are Raman allowed are infrared forbidden, and *vice versa.* Isomeric molecules such as *cis*-dichloroethene (C_{2v}) and *trans*-dichloroethene (C_{2h}) can be identified easily by the fact that for the latter molecule, the mutual exclusion principle holds.

2.6 Discussion of selected small molecules

2.6.1 Tetrahedral molecules: carbon tetrachloride, CCl_4, and methane, CH_4

The final section in this chapter presents a detailed derivation of the allowed vibrational transitions for tetrahedral molecules (CCl_4, CH_4), followed by a summary of the vibrations of small molecules with symmetry lower than tetrahedral. Such molecules (chloroform and methyl chloride of C_{3v} symmetry, dichloromethane of C_{2v} symmetry, and methylene chloride-d1 of C_s symmetry) give important insight into how symmetry influences the observed vibrational spectra, and also, introduce the vibrations of very important moieties such as the methyl (—CH_3) and methylene (—CH_2—) groups.

This discussion starts with a detailed derivation of the reducible representation of the Cartesian displacement coordinates of a tetrahedral molecule such as carbon tetrachloride (CCl_4) or methane (CH_4), shown schematically in Figure 2.9(a). This figure shows the atomic numbering scheme used in the discussion to follow, and also shows the molecule in two different orientations that demonstrate different symmetry properties. Figure 2.9(b) depicts four of the five symmetry operations (except *E*) allowed for the T_d point group, for which the character table is listed in Eq. 2.53.

T_d	E	$8C_3$	$3C_2$	$6S_4$	$6\sigma_d$		
A_1	1	1	1	1	1		$x^2+y^2+z^2$
A_2	1	1	1	−1	−1		
E	2	−1	2	0	0		$(2z^2-x^2-y^2, x^2-y^2)$
T_1	3	0	−1	1	−1	(R_x,R_y,R_z)	
T_2	3	0	−1	−1	1	(x,y,z)	(xy, xz, yz)

(2.53)

Following the procedure defined earlier (Eqs. 2.31–2.34), the transformation matrices of the Cartesian displacement vectors for the five symmetry classes are defined next. In order to save space, this procedure will be presented in a somewhat abbreviated form: rather than listing all 15 Cartesian displacement coordinates

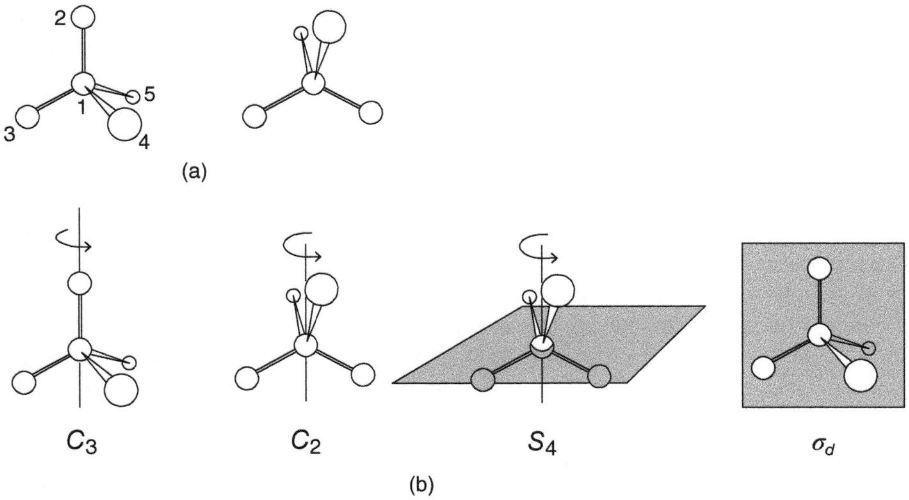

Figure 2.9 *Structure and symmetry operations for a tetrahedral molecule*

for all five atoms, they will be grouped as the coordinates ξ_i, with $i = 1, \ldots, 5$, where ξ denotes x, y, and z displacements. Thus, instead of listing the transformation matrix as was done previously (Eq. 2.31), this matrix is represented as follows:

$$\begin{pmatrix} \xi_1' \\ \xi_2' \\ \xi_3' \\ \xi_4' \\ \xi_5' \end{pmatrix} = \begin{pmatrix} M & 0 & 0 & 0 & 0 \\ 0 & M & 0 & 0 & 0 \\ 0 & 0 & M & 0 & 0 \\ 0 & 0 & 0 & M & 0 \\ 0 & 0 & 0 & 0 & M \end{pmatrix} \begin{pmatrix} \xi_1 \\ \xi_2 \\ \xi_3 \\ \xi_4 \\ \xi_5 \end{pmatrix} \tag{2.54}$$

where ξ_i denotes x_i, y_i, and z_i, the primed and unprimed quantities denote coordinates after and before a symmetry operation was applied, respectively, and M stands for a 3×3 transformation matrix specific to a given symmetry operation. These matrices were introduced in Eqs. 2.8 and 2.19.

For the identity operation, the overall transformation matrix is given by Eq. 2.54, with M given by

$$M = \begin{pmatrix} 1 & 0 & 0 \\ 0 & 1 & 0 \\ 0 & 0 & 1 \end{pmatrix} \tag{2.55}$$

which gives a trace of the transformation matrix of 15.

For each of the eight C_3 operations (along all four chemical bonds, both clockwise and counterclockwise) the operations can be presented by Eq. 2.56, which describes a counterclockwise rotation about the 1–2 bond direction. For this operation, atoms 1 and 2 (see Figure 2.9) stay in place, whereas atom 3 goes to the position of atom 4, atom 4 to atom 5, and atom 5 to atom 3. Thus, the transformation matrix has the form:

$$\begin{pmatrix} \xi_1' \\ \xi_2' \\ \xi_3' \\ \xi_4' \\ \xi_5' \end{pmatrix} = \begin{pmatrix} M & 0 & 0 & 0 & 0 \\ 0 & M & 0 & 0 & 0 \\ 0 & 0 & 0 & 0 & M \\ 0 & 0 & M & 0 & 0 \\ 0 & 0 & 0 & M & 0 \end{pmatrix} \begin{pmatrix} \xi_1 \\ \xi_2 \\ \xi_3 \\ \xi_4 \\ \xi_5 \end{pmatrix} \tag{2.56}$$

and the transformation matrix M has the form

$$M = \begin{pmatrix} -0.5 & -0.866 & 0 \\ 0.866 & -0.5 & 0 \\ 0 & 0 & 1 \end{pmatrix} \tag{2.57}$$

The trace of the C_3 transformation matrix for the entire molecule, therefore, is zero. Similarly, for all C_2 operations, four atoms (2 and 3, and 3 and 4) will change places, and only the central atom stays stationary. Thus, the overall transformation matrix for C_2 operations is

$$\begin{pmatrix} \xi_1' \\ \xi_2' \\ \xi_3' \\ \xi_4' \\ \xi_5' \end{pmatrix} = \begin{pmatrix} M & 0 & 0 & 0 & 0 \\ 0 & 0 & M & 0 & 0 \\ 0 & M & 0 & 0 & 0 \\ 0 & 0 & 0 & 0 & M \\ 0 & 0 & 0 & M & 0 \end{pmatrix} \begin{pmatrix} \xi_1 \\ \xi_2 \\ \xi_3 \\ \xi_4 \\ \xi_5 \end{pmatrix} \tag{2.58}$$

with

$$M = \begin{pmatrix} -1 & 0 & 0 \\ 0 & -1 & 0 \\ 0 & 0 & 1 \end{pmatrix} \tag{2.59}$$

Thus, the trace of the transformation matrix given in Eq. 2.59 is −1.

The changes in atomic positions for one of the S_4 operations are shown in Figure 2.10.

The overall transformation matrix is

$$\begin{pmatrix} \xi_1' \\ \xi_2' \\ \xi_3' \\ \xi_4' \\ \xi_5' \end{pmatrix} = \begin{pmatrix} M & 0 & 0 & 0 & 0 \\ 0 & 0 & 0 & M & 0 \\ 0 & 0 & 0 & 0 & M \\ 0 & 0 & M & 0 & 0 \\ 0 & M & 0 & 0 & 0 \end{pmatrix} \begin{pmatrix} \xi_1 \\ \xi_2 \\ \xi_3 \\ \xi_4 \\ \xi_5 \end{pmatrix} \tag{2.60}$$

Since the M-matrix for an S_4 operation is

$$M = \begin{pmatrix} 0 & 1 & 0 \\ -1 & 0 & 0 \\ 0 & 0 & -1 \end{pmatrix} \tag{2.61}$$

the trace of the matrix given by Eq. 2.61 will be −1. Finally, for reflection by a mirror plane, three atoms remain stationary and only two atoms exchange positions, as shown in Figure 2.11

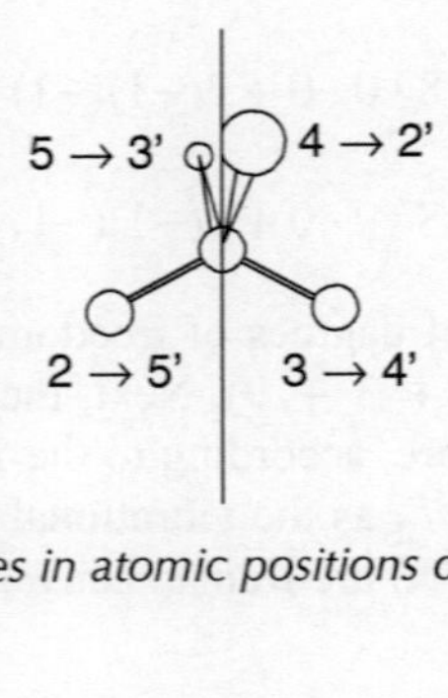

Figure 2.10 *Changes in atomic positions during an* S_4 *operation*

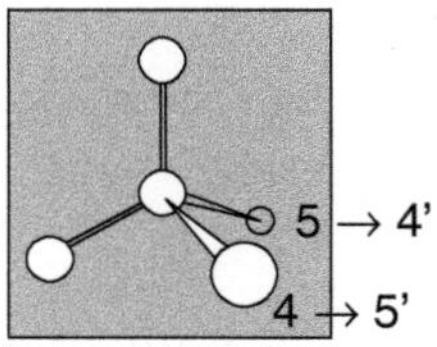

Figure 2.11 *Changes of atomic positions for a reflection by a mirror plane*

Consequently, the overall transformation matrix is

$$\begin{pmatrix} \xi'_1 \\ \xi'_2 \\ \xi'_3 \\ \xi'_4 \\ \xi'_5 \end{pmatrix} = \begin{pmatrix} M & 0 & 0 & 0 & 0 \\ 0 & M & 0 & 0 & 0 \\ 0 & 0 & M & 0 & 0 \\ 0 & 0 & 0 & 0 & M \\ 0 & 0 & 0 & M & 0 \end{pmatrix} \begin{pmatrix} \xi_1 \\ \xi_2 \\ \xi_3 \\ \xi_4 \\ \xi_5 \end{pmatrix} \tag{2.62}$$

with

$$M = \begin{pmatrix} -1 & 0 & 0 \\ 0 & 1 & 0 \\ 0 & 0 & 1 \end{pmatrix} \tag{2.63}$$

The trace of the matrix in Eq. 2.62 is +3. The reducible representation of the Cartesian displacement coordinates that form the basis of the displacement vectors of the normal coordinates is

$$\begin{matrix} & E & C_3 & C_2 & S_4 & \sigma_d \\ \Gamma_{T_d} & 15 & 0 & -1 & -1 & 3 \end{matrix} \tag{2.64}$$

Next, one decomposes this reducible representation into its contributions of the irreducible representations, using

$$n_i = \left(\frac{1}{h}\right) \sum\nolimits_\alpha g_\alpha X_\alpha(\Gamma_i)\, X_\alpha(\Gamma_{\mathrm{Td}}) \tag{2.26}$$

For the T_d point group, $h = 24$ (see Eq. 2.24), and the reduction yields

$$n(A_1) = \left(\frac{1}{24}\right)[1 \cdot 15 \cdot 1 + 8 \cdot 0 \cdot 1 + 3 \cdot (-1)1 + 6 \cdot (-1) \cdot 1 + 6 \cdot 3 \cdot 1] = 1$$

$$n(A_2) = \left(\frac{1}{24}\right)[1 \cdot 15 \cdot 1 + 8 \cdot 0 \cdot 1 + 3(-1)1 + 6(-1)(-1) + 6 \cdot 3(-1)] = 0$$

$$n(E) = \left(\frac{1}{24}\right)[1 \cdot 15 \cdot 2 + 8 \cdot 0(-1) + 3(-1)2 + 6(-1) \cdot 0 + 6 \cdot 3 \cdot 0] = 1$$

$$n(T_1) = \left(\frac{1}{24}\right)[1 \cdot 15 \cdot 3 + 8 \cdot 0 \cdot 0 + 3(-1)(-1) + 6(-1) \cdot 1 + 6 \cdot 3(-1)] = 1$$

$$n(T_2) = \left(\frac{1}{24}\right)[1 \cdot 15 \cdot 3 + 8 \cdot 0 \cdot 0 + 3(-1)(-1) + 6(-1)(-1) + 6 \cdot 3 \cdot 1] = 3 \tag{2.65}$$

or $1A_1 + E + T_1 + 3\,T_2$. The number of degrees of freedom expressed by these irreducible representations does, indeed, add up to 15: (1 + 2 + 3 + 9). Next, the three translational and the three rotational degrees of freedom are subtracted, which are, according to the T_d character table above, one T_1 and one T_2 representation. This leaves $1A_1 + E + 2\,T_2$ as the vibrational representations, which, indeed, account for $3\,N - 6 = 9$ degrees of freedom. Of these, all are Raman active, since polarizability tensor terms appear for

Table 2.1 Observed IR and Raman transitions for carbon tetrachloride[a]

Designation	Observed transition (cm^{-1})	Symmetry	Activity	Mode description
ν_1	215	E	Ra	Deformation
ν_2	314	T_2	Ra, IR	Deformation
ν_3	458	A_1	Ra	Symmetric stretching
ν_4	790	T_2	Ra, IR	Antisymmetric stretching

[a]All vibrational frequencies and assignments shown in Tables 2.1–2.6 were taken from this reference.

all three irreducible representations, but only the two T_2 modes are infrared active. Thus, due to degeneracies and restrictive selection rules, only two bands are observed in the infrared and only four in the Raman spectrum; the two infrared-active vibrations are observed in Raman spectroscopy as well.

Next, it is instructive to look at the displacement coordinates of the atoms for the normal modes of vibration. This is best accomplished by considering a distinct example of a molecule with T_d symmetry, such as carbon tetrachloride (CCl_4). The vibrational modes, along with observed vibrational frequencies, are listed in Table 2.1.

The atomic displacement vectors are represented schematically in Figure 2.12, but one should keep in mind that the absolute amplitude of the displacement vectors is indeterminate, and only the relative motion of

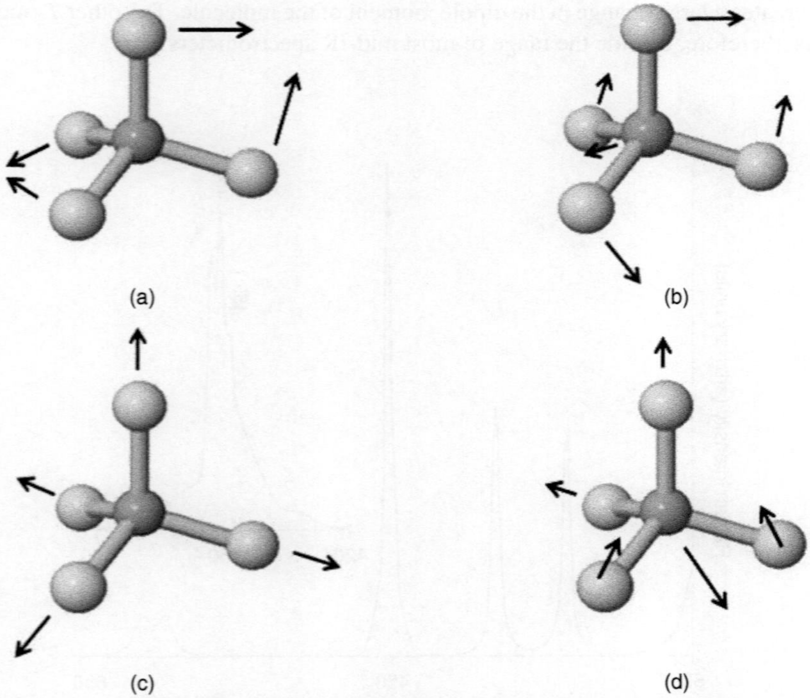

Figure 2.12 Approximate atomic displacement vectors for the four normal modes observed for carbon tetrachloride. (a) ν_1 at 215 cm^{-1}; (b) ν_2 at 314 cm^{-1} ; (c) ν_3 at 458 cm^{-1}; and (d) ν_4 at 790 cm^{-1}

the atoms with respect to each other is relevant. The actual motion can also be visualized by computer animation; for example, the website www.chemeddl.org has an easy to use graphic interface that permits for the inspection of the molecular vibrational motions. Other alternatives to visualize the atomic motions is by quantum mechanical computational packages such as Gaussian 2009 [3] (or earlier versions) or HyperChem [4]. It is also easy to see that for the E_2 mode, for example, there are different ways to construct deformation coordinates that describe similar motions, but utilize different atom pairs, since the mode is twofold degenerate.

The Raman spectrum of CCl_4 is shown in Figure 2.13. This spectrum shows the expected bands at about 215, 314, 458, and 790.5 cm^{-1}; in addition, a fifth peak is observed at 762 cm^{-1}. This latter peak is an overtone that is Fermi resonance enhanced (see Chapter 5). The inset of Figure 2.13 shows an expanded view of the strong A_1 mode of CCl_4; at higher spectral resolution, this band shows a resolved fine structure, which is due to isotopic splitting. Chlorine occurs naturally as a mixture of about 75% ^{35}Cl and 25% ^{37}Cl; hence, its average atomic mass is reported in most periodic tables to be about 35.5. Owing to the natural abundance of the chlorine isotopes, the most likely species to be found in a sample of CCl_4 is $C^{35}Cl_3{}^{37}Cl$. Therefore, the Raman scattered intensity of the A_1 mode of this isotopic species is largest, and corresponds to the peak at 458.4 cm^{-1}. The second most likely species is $C^{35}Cl_4$; since it corresponds to a lighter species, its A_1 mode frequency occurs at a slightly higher frequency, 461.5 cm^{-1}. The third most abundant species is $C^{35}Cl_2{}^{37}Cl_2$, and its symmetric stretching mode is observed at 455.1 cm^{-1}. At room temperature, the corresponding vibration of the other two species, $C^{35}Cl^{37}Cl_3$ and $C_3{}^{37}Cl_4$, are not resolved, but they can be observed at low temperature or in the gas phase. The low frequency side of this peak also contains a Fermi-resonance-enhanced overtone band of ν_3 and $2\nu_1$. The corresponding infrared absorption spectrum contained two fundamentals. The antisymmetric stretching mode, ν_4 at 790 cm^{-1} of T_2 symmetry, is extremely strong in the infrared spectrum because it creates a large change in the dipole moment of the molecule. The other T_2 mode occurs below 400 cm^{-1}, and is, therefore, outside the range of most mid-IR spectrometers.

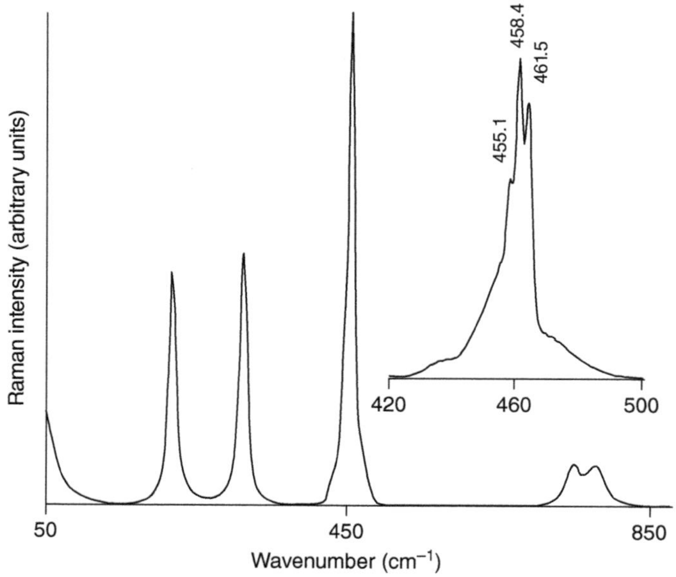

Figure 2.13 *Raman spectrum of liquid carbon tetrachloride*

Table 2.2 Observed IR and Raman transitions for methane

Designation	Observed transition (cm^{-1})	Symmetry	Activity	Mode description
ν_1	1306	T_2	Ra	Deformation
ν_2	1534	E	Ra, IR	Deformation
ν_3	2917	A_1	Ra	Symmetric stretching
ν_4	3019	T_2	Ra, IR	Antisymmetric stretching

Methane, CH_4, also has tetrahedral symmetry. Therefore, the vibrational irreducible representation is $1A_1 + E + 2\,T_2$ as well, and only the two T_2 representations are allowed in infrared absorption. The vibrational fundamentals for gaseous CH_4 are listed in Table 2.2.

For one of these, the antisymmetric stretching mode ν_4 observed at 3019 cm^{-1}, the spectral trace is shown in Figure 2.14. In addition to the strongest peak at 3019 cm^{-1}, a number of nearly equidistant "side bands" are observed, which are due to the interaction of vibration and free rotation of the molecule. Such "rot-vibrational" spectra are observed in both Raman and infrared spectroscopy of gaseous molecules. This book will not deal with pure molecular rotational spectroscopy at all, since this subject is outside the intended scope of discussion. However, the interaction between vibration and rotational transition is observed frequently (e.g., for atmospheric water and CO_2), and will be introduced in Chapter 5.

The rot-vibrational bands shown in Figure 2.14 result from transitions in which both

$$\Delta n = +1 \tag{2.66}$$

and

$$\Delta J = \pm 1 \tag{2.67}$$

that is, the molecules undergo transitions to a higher vibrational energy level, and to a lower or higher rotational energy level, indicated by the change in rotational quantum number J. As will be discussed later

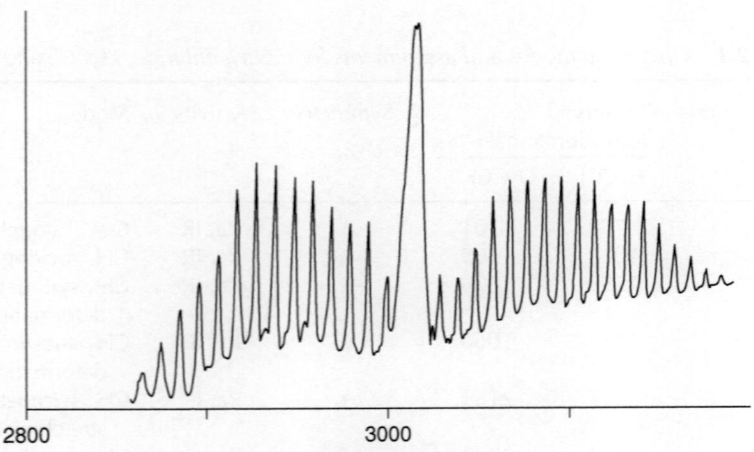

Figure 2.14 Rot-vibrational details of the ν_4 mode of methane

(Section 5.3), the rotational energy of a spherical molecule is given by

$$E_J = BJ(J + 1) \tag{2.68}$$

where B is the rotational constant, related to the moment of inertia of the molecule. Equations 2.67 and 2.68 lead to an energy level diagram shown in Figure 5.5, with the allowed transitions indicated by the solid arrows.

Returning to Figure 2.14, it is interesting to note that methane, as a tetrahedral molecule, does not possess a permanent dipole moment. Since pure rotation of a molecule does not alter a molecular dipole moment, one of the *sine-qua-non* conditions of rotational absorption spectroscopy is that the molecule must possess a permanent dipole moment. Thus, it may seem surprising, at first, to see a rot-vibrational spectrum for the ν_4 mode of CH_4. A plausible reason for this may be found by the following argument: the vibrational excited state of the ν_4 mode is no longer symmetric and possesses an albeit small dipole moment, which makes the transition allowed in rot-vibrational spectroscopy.

2.6.2 Chloroform and methyl chloride

Next, two molecules are discussed that result from the tetrahedral molecules discussed above by substitution of one of the atoms: when one chlorine atom in carbon tetrachloride is substituted by a hydrogen atom, one obtains chloroform, $HCCl_3$. Similarly, when one H-atom in methane is substituted by a chlorine atom, one obtains methyl chloride, H_3CCl. Both these molecules have C_{3v} symmetry. The discussion of these two species demonstrates how lowering the symmetry will affect the observed spectra. The latter of the two molecules is of particular interest because it serves as a model for the methyl group, —CH_3, which is found in many larger molecules, and whose individual vibrations can be identified in many molecules containing a methyl group.

The vibrational modes allowed in infrared and Raman spectroscopy are derived for these molecules as was demonstrated for carbon tetrachloride above, using the transformation properties of the Cartesian displacement coordinates and the character table for C_{3v}. For both molecules, one finds that the vibrational modes fall into $3A_1$ and $3E$ irreducible representations accounting for nine degrees of freedom. For methyl chloride, these modes are listed in Table 2.3. The listed frequencies, however, are not exactly the observed frequencies, since the symmetric stretching mode ν_5 is strongly influenced by Fermi resonance. This subject is discussed in more detail in Chapter 5; for the discussion here, may it suffice to state that the overtone of the antisymmetric

Table 2.3 *Vibrational modes and assignments for methyl chloride, H_3CCl, and D_3CCl*

Designation	Observed transition (cm^{-1})		Symmetry	Activity	Mode
	H_3CCl	D_3CCl			
ν_1	732	701	A_1	Ra, IR	C—Cl stretching
ν_2	1017	768	E	Ra, IR	CH_3 rocking
ν_3	1355	1029	A_1	Ra, IR	CH_3 symmetric deformation
ν_4	1452	1060	E	Ra, IR	CH_3 antisymmetric deformation
ν_5	2937	2160	A_1	Ra, IR	CH_3 symmetric stretching
ν_6	3039	2283	E	Ra, IR	CH_3 antisymmetric stretching

deformation mode ν_4, which is expected to occur around 2900 cm^{-1}, interacts, or strongly mixes, with ν_5. This mixing has two consequences: first, the intensities of the overtone and the fundamental become very similar; second, the splitting between these modes is increased. Thus, the C—H stretching region of methyl chloride does not show two peaks as expected (the symmetric and antisymmetric methyl stretching modes, of A_1 and E symmetry, respectively), but exhibits three peaks at 2879, 2968, and 3039 cm^{-1}. The first two of them are the Fermi resonance-coupled $2\nu_4 + \nu_5$ pair. From the theory of Fermi resonance (see Chapter 5), the unperturbed frequency of the ν_5 mode can be estimated and yields the value of 2937 cm^{-1} listed in Table 2.3.

Under C_{3v} symmetry, all the A_1 and E modes are allowed in both infrared and Raman spectroscopy; thus each spectrum should show six fundamental bands. Of these, all but ν_1 are characteristic "group frequencies" of the methyl group. This statement implies that most molecules containing methyl groups will exhibit these five vibrations relatively unchanged, even if the overall symmetry of the molecule is lower (e.g., in toluene of C_2 symmetry). For molecules of lower symmetry that do not support degeneracy, the E modes of the methyl group are split into two modes separated by a few wavenumbers. A schematic representation of the atomic displacement vectors, obtained by normal coordinate calculations (see Chapter 1), for the three stretching modes is shown in Figure 2.15.

This example demonstrates that lowering the symmetry increases the number of bands observed in the infrared and Raman spectra: while for the tetrahedral molecules discussed above only two fundamentals are observed in absorption, six bands are observed in C_{3v} molecules such as methyl chloride.

Next, the vibrational spectra of chloroform, $HCCl_3$, is discussed. Similarly to methyl chloride, chloroform has C_{3v} symmetry (ignoring chlorine isotopic effects) and exhibits the same symmetry representations, $3A_1$ and $3E$ modes. The observed vibrational frequencies are listed in Table 2.4.

Infrared and Raman spectra of chloroform in the mid-range (400–1600 cm^{-1}) region are shown in Figure 2.16. These spectral traces illustrate a number of important facts of vibrational spectroscopy. The two low frequency modes (at 261 and 363 cm^{-1}) of chloroform are outside the range of most modern infrared instruments, but can clearly be observed in Raman spectroscopy (*cf.* Figure 2.17). The symmetric stretching mode at 680 cm^{-1} is very strong in the Raman spectrum and relatively weak in the IR, whereas the antisymmetric stretching mode at 774 cm^{-1} is extremely strong in the IR, but weak in the Raman spectrum. It is generally true that modes that strongly change the dipole moment, such as an anti-symmetric stretching vibration, exhibit strong infrared bands, whereas vibrations with little change in the dipole moment, such as the symmetric stretching vibrations, are very weak in the infrared, but strong in the Raman spectra. This fact makes infrared and Raman spectroscopy two mutually complimentary tools. The small spectral features observed in the infrared (at about 20 μm sample path length) are overtones and combination bands that mostly have been assigned.

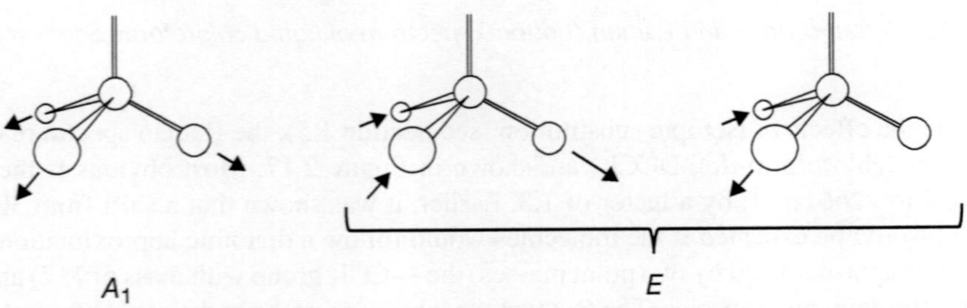

Figure 2.15 *Displacement vectors for the ν_5 and ν_6 vibrations of methyl chloride or a methyl groups under C_{3v} symmetry*

Table 2.4 *Vibrational modes and assignments for chloroform, $HCCl_3$, and $DCCl_3$*

Designation	Observed transitions (cm^{-1})		Symmetry	Activity	Mode description
	$HCCl_3$	$DCCl_3$			
ν_1	261	262	E	Ra, IR	CCl_3 antisymmetric deformation
ν_2	363	369	A_1	Ra, IR	CCl_3 symmetric deformation
ν_3	680	659	A_1	Ra, IR	CCl_3 symmetric stretching
ν_4	774	749	E	Ra, IR	CCl_3 antisymmetric stretching
ν_5	1220	914	E	Ra, IR	CH bending
ν_6	3034	2266	A_1	Ra, IR	CH stretching

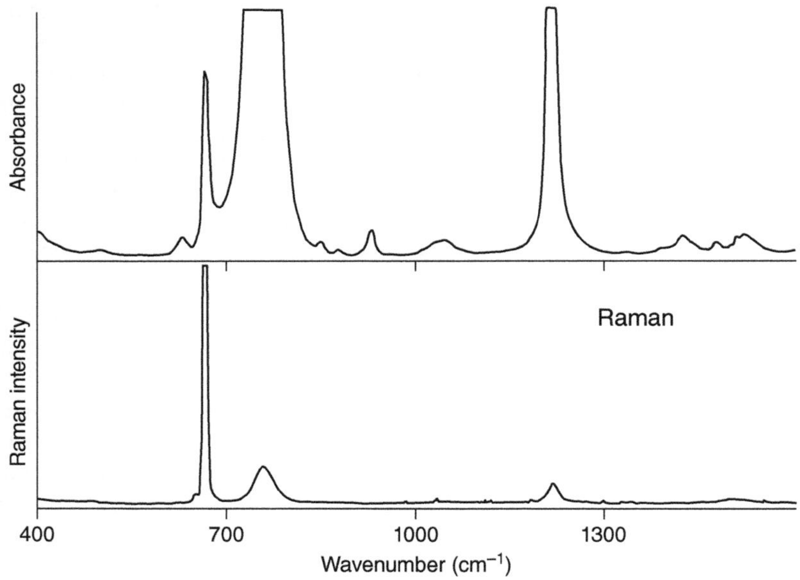

Figure 2.16 *Infrared (top) and Raman (bottom) spectrum of liquid chloroform. See text for details*

To demonstrate the effects of isotopic substitution (see Section 1.5), the Raman spectra of chloroform and deuterochloroform (chloroform-d_1), $DCCl_3$, are shown in Figure 2.17. Most obvious is the shift of the ν_6 mode, from 3034 to 2266 cm^{-1}, by a factor of 1.3. Earlier, it was shown that a shift from 3034 to 2155 (by a factor of 1.41) would be expected if the molecules would follow a diatomic approximation – that is, if the molecules could be approximated by two point masses (the —CCl_3 group with mass of 117) and 1 for the light species and 2 for the deuterated species. The fact that the vibrations of the hydrogen/deuterium do couple with the rest of the molecule is indicated by the deviation from the predicted values above, and from the fact that the deuteration slightly affects the vibrational frequencies of the heavy atom modes (see Table 2.4). The C–D

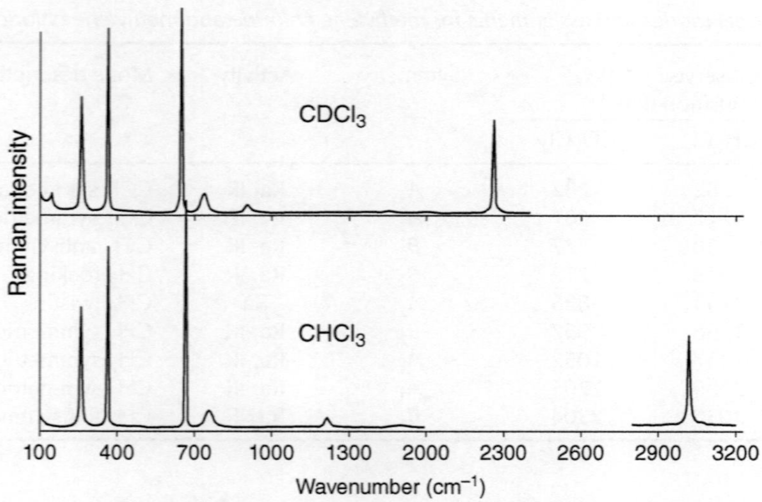

Figure 2.17 *Raman spectra of chloroform and deuterochloroform. Note that the wavenumber scale above 2000 cm^{-1} is compressed by a factor of 2*

deformation shifts from 1220 to 850 cm^{-1} or by a factor of 1.43. The fact that chloroform-d_1 has a spectrum devoid of fundamental transitions in the 900–2200 cm^{-1} region makes it a very useful solvent for vibrational spectroscopy.

2.6.3 Dichloromethane (methylene chloride), CH_2Cl_2

This molecule of C_{2v} symmetry incorporates a methylene (—CH_2—) group that exhibits, similarly to the methyl group discussed earlier, very characteristic group frequencies. In fact, the group frequencies of the methyl and methylene moieties dominate the vibrational spectra of alkanes as well as the side chains of fatty acids and are rather diagnostic for many of their physical properties. The conformation of phospholipids in biological membranes, for example, can be probed with vibrational spectroscopy. Here, as in many areas of IR and Raman spectroscopy, the insertion of —CD_2— groups adds another dimension for structural sensitivity [5].

The irreducible representation of methylene chloride reveals that its nine vibrational modes are given by $4A_1$, $1A_2$, $2B_1$, and $2B_2$ species. Since there are no longer any degeneracies at this low level of symmetry, there are nine distinct normal modes, eight of which are infrared and all nine are Raman active. The vibrational frequencies are listed in Table 2.5. This example further demonstrates that lowering the symmetry increases the number of bands allowed in infrared and Raman spectra, since degeneracy no longer exists and the selection rules do not prohibit any but one vibrational mode in the infrared.

Although the symmetric and antisymmetric C—H stretching frequencies listed above for methylene chloride occur at 2999 and 3030 cm^{-1}, respectively, the different masses attached to the carbon atom in (—CH_2—)$_n$ chains shift the observed C—H stretching frequencies in fatty acids and alkanes at about 2852 and 2922 cm^{-1}. These modes are quite strong in both Raman and infrared spectra, and can serve as probes for cis–trans isomerization, or order, in biomembranes [6]. The methylene stretching modes occur quite well separated from those of the methyl symmetric and antisymmetric modes that occur in alkanes at 2872 (A_1) and 2962 cm^{-1} (E). Again, in the smallest molecule containing a methyl group, H_3CCl, these modes occur at 2937 and 3039 cm^{-1}, respectively (see Table 2.3). A comparison of vibrational frequencies for methyl and methylene modes in alkanes is shown in Table 2.6.

Table 2.5 *Vibrational modes and assignments for methylene chloride and methylene chloride-d$_2$*

Designation	Observed transition (cm^{-1})		Symmetry	Activity	Mode description
	CH_2Cl_2	CD_2Cl_2			
ν_1	282	282	A_1	Ra, IR	CCl_2 symmetric deformation
ν_2	717	687	A_1	Ra, IR	CCl_2 symmetric stretching
ν_3	758	727	B_2	Ra, IR	CCl_2 antisymmetric stretching
ν_4	898	712	B_2	Ra, IR	CH_2 rocking
ν_5	1153	826	A_2	Ra	CH_2 twisting
ν_6	1268	957	B_2	Ra, IR	CH_2 symmetric deformation (wag)
ν_7	1467	1052	A_1	Ra, IR	CH_2 symmetric deformation
ν_8	2999	2205	A_1	Ra, IR	CH_2 symmetric stretching
ν_9	3030	2304	B_1	Ra, IR	CH_2 antisymmetric stretching

Table 2.6 *Summary of methyl and methylene stretching frequencies in alkanes (cm^{-1})*

Methyl		Methylene	
ν_{sym} CH_3	2872	ν_{sym} CH_2	2852
$\nu_{antisym}$ CH_3	2962	$\nu_{antisym}$ CH_2	2922

2.6.4 Dichloromethane-d$_1$ (methylene chloride-d_1), $CHDCl_2$

The discussion of small molecules in this chapter presented several pentatomic molecules in order of decreasing symmetry, T_d, C_{3v}, and C_{2v}. The next molecule introduced here is of even lower symmetry: $CHDCl_2$ of C_s symmetry. In this symmetry group, only the identity element and a mirror plane are allowed, and two irreducible representations, A' and A''. The vibrational modes are distributed among these two representations as follows: $6A'$ and $3A''$, and are listed in Table 2.7.

The vibrations of the CCl_2 moiety are nearly unchanged from those in dichloromethane. The CH and CD stretching modes are observed in the expected frequency ranges. Most interesting is the behavior of the CH and CD deformations: the methylene symmetric deformation (ν_6 in CH_2Cl_2, see Table 2.5) is observed as the two bending modes of the A'' symmetry (ν_5 and ν_6) modes in $CHDCl_2$, whereas the twisting and rocking motions of the CH_2 group occur in the symmetry plane (the A' bending modes).

If the symmetry is lowered even more to the C_1 point group, which only accommodates the identity element, all vibrations belong to the same irreducible representation, A, which are allowed in both infrared and Raman spectra. Most macromolecules to be discussed in Chapter 10 belong to this very low symmetry group, although ideal helices have certain symmetry properties. Similar, the catalytically active iron porphyrin moieties in hemoglobin and some cytochromes possess locally higher symmetry, but these groups are generally embedded in an asymmetric environment, and yet maintain some properties of their local symmetry.

The discussion of the small molecules presented above, in Section 2.6, is the only foray toward small molecules in this book, since Part II is devoted to large biological molecular systems. The reader is referred to extensive literature of the 1950s–1970s for normal coordinate analysis and vibrational assignments of hundreds of small molecules. In fact, the present understanding of the symmetries and shapes of small molecules

Table 2.7 *Vibrational modes and assignments for* dichloromethane-d$_1$

Designation	Observed transition (cm^{-1})	Symmetry	Activity	Mode description
ν_1	283	A'	Ra, IR	CCl_2 symmetric deformation
ν_2	692	A'	Ra, IR	CCl_2 symmetric stretching
ν_3	738	A''	Ra, IR	CCl_2 antisymmetric stretching
ν_4	778	A'	Ra, IR	CD bend
ν_5	890	A''	Ra, IR	CD bend
ν_6	1223	A''	Ra, IR	CH bend
ν_7	1282	A'	Ra, IR	CH bend
ν_8	2249	A'	Ra, IR	CD stretching
ν_9	3024	A'	Ra, IR	CH stretching

that are taught in introductory chemistry courses, *via* valence shell electron pair repulsion theory (VSEPR), is due in part to an analysis of the vibrational spectra and their symmetry properties.

References

1. Cotton, F.A. (1990) *Chemical Applications of Group Theory*, 3rd edn, John Wiley & Sons, Inc., New York.
2. Engel, T. and Reid, P. (2010) *Physical Chemistry*, 2nd edn, Pearson Prentice Hall, Upper Saddle River, NJ.
3. Gaussian http://www.gaussian.com (accessed 3 January 2015).
4. Hypercube, Inc. http://www.hyper.com (accessed 3 January 2015).
5. Mendelsohn, R., Davies, M.A., Brauner, J.W. *et al.* (1989) Quantitative determination of conformational disorder in the acyl chains of phospholipid bilayers by infrared spectroscopy. *Biochemistry*, **28**, 8934–8939.
6. Lee, C.D. and Chapman, D. (1986) Infrared spectroscopic studies of biomembranes and model membranes. *Biosci. Rep.*, **6** (3), 235–256.
7. Barth, A. (2007) Infrared spectroscopy of proteins. *Biochim. Biophys. Acta*, **1767**, 1073–1101.

3

Infrared Spectroscopy

In this and the following chapter, the two main modalities for observing molecular vibrational spectra are discussed, starting with infrared (IR) spectroscopy. This field started in the late 1930s, using mostly home-built instruments that incorporated prism monochromators and thermocouple detectors and the most basic electronic amplification circuitry, because solid-state (or even digital) electronics was still decades in the future. Nevertheless, the value of IR spectroscopy as a qualitative structural tool was soon realized, and the methodology soon migrated from physics laboratories, where it was originally developed, into chemical laboratories, both in academe and industry, and soon became one of the first physical methods to aid in structure determination. Except for very high resolution studies and astrophysical applications, IR spectroscopy is now being used and developed largely in physical chemistry research, since this technique offers a phenomenally sensitive probe for molecular structure, conformation, interactions, and dynamics.

The basic concepts of the theory of IR spectroscopy were introduced in Sections 1.2–1.5, and fundamental aspects of the observation of the effect were discussed in Section 1.6. The selection rules based on molecular symmetry were introduced in Chapter 2, and a few IR absorption spectra of small molecules were shown in Section 2.6. In the following sections, basic optical concepts of the interaction of IR radiation with molecular samples are introduced, followed by a discussion of instrumentation for the observation of IR spectra. Finally, data collection strategies (i.e., transmission, reflection, attenuated total reflection (ATR), etc.) are discussed.

3.1 General aspects of IR spectroscopy

IR spectra can be observed using several sampling methods and geometries that will be discussed in turn. However, in the most basic form of this technique, an IR photon is absorbed by the sample, causing excitation of one normal mode of vibration into a higher vibrational energy level. As pointed out earlier, such a process can be described by Eq. 1.114, which states that the quantum mechanical transition moment determines the molar extinction coefficient $\varepsilon(\tilde{\nu})$

$$D_{01} \propto \langle \psi_1 | \mu | \psi_0 \rangle^2 = \frac{1}{\tilde{\nu}_0} \int_{\text{band}} \varepsilon(\tilde{\nu})\, d\tilde{\nu} \qquad (1.114)$$

Modern Vibrational Spectroscopy and Micro-Spectroscopy: Theory, Instrumentation and Biomedical Applications, First Edition.
Max Diem.

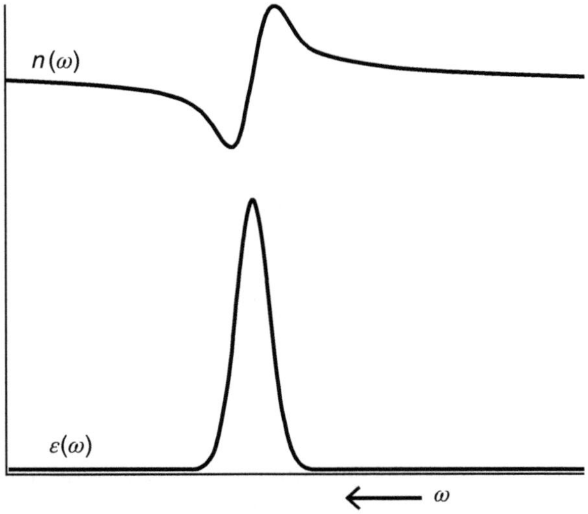

Figure 3.1 *Dispersion of the refractive index (top) within an absorption peak (bottom)*

which, in turn, determines the likelihood that a photon of wavenumber $\tilde{\nu}$ is absorbed by the sample. A plot of the molar extinction coefficient against the wavenumber of radiation gives the desired IR spectrum.

In general, the interaction of a sample with a medium is governed by the complex index of refraction η, defined as

$$\eta = n - i\kappa \tag{3.1}$$

in which n is the real refractive index familiar from classical optics and κ is known as the absorption index, which is related to the extinction coefficient by

$$\kappa = \frac{2.3\ \varepsilon\ \lambda C}{4\pi} \tag{3.2}$$

In Eq. 3.2, λ represents the wavelength of light and C the molar concentration of the sample. Equation 3.1 teaches one of the fundamental aspects of spectroscopy: that the refraction and absorption of light are coupled processes and that absorption of light is always accompanied by changes in refractive index of the medium. Interestingly, the connection expressed in Eq. 3.1 is often ignored in basic courses in geometric optics since optical materials from which lenses and prisms are produced generally do not have any (vibrational or electronic) absorptions in the spectral region for which they are manufactured. Therefore, the refractive index often is treated as independent of wavelength, or at best, slowly varying with wavelength. However, at the wavelength at which any absorption process occurs, the refractive index undergoes what is known as "anomalous dispersion," indicated by a deflection point in the otherwise smooth $n(\lambda)$ curve. This is shown schematically in Figure 3.1. At the center of the absorption curve (or at the deflection point of the dispersive curve), the phase between the electromagnetic radiation and the response of the medium changes as well. These effects are well understood, and discussed in books on classical optics, such as the monograph by Born and Wolf [1].

The coupling of absorption and dispersion is expressed mathematically by the Kramers–Kronig transform, written here in terms of the angular frequency ω of the light, rather than its wavelength λ:

$$n(\omega) = \frac{1}{\pi}\int_{-\infty}^{\infty} \frac{\varepsilon(\omega)}{\omega - \omega_0}\, d\omega \tag{3.3}$$

$$\varepsilon(\omega) = -\frac{1}{\pi}\int_{-\infty}^{\infty}\frac{n(\omega)}{\omega - \omega_0}\,d\omega \tag{3.4}$$

In these equations, ω_0 describes the frequency of light at which a transition occurs. Equations 3.3 and 3.4 indicate that knowing one of the quantities, for example, $n(\omega)$, uniquely defines the other, $\varepsilon(\omega)$, and *vice versa.* The integrations in Eqs. 3.3 and 3.4 are over a singularity at $\omega = \omega_0$, which requires that the principal value of this Cauchy integral is evaluated. The relationship between refractive index and extinction coefficient is also well known in other areas of spectroscopy: optical rotatory dispersion, which is based on the differential refractive index of a sample toward left and right circularly polarized light, is related by the Kramers–Kronig transform to circular dichroism, which is the differential absorption of a sample toward left and right circularly polarized light.

For spectroscopy, the coupled changes in n and ε have far-reaching consequences. This coupling implies that an IR absorption spectrum can actually be obtained from a reflectance spectrum after suitable mathematical manipulation but also that their interference can confound the observed spectra. In Fourier transform (FT) IR spectroscopy, for example, instrumental software (the so-called "phase correction") is required to insure that the refractive and the absorptive components (i.e., those resulting from changes in $n(\omega)$ and $\varepsilon(\omega)$, respectively) are properly separated. This is discussed in Sections 3.3 and 3.4. Furthermore, any observation methods in IR spectroscopy that explicitly involve the refractive index are subject to perturbations from the dispersion of the refractive index.

3.2 Instrumentation

Notwithstanding the discussion in the previous section, the following discussion assumes, to a first approximation, that the IR spectrum is caused by the changes in extinction coefficients due to the $3N-6$ normal vibrations of a molecule, and that an IR spectrum of a sample can be obtained by shining IR radiation into a sample and measuring the intensity of the light transmitted by it. This is shown schematically in Figure 3.2.

Figure 3.2, which is similar to Figure 1.13, schematically shows the main components of an IR spectrometer: a source (S), a "color"-selecting device, either a monochromator (MC) or an interferometer (IF), the sample (SA), and a detector (D). Notice that the sample can be placed before or after the color-selecting device. Recently, tunable IR laser sources have been incorporated into IR spectrometers (Figure 3.2(c)); although still the most expensive method to excite an IR spectrum, this approach offers the highest light throughput and may eventually replace the more conventional designs shown in Figure 3.2(a) and (b). The optical components listed in Figure 3.2 are discussed in turn.

3.2.1 Sources of infrared radiation: black body sources

IR radiation, first described by Herschel in 1800, is the long wavelength continuation of visible light. One distinguishes the near IR (0.8–2.5 μm in wavelength or 12,500 to 4000 cm^{-1}), mid-IR (2.5–25 μm or 4000 to 400 cm^{-1}), and the far-IR regions (25–250 μm or 400–40 cm^{-1}), before reaching the terahertz spectral range. Mid-IR radiation is typically produced by a hot filament or rod that is heated to orange or white glow by an electric current. This kind of source is typically referred to as a "black body" source, since its emitted intensity distribution, as a function of wavelength and temperature, obeys the black body curve shown in Figure 3.3. In this graph, the intensity of the electromagnetic energy is shown for several temperatures between 1000 and 5000 K as a function of wavelength. The shape of the black body curve is described by Eq. 3.5 [2]:

$$I(\nu, T) = \frac{2h\nu^3}{c^2}\frac{1}{\left(e^{\frac{h\nu}{kT}} - 1\right)} \tag{3.5}$$

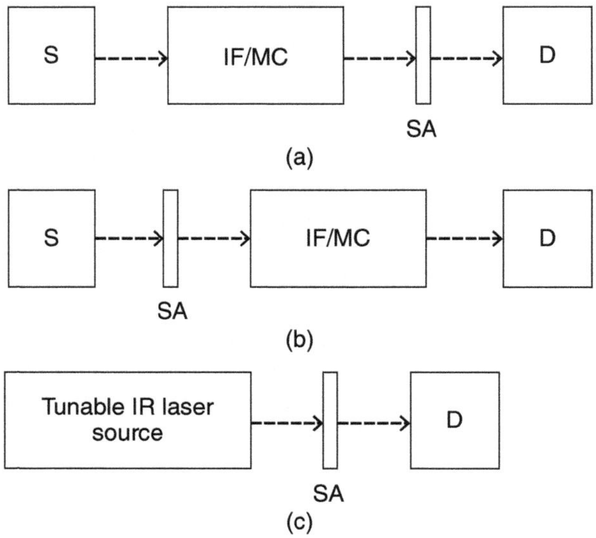

Figure 3.2 *Block diagrams of different configurations of IR spectrometers. S: source, IF/MC: interferometer or monochromator, SA: sample, and D: detector*

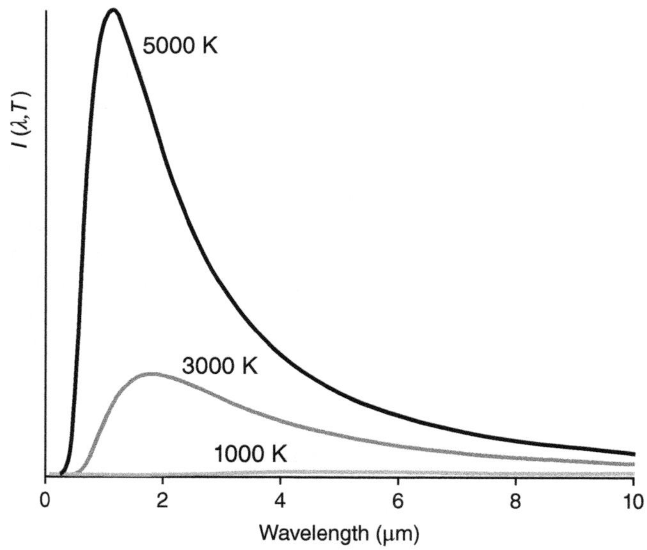

Figure 3.3 *Plot of black body emission curve*

The graph shows that the overall emitted energy increases with increasing temperature, and that the peak wavelength of maximum intensity shifts toward lower wavelength (Wien's law). The total energy radiated by such a source per unit area and unit time (the irradiance), integrated over all wavelengths, is proportional to the absolute temperature to the fourth power:

$$\int_0^\infty W\, d\lambda = \sigma T^4 \text{ (Stefan–Boltzmann law)} \tag{3.6}$$

The irradiance is expressed in units of $\left[\frac{\mathrm{W}}{\mathrm{m}^2\ \mathrm{s\ sr}}\right]$ or $\left[\frac{\mathrm{photons}}{\mathrm{m}^2\ \mathrm{s\ sr}}\right]$.

Inspection of Figure 3.3 reveals that higher power density in the IR spectral region is obtained at higher temperatures, but that proportionally less IR radiation is produced as compared to the total energy emitted. Thus, one usually compromises and utilizes sources between about 1400 and 2000 K. Such sources still produce a broadband visible emission in addition to IR radiation, but the overall energy consumption can be relatively low (<100 W). The problem with black body sources is that the emitting material is quite hot, and therefore, will readily react with atmospheric oxygen. Therefore, the hot filament in a standard incandescent light bulb is enclosed in a glass envelope that is filled with an inert gas such as argon or nitrogen. However, a glass envelope cannot be utilized in IR spectroscopy since glass is a very strong absorber of IR radiation. Thus, sources must be used that are either inert to oxygen, or sources must be constructed that operate in an inert atmosphere and use an IR-transparent window in the path of the beam.

Among materials that are relatively inert to atmospheric oxygen in the 1400–2000 K temperature range are certain noble metals such as Pt, but the high price of this metal makes it a poor choice. Alloys of nickel and chromium (known as "nichrome" sources) are relatively inert at these temperatures, although a thick coat of oxides may form on such a source after extended use. "Globars" fabricated from a 1:1 mixture of Si and C atoms in a diamond lattice are relatively stable black body IR sources; other doped Si alloys are in use as well. Finally, a semiconductor source known as a "Nernst" glower can be used. These sources are made of zirconium oxide (ZrO_2), doped with rare earth oxides (Y_2O_3 and Er_2O_3); this material is non-conductive at room temperature. After pre-heating these sources radiatively to about 800 K, they become conductive. They need to be operated with a resistive element in series, since a Nernst glower's electric resistance decreases with increasing temperature. They can be operated up to about 2500 K, and are excellent broadband IR sources.

3.2.2 Sources of infrared radiation: quantum-cascade lasers, nonlinear devices

The source described in the previous paragraphs are broadband emitters and therefore, require a device that separates the IR "colors" used to illuminate the sample. This can be avoided using tunable monochromatic IR sources based on laser technology. IR lasers have been used as sources in IR spectroscopy for some time; the CO_2-laser was one of the early coherent and monochromatic sources used in IR spectroscopy. However, the CO_2-laser does not produce continuously tunable IR radiation, but only fixed wavelengths in the 9.4 and 10.6 µm range due to distinct rot-vibrational transitions of CO_2 [3].

Tunable solid state lasers for the IR spectral region have been available for some time, but their narrow tuning range precluded them being used except in cases of very high resolution spectroscopy. Only recently have broadband tunable IR lasers become commercially available, albeit at very high cost. Most promising for the near future are quantum-cascade lasers (QCLs) that are semiconductor devices based on semiconductor superlattice structures produced from materials such as InGaAs/InAlAs. Tuning ranges of over 200 cm^{-1} for one semiconductor chip have been achieved; the combination of several chips of different materials has resulted in tuning ranges of QCL lasers over 600 cm^{-1}. This technology appears highly promising for applications such as IR microspectroscopy (IR microscopy) to be discussed in Chapter 11.

Another approach to broad tuning range, high power lasers is *via* nonlinear optical processes such difference frequency generation (DFG) or optical parametric oscillators (OPOs) [4], to be discussed in Section 6.3.2. In the former case, two laser pulses at frequency ω_1 and ω_2 are combined in a crystal material with nonlinear response, such as $LiNbO_3$, to produce a photon with the difference frequency $\omega_{DFG} = \omega_1 - \omega_2$. By tuning the laser frequency ω_1 or ω_2 (or both), pulsed IR radiation at different frequency can be obtained. In OPOs, an incident photon at a pump frequency ω_p is decomposed inside a crystal of nonlinear material into two photons with frequencies $\omega_1 + \omega_2$. If a visible frequency ω_p is utilized, both photons produced may be IR photons. OPOs are mature, broad-range tunable IR sources that are now commercially available.

3.2.3 Transfer optics

The optical elements directing the IR radiation from the source to the monochromator/interferometer, and from there to the sample and the detector, are generally based on reflective optics (mirrors). While in spectrometers using visible light, these optical elements are often based on fiber optics or refractory optics such as lenses, the restrictions of IR transmitting materials dictate optical designs using mirrors. This is, as pointed out in Section 1.6 above, because glass or quartz has strong molecular absorptions in the mid-IR region and cannot be used for lens materials, and IR-transparent materials such as NaCl, KCl, KBr, AgCl, AgBr, CaF_2, Si, Ge, KRS-5 (a mixed thallium halide), or diamond are either hygroscopic, soft, light sensitive, toxic, and expensive, or exhibit prohibitively high refractive indices that restrict usage due to high reflective losses (see Section 3.4 and Table 3.2). Thus, light collection is usually carried out by spherical or ellipsoidal mirrors, preferably gold-coated ones. The use of mirrors also eliminates chromatic aberrations one would encounter using lenses for light collection and focusing. Chromatic aberrations for single lenses in the IR regions are aggravated by the fact that the IR spectral range extends over more than an order of magnitude in wavelengths (2–25 μm), whereas the visible range barely covers a factor of 2 in wavelength range.

3.2.4 Color sorting devices: monochromators

Prior to the mid-1970s, nearly all IR spectroscopy was carried out using instruments that incorporated a monochromator to separate the broadband radiation from the IR source into narrow wavenumber bands that consecutively expose the sample to the different IR "colors." These monochromators utilized gratings or prisms to disperse the light, and slits of appropriate widths to select a desired wavenumber band to irradiate the sample. Since only one spectral element (band) was sampled at a time, acquisition of spectral data often took hours.

In the following paragraphs, the operation of grating-based monochromators is introduced. As an example, the commonly used Czerny–Turner design will be elaborated upon, although there are several other common designs. In Chapter 4, the use of a Czerny–Turner monochromator in a spectrograph configuration, useful for multichannel Raman instruments, is discussed.

Figure 3.4 shows the basic design of a Czerny–Turner monochromator. Polychromatic light enters at the slit S_1, whose function it is to define the image size. The light is collimated by a mirror of focal length F placed at a distance F from the slit; therefore, the light reflected by mirror M_1 is collimated. This light impinges

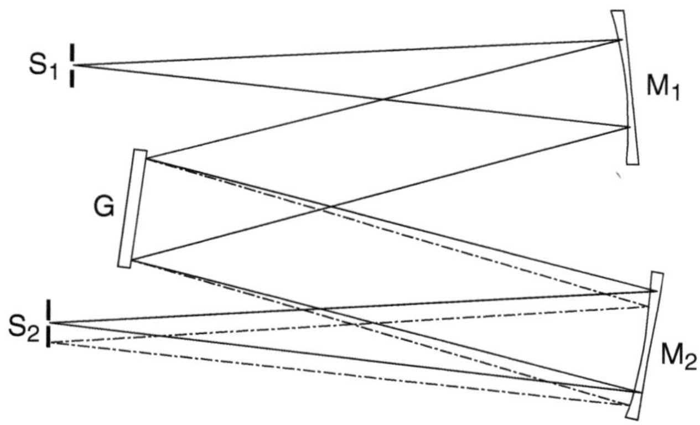

Figure 3.4 Basic design of a Czerny–Turner monochromator. S_1, S_2: entrance and exit slits, respectively; G: grating; and M_1, M_2: collection and focusing mirror, respectively

on the grating G, where different colors are diffracted at different angles. The grating is a reflective surface with parallel lines, or grooves, etched in the surface (perpendicular to the drawing plane in Figure 3.4). These lines act as multiple slits and diffract the light into different colors. The process of manufacturing gratings is quite involved, since the lines need to be perfectly uniform and equidistant. In the past, these lines were scratched into the surface of a suitable substrate by a diamond stylus. To date, nearly all gratings are produced by a holographic process in which the grating surface is coated with a photosensitive film, and an image of the grating is produced on this film holographically. After developing the film and chemically etching off the light exposed areas, nearly perfect grating surfaces are obtained. As indicated in Figure 3.5, the profile of the grating grooves resembles an asymmetric saw tooth function to maximize the diffracted intensities for certain center wavelength. This is referred to as the "grating blaze."

The diffracted light is focused by mirror M_2 onto the exit slit (solid path in Figure 3.4) whose width determines the transmitted bandpass. Different colors are diffracted at a different angle by the grating, and consequently, are focused by mirror M_2 such that the light hits the slit jaws, and is not propagated (dashed path). By rotating the grating, rays of different colors can pass the exit slit.

The diffraction of light by a grating is described by the grating equation [2] and is explained in Figure 3.5:

$$d\,(\sin\,\alpha\,+\,\sin\,\beta) = n\,\lambda \tag{3.7}$$

In Eq. 3.7, α denotes the angle of incidence and β the diffracted angle (both measured with respect to the grating normal, represented by the dashed line in Figure 3.5). In Eq. 3.7, λ is the wavelength, d the spacing between the lines on the surface of the grating, and n the diffraction order ($n = 0, 1, 2, \ldots$). Equation 3.7 is an extension of the equation governing diffraction from a single slit. A more detailed view of the grating surface and the diffracted angles for two wavelengths is presented in Figure 3.5. This figure shows a vastly enlarged cross section through a grating, at left, with the grooves perpendicular to the drawing plane. The profile of the grooves is not symmetric, but etched at a certain angle (the "blaze" angle), similar to a saw tooth to be discussed below. Light is incident on the grating at a given angle to the grating normal. This angle, here assumed to be 20°, is determined by the physical design of the monochromator, that is, the distance between mirrors M_1 and M_2 in Figure 3.4. For the discussion to follow, it is assumed that the incident light is in the visible region (450–750 nm), but the principles equally hold for IR light. The distance d between adjacent grooves is chosen roughly to coincide with the wavelength of light; for the example discussed next, a grating with 1800 lines/mm is used. This is a common groove density for gratings to be used in the visible region of the spectrum and corresponds to a spacing between lines of 1/1800 mm, or $d = 556$ nm. Polychromatic light incident on the grating will be diffracted into different colors, and it is easy to show that under the above assumptions, two rays of diffracted light with wavelength of 500 and 510 nm will be diffracted at angles β_1 and β_2 of 33.9° and 35.1°, respectively, as shown in Figure 3.5.

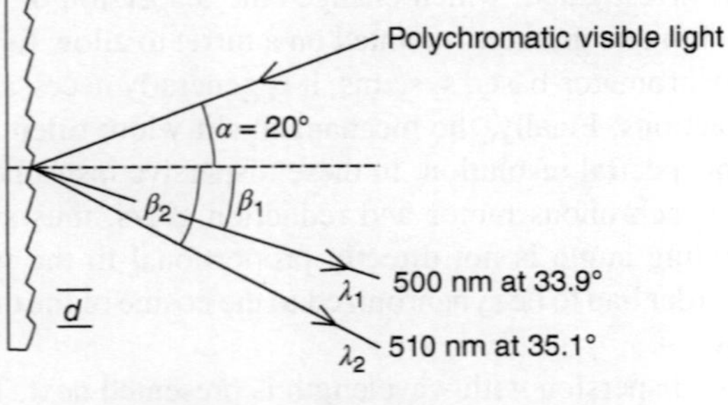

Figure 3.5 *Diffraction by a grating*

If a monochromator is being used in the IR spectral region, the grating groove density will be much lower since the wavelength is so much larger. For radiation at 1000 cm^{-1}, corresponding to a wavelength of 10 μm, grating groove densities of 100 lines/mm (d = 10 μm) are adequate. Gratings are "blazed" for maximum efficiency at a given wavelength. This is accomplished by varying the profile of the grooves to maximize the diffracted intensity at a given wavelength. A general rule of thumb for the use of blazed gratings is that they are good in a wavelength range from 2/3 to 3/2 of the blaze wavelength.

In order to calculate the exit slit width required to obtain a certain resolution (bandpass), one needs to calculate the linear dispersion $d\lambda/dx$ of the monochromator at the exit slit plane. This quantity specifies the change in wavelength of the diffracted light in the exit plane. This is obtained by differentiation of Eq. 3.7 with respect to the diffracted angle β to obtain

$$\frac{d\lambda}{d\beta} = \frac{(d \cdot \cos\beta)}{n} \tag{3.8}$$

where the quantity $d\lambda/d\beta$ is referred to as the angular dispersion. The angular dispersion can be converted to the linear dispersion $d\lambda/dx$ in the exit plane using

$$dx = F\,d\beta \tag{3.9}$$

where F is the focal length of the monochromator. Combining Eqs. 3.8 and 3.9 yields the equation for the linear dispersion of a monochromator

$$\frac{d\lambda}{dx} = \frac{(d \cdot \cos\beta)}{nF} \quad \text{or} \quad \frac{d\lambda}{dx} = \frac{\cos\beta}{n\,g\,F} \tag{3.10}$$

where the groove density g was used as $g = 1/d$.

The important consequence of Eq. 3.10 is that the linear dispersion of a monochromator in the exit plane is not constant, but varies with the cosine of the diffracted angle, β. Thus, at different incident angles, α, corresponding to different grating rotations, the bandpass at constant mechanical slit width can vary substantially.

In dispersive IR instruments, monochromators with 20–50 cm focal length are used with gratings ruled at 50–300 lines/mm. This produces optical resolutions on the order of 0.5–5 cm^{-1} at a mechanical slit width of 0.1–1 mm. Typical IR spectral measurements are made over a range of 400 to nearly 4000 cm^{-1} (25 to 2.5 μm). Inspection of Eq. 3.10 shows that the linear dispersion of the monochromator depends on the cosine of the angle of diffraction. Consequently, one finds that it is impossible to achieve a reasonably constant resolution over the wavelength range at a constant mechanical slit width with one and the same grating in the same order. In the past, instrument designers avoided this problem by using the grating in first and second order, depending on the wavelength investigated, which changes the dispersion by a factor of 2. Also, dispersive instruments often used more than one gratings mounted on a turret to allow for automatic grating change at a certain wavelength. For monochromator-based systems, it is generally necessary to include a band-pass filter to exclude higher order diffractions. Finally, the mechanical slit width often was adjusted automatically to produce a reasonably constant spectral resolution. In these dispersive instruments, a spectrum was collected by rotating the grating by a synchronous motor and reduction gears, thus sweeping different wavelength by the exit slit. Since the grating angle is not directly proportional to the wavelength transmitted by the monochromator, the chart recorder had to be synchronized to the cosine of the diffracted angle, which required elaborate mechanical systems.

An example of the change in dispersion with wavelength is presented next. Table 3.1 shows the dispersion and resolution for a 320 mm focal length Czerny–Turner monochromator with a 120 lines/mm grating at a

Table 3.1 *Example of optical band path of monochromator system*

Wavelength (μm)	Wavenumber (cm^{-1})	Angle of diffraction (°)	Linear dispersion (nm mm^{-1})	Bandpass at 2 mm slit width (cm^{-1})
5.56	1800	31.9	22.5	14.2
6.25	1600	34.6	22.1	10.9
7.14	1400	38.0	21.5	8.0
8.33	1200	42.8	19.1	5.5
10.0	1000	49.8	16.8	3.4

constant mechanical slit width of 2 mm. This monochromator was used in an IR CD (vibrational circular dichroism VCD) spectrometer designed and constructed by Diem *et al.* [5].

In modern monochromators, the rotation of the grating is achieved by computer controlled stepping motors. This allows a step-wise motion of the grating, where the grating is moved to a particular wavelength position, stopped, and a spectral data point is measured, before the process is repeated. Thus, it is easy to incorporate in the design some software that rotates the grating by varying angles to keep the wavenumber increments between subsequent data points constant. This method also allows for an automatic adjustment of the mechanical slit for each grating position. However, the main disadvantage of a monochromator/single detector system is that only one spectral element is measured at a time, and that acquisition of hundred or even thousand spectral elements can be a time-consuming process.

Furthermore, in dispersive systems, increased spectral resolution is associated with a steep price in signal-to-noise ratio. In order to change the optical resolution from 4 to 2 cm^{-1}, for example, the mechanical slit width needs to be reduced by a factor of 2; yet, the intensity of light reaching the detector depends on the square of the mechanical slit width. Thus, the necessary change in mechanical slit width causes a decrease in light intensity at the detector by a factor of 4. This, in turn, degrades the signal-to-noise ratio of the measurement.

3.2.5 Color encoding devices: interferometers

In a dispersive system incorporating a grating and slits, only one spectral element is sampled by the detector at a time. In contrast, in interferometric methods, all wavelengths are examined simultaneously. In order to obtain the desired spectral distribution $I(\tilde{\nu})$ *vs.* $(\tilde{\nu})$, one modulates the interference pattern of the light reaching the detector and takes the FT of the "interferogram" obtained in this manner. The principle of the central apparatus in FT spectroscopy, the interferometer, is discussed next, followed by a short review of the interference process taking place in such an interferometer, and the mathematical foundation to understand the process.

A schematic of a Michelson interferometer is shown in Figure 3.6. Light from a source is collimated by a lens or mirror and enters the interferometer as a parallel beam. This beam of light impinges on a beam splitter such that half the light intensity is reflected and half is transmitted. The transmitted light is reflected by a mirror (fixed mirror) and impinges on the beam splitter again.

The portion of the light that was originally reflected by the beam splitter is reflected by a mirror marked "movable mirror" in Figure 3.6, which may be moved in the direction indicated by the arrow. The light reflected by this movable mirror reaches the beam splitter where the two beams are recombined. Again, at the beam splitter half the intensity is directed back to the source and half of it is directed toward the detector. The two beams that are propagating toward the detector – one originating from the "fixed mirror" path and

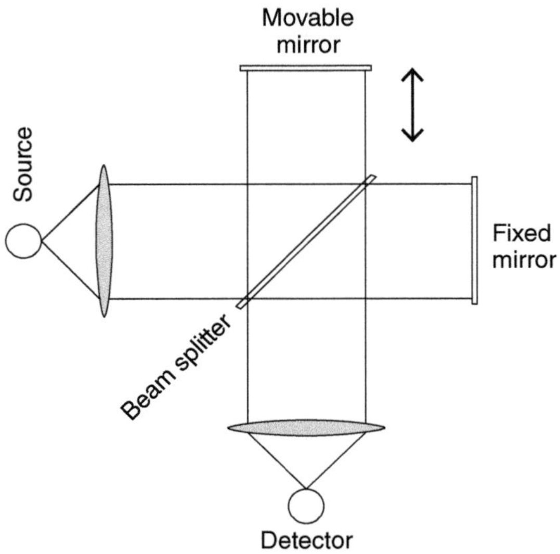

Figure 3.6 *Schematic of a Michelson interferometer. See text for details*

the other from the "movable mirror" path – undergo constructive or destructive interference depending on their path difference or phase shift. This path difference depends, of course, on the momentary position of the movable mirror. Thus, the detector records an intensity pattern produced by the motion of the mirror. This pattern is referred to as the interferogram, $J(x)$, where x is the mirror position.

The function of the interferometer can be understood from the following discussion. First, if the movable mirror is in a fixed position, and the light entering the interferometer is purely monochromatic, the two waves that interfere constructively or destructively can be described by

$$\mathbf{E}_1 = E_0(\sin \omega t - \alpha_1)$$
$$\mathbf{E}_2 = E_0(\sin \omega t - \alpha_2) \tag{3.11}$$

Here, α denotes the phase angles, ω the frequency of the light waves, $\mathbf{E}_1$ and $\mathbf{E}_2$ the instantaneous values of the electric vectors of each wave, and E_0 is the amplitude, which is equal for both waves since it is assumed that exactly one-half of the total intensity is reflected and one-half is transmitted by the beam splitter. The total amplitude E_T of the wave resulting by adding the two components of Eq. 3.11 can be written as

$$E_T = E_0[(\sin \omega t - \alpha_1) + (\sin \omega t - \alpha_2)] \tag{3.12}$$

Using the identity

$$\sin(\alpha - \beta) = \sin \alpha \cos \beta - \cos \alpha \sin \beta \tag{3.13}$$

Eq. 3.12 can be rewritten as

$$\begin{aligned} E_T &= E_0(\sin \omega t \ \cos \alpha_1 - \cos \omega t \ \sin \alpha_1 + \sin \omega t \ \cos \alpha_2 - \cos \omega t \ \sin \alpha_2) \\ &= (\cos \alpha_1 + \cos \alpha_2)\, E_0(\sin \omega t) - (\sin \alpha_1 + \sin \alpha_2)\, E_0(\cos \omega t) \end{aligned} \tag{3.14}$$

For a fixed mirror position, E_T depends on the mirror position (that determines the phase of the two waves) and the frequency of the radiation.

The sum of two sine (cosine) functions of equal frequency is another sine (cosine) function with amplitude A:

$$(\cos\alpha_1 + \cos\alpha_2)\, E_0 = A\ \cos\Theta$$

$$(\sin\alpha_1 + \sin\alpha_2)\, E_0 = A\ \sin\Theta \tag{3.15}$$

where $\Theta = \Theta(\alpha_1, \alpha_2)$.

Thus, Eq. 3.14 can be written as

$$E_\mathrm{T} = A\ \cos\Theta(\sin\omega t) - A\ \sin\Theta(\cos\omega t) \tag{3.16}$$

Squaring and adding the time-independent amplitudes gives

$$A^2\cos^2\Theta = {E_0}^2(\cos^2\alpha_1 + \cos^2\alpha_2 + 2\cos\alpha_1\cos\alpha_2)$$

$$A^2\sin^2\Theta = {E_0}^2(\sin^2\alpha_1 + \sin^2\alpha_2 + 2\sin\alpha_1\sin\alpha_2) \tag{3.17}$$

$$A^2(\cos^2\Theta + \sin^2\Theta) = A^2$$

$$= {E_0}^2[1 + 1 + 2(\cos\alpha_1\cos\alpha_2 + \sin\alpha_1\sin\alpha_2)] \tag{3.18}$$

$$A^2 = 2{E_0}^2[1 + (\cos\alpha_1\cos\alpha_2 + \sin\alpha_1\sin\alpha_2)] \tag{3.19}$$

Using $\cos(\alpha - \beta) = \cos\alpha\cos\beta + \sin\alpha\sin\beta$, one obtains from Eq. 3.19

$$A^2 = 2{E_0}^2 + 2{E_0}^2\cos(\alpha_1 - \alpha_2) = 2{E_0}^2(1 + \cos\delta) \tag{3.20}$$

where δ is the phase shift (retardation) between the two waves,

$$\delta = \alpha_1 - \alpha_2 \tag{3.21}$$

The detailed derivation of Eqs. 3.11–3.16 can be found in texts on classical optics, for example, Jenkins and White [2]. Since the intensity of light is proportional to the square of the amplitude, one substitutes $I(\delta)$ for A^2 and I_0 for ${E_0}^2$:

$$I(\delta) = 2\, I_0(1 + \cos\delta) \tag{3.22}$$

If the movable mirror is scanned at constant velocity, the phase difference δ can be expressed in terms of the wavelength λ of the light and the mirror position x, as follows:

$$\delta = \frac{2\pi x}{\lambda} = 2\pi x\tilde{\nu} \tag{3.23}$$

where $\tilde{\nu}$ is the wavenumber of the radiation.

Thus,

$$I(x) = 2\, I_0[\ 1\ +\ \cos(2\pi\tilde{\nu}\, x)] \tag{3.24}$$

Equation 3.24 gives the interference pattern for a monochromatic light source as a function of the mirror position, x. If the light source is not monochromatic but has a spectral distribution $S(\tilde{\nu})$, one obtains instead of Eq. 3.24

$$I(x) = \int 2\, S(\tilde{\nu})\{\ 1 + \cos\,(2\pi\tilde{\nu}x)\}d\tilde{\nu}$$

$$= \int 2\, S(\tilde{\nu})d\tilde{\nu} + \int 2\, S(\tilde{\nu})\cos\,(2\pi\tilde{\nu}\ x)d\tilde{\nu} \tag{3.25}$$

where the integration is from zero to infinity. The first integral in Eq. 3.25 can be evaluated at the zero path difference (ZPD) position ($x = 0$) to give

$$I(0) = 4 \int_0^\infty S(\tilde{\nu}) d\tilde{\nu} \tag{3.26}$$

Thus, $I(x)$ can be written as

$$I(x) = \frac{1}{2} I(0) + 2 \int_0^\infty S(\tilde{\nu}) \cos(2\pi\tilde{\nu}\, x) d\tilde{\nu} \tag{3.27}$$

After extending the integral symmetrically from $-\infty$ to ∞, the interferogram $J(x)$ can be defined as

$$I(x) - \frac{1}{2} I(0) = J(x) = \int_{-\infty}^\infty S(\tilde{\nu}) \cos\,(2\pi\tilde{\nu}\, x) d\tilde{\nu} \tag{3.28}$$

$$S(\tilde{\nu}) = \int_{-\infty}^\infty J(x) \cos\,(2\pi\tilde{\nu}\, x) dx \tag{3.29}$$

Equations 3.28 and 3.29 are said to be a Fourier pair. Fourier series and FTs are common ways to analyze harmonic or periodic signals, and are discussed in many books on advanced calculus and are reviewed in Appendix D.

Here, one can think of the FT as a procedure to take a signal, observed in one domain (the mirror position of the interferometer), to another domain, the frequency (or time) domain. In a Fourier pair of relationships, knowledge of one of the quantities, say $J(x)$, uniquely defines the other, here $S(\tilde{\nu})$, similarly to the relationship between the refractive index n and the absorption coefficient ε discussed in Eqs. 3.3 and 3.4. In order to get from the interferogram $J(x)$, which is an intensity distribution as a function of the position, x, of the movable mirror, to the spectrum $S(\tilde{\nu})$, one has to take the FT of $J(x)$. The background of harmonic analysis, Fourier series, FTs, and methods to perform the FT computationally are presented in Appendix D. An interferogram obtained by scanning the movable mirror from –L to +L over the ZPD position is shown in Figure 3.7.

After the color selection device – be it a monochromator or an interferometer – the sample is placed into the optical beam as indicated in Figure 3.1. In principle, the sample can also be placed before the monochromator/interferometer. In general, the former arrangement – the sample after the color sorting device – is preferred since the total radiation flux can be reduced to avoid sample heating. This is particularly so since the interferometer usually contains filters or other optical elements that limit the bandpass by cutting out the visible/near IR radiation emitted by the source. There are several sampling techniques, from the simplest transmission measurements to sophisticated modulated reflectance measurements, to collect an IR spectrum. An entire Section 3.4 is devoted to this subject.

The operational details of IR interferometry also requires more detailed discussion, which is presented in Section 3.3 to include subjects such as optical resolution, phase correction, apodization, zero filling, and some data processing in Fourier space. Thus, the present discussion on IR instrumentation will continue with the next major component of an IR instrument, the detector.

3.2.6 Detectors

IR photons have only between a quarter and a tenth of the energy of a visible photon. Therefore, IR spectroscopy utilizes detector technology quite different from that of UV or visible spectroscopy. Detectors sensitive in the IR region can be classified as thermal and semiconductor detectors. Thermal detectors directly measure a heating effect due to the IR radiation. These include thermocouples, bolometers (thermal resistor), and pyroelectric detectors.

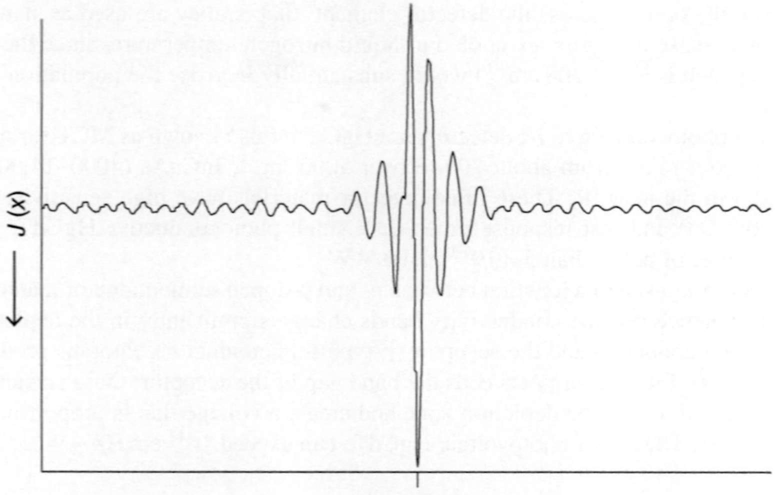

Figure 3.7 *Example of an interferogram. ZPD: zero path difference position of interferometer. The magnitude of the signal shown is inverted*

Among these, thermocouples have been used for the detection of IR radiation for over half a century [6]. Thermocouples are constructed from two dissimilar metals or alloys A and B to have two junctions of the metals, for example, A–B–A. The left A–B junction serves as the measuring point, while the right B–A junction is at a constant reference temperature. If junction A–B is at a different temperature than the B–A junction, a voltage develops between the two points A. The sensitivity of such a thermocouple is about 50 μV/°. Here, the intensity of the IR radiation impinging on the A–B junction is measured as the change in voltage of the entire thermocouple. To reduce the response time of a thermocouple detector, their thermal mass is generally very low; thus, the actual junction that is heated by the radiation is often just a connection of small wires. Thermocouple technology is very mature, and temperature sensors based on this methodology are readily available from many manufacturers.

Bolometers also are thermal detectors in that they utilize a small, low heat capacity metal or semiconductor detector element that is heated by the IR radiation. This detector is connected by a wire to a temperature sink; as the detector heats, the resistance of the wire changes. This changed resistance can be observed as a voltage drop between the detector and the heat sink.

Deuterated triglycine sulfate (DTGS) is an example of a pyroelectric material in which the dielectric constant depends on the temperature. Thus, DTGS is used as a capacitor material, and the change in capacitance, induced by the temperature change, is measured as a change in voltage across the capacitor. All the thermal detectors have a flat wavelength response (typically between 400 and 4000 cm^{-1}) and are reasonably sensitive. Their main drawback is the slow frequency response; thus, DTGS detectors need to be used with slower interferometer scan speeds to prevent the Fourier frequencies (*cf.* Section 3.4) to exceed the detector's frequency response. Nevertheless, DTGS detectors are the most commonly used detectors in low-cost, medium-performance instruments and offer the further advantage that they can be used at room temperature.

Semiconductor detectors are used either in photoconductive or photovoltaic modes. In the photoconductive mode, materials with band gaps between 400 and 800 cm^{-1} (~0.05–0.1 eV) are used. When exposed to IR radiation, electrons are promoted form the valence band into the conductivity band, thereby reducing the resistance of the detector. These detector elements are biased at a constant voltage, and the change in resistance causes a variation in the current across the detector element; that is, they are used as standard photodiodes. For best performance, these detectors are cooled to liquid nitrogen temperature, since the thermal energy at room temperature (which is about 208 cm^{-1}) would substantially increase the population of electrons in the conductivity band.

The most common photoconductive IR detectors are HgCdTe (also known as MCT for mercury-cadmium-telluride) with a spectral range from about 700 to over 5000 cm^{-1}, InGaAs (4000–14,000 cm^{-1}) and InSb (2000–10,000 cm^{-1}) in the near IR. These three detector materials have high sensitivity (expressed as the "specific detectivity" D^*) and fast response times (μs); small photoconductive HgCdTe detector elements typically have D^* values of better than 5×10^{10} cm $Hz^{1/2}$ W^{-1}.

Photovoltaic detectors consist of a junction between n- and p-doped semiconductor materials. In these junctions, the energy of valence and the conductivity bands change significantly in the depletion zone between the donor (n-type semiconductor) and the acceptor (p-type semiconductor). Photons produce carriers (electrons and electron holes) if their energy exceeds the band gap of the acceptor; these carriers are immediately separated by the potential across the depletion zone and create a voltage that is proportional to the intensity of the incident radiation. The D^* of photovoltaic HgCdTe can exceed 10^{11} cm $Hz^{1/2}$ W^{-1}.

3.2.7 Read-out devices

All modern instruments – interferometric or grating-based – are operated by laptop or bench top computers, or equivalent computers build into the instrument. Thus, the signal from the detector must be converted from an analog voltage or current to a digital signal that can be read and stored digitally. This is accomplished as follows. First, the signal is amplified by a current-to-voltage converter or an operational amplifier by about million-fold to produce a full-scale signal in the 0–1 V range, which is subsequently converted to a digital (binary) number by an analog-to-digital (A/D) converter. For interferometry, the dynamic range of the A/D converter is highly important: the center burst (the signal at the ZPD, see Figure 3.7) can be 100-fold stronger than the small fringes away from the ZPD; both these signals have to be converted to a digital number with high fidelity. A 10 bit A/D converter has a range from 1 to 1024 (2^{10}) and therewith, a digitization limit of about 0.1%. For a 12 bit converter, the range is 1–4096 with a digitization limit of about 0.02%, and a 14 bit converter has a range of 1–16,384 or a limit of 0.006%. Thus, most current instruments utilize either 14 or 16 bit A/D conversion. Interferograms and spectral vectors are generally stored as 32 bit floating point numbers for further processing. Some concepts of data processing, such as smoothing, derivatization, and so on, are introduced in Chapter 12.

3.3 Methods in interferometric IR spectroscopy

3.3.1 General instrumentation

At the time of writing of this book (2013), nearly all commercial IR spectrometers are based on interferometric data acquisition methods. One notable exception to this statement will be discussed at the end of this chapter (planar array infrared spectroscopy). Thus, it is appropriate to devote some discussion to the design and operation of a modern FT-based IR spectrometer. The principles of the core technology, the interferometer, were discussed above (Section 3.2), and the theory of FT and fast Fourier transform (FFT) are introduced in Appendix D.

A schematic layout of an FTIR spectrometer is shown in Figure 3.8. The source typically is a broad-band black body source, and the radiation is typically detected by a DTGS or semiconductor detector, the latter generally at liquid nitrogen temperature. Care has to be taken that the frequency response of the detector/preamplifier is sufficiently high to follow the Fourier frequencies, which result from the rapid scanning

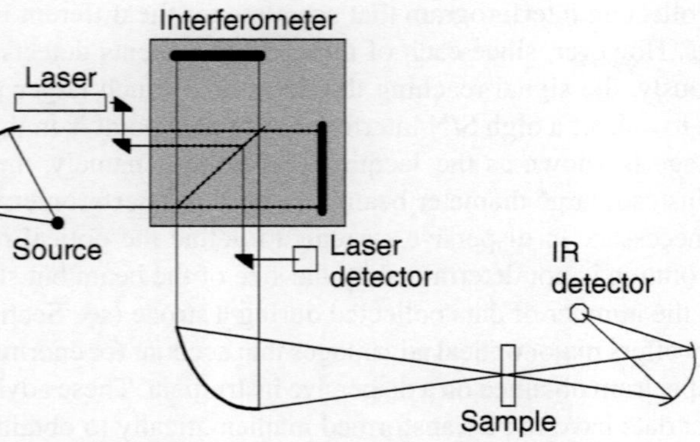

Figure 3.8 *Schematic of an FTIR spectrometer*

of the movable mirror. Typical scan speeds are on the order of several centimeters per second, and typical Fourier frequencies are in the low kilohertz range.

The path difference x, that is, the difference in the path the light travels between the movable and the fixed mirror, needs to be known very accurately for interferometric methods to work properly. This is accomplished *via* the "reference laser" shown in Figure 3.8. Data acquisition of the interferometer is synchronized to the fringe pattern produced at the "laser detector" (see Figure 3.8) when the interferometer is scanned. The reference laser beam, most commonly from a He—Ne laser, is passed through a portion of the interferometer, typically the corners or periphery of the mirrors. A diode detector measures the laser intensity variations produced when the interferometer is scanned. As discussed in Eq. 3.24, this laser intensity pattern is a shifted (raised) sine or cosine wave. The signal of the laser diode is passed through a DC filter that cuts out the offset, and a zero crossing of this laser signal is observed whenever the movable mirror has moved 1/2 laser wavelength. For a He—Ne laser ($\lambda = 632.8$ nm), the laser zero crossing points are spaced by 316.4 nm. Whenever the laser signal crosses the zero level, data acquisition is triggered: the signal from the IR detector is read, digitized, and stored as a point of the interferogram.

The laser signal is periodic in the motion of the movable mirror; yet, since it is monochromatic, no center burst is observed for the laser reference signal. Thus, a method is needed to determine the ZPD of the interferometer to the same accuracy as the fringe spacing. This is accomplished using the "white light interferogram" (i.e., the center burst of the broadband IR radiation interferogram) or the most intense peak of the IR detector (see Figure A.4(b)). Once this "ZPD" (see Figure 3.7) is found, all subsequent measurements are triggered by the zero crossing of the reference laser signal. Thus, the path difference between the two arms of the interferometer is accurately known.

Since the wavelength of a He—Ne laser is very stable, and very accurately known, data can be acquired at very precise path differences, even if there is some jitter in the velocity of the moving mirror. Thus, the accuracy of the interferometric data is very high, allowing for repetitive scans of the interferometer to be very well aligned with respect to each other. Furthermore, the wavenumber accuracy of FTIR spectroscopy is higher than in dispersive IR spectroscopy since it is determined by a very accurately known and highly reproducible visible laser wavelength. This is known as the Conne advantage of Fourier transform spectroscopy.

In addition to the Conne advantage, there are two other major improvements in interferometric over dispersive IR spectrometers. One of these is the multiplexing advantage, which is also known as the Fellgett advantage. Multiplexing here refers to the principle that all wavelengths are measured simultaneously in interferometry, whereas in dispersive systems only one spectral element passes the exit slit at a time. It is true that in interferometry, many consecutive measurements have to be taken as well at different positions of the movable mirror to collect an interferogram that accounts for the different interference patterns produced by different phase shifts. However, since each of these measurements detects an interference pattern of all wavelengths simultaneously, the signal reaching the detector is much larger than that for a single spectral element. Thus, the time to collect a high S/N interferometric data point is in the microsecond regime.

A third major advantage is known as the Jacquinot advantage, namely, the absence of beam restricting elements such as slits. Instead, large diameter beams are used in interferometry, and the intensity losses due to the slits, which are necessary in dispersive systems to define the optical resolution, are avoided. In FT instrumentation, the resolution is not determined by the size of the beam but strictly by the stroke (travel) of the movable mirror and the number of data collected during a stroke (see Section 3.3.2).

Thus, FT spectroscopy offers major optical advantages that account for enormously reduced data acquisition times as compared to a spectrum obtained on a dispersive instrument. These advantages more than compensate for the disadvantage that data have to be transformed mathematically to obtain the desired spectra. With the recent increase in the computational power of personal computers, it is possible to carry out an 8k transform in a few milliseconds *via* software FFT.

Next, a number of mathematical procedures that are necessary in FTIR spectroscopy are discussed. These procedures, phase correction, apodization, and zero filling are very well established, and are generally embedded in the instrument's operation software and hidden from the user's view. However, an understanding of these procedures is of utmost importance to avoid confounded spectra being recorded.

3.3.2 Optical resolution

In interferometric IR spectroscopy, the highest frequency signal observed is determined by the reference laser wavelength. This is because the signal is sampled at intervals determined by the zero crossing of the laser fringe pattern. Using a He—Ne laser ($\lambda = 632.8$ nm or $\tilde{\nu} = 15{,}803\ \text{cm}^{-1}$), it follows that the entire frequency space sampled by the interferogram extends from 0 to 15,803 cm^{-1}, even if the detector or various filters restrict the band path to a lower range. The number of data points collected during a stroke of the interferometer is an integer power of 2 (1024, 2048, 4096, 8192, and so on, abbreviated as 1, 2, 4, 8k) for computational reasons (the FFT algorithm only can be applied to datasets of dimension of integer powers of 2). Thus, for a 1k data set, 1k data points are obtained after FFT that span the wavenumber range from 0 to 15,803 cm^{-1}. Consequently, data points are spaced by $15{,}803/1024 = 15.4\ \text{cm}^{-1}$. Here, the interferometer stroke is only $632.8 \times 1024/2 = 0.323$ mm. To increase the optical resolution to about 2 cm^{-1} (which is the typical optical resolution required for most work), one would collect 8k data points and perform an 8k FFT, resulting in a data point spacing of 1.93 cm^{-1}. This requires that the interferometer stroke is increased to about 2.6 mm.

Increasing the optical resolution does not come with an intensity penalty, as discussed before for a dispersive system where the slit width needs to be decreased, and consequently, the transmitted intensity is reduced. However, it still comes with a penalty in signal-to-noise ratio, since changing the optical resolution from 4 to 2 cm^{-1} in interferometry requires twice the scanning time, and affects the signal-to-noise ratio adversely, as discussed later.

3.3.3 Zero filling and fourier smoothing

Upon inspection of the interferogram of a real IR spectrum, it becomes obvious that the "high frequency" components of the interferogram become very small. This is shown in Figure 3.9, which depicts the first 1024 points of a 2k FT of a data set. The amplitude of the fringe pattern between points 600 and 1000 become very small (see inset) and above point 1500, the signal may be buried by detector noise. Thus, scanning the interferometer past the 1k mark will, indeed, increase the resolution of the spectrum since a 2k transform,

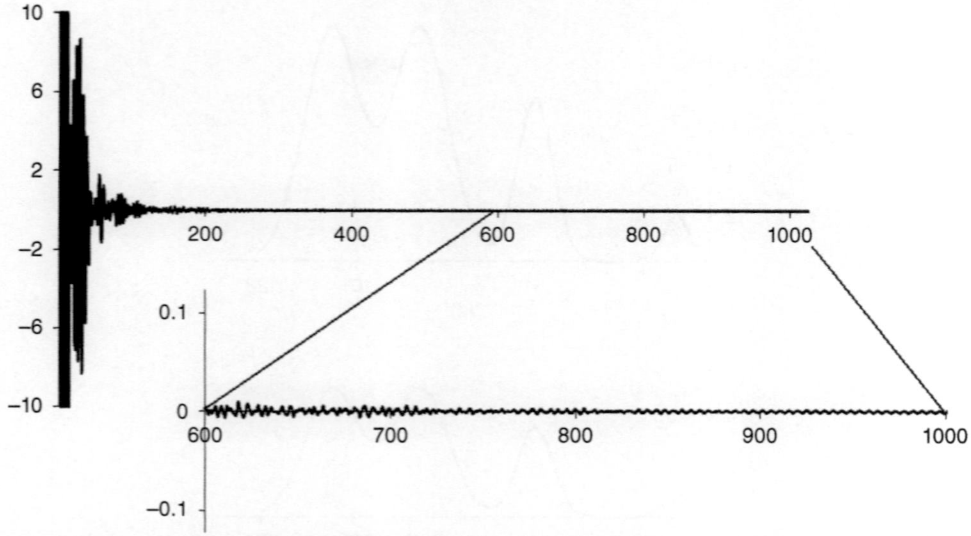

Figure 3.9 *Sections of the real part of a 2k interferogram (plotted against the number of data points) of a biological sample, demonstrating the decrease in fringe intensity far removed from the center burst*

rather than a 1k transform can be carried out, but the resulting spectrum becomes noisier since the high frequency (noise) components are Fourier transformed into the spectrum.

Thus, the question arises: What would happen to a spectrum if the interferogram, rather than being scanned to the next integer power of 2 data point, is augmented by the same number of zeros? This process is known as zero filling. The argument for zero filling is as follows: most of the spectral information is contained, for example, in the first 1024 points of the interferogram; thus, scanning another 1024 points with their inherent noise will, indeed, enhance the spectral resolution, but at a cost in signal-to-noise ratio. Adding instead 1024 zeros after the initial 1024 data points allows a 2k transform to be carried out without increasing the noise. Thus, it appears that one can gain resolution "for nothing." This apparent paradox is further explained using Figure 3.10.

In Figure 3.10(b), a 256 data point spectrum is shown that consists of four Gaussian bands (see Eq. 1.115). When this spectrum is forward and subsequently reverse Fourier transformed, an identical spectrum is obtained. If, however, the spectrum is transformed into Fourier space and zero-filled with 256 points to give a 512 point spectral vector that is subsequently reverse transformed, a new spectrum is obtained as shown in Figure 3.10(c). This spectrum now has 512, rather than 256 data points, but its amplitude is reduced by a factor of 2. This is because the amplitude of the interferogram is divided over twice as many data points. If the original spectrum is Fourier transformed, and the transformed vector truncated to 128 data points and reverse transformed, the spectrum shown in Figure 3.10(a) is obtained. Notice that this spectrum has twice the amplitude as compared to the original. What is not obvious from Figure 3.10 is the fact that the truncated spectrum has the high frequency components removed. This is demonstrated in Figure 3.11(a), which shows a noisy 256 point spectrum. Upon FT, truncation of the interferogram to 128 points, and upon reverse FT, the trace shown in Figure 3.11(b) is obtained. This "smoothed" trace obviously has reduced high frequency noise. This may be considered a primitive approach to "Fourier smoothing."

The procedure of smoothing a spectrum by just truncating the high frequency half of the interferogram is a rather non-elegant approach since the truncation is equivalent to subjecting the interferogram to a box-car function. The effect of multiplying a spectrum with a box-car function is discussed later in this section (under

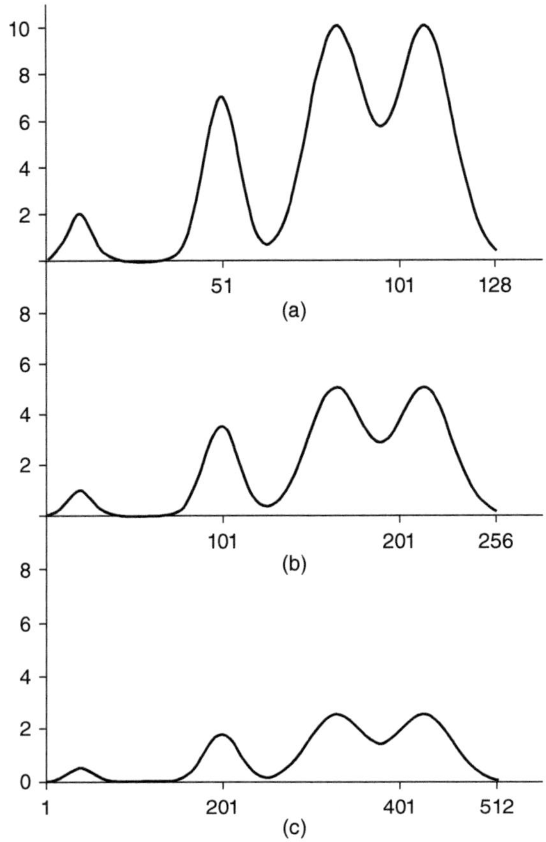

Figure 3.10 *Zero filling and truncation in Fourier transform spectroscopy. See text for details*

Apodization). A more elegant way to carry out Fourier smoothing was introduced by Kauppinen *et al.* [7], and involves multiplying the interferogram with the sinc function,

$$\text{sinc}\,(x) = \frac{\sin\,(x)}{x}. \tag{3.30}$$

Multiplying the interferogram by this function produces the best improvement of the signal-to-noise ratio with minimal distortion of the original band shapes.

3.3.4 Phase correction

In the early years of FT IR spectroscopy (and for this matter, FT-NMR spectroscopy as well), it was observed that a chirped, that is, asymmetric, interferogram produces a spectrum, upon FT, that is contaminated by dispersive line shapes, as shown in Figure 3.12(c) [8]. This is due to the fact that a perfectly symmetric interferogram can be transformed from mirror displacement domain $J(x)$ into wavenumber domain $S(\tilde{\nu})$ by a cosine FT that only transforms the real part of the interferogram, as indicated by Eqs. 3.28 and 3.29. However, a chirped interferogram needs to be transformed by a complex FT that explicitly includes a "phase

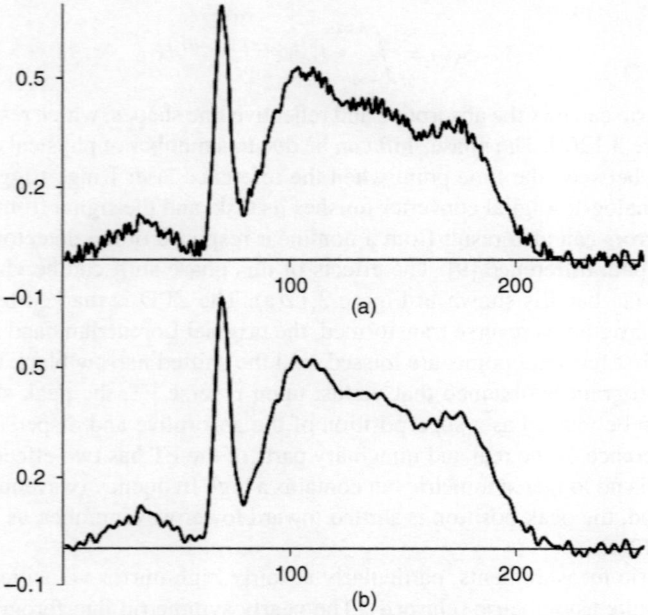

Figure 3.11 *Simple Fourier smoothing via truncation of interferogram. See text for details*

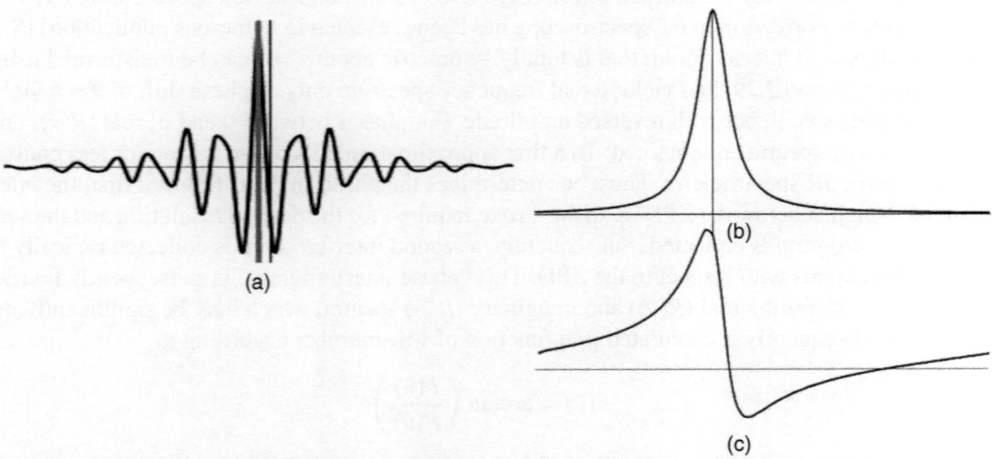

Figure 3.12 *(a) Real part of an interferogram. (b) Reverse FT of interferogram in (a) if the zero points of interferograms are assumed to coincide with the ZPD, indicated by the heavy gray line. (c) Reverse FT of interferogram in A if the zero point of the interferogram is shifted from the ZPD to the position indicated by the thin black line*

angle" θ according to Eq. 3.31 [8], where the imaginary part may contain some dispersive line shapes as indicated in Figure 3.1:

$$S(\tilde{\nu}) = \int_{-\infty}^{+\infty} J(x)\, e^{2\pi i\, \tilde{\nu}(x-\theta)} dx \tag{3.31}$$

This complex transform can mix the absorptive and reflective line shapes, which results in the distorted band shape depicted in Figure 3.12(c). The phase shift can be due to a number of physical effects, for example, due to varying time delays between the time point when the reference laser fringe triggers data acquisition and the time at which the analog-to-digital converter finishes its task, and the signal from the detector is recorded as a digital number. Errors can also result from a nonlinear response of the detector (detector saturation) in the region of the zero path difference [9]. The effects of this phase shift can be visualized as follows: The FT of a single Lorentzian band is shown in Figure 3.12(a). The ZPD is marked by the heavy gray line in this panel. If this interferogram is reverse transformed, the original Lorentzian band is recreated, as shown in Figure 3.12(b). If the first few data points are missed, and the shifted narrow black line is taken as the ZPD, a phase shifted interferogram is obtained that yields, upon reverse FT, the peak shown in Figure 3.12(c). This distorted peak can be viewed as a superposition of the absorptive and dispersive line shapes shown in Figure 3.1. This interference of the real and imaginary parts of the FT has two effects.

First, the band shape is no longer symmetric but contains a high frequency (wavenumber) negative intensity contribution, and second, the peak position is shifted toward lower wavenumber, as indicated by the vertical black line in Figure 3.12(c).

In most interferometric measurements, particularly at fairly high mirror velocities, the collected interferograms are normally quite asymmetric (chirped). The nearly symmetric interferogram shown in Figure 3.7 above is actually quite an exception and was collected with a step-scanning interferometer. (The concept of step-scanning is introduced in Chapter 8; suffice it to say at this point that this is a methodology in which interferogram points are collected at stationary mirror positions.) For all other cases, a phase correction needs to be applied that accounts for the chirped interferogram to produce undistorted spectra upon FT.

The theory of phase correction in FT spectroscopy has been presented in numerous publications [8, 10–13]. As pointed out above, an interferogram that is totally symmetric about $x = 0$ can be transformed using a real (cosine) FT (Eqs. 3.28 and 3.29) and yields a real frequency spectrum only. A phase shift of $\theta = \pi$ yields a real frequency spectrum as well, but with reversed amplitude. For phases between 0 and π, real ($R(\tilde{\nu})$) and imaginary ($I(\tilde{\nu})$) frequency spectra are obtained. To a first approximation, this phase is more or less constant for a given interferometric IR spectrometer. Thus, one determines the phase angle as follows: first, the interferometer is scanned from just below the ZPD up to the stroke required for the desired resolution and the sample (or background) interferogram is collected. Subsequently, a second interferogram is collected typically between −128 and 128 data points with respect to the ZPD. This "phase interferogram" is subsequently Fourier transformed to give low-resolution real ($R(\tilde{\nu})$) and imaginary ($I(\tilde{\nu})$) spectra, which may be significantly distorted. The phase angle subsequently is calculated as a function of wavenumber according to

$$\theta(\tilde{\nu}) = \arctan\left(\frac{I(\tilde{\nu})}{R(\tilde{\nu})}\right) \tag{3.32}$$

This phase angle then is used to correct the real and imaginary parts of the sample or background interferograms according to

$$\begin{aligned} R'(\tilde{\nu}) &= R(\tilde{\nu})\cos\theta(\tilde{\nu}) + I(\tilde{\nu})\sin\theta(\tilde{\nu}) \\ I'(\tilde{\nu}) &= -R(\tilde{\nu})\sin\theta(\tilde{\nu}) + I(\tilde{\nu})\cos\theta(\tilde{\nu}) \end{aligned} \tag{3.33}$$

Thus, one assumes that the phase does not change between the time the sample (or background) interferogram and the time the phase interferogram are collected. The phase correction is generally carried out automatically by algorithms embedded deeply within the instrument data acquisition software, and is barely accessible to the end user. Details of the procedure outlined here can be found in by Griffiths and de Haseth [8]. The original phase correction methods were published in the 1960s [10, 11].

3.3.5 Apodization

In addition to phase correction discussed above, another correction function is automatically performed by the instrument control software, nearly invisible to the user. This correction is called "apodization" and is required since the interferogram is truncated during data acquisition at the end of the stroke of the interferometer. The equations for FT (Eqs. 3.28 and 3.29), in principle, require integration from $-\infty$ to $+\infty$, which would require an infinite interferometer stroke. The truncation at a certain finite mirror travel, however, introduces a distortion of the band shapes, as follows.

One can visualize that an interferogram collected at a finite mirror stroke is obtained by multiplying a hypothetical interferogram collected at infinite mirror stroke by a box-car function as shown in Figure 3.13(a). Thus, if a 2k interferogram is collected, all points between the ZPD and the 2kth point can be thought to be multiplied by 1, whereas all higher points are multiplied by zero. The interferogram, therefore, needs be thought of as the product the infinite interferogram and the box-car function.

The FT of the product of two functions is the convolution of the FTs of each function. A box-car function, however, has its own FT, as shown in Figure 3.13(b). The FT of a finite interferogram, therefore, will have the FT of the box-car function superimposed on each band shape. This is shown in Figure 3.13(d), which depicts the result for a single Lorentzian band (Figure 3.13(c)) whose FT was multiplied with a tight box-car function before reverse FT. The resulting band profile, shown in Figure 3.13(d), is unacceptable due to the side lobes resulting from the convolution with the box-car function. As indicated, every peak and band in the spectrum will exhibit these side lobes.

This problem can be alleviated by the aforementioned process of apodization, which consists of multiplying the collected interferogram with a function that does not exhibit as drastic an oscillatory pattern as the box-car function. Several functions have been proposed that have band shapes, upon FT, that approximate the actual

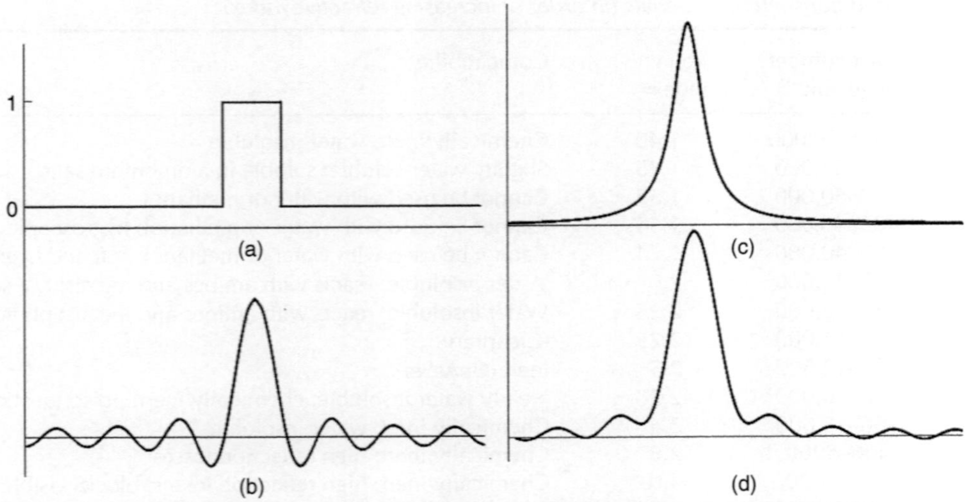

Figure 3.13 *(a) Box-car function and its Fourier transform (b). (c) Lorentzian band shape and its Fourier transform (d) when band shape is convoluted with box-car function (d)*

band shapes of the spectral features to be observed. These functions usually have the value 1 at the ZPD, and gradually decrease to zero at the end of the interferometer stroke. A simple straight line apodization (known as the Bartlett triangular apodization function), a simple cosine or the sum of several cosine functions (Blackmann), a cosine square function (Hanning), and several others are being used routinely. These functions do reduce (but not eliminate) the side lobes to an acceptably low level. The reader is referred to the literature for more details [14].

3.4 Sampling strategies

The majority of routine IR spectra used to be acquired in transmission mode, with the sample contained in a sample cell fitted with transparent windows. While this sampling method is certainly the easiest to understand, several other sampling arrangements are now in common use, often in order to facilitate and accelerate sample preparation. Some of these methods are discussed in turn.

3.4.1 Transmission measurement

Here, the sample is contained in a suitable sample cell that accommodates the physical state and property of the sample. In transmission measurements, the attenuation of the IR beam within the sample is observed directly; this attenuation follows the Lambert–Beer law (Eq. 1.113). The most accurate spectral information is obtained if the sample path or concentration is adjusted such that the peaks of interest have an absorbance between 0.01 and 1, corresponding to percent transmission values between 97.7 and 10%. IR-transparent material for the windows of either liquid or gas-phase sample cells are listed in Table 3.2.

In this table, both the transmission range (cm^{-1}) and the refractive index of the window material are listed. The refractive index is highly important for the selection of sample cell windows, since it determines the amount of light lost by reflection from the cell interfaces. This aspect is discussed in the next section.

Table 3.2 *Infrared transmitting materials (in order of increasing refractive index)*

Material	Wavenumber range (cm^{-1})	Refractive index[a]	Compatibility
CaF_2	1100–66,000	1.40	Chemically inert, water insoluble
BaF_2	800–66,000	1.45	Slightly water soluble; soluble in ammonium salts
NaCl	500–50,000	1.52	Cannot be used with water or methanol
KBr	400–48,000	1.54	Cannot be used with water or methanol; hygroscopic
CsI	200–40,000	1.74	Cannot be used with water or methanol; soft and hygroscopic
AgCl	435–25,000	2.0	Water insoluble; reacts with amines and metals; UV sensitive
AgBr	285–22,000	2.25	Water insoluble; reacts with amines and metals; photosensitive
ZnS	700–27,000	2.25	(Cleartran)
Amtir-1	750–12,500	2.5	Inert (GeAsSe)
KRS-5	250–16,000	2.38	Nearly water insoluble; chemically inert; poisonous!!!
ZnSe	500–18,000	2.4	Chemically inert; water insoluble
CdTe	400–5000	2.6	Chemically inert; high reflection losses
Ge	600–5000	4.0	Chemically inert; high reflection losses; blocks visible light

[a] At 2000 cm^{-1} (5 μm).
http://www.infraredtraininginstitute.com/infrared-transparent-materials/.

Figure 3.14 Picture of a 10 cm path length demountable sample cell for gases. PikeTech. http://www.piketech.com/GS-Short-Path-Gas-Cells.html

Gaseous samples with 10–50 mm Hg vapor pressure at room temperature can be studied at these pressures with gas cells with 50–100 mm path lengths. A typical gas transmission cell is shown in Figure 3.14. For samples with lower vapor pressures, correspondingly longer path lengths need to be employed. Gas cells with up to 10 m path lengths are commercially available. For these long paths, sample cells with internal mirrors to fold the optical path are utilized. Gas phase spectra give the most accurate spectral data, because molecules in a gaseous sample are normally monomeric, and undergo only minimal interactions with neighboring molecules or a solvent. The rotational fine structure (*cf.* Figure 2.14 and Chapter 5) can sometimes be used to obtain very accurate structural data and can aid in the vibrational assignment of small molecules. If the rotational structure obscures vibrational peaks, the addition of a few atmospheres of inert gas often sufficiently broadens the rotational–vibrational lines into broad envelopes to permit observation of the vibrational peaks.

Neat liquids are sampled most often with sample path lengths between 10 and 100 µm, between windows of materials described in Table 3.2. Several sample cell designs exist, from demountable cells consisting of two polished, IR-transparent windows with suitable spacers, to sealed flow-through cells and variable path length cells (0.01–5 mm); see Figure 3.15. Even very short path lengths, such as 10 µm, often will produce very intense absorption spectra. Therefore, many routine spectral data acquisitions are carried out with the sample dissolved in a suitable solvent. In dilute solution spectra, monomeric solute spectra can be obtained, and interactions between solute molecules (e.g., hydrogen-bonded dimers) or between solvent and solute molecules can be observed. Vibrational frequencies of gaseous and solution phase species are usually shifted due to solvent interactions.

Low solubility of the solute in a given solvent may require long path lengths to be used, which may create problems, since even for a 0.5 mm path, most organic liquids have enormously strong overtone and combination bands in the mid-IR spectrum. Therefore, many organic solvents, such as acetone, dimethylsulfoxide (DMSO), or acetonitrile, can only be used in very narrow spectral regions. Acceptable solvents for the spectral range above about 1200 cm^{-1} are CCl_4, CS_2, chloroform-d_1 ($CDCl_3$, see Section 2.6), and bromoform-d_1 ($CDBr_3$).

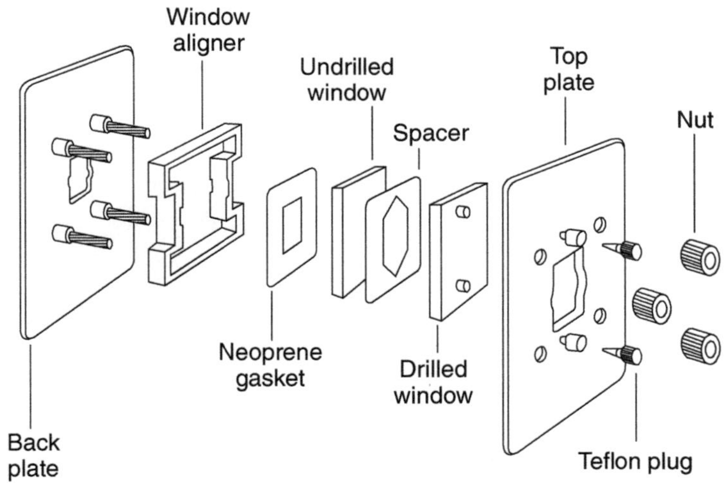

Figure 3.15 *Picture of a demountable sample cell for liquids. PerkinElmer. Reproduced with permission from PerkinElmer, Inc*

Biological molecules and ionic compounds often are insoluble in any solvent except water and heavy water (deuteroxide). Both these solvents have enormously intense IR absorptions, and it is virtually impossible to utilize path lengths of more than 20 μm with H_2O and 100 μm with D_2O. Therefore data acquisition from these solvents in transmission spectroscopy is challenging. Absorption spectra of H_2O and D_2O are shown in Figure 3.16. For IR spectroscopy of liquids and solutions, one has to ascertain that no chemical interaction occurs between the aqueous solvent and the window materials: NaCl or KBr windows are incompatible with aqueous solutions, as are amines and silver halide windows.

Owing to problems in dissolving samples, many routine collections of IR spectra are carried out in the solid phase. This is accomplished by finely grinding the solid sample, and mixing it with inert, finely ground salts (such as KBr or KCl) and pressing the powders to produce clear pellets. However, solid phase spectra are more difficult to interpret, since several effects perturb the spectral features of the isolated molecules. Among them are strong interactions, such as hydrogen bonding, and site interaction. This latter effect is due to the fact that more than one molecule, occupying different sites, may be found in a unit cell. Since the unit cell, and not the molecule, is the smallest repeating unit in a crystalline solid, the observed vibrational spectrum is that of the unit cell. Thus, some vibrations may be split into multiplet structures depending on the number of molecules per unit cell. In principle, the site splitting can be analyzed by group theoretical methods, since the sites are related by symmetry. However, most qualitative IR work totally ignores these effects.

3.4.2 Specular reflection

As pointed out above, any sample exhibiting an absorption spectrum in a given frequency range exhibits anomalous dispersion of the refractive index at the wavelengths of absorption maxima. Thus, measuring the reflectivity of a sample as a function of wavenumber produces a reflectance spectrum that can be transformed to an absorption spectrum according to the Kramers–Kronig transform (Eqs. 3.3 and 3.4). The amount of radiation reflected by a surface (in regions with no absorptions) is given by

$$R = \frac{(n-1)^2}{(n+1)^2} \tag{3.34}$$

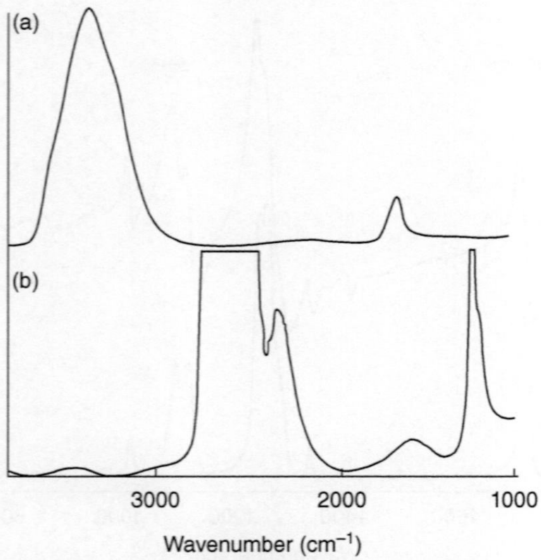

Figure 3.16 *(a) Liquid phase absorption spectrum of H_2O, 5 µm path length. (b) Liquid phase absorption spectrum of D_2O, 50 µm path length*

where n is the ratio of the refractive index of the medium and air (which is assumed to be 1.0). Thus, for a material with a high refractive index, such as ZnSe (see Table 3.2), the reflection losses per interface are substantial:

$$R = \frac{(2.4 - 1)^2}{(2.4 + 1)^2} = 0.17 \text{ or } 17\%! \tag{3.35}$$

For this reason, sample cell windows should be made of materials with low refractive indices, or be anti-reflection coated.

Returning to the discussion of reflectance measurements, one needs to modify Eq. 3.34 as follows if the sample exhibits absorption in the wavelength range of interest:

$$R = \frac{(n - 1)^2 + \kappa^2}{(n + 1)^2 + \kappa^2} \tag{3.36}$$

where the absorption index κ was defined in Eq. 3.2. Thus, light reflected from a sample interface contains information on its absorption spectrum. The amount of radiation reflected, however, is generally quite small, and reflectance measurements (also referred to as "specular reflection") are, in general, noisier than absorption measurements. A specular reflection spectrum from a single reflection of an incident beam from a thick sample of polyethylene terephthalate, (PET) is shown in Figure 3.17, along with the corresponding absorption trace, obtained by Kramers–Kronig transform of the reflectance spectrum [15]. The example reproduced here follows a long line of experimental efforts [16–18] to determine the "optical constants" n and κ, which are the fundamental parameters describing the interaction of light with a medium.

3.4.3 Diffuse reflection

In diffuse reflectance spectroscopy, also referred to as DRIFTS (diffuse reflection infrared Fourier transform spectroscopy), a spectrum of a polycrystalline sample is collected by illuminating the sample surface with a

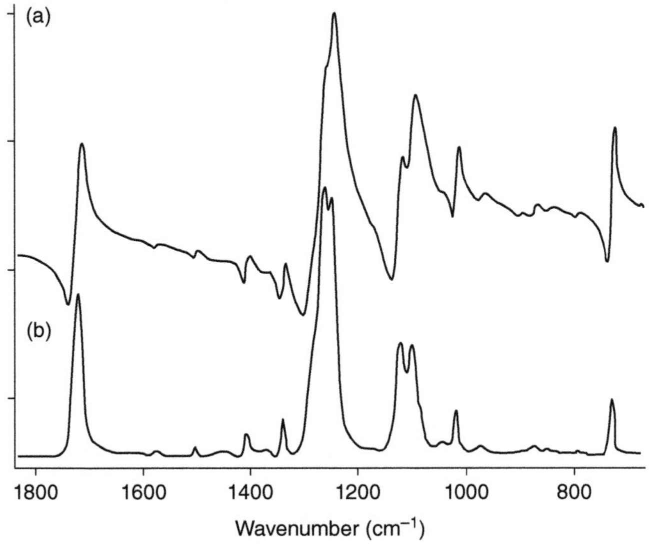

Figure 3.17 *(a) Pure reflectance of PET and (b) corresponding absorption spectrum obtained by Kramers–Kronig transform of trace (a). Everall* et al. *[15]. Reproduced with permission from Elsevier*

focused IR beam, and collecting the light that is reflected from the lower layers of the sample and absorbed by passing through sample crystals between the surface and the reflecting layer. The diffusely reflected light is usually collected using an integrating sphere. In general, one tries to avoid the light from the specular reflection since this light has dispersive line shapes and it is easy to see that in diffuse reflectance spectroscopy distorted band shapes can occur if the diffusely reflected light mixes with light undergoing pure reflectance. Such mixing of the absorptive and dispersive line shapes can produce spectra that are nearly un-interpretable, as shown in Figure 3.18 [19].

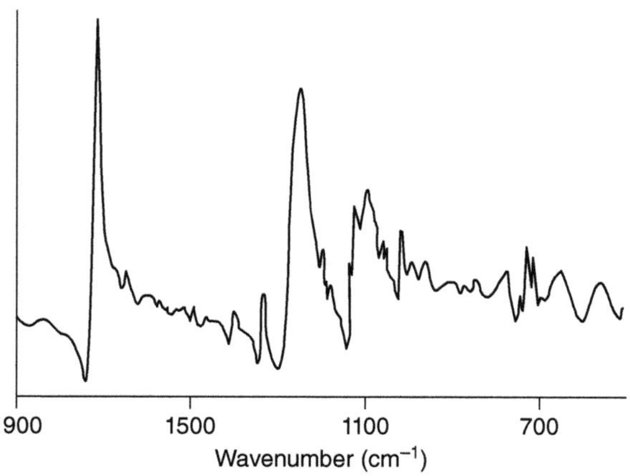

Figure 3.18 *Diffuse reflectance spectrum of PET (see Figure 3.17). Here, the pure reflectance spectrum is contaminated by a specular reflection component, confounding the observed line shapes. Chalmers and MacKenzie, 1985, [19]. Reproduced with permission from Society for Applied Spectroscopy*

Diffuse reflectance spectra should provide information comparable to absorption spectra once the reflected intensities are converted to R_∞ (the reflectance of an infinitely thick sample) *via* the Kubelka–Munk relationship [20]. Diffuse reflectance works best for powdery substances or very fine crystalline samples. Often, the sample needs to be diluted in an inert substance, for example, by finely grinding it with KBr, to reduce the observed absorbance values.

3.4.4 Attenuated total reflection

ATR, also referred to as total internal reflection spectroscopy, is carried out by having the IR beam confined in a material of high refractive index that is in contact with the sample. The IR beam impinges on the interface between the optically dense material and the sample at an angle higher than the critical angle, such that total internal reflection of the IR beam occurs. This is illustrated in Figure 3.19.

This figure reviews the principles of reflection and refraction when a beam of light passes through an interface. Figure 3.19(a) depicts reflection of a beam of light traveling in a medium with low refractive index, and encountering an interface with a material of higher refractive index (i.e., $n_2 > n_1$). In this case, some of the light is being reflected at the same angle as the angle of incidence, that is, $\theta_i = \theta_r$. A portion of the beam is refracted into the medium of higher refractive index. Here, the angle of refraction, θ_f, is given by Snell's law:

$$\frac{\sin\theta_i}{\sin\theta_f} = \frac{n_2}{n_1} \quad (3.37)$$

When a beam of light is traveling from a denser into the less dense medium, and impinges onto the interface at a small angle (as measured from the normal), light again is reflected and refracted, as shown in Figure 3.19(b). However, as the angle of incidence increases, more light is reflected back into the optically dense medium, and less light is being refracted into the less dense medium. At a limiting angle, called the critical angle, no light is being refracted into the less dense medium, and total internal reflection occurs. The critical angle is given by

$$\sin\theta_c = \frac{n_1}{n_2} \quad (3.38)$$

At this angle, the refracted beam is confined to the surface of the optically dense medium, and is polarized in the direction of the surface plane. This surface-confined radiation, also known as the "evanescent" wave, gives rise to surface plasmon resonance spectroscopy and other surface effects, such as ATR. In ATR, it is this evanescent wave that interacts with, and is attenuated by, the sample; thus, the light totally reflected at the interface is attenuated similarly to the attenuation that occurs by straightforward absorption. Since the

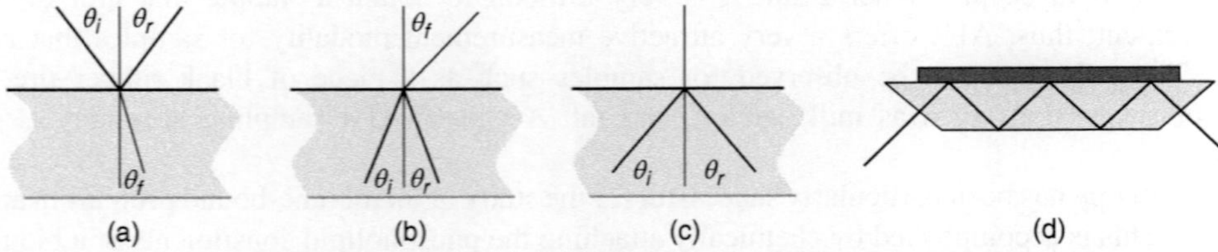

Figure 3.19 *(a) Reflection and refraction of beam of light traveling from a medium of low into a medium of high refractive index (gray shaded regions). (b) Reflection and refraction of beam of light traveling from medium of high into a medium of low refractive index at an angle of incidence below the critical angle. (c) Total internal reflection at or above the critical angle. (d) Multiple reflection IRE for ATR measurement. The dark gray rectangle represents the sample. See text for details*

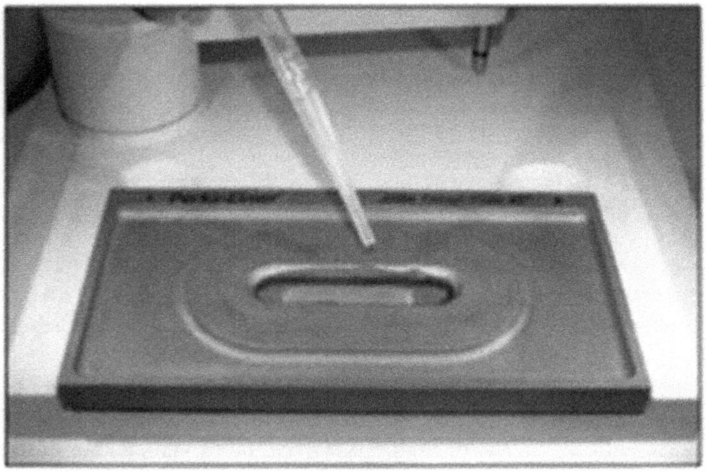

Figure 3.20 *ATR sampling accessory. The IRE crystal is visible under the pipette tip*

amplitude of the electric field of the evanescent wave drops off exponentially with the distance from the surface, the attenuation experienced by the reflected beam corresponds to an "absorption" path length of about 2 µm. Therefore, only species very close to the interface can be observed *via* ATR but due to the short equivalent path length, highly absorbing materials can be studied.

ATR spectroscopy is carried out by confining the IR beam within an internal reflection element (IRE) made of an optically dense, but IR-transparent material such as ZnSe, diamond, or KRS-5. Frequently, the IRE is designed such that multiple internal reflections occur in a crystal with trapezoidal cross section; see Figure 3.19(d). At the interface of the IRE and the sample, the penetration depth d_p of the evanescent wave can be estimated to be

$$d_p = \frac{\lambda}{2\pi n_c\,[\sin^2\theta - n_{sc}^2\,]^{1/2}} \tag{3.39}$$

In this equation, λ is the wavelength of the light, n_c the refractive index of the IRE material, θ the angle of incidence, and n_{sc} the ratio of the refractive indices of the sample and the IRE material [21].

If ZnSe is used as the IRE material ($n_c = 2.2$), the ratio $n_{sc} = n_s / n_c \approx 0.6$ for typical organic liquids. In this case, the critical angle is about 37°; thus, for any angle of incidence larger than this value, total reflection will occur. Assuming an angle of incidence of 45°, one can estimate the penetration depth at 10 µm wavelength to be just under 2 µm. It is very difficult to obtain a sample that thin for transmission measurement; thus, ATR offers a very attractive measurement modality for samples that are highly absorbing. Thus, IR data can be observed for samples such as a piece of black rubber tire, aqueous solutions, or suspensions such as milk, grease, and fat. A typical ATR sampling accessory is shown in Figure 3.20.

ATR spectroscopy has been particularly successful for the study of membrane-bound proteins in an aqueous environment. This is accomplished by chemically attaching the phospholipid constituents of a biomembrane to the IRE, and subsequently incorporating proteins into the membrane. The entire assembly can subsequently be immersed in water, and membrane-bound proteins can be studied in their native environment, and can even undergo catalytic reactions [22]. Another example of observing the spectral signatures of aqueous solutions of sucrose is shown in Figure 3.21 [23]. Owing to the high absorption of water, it is nearly impossible to measure these spectra in transmission.

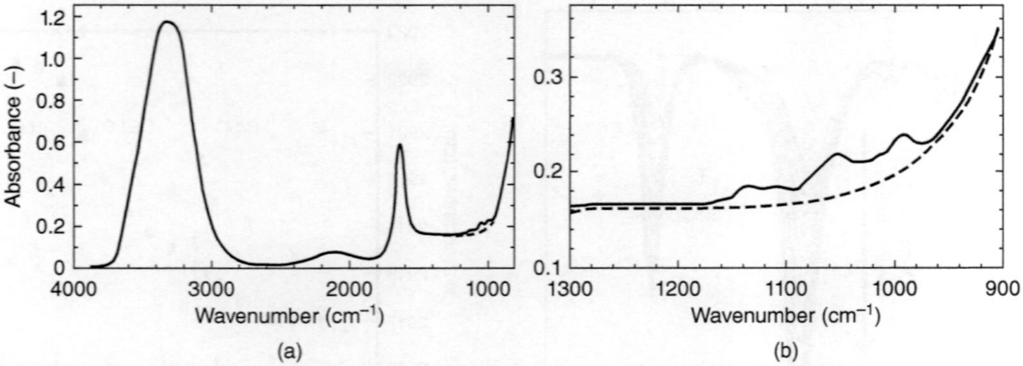

Figure 3.21 *(a) Example of an ATR spectrum of an aqueous solution of sucrose (3% by weight). (b) Expanded view in the low frequency region. In both graphs, the dashed traces are for pure water (see Figure 3.16) and the solid traces for the solution. Hashimoto and Keameoka, 2005, [23]. Reproduced with permission from Society for Applied Spectroscopy*

ATR spectroscopy also has found application in first responder teams where it may be essential to analyze an unknown sample rapidly, and without sample preparation. ATR-based, portable IR instruments exist in which the unknown sample is pressed into a sample pit with an ATR crystal at its bottom. This allows very rapid identification and classification of unknown samples.

3.4.5 Infrared reflection absorption spectroscopy (IRRAS)

Infrared reflection absorption spectroscopy (IRRAS) is another reflection-based measurement technique that is particularly suited for studying monolayers of interfacial species, such as phospholipids on an aqueous sub-phase. IRRAS is carried out by detecting the small amount (~6%) of light being reflected from an aqueous surface at angles of incidence between 40° and 70°. The reflected light is measured at an angle of reflection equal to the angle of incidence. The reflectance–absorbance (RA) is defined as $-\log(R/R_0)$ in analogy to the definition of absorbance, where R_0 is the reflectance of the pure sub-phase and R the reflectance of the film-covered surface. Furthermore, the state of polarization of the incident light can be varied to be parallel or perpendicular to the reflection plane.

For measurements using water as the sub-phase, the contamination of the reflectance spectra by water vapor contributions is a major problem, which has been overcome by either of two methods: a "shuttle" approach where the sample is divided into two areas, one with and one without a surface monolayer and the measurement spot is "shuttled" between the two areas. Here, it is assumed that the water vapor contribution is the same in both sides. The other involves a polarization modulation method, where the water vapor spectral components are cancelled between the two polarization measurements since the vapor components are the same for both polarization directions [24].

IRRAS intensities depend on the real and imaginary optical constants of the film and sub-phase, as well as the orientation of the transition moment with respect to the plane of reflection, and can be positive or negative. Therefore, RA intensities permit determination of the orientation of molecules deposited at the surface. In particular, phase transitions of lipids, as a function of surface pressure or molecular surface area, can be monitored *via* changes in the alignment of a transition moment with respect to the surface. This is shown in Figure 3.22 for a monolayer of a model lipid, 1,2-dipalmitoyl-sn-glycero-3-phosphocholine (DPPC) on an aqueous sub-phase.

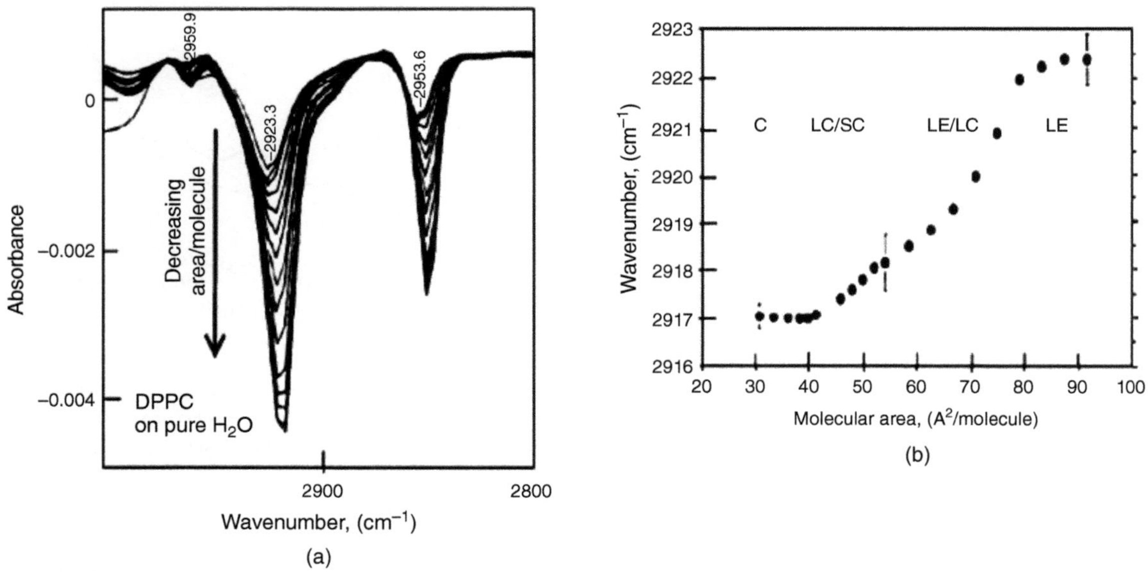

Figure 3.22 *(a) IRRAS spectra of a monolayer of 1,2-dipalmitoyl-sn-glycero-3-phosphocholine (DPPC) on water. The symmetric (2863 cm^{-1}) and antisymmetric (2923 cm^{-1}) methylene stretching modes (cf. Table 2.6) are shown as a function of molecular surface area. (b) Dependence of the antisymmetric methylene stretching frequency on the molecular area. The phases are marked "C" for solid condensed, "LC" for liquid condensed, and "LE" for liquid expanded phases Mendelsohn, Mao and Flach, 2010, [24]. Reproduced with permission from Elsevier*

3.4.6 Fourier transform photoacoustic spectroscopy (FT-PAS)

A different method to detect the interaction of IR radiation with the sample is provided by photoacoustic spectroscopy. Here, the instrumentation for exciting an IR spectrum is basically the same as for (interferometric or dispersive) absorption measurements, but rather than detecting the number of photons transmitted by the sample, as in standard transmission measurement, one detects the effect of sample heating by the absorbed IR radiation. The local heating of the sample at the focus of the IR radiation causes the sample to expand. If the radiation is modulated at a frequency f, for example, by the interferometer Fourier frequencies, the heating and cooling of the sample occurs at these frequencies, creating temperature oscillations within the sample. These oscillations are the origin of acoustic waves that propagates through the sample. Upon approaching the surface of the sample, the gas in contact with the sample surface is heated and cooled, and propagates an acoustic signal in the gas that can be detected with a microphone. This method is particularly suitable for samples of very high absorbance (and therefore, low transmitted intensities), where the high absorbance causes strong local heating, and therefore, a strong acoustic signal [25].

The sampling depth in Fourier transform photoacoustic spectroscopy (FT-PAS) depends inversely on the modulation frequency of the IR beam, since low frequency excitation allows a longer time for the thermal waves to propagate from deeper within the sample. Sampling depths of more than 100 μm can be achieved, far surpassing the sampling depth of reflective techniques such as ATR discussed above. Thus, FT-PAS is an ideal method to probe layered samples. An example of FT-PAS spectra is shown in Figure 3.23 for a thin (6 μm) layer of PET (see Figure 3.17) on a polycarbonate substrate. FT-PAS evidently can independently measure the spectra of the individual layers of the sample. Such depth-resolved measurements can easily be

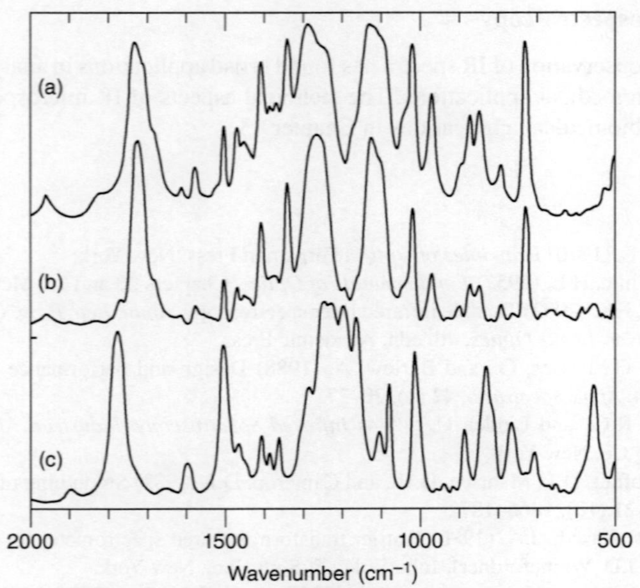

Figure 3.23 *(a): FT-PAS spectrum of pure PET (see Figures 3.17 and 3.18). (b) FT-PAS spectrum of a 6 μm thick layer of PET on a polycarbonate substrate. (c) FT-PAS spectrum of pure polycarbonate substrate. McClelland, Jones and Bajic, 2002 [25]. Reproduced with permission from J.Wiley & Sons*

carried out in confocal Raman microspectroscopy (see Chapter 11), but not in IR microspectroscopy due to the restricted confocal depth resolution in the latter technology.

3.4.7 Planar array infrared spectroscopy (PA-IRS)

This technique could be construed as a third experimental approach to IR spectroscopy. It utilizes a continuum source and dispersive optics (a prism or a grating), but rather than re-collimating the diffracted light and focusing it onto the exit slit of a monochromator, a planar array detector is placed at the exit plane to collect all diffracted wavelength simultaneously. This is, in principle, quite similar to the arrangement used in multichannel Raman instrumentation (*cf.* Section 4.5). Typical detectors used for this purpose are 256×256 element HgCdTe focal plane array (FPA) detectors. An image of the monochromator entrance slit is projected onto one dimension of the FPA, while the diffracted wavelengths are projected in the other dimension. By dividing the slit image equally between a reference and a sample beam path, these two spectral components can be measured simultaneously and in real time. Since the optics of a PA-IR instrument can be designed such that the optical paths of sample and reference beams are exactly equal in length, this method can compensate for the atmospheric water and carbon dioxide compensation. Furthermore, the high sensitivity and fast read-out of an FPA allow spectral data acquisition in the low millisecond time frame. Thus, this experimental approach permits time-resolved spectral information to be collected [26].

3.4.8 Two-dimensional FTIR

This technique is discussed along with other data analysis methods in Chapter 12.

3.4.9 Infrared microspectroscopy

The microscope-based observation of IR spectra has found broad applications in analytical chemistry, material science as well as in biomedical applications. The technical aspects of IR microspectroscopy are discussed in Chapter 11, and the biomedical applications in Chapter 13.

References

1. Born, M. and Wolf, E. (1970) *Principles of Optics*, Pergamon Press, New York.
2. Jenkins, F.A. and White, H.E. (1957) *Fundamentals of Optics*, Chapters 23 and 25, McGraw-Hill, New York.
3. Curl, R.F. and Tittel, F.K. (2002) Tunable infrared laser spectroscopy. *Annu. Rep. Prog. Chem. Sect. C*, **98**, 219–272.
4. Boyd, R.W. (2008) *Non-linear Optics*, 4th edn, Academic Press.
5. Diem, M., Roberts, G.M., Lee, O., and Barlow, A. (1988) Design and performance of an optimized dispersive infrared dichrograph. *Appl. Spectrosc.*, **42** (1), 20–27.
6. Barnes, R.B., Gotr, R.C., and Liddel, U. (1944) *Infrared Spectroscopy. Industrial Applications and Biography*, Reinhold Publishing Co, New York.
7. Kauppinen, J.K., Moffatt, D.J., Mantsch, H.H., and Cameron, D.J. (1982) Smoothing of spectral data in the Fourier domain. *Appl. Opt.*, **21** (10), 1866–1872.
8. Griffiths, P.R. and de Haseth, J.A. (1986) Fourier transform infrared spectrometry, in *Chemical Analysis*, vol. **83** (eds P.J. Elving and J.D. Winnefordner), John Wiley & Sons, Inc., New York.
9. Chase, D.B. (1984) Nonlinear detector response in FT-IR. *Appl. Spectrosc.*, **38** (4), 491–494.
10. Mertz, L. (1967) Auxiliary computation for Fourier spectrometry. *Infrared Phys.*, **7** (1), 17–23.
11. Forman, M.L., Steel, W.H., and Vanasee, G.A. (1966) Correction of asymmetric interferograms obtained in Fourier transform spectroscopy. *J. Opt. Soc. Am.*, **56**, 59–67.
12. Chen, C.N. and Kan, L.S. (1988) An iterative phase correction program for nuclear magnetic resonance (NMR) spectra. *Comput. Meth. Programs Biomed.*, **26** (1), 81–84.
13. Montigny, F., Elbayed, K., Brondeau, J., and Canet, D. (1988) Automatic phase correction of Fourier transform nuclear magnetic resonance spectroscopy data and estimation of peak area by fitting to a Lorentzian shape. *Anal. Chem.*, **62**, 864–867.
14. Weisstein, E. (1996) Apodization.
15. Everall, N.J., Chalmers, J.M., and Local, A. (1996) Measurement of surface orientation in uniaxial poly(ethylene terephthalate) films using polarised specular reflectance FT infrared microscopy. *Vib. Spectrosc.*, **10**, 253–259.
16. Bertie, J.E. and Zhang, S. (1992) Infrared intensities of liquids. IX. The Kramers-Kronig transform, and its approximation by the finite Hilbert transform via fast Fourier transforms. *Can. J. Chem.*, **70**, 520–531.
17. Yamamoto, K. and Ishida, H. (1994) Optical theory applied to infrared spectroscopy. *Vib. Spectrosc.*, **8**, 1–36.
18. Bertie, J.E., Zhang, S.L., and Keefe, C.D. (1995) Measurement and use of absolute infrared absorption intensities of neat liquids. *Vib. Spectrosc.*, **8**, 215–229.
19. Chalmers, J.M. and MacKenzie, M.W. (1985) Some industrial applications of FT-IR diffuse reflectance spectroscopy. *Appl. Spectrosc.*, **39** (4), 634–641.
20. Kubelka, P. and Munk, F. (1931) Ein Beitrag zur Optik der Farbanstriche. *Z. Tech. Phys.*, **11a**, 593–601.
21. Harrick, N.J. (1967) *Internal Reflection Spectroscopy*, John Wiley & Sons, Inc., New York.
22. Schartner, J., Güldenhaupt, J., and Mei, B. (2013) Universal method for protein immobilization on chemically functionalized germanium, investigated by ATR-FTIR difference spectroscopy. *J. Am. Chem. Soc.*, in press.
23. Hashimoto, A. and Keameoka, T. (2000) Mid-IR spectroscopic determination of sugar contents in plant-cell culture media using an ATR method. *Appl. Spectrosc.*, **54** (7), 1005–1011.
24. Mendelsohn, R., Mao, G., and Flach, C.R. (2010) Infrared reflection-absorption spectroscopy: principles and applications to lipid-protein interactions in Langmuir films. *Biochim. Biophys. Acta*, **1798**, 788–800.
25. McClelland, J.F., Jones, R.W., and Bajic, S.J. (2002) FTIR photoacoustic spectroscopy, in *Handbook of Vibrational Spectroscopy*, vol. **2** (eds J.M. Chalmers and P.R. Griffiths), John Wiley & Sons, Ltd, Chichester, pp. 1231–1251.
26. Pelletier, I., Pellerin, C., and Chase, D.B. (2005) New developments in planar array infrared spectroscopy. *Appl. Spectrosc.*, **59** (2), 156–163.

4
Raman Spectroscopy

As indicated in Section 1.6, vibrational spectral information can be collected by direct absorption of infrared (IR) photons or by light scattering *via* the Raman effect. Before discussing this conceptually somewhat more complicated form of vibrational spectroscopy, a quick interpretation of the term "light scattering" is appropriate, since this terminology is used both for macroscopic and microscopic effects. Macroscopic light scattering – for example, the scattering of light from aerosols – occurs when the scattering particles are about the same size (about 0.5 μm) as the wavelength of light. This scattering depends on the fourth power of the frequency of light; thus, suspensions of colloid particles within this size range may appear bluish, since blue visible light is scattered more efficiently than red light. Suspended colloid particles are typical examples of media that exhibit macroscopic light scattering (the Tyndall effect).

However, light scattering also occurs at the molecular level. A strong, focused visible light laser beam traveling through a clean, transparent liquid sample can be seen with the naked eye since the molecules in the sample scatter photons in the direction of the observer. This scattering occurs at the same wavelength as the incident light and involves a change in the direction but not the momentum and energy of the scattered photon. Therefore, this process is often termed elastic scattering. The scattering cross section (efficiency) of molecular scattering depends on the fourth power of the frequency of the incident light as well. This accounts, for example, for the blue sky observed on a clear day: the (mostly) N_2 molecules of the atmosphere scatter blue light more efficiently than red light. Thus, a casual observer standing on the surface of the earth will observe more blue than red photons. The orange-red color of the sky at sunset results from light that has been depleted of blue photons. Scattering is a relatively weak process and only about 1 in 10^6 photons traveling through a medium will undergo elastic scattering that is also referred to as Rayleigh scattering. Owing to the low scattering efficiency, the intensity of the blue sky is orders of magnitude lower than that of the light directly reaching the surface of the earth from the sun.

In addition to this molecular form of elastic scattering, inelastic scattering also may occur, but with an even lower efficiency than elastic scattering. In this inelastic scattering, both the direction and energy of the photon change. This scattering process is known as Raman scattering, named after Sir Chandrashekhar Raman who first observed this effect experimentally in 1928 after it had been predicted by A. Smekal in 1923.

Since the advent of commercial lasers in the late 1960s, Raman spectroscopy has become a powerful technique that rivals FTIR methods in the ease and speed of data acquisition, and in information content of the

Modern Vibrational Spectroscopy and Micro-Spectroscopy: Theory, Instrumentation and Biomedical Applications, First Edition.
Max Diem.

collected spectra. In addition, high power pulsed laser sources have allowed the development of several new Raman techniques based on nonlinear dependence of the induced molecular polarization, which have opened new avenues in molecular spectroscopy (see Chapter 6). This chapter introduces a phenomenological description of Raman scattering.

4.1 General aspects of Raman spectroscopy

In Section 1.6, Raman spectroscopy was introduced from the viewpoint of interference between molecular vibrational frequencies ω_k (which modulate the molecular polarizability, α, see Eq. 1.118) and the angular frequency of the incident laser light, ω_L. This interference creates beat frequencies at ω_L and $(\omega_L \pm \omega_k)$. In the energy level diagram shown in Figure 1.13(a), it was also demonstrated that in Raman spectroscopy, the incident photons are far from the resonance condition

$$\Delta E_{\text{molecule}} = E_{\text{photon}} \tag{1.91}$$

and, in general, have much higher energy than the energy difference between vibrational states. Thus, Raman spectroscopy is an "off-resonance" spectroscopic effect, mediated by the polarizability, α. It will be demonstrated later (Section 4.2) that the polarizability actually contains the "off-resonance" conditions explicitly.

First, however, a classical view of the polarizability is presented. As pointed out above, the oscillating electromagnetic field of light induces a dipole moment μ_{ind}, which is proportional to the field strength E of the electromagnetic field:

$$\mu_{\text{ind}} = \alpha E \tag{1.116}$$

where the proportionality constant α is referred to as the polarizability. The induced dipole radiates energy at the frequency of the exciting light, and at the beat frequencies between the molecular frequency and the frequency of the incident radiation. However, the directions of the induced dipole moment and the electric vector of the incident radiation need not necessarily coincide. This is expressed by writing Eq. 1.116 as a vector equation:

$$\boldsymbol{\mu}_{\text{ind}} = \overline{\alpha}\mathbf{E} \tag{4.1}$$

where both μ and $\mathbf{E}$ are vectors and $\overline{\alpha}$ is a tensor, represented by a 3×3 matrix:

$$\begin{pmatrix} \mu_X \\ \mu_Y \\ \mu_Z \end{pmatrix} = \begin{pmatrix} \alpha_{XX} & \alpha_{XY} & \alpha_{XZ} \\ \alpha_{YX} & \alpha_{YY} & \alpha_{YZ} \\ \alpha_{ZX} & \alpha_{ZY} & \alpha_{ZZ} \end{pmatrix} \begin{pmatrix} E_X \\ E_Y \\ E_Z \end{pmatrix} \tag{4.2}$$

Here, the coordinates X, Y, and Z represent a laboratory Cartesian coordinate system. Equation 4.2 implies that light traveling in the Z-direction and polarized along the X-direction will induce dipole moment components not only along the X, but also along the Y- and Z-directions, according to

$$\begin{pmatrix} \mu_X \\ \mu_Y \\ \mu_Z \end{pmatrix} = \begin{pmatrix} \alpha_{XX} & \alpha_{XY} & \alpha_{XZ} \\ \alpha_{YX} & \alpha_{YY} & \alpha_{YZ} \\ \alpha_{ZX} & \alpha_{ZY} & \alpha_{ZZ} \end{pmatrix} \begin{pmatrix} E_X \\ 0 \\ 0 \end{pmatrix} = \begin{pmatrix} \alpha_{XX}E_X & 0 & 0 \\ \alpha_{YX}E_X & 0 & 0 \\ \alpha_{ZX}E_X & 0 & 0 \end{pmatrix} = \begin{pmatrix} \alpha_{XX}E_X \\ \alpha_{YX}E_X \\ \alpha_{ZX}E_X \end{pmatrix} \tag{4.3}$$

The consequences of Eq. 4.3 are discussed below when the subject of polarization properties of Raman scattering is further explored. Next, a simple quantum mechanical explanation for the polarizability will be presented.

4.2 Polarizability

So far, the polarizability was introduced simply as a proportionality between induced dipole moment and the incident electric field. Quantum mechanically, this interaction can be described as a sum of all transition moments of a molecular system, each one weighted by the frequency difference between the energy of the exciting field, and the actual transition frequency. A description of these steps can be found in Kauzmann's book, *Quantum Chemistry* [1]. In short, this derivation proceeds as follows.

One writes the induced dipole moment, μ_{ind}, as the expectation value of the operator, μ:

$$\mu_{\text{ind}} = \int \Psi^* \mu \Psi \, d\tau \tag{4.4}$$

where Ψ denotes the time-dependent wavefunction defined in Eq. 1.84:

$$\Psi(x,t) = \sum\nolimits_k c_k(t)\Psi(x,t) = \sum\nolimits_k c_k(t)\varphi(t)\,\psi(x) \tag{1.84}$$

As before, the wavefunctions $\psi(x)$ are the stationary state wavefunctions, $\varphi(t)$ the time-dependent wavefunctions, and $c_k(t)$ the time-dependent expansion coefficients that describe how the system transitions from one state to another. Assuming that the time dependence of the electric field can be written as

$$E(t) = E_0 \cos \omega t \tag{4.5}$$

the induced dipole moment is given by

$$\mu_{\text{ind}} = \mu_0 + \frac{2E_0}{\hbar} \cos \omega t \sum\nolimits_m \frac{\omega_{0m} |\langle \psi_m|\mu|\psi_0\rangle|^2}{\omega_{0m}^2 - \omega^2} - \frac{2E_0}{\hbar} \sum\nolimits_m \frac{|\langle \psi_m|\mu|\psi_0\rangle|^2}{\omega_{0m}^2 - \omega^2} \omega_{0m} \cos \omega_{0m} t \tag{4.6}$$

Here, one assumes that the molecule is in the ground state $|\psi_0\rangle$ before any interaction with the radiation occurs. The first term in Eq. 4.6 describes the permanent dipole moment, the second term a component of the induced dipole that oscillates at the same frequency as the incident light, and the third term an induced dipole that oscillates at the transition frequency ω_{0m} and is clearly not in phase with the frequency of the light inducing the dipole moment. The second term is called the polarizability, and a comparison between Eqs. 4.5 and 4.6 reveals that

$$\alpha = \frac{2}{\hbar} \sum\nolimits_m \frac{\omega_{0m}}{\omega_{0m}^2 - \omega^2} |\langle \psi_m|\mu|\psi_0\rangle|^2 \tag{4.7}$$

Equation 4.7 implies that any medium exposed to electromagnetic radiation will undergo some, albeit small, change, even when the frequency of the radiation is far from any transition frequency, ω_{0m}. This change can be viewed as a small, induced oscillatory motion of the electrons in a molecule; tightly bound inner electrons will respond to a lesser degree than more loosely bound electrons such as π-electrons in a double bond.

The polarizability then can be viewed as the sum of all the possible transitions in a molecule, each one weighted by an energy term in the denominator. This weighting term gets very small if the frequency of the light ω approaches the energy difference between two real stationary states ($\omega_{0m}^2 = \omega^2$) and the corresponding term in the sum expressed by Eq. 4.7 gets very large. This qualitatively describes the concept of resonance Raman spectroscopy (see Chapter 6).

The contribution of a given transition moment is small when the exciting light is far from the energy of a transition. All possible excited states must be summed up to yield the polarizability, which thus can be viewed as an attempt of the system to undergo a transition, but the energy of the photon is insufficient, or too large, to actually permit the transition. Equation 4.7 holds in the so-called "far from resonance" (FFR) case in which the photons of the incident light do not have energies close to those of an electronic transition. A more general discussion applicable to both the FFR and the "single electronic state" (SES) resonance Raman cases can be found in the Ref. [2], and is touched upon briefly in Chapter 6.

The polarizability, as discussed before, is an off-resonance effect that does not require a photon to be absorbed, but rather, indicates how the electron distribution in a molecule is affected by an applied static or dynamic electromagnetic field. As such, it is related to the refractive index n of a medium by

$$\alpha = \frac{3}{4\pi N}\frac{n^2 - 1}{n^2 + 2} \tag{4.8}$$

Here, N is the Avogadro number. Thus, one finds that off-resonance spectroscopic properties, such as the dielectric constant or the refractive index, are related to the transition moment *via* the polarizability. The results of this statement are that the proximity of an electronic transition will increase the refractive index. In the visible region of the spectrum, one can, therefore, predict that acetone has a higher refractive index than water, since its closest UV transition occurs around 280 nm, whereas the closest transition for water lies below 200 nm in the far UV region.

In general, an oscillating dipole radiates light intensity I into all space according to

$$I = \frac{16\pi^4 \tilde{\nu}^4}{3c^3}\mu^2 \tag{4.9}$$

Equation 4.9 introduces the dependence of the scattered light intensity on the fourth power of the frequency. The intensity scattered from an induced dipole depends on the square of the electric field and the polarizability; however, since the polarizability is not a scalar, but a tensor, the dependence on the polarizability is somewhat more complicated and is discussed in Section 4.3. Equation 4.9 is responsible for the Rayleigh scattering introduced in Chapter 1.

The expression for the polarizability, Eq. 4.7, describes a general scenario for the light-induced response of a medium FFR. In order to explain the Raman scattered intensities, one has to account for the fact that the molecule, after a Raman scattering event, is left in a state different from the original state. This is accomplished by writing the ground and excited vibrational states explicitly as vibronic wavefunctions ψ_{ev} as follows:

$$\alpha_{\alpha\beta} = \frac{2}{\hbar}\sum_m \frac{\omega_{0m}}{\omega_{0m}^2 - \omega^2}\langle\psi_{ev}|\mu_\alpha|\psi_{e'v'}\rangle\langle\psi_{e'v'}|\mu_\beta|\psi_{ev'}\rangle \tag{4.10}$$

Here, the primed quantities represent excited states and the unprimed quantities the ground state. This equation implies that a Raman transition starts at the vibrational ground state of the electronic ground state $\langle\psi_{ev}|$ and ends up in the vibrationally excited state of the electronic ground state, $|\psi_{ev'}\rangle$. The intermediate virtual state, shown by the dashed line in Figure 4.1, is created by the photon and is mediated by the polarizability. The polarizability then can be viewed as an interaction of the incident photon with all real excited states; this interaction is given by the transition moment between the vibronic ground and excited states, weighted by an energy denominator indicating the proximity of the transition energy into the intermediate state and the energy of the exiting light.

The Raman intensities are determined by the derivatives of the tensor components with respect to the normal coordinates, Q_k. The subscripts α and β in Eq. 4.10 indicate the Cartesian coordinates of the polarization tensor. Equation 4.10 clearly shows the vibronic nature of Raman intensities. This equation also indicates that Raman spectroscopy is a two-photon process, since there are two electronic transition moment expressions contained in the definition of the polarizability. Thus, the selection rules for the Raman effect, as discussed in Section 2.5, depend on the binary components $x{\cdot}x$, $x{\cdot}y$, $x{\cdot}z$, ..., of the Cartesian displacement components, which are listed in the character tables.

The equations presented so far describe Rayleigh and Stokes Raman scattering, in which the molecule, after the scattering event, is left in the same ground state, or a vibrationally excited state, respectively. Anti-Stokes

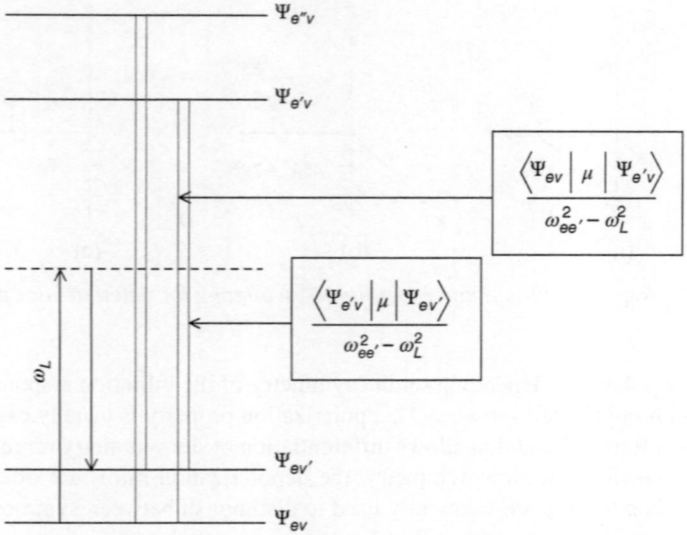

Figure 4.1 *Energy level diagram to model polarizability in terms of sum of excited state transitions. See text for details*

Raman spectroscopy occurs if the scattering event originates from a molecule originally in a vibrationally excited state and ends in the vibrational ground state. In this case, the scattered photon occurs at a frequency ($\omega_L + \omega_k$), where, as defined in Eq. 1.120, ω_L denotes the frequency of the laser photon, and ω_k the frequency of the kth normal vibration. The anti-Stokes radiation is blue-shifted with respect to the incident laser beam, and the anti-Stokes Raman frequencies appear as a mirror image of the Stokes Raman frequencies. However, anti-Stokes Raman intensities are lower than those of the Stokes spectrum, since the former depend on the population of the vibrationally excited state from which they originate. This population depends, *via* the Boltzmann equation (Eq. 1.103), on the absolute temperature and the energy of the vibrational level. Thus, anti-Stokes intensities are highest at high temperature and for those vibrational energy levels with lowest vibrational frequencies.

The description of the polarizability presented so far is applicable not only to the standard Raman effect but describes qualitatively the resonance enhancement exhibited by the Raman effect when the exciting wavelength is close to a molecular absorption. In this case, the denominator in one term in the sum in Eq. (1.4.7) becomes very small, and the corresponding polarizability term becomes very large. A more detailed description, using the Raman scattering tensor, is introduced in Chapter 6.

The review of the theory of Raman spectroscopy was presented here in a somewhat cursory manner. For an excellent review of the theory of spontaneous (linear) Raman, as well as resonance, surface-enhanced, and nonlinear Raman spectroscopy, the reader is referred to an exhaustive discussion of the subject in a book by Long [3].

4.3 Polarization of Raman scattering

In infrared absorption spectroscopy, a polarization response of the absorption intensities toward linearly polarized light is observed only for oriented, non-isotropic samples. Raman scattering, on the other hand, is

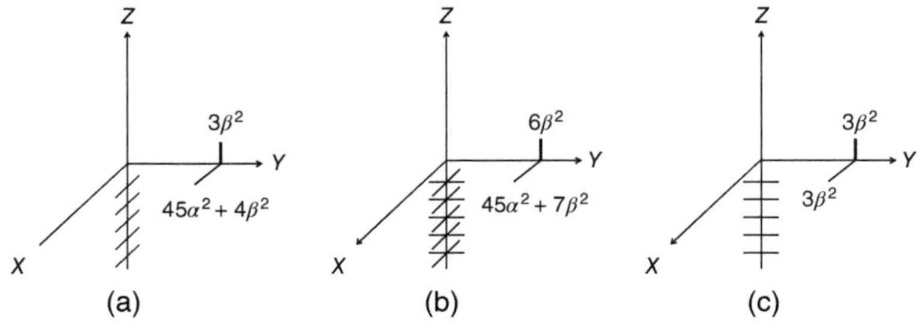

Figure 4.2 *Raman scattered intensities for 90° scattering for different laser polarization*

completely or partially polarized, depending on the symmetry of the vibration responsible for the scattering event for isotropic and non-oriented samples. This polarization property is usually expressed in terms of the depolarization ratio of a Raman band that allows differentiation of the symmetry representation of the group. Moreover, even for molecules with low symmetry, the depolarization ratios are often very helpful for the assignment of spectral bands: they are frequently used to distinguish between symmetric and antisymmetric vibrations of a locally symmetric part of a molecule, such as a methyl group.

The derivation of depolarization ratios can be carried out using the classical description of the scattering phenomenon and averaging the scattered intensities over a large ensemble of scattering species. This derivation again starts with Eq. 4.2:

$$\begin{pmatrix} \mu_X \\ \mu_Y \\ \mu_Z \end{pmatrix} = \begin{pmatrix} \alpha_{XX} & \alpha_{XY} & \alpha_{XZ} \\ \alpha_{YX} & \alpha_{YY} & \alpha_{YZ} \\ \alpha_{ZX} & \alpha_{ZY} & \alpha_{ZZ} \end{pmatrix} \begin{pmatrix} E_X \\ E_Y \\ E_Z \end{pmatrix} \tag{4.2}$$

where the capital subscripts X, Y, and Z denote, as before, space fixed (laboratory) coordinates, and the polarizability tensor elements in Eq. 4.2 can be construed to be either the static polarizability or the polarizability derivatives $(\partial\alpha_{\alpha\beta}/\partial Q)$. The first case describes the dipole moment induced at the laser frequency only, and the corresponding emission process is Rayleigh scattering. In the second case, the induced dipole moment is modulated by the normal vibration as discussed before in Eq. 1.120 and leads to Raman scattering. Since the formalism presented below for averaging the tensor elements over all orientations is the same in both cases, the tensor elements themselves, rather than their derivatives, are used in the following discussion.

Consider a non-rotating molecule, located at the origin of a coordinate system (X, Y, Z), exposed to light from a laser beam, traveling in the positive Z-direction and polarized in the XZ-plane, as shown in Figure 4.2(a). For this laser beam, $E_Y = E_Z = 0$, and only $E_X \neq 0$. In this case of linearly polarized incident light, the induced dipole moments, expressed in a space-fixed, laboratory coordinate system are (*cf.* Eq. 4.3) given by

$$\begin{pmatrix} \mu_X \\ \mu_Y \\ \mu_Z \end{pmatrix} = \begin{pmatrix} \alpha_{XX}E_X \\ \alpha_{YX}E_X \\ \alpha_{ZX}E_X \end{pmatrix} \tag{4.11}$$

Equation 4.11 indicates, as pointed out before, that a dipole moment is induced in all three Cartesian directions, even though the electric field oscillates along the X-direction only. This is due to the fact that the polarizability is a tensor, which relates vectors of cause and effect that are not collinear.

According to Eq. 4.9, this induced oscillating dipole will radiate total light intensity I^T in the Y-direction within the solid angle $\sin\theta\, d\theta\, d\varphi$ given by

$$I^T = \frac{2\pi^3 v^4}{c^3}\left(\alpha_{XX}^2 + \alpha_{XZ}^2\right) E_X^2 \tag{4.12}$$

Furthermore, one can analyze the light scattered in the Y-direction through a polarizer that transmits light polarized along the $Z(I^Z)$ and $X(I^X)$ directions separately:

$$I^X = \frac{2\pi^3 \nu^4}{c^3} \alpha_{XX}^2 E_X^2 \tag{4.13}$$

$$I^Z = \frac{2\pi^3 \nu^4}{c^3} \alpha_{XZ}^2 E_X^2 \tag{4.14}$$

In Raman spectroscopy of single crystals that are oriented properly with respect to the laser propagation and polarization directions, these components of the polarizability tensor derivatives can be measured individually. However, in non-oriented molecules in the liquid and gaseous phases, as well as for randomly oriented molecules in the solid phase, it is necessary to transform from the space-fixed (laboratory) Cartesian coordinate system X, Y, and Z to a molecule-fixed coordinate system, and express the polarizability tensor in this new coordinate system. Such a coordinate transformation is a simple matrix multiplication process, known in linear algebra as a *similarity transformation.* Then, the polarizability tensor (or the derivatives of the polarizability tensor with respect to the normal coordinates) can be expressed as a diagonal matrix:

$$\begin{pmatrix} \alpha_{XX} & \alpha_{XY} & \alpha_{XZ} \\ \alpha_{YX} & \alpha_{YY} & \alpha_{YZ} \\ \alpha_{ZX} & \alpha_{ZY} & \alpha_{ZZ} \end{pmatrix} \rightarrow \begin{pmatrix} \alpha_1 & 0 & 0 \\ 0 & \alpha_2 & 0 \\ 0 & 0 & \alpha_3 \end{pmatrix} \tag{4.15}$$

The molecule-fixed coordinate system in which the polarizability is in diagonal form is known as the principal axes of polarizability.

The similarity transformation that converts the polarizability tensor from the space-fixed coordinates X, Y, and Z (henceforth referred to as the α_F matrix) to the molecule-based coordinates (the α_g matrix) is given by

$$\boldsymbol{\alpha_F} = \mathbf{S}\,\boldsymbol{\alpha_g}\,\mathbf{S}^T \tag{4.16}$$

where

$$\mathbf{F} = \mathbf{S}\,\mathbf{g} \tag{4.17}$$

In Eqs. 4.16 and 4.17, bold quantities denote matrices and the superscript T denotes the transpose of the matrix. Here, the second equation describes the relationship between the two coordinate systems, whereas the first equation describes how a physical quantity, such as the polarizability, can be transformed from one into another coordinate system.

Before carrying out this coordinate transformation, and the subsequent averaging over all possible orientations, a number of tensor properties that are maintained under a similarity or any other linear transformation need to be discussed. In other words, there are properties of the tensor that are independent of the coordinate system in which the tensor is expressed and, therefore, are known as the *tensor invariants*. There are three of these invariants:

The *first invariant* (trace or tensor isotropic part) is defined as the sum of the diagonal terms:

$$\alpha_{XX} + \alpha_{YY} + \alpha_{ZZ} = \alpha_{xx} + \alpha_{yy} + \alpha_{zz} \tag{4.18}$$

and is invariant (constant) under a similarity transform. Here, the uppercase subscripts denote the laboratory coordinate system, whereas the lowercase subscripts denote molecule-fixed coordinates. (The reason that the traces of the transformation matrices were used to determine a molecule's reducible representation in Chapter 2 can be understood from the fact that these traces represent the invariants of the coordinate transform before and after a symmetry operation.)

The *second invariant* (also known as the first anisotropy) is the sum of all minors:

$$\begin{vmatrix}\alpha_{XX} & \alpha_{XY}\\ \alpha_{YX} & \alpha_{YY}\end{vmatrix} + \begin{vmatrix}\alpha_{XX} & \alpha_{XZ}\\ \alpha_{ZX} & \alpha_{ZZ}\end{vmatrix} + \begin{vmatrix}\alpha_{YY} & \alpha_{YZ}\\ \alpha_{ZY} & \alpha_{ZZ}\end{vmatrix} = \text{constant} \tag{4.19}$$

When the determinants in Eq. 4.19 are expanded and added up, one obtains for the first anisotropy (with $\alpha_{XY} = \alpha_{YX}$)

$$\alpha_{XX}\alpha_{YY} + \alpha_{XX}\alpha_{ZZ} + \alpha_{YY}\alpha_{zz} - \alpha_{XY}^2 - \alpha_{XZ}^2 - \alpha_{YZ}^2 \tag{4.20}$$

which, again, is invariant under a similarity transform.

The third invariant, also known as the second anisotropy, is the determinant of the tensor:

$$\text{Det}\begin{vmatrix}\alpha_{XX} & \alpha_{XY} & \alpha_{XZ}\\ \alpha_{YX} & \alpha_{YY} & \alpha_{YZ}\\ \alpha_{ZX} & \alpha_{ZY} & \alpha_{ZZ}\end{vmatrix} \tag{4.21}$$

After this introduction of tensor invariants, the discussion will return to the transformation referred to in Eq. 4.17 which involves rotation of the coordinate axes by the angles θ between the X, Y, Z coordinates and the principal axes 1, 2, 3. For this case, Eq. 4.16 can be written as

$$\alpha_{FF'} = \sum_{i=1}^{3} \alpha_i \Phi_{Fi} \Phi_{Fi'} \tag{4.22}$$

where F and F' are any combination of the components X, Y, or Z and the elements Φ_{Fi} are the directional cosine matrix elements between the two coordinate systems. Thus, the two elements, α_{XX} and α_{XY} of the space-fixed polarizability tensor elements in Eqs. 4.13 and 4.14, may be expressed as

$$\alpha_{XX} = \alpha_1 \Phi_{X1}^2 + \alpha_2 \Phi_{X2}^2 + \alpha_{31} \Phi_{X3}^2 \tag{4.23}$$

and

$$\alpha_{XZ} = \alpha_1 \Phi_{X1} \Phi_{Z1} + \alpha_2 \Phi_{X2} \Phi_{Z2} + \alpha_3 \Phi_{X3} \Phi_{Z3} \tag{4.24}$$

Since the polarizability tensor elements in Eqs. 4.13 and 4.14 occur as second powers, α_{XX}^2 contains terms in $\alpha_1^2, \alpha_1\alpha_2, \ldots,$ and Φ_{X1}^4, Φ_{X1}^2, Φ_{X2}^2, ... The directional cosine matrix elements Φ_{Fi} now need to be averaged over all orientations to account for the random orientation, caused by the tumbling motion, of the molecules in liquid and gaseous phases. This integration is given in detail in Chapter 3.6 of Ref. [4].

For the term α_{XX}^2 in Eq. 4.13 the averages over all possible orientations can be written as

$$\overline{\alpha_{XX}^2} = \frac{1}{5}(\alpha_1 + \alpha_2 + \alpha_3) + \frac{2}{15}(\alpha_1\alpha_2 + \alpha_1\alpha_3 + \alpha_2\alpha_3). \tag{4.25}$$

A similar expression holds for α_{XZ}^2.

Setting all the constants in Eqs. 4.13 and 4.14 to equal k, these equations can then be written as

$$I^X = k\left[3\sum_{i=1}^{3} \alpha_i^2 + 2\sum_{i<j}^{3} \alpha_i\alpha_j\right] \tag{4.26}$$

$$I^Z = k\left[3\sum_{i=1}^{3} \alpha_i^2 - \sum_{i<j}^{3} \alpha_i\alpha_j\right] \tag{4.27}$$

Using the tensor invariants defined above (Eqs. 4.19 and 4.20) one defines a spherical polarizability α^2 as the first invariant of the polarizability tensor, and β^2 as the first anisotropy of the polarizability tensor to obtain

$$I^X = k[45\,\alpha^2 + 4\,\beta^2] \tag{4.28}$$

$$I^Z = k[3\,\beta^2] \tag{4.29}$$

Equations 4.28 and 4.29 hold for the "FFR" condition only and need to be modified for resonance Raman spectroscopy [2]. If the polarizability tensor is expressed in any system of molecule-fixed axes that are not the principal axes of polarizability, the first anisotropy is written as

$$\beta^2 = (\alpha_{xx} - \alpha_{xx})^2 + (\alpha_{xx} - \alpha_{zz})^2 + (\alpha_{yy} - \alpha_{zz})^2 + 6(\alpha_{xy}^2 + \alpha_{xz}^2 + \alpha_{yz}^2) \tag{4.30}$$

Equations 4.28 and 4.29 are the take-home message from the derivation presented in this section. Starting with a molecule defined in a space-fixed coordinate system the Raman scattered intensity components, I^X and I^Z, were obtained. Initially, these scattered intensity components were expressed in terms of the polarizability components in space-fixed coordinates (Eqs. 4.13 and 4.14). After a suitable coordinate transformation and averaging over all possible orientations, these same scattered intensity components were derived in terms of polarizability components written in a coordinate system attached to, and rotating with, the molecule. Finally, the scattered intensities were expressed in terms of the tensor anisotropies, α and β.

The two intensity components, I^X and I^Z, can readily be observed separately by placing a polarization analyzer in the scattered beam. For an isotropic sample, the scattered intensities depend on whether the anisotropic part of the polarizability tensor or a combination of the isotropic and the anisotropic parts were responsible for the scattering process. These two possibilities arise from different irreducible representations of the symmetry group to which a molecule belongs, and are discussed next.

4.4 Dependence of depolarization ratios on scattering geometry

For the 90° scattering discussed so far, the depolarization ratio of a Raman active vibration is defined as the quotient of the intensity of scattered light polarized perpendicularly (I_Y^Z) and the intensity of scattered light polarized parallel (I_Y^X) to the incident laser polarization:

$$\rho = \frac{I_Y^Z}{I_Y^X} = \frac{3\beta^2}{45\alpha^2 + 4\beta^2} \tag{4.31}$$

Here, the subscript Y was added to the scattered intensities I^X and I^Z to indicate that these intensities were defined for right angle scattering in the Y-direction, as depicted in Figure 4.2(a); both intensities are observed through a polarization analyzer oriented to transmit light polarized in the X- or Z-direction.

This depolarization ratio allows one to assess the symmetry of the vibration giving rise to the scattering. In a *totally symmetric* vibration of a molecule belonging to a spherical point group (T_d, O_h), all the Raman scattered intensity is contributed by the isotropic (spherical) part, α^2, of the polarizability, and $\boldsymbol{\beta}^2$ is zero. Thus, I_Y^X is $45\alpha^2$, and I_Y^Z is zero; consequently, ρ is zero. Consequently, the vibration maintains the polarization of the laser beam, and the light scattered is linearly polarized in the same direction as the incident laser.

For a vibration that is not due to the most symmetric irreducible representation of the group, the isotropic scattering is zero, and all scattered intensity is contributed by the anisotropic part β^2. Therefore, the depolarization will be

$$\rho = \frac{I_Y^Z}{I_Y^X} = \frac{3\beta^2}{4\beta^2} = 0.75 \tag{4.32}$$

Thus, the polarization properties of the incident laser beam are not maintained, and the vibration acts as a partial polarization scrambler. When unpolarized laser radiation is used to excite the sample, the depolarization ratio varies between 0 and 6/7, as indicated in Figure 4.2(b).

The limiting values for the depolarization ratios, 0 and 0.75, are reached only for molecules belonging to spherical point groups (T_d, O_h). In CCl_4 , for example, the depolarization ratio for the symmetric stretching

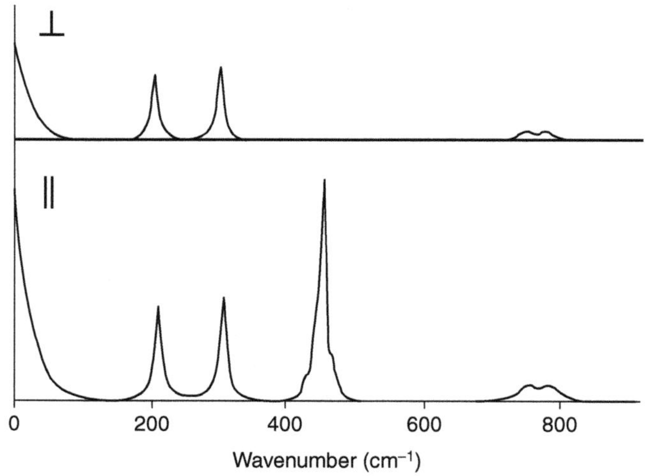

Figure 4.3 *Example of polarized Raman spectra of CCl_4 (see also Figure 2.13). Top trace: Perpendicularly polarized spectrum. Bottom trace: Parallel polarized spectrum. Kiefer and Topp, 1974, [5]. Reproduced with permission from Society for Applied Spectroscopy*

mode at 459 cm^{-1} (*cf.* Chapter 2) is smaller than 0.01, where the deviation from exactly zero is mostly due to the isotopic composition of CCl_4. The other three Raman-active modes show depolarization ratios very close to 0.75 (see Figure 4.3). For molecules with lower symmetry, the experimentally observed depolarization ratios may vary typically between 0.01 and 0.6. These depolarization ratios are very valuable even in molecules of lower symmetry, because the most polarized vibrations originate from the vibration with the most symmetric representation in the point group of the molecule.

In right angle (90°) scattering, the accurate measurement of the depolarization ratios is by no means a trivial problem. The most straightforward approach to measure the depolarization ratio is to place the polarization analyzer into the scattered beam, and to carry out two measurements, one with the polarizer transmitting along the Z- and another one with it transmitting in the X-directions, respectively. The observed depolarization ratios are given by Eq. 4.32 and will vary from 0 to 0.75. However, the rotation of a polarization analyzer in the scattered beam between the two perpendicular positions bears some disadvantages. First, when using a Polaroid sheet polarizer, its extinction ratio and exact alignment are the limiting factors determining the accuracy of the depolarization measurement. If a polarizing prism is used, a slight beam offset must be compensated for when the polarizer is rotated by 90°. Second, and more damaging, is the fact that a grating-based monochromator that disperses the scattered light has a different response for the two light components polarized at right angles to each other. The grating in the monochromator is the main culprit: the diffraction efficiency varies drastically between light components polarized parallel or perpendicularly to the ruling direction of the grating. Thus, a polarization scrambler should be placed in front of the monochromator entrance slit (after the polarization analyzer), which removes the polarization properties of the light.

Therefore, another approach for depolarization measurement may be employed. This approach involves the rotation of the plane of polarization of the incident laser beam and observing the depolarization ratios through a fixed polarization analyzer, aligned in the X-direction, in the scattered light. Inspection of Figure 4.2(a) and (b) reveals that the depolarization ratio observed by rotating the polarization direction of the laser beam is given by Eq. 4.32 as well. This latter method lends itself for an automatic measurement of ρ, since the laser plane of polarization may be rotated by 90° with an electro-optic modulator or half-wave plate, and the two scattered intensities can be measured and ratioed by synchronous detection methods. This approach

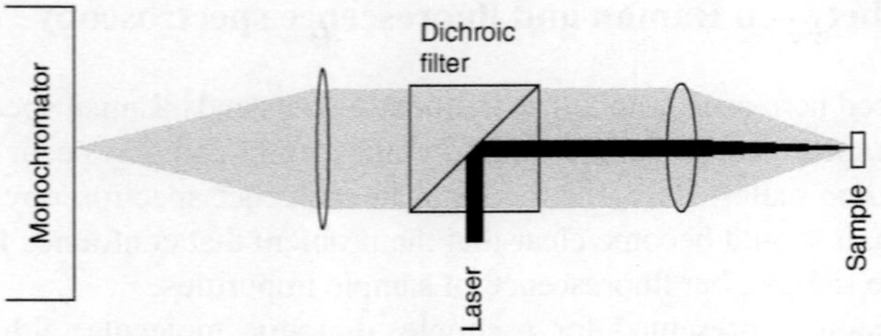

Figure 4.4 *Backscatter (180°) geometry. The gray areas represent the Raman backscattered light*

eliminates the problems with the alignment of the polarization analyzer in the scattered beam [5]. Figure 4.3 depicts an automatically scanned polarized/de-polarized spectral set for CCl_4. Here, it was found that only the symmetric stretching mode of the species containing four ^{35}Cl species exhibits a depolarization ratio of 0.002. The remaining intensity in the parallel polarization results most likely from the non-zero degree solid collection angle. All other vibrational modes of this molecule exhibit depolarization ratios of 0.75, as predicted by theory.

When the laser beam is polarized as shown in Figure 4.2(b), that is, it is unpolarized or circularly polarized, the total scattered intensity, $I_Y^X + I_Y^Z$, is given by $45\alpha^2 + 7\beta^2$, that is, a sum of the scattered intensity components depicted in Figure 4.2(a) and (c). When the polarization analyzer is rotated as shown in Figure 4.2(b), the total scattered intensity is $6\,\beta^2$.

Recently, different scattering geometries for Raman spectroscopy have been utilized successfully, particularly a geometry known as Raman backscattering (or 180° scattering). The reasons for the success of this scattering geometry are elaborated upon below in more detail; suffice it here to state that this geometry is particularly successful since the availability of holographic notch or low-pass optical filters. A typical backscattering geometry is shown in Figure 4.4. The laser beam is focused through a dichroic mirror and a lens into the sample. The same lens also collects and collimates the scattered Raman radiation and directs it toward the dichroic mirror that has maximal reflectivity at the laser wavelength, but low reflectivity for light at lower frequency than the laser. Thus, the incident laser is nearly completely reflected by the dichroic, but the Raman scattered light, since it is red-shifted compared to the incident laser light, passes through the dichroic mirror and is focused by a second lens onto the entrance slit of the monochromator.

This allows for very efficient collection of the Raman scattered light, and makes alignment of the entire optical system less problematic than in right angle scattering, where the image of the laser beam in the sample needs to be focused very accurately onto the exit slit of the monochromator. This geometry is employed in Raman microscopy (see Chapter 11), where the laser beam is focused through, and the Raman scattered light collected by, the same microscope objective.

In backscattering of linearly polarized light, the total scattered intensity is also given by $45\boldsymbol{\alpha}^2 + 7\boldsymbol{\beta}^2$, and the depolarization ratio varies from 0 to 6/7. For circularly polarized incident light, the meaning of the depolarization ratio changes since backscattered light can be circularly or elliptically polarized, depending on the symmetry species of the vibrational mode. This is expressed by the "reversal ratio" P, defined as

$$P = \frac{6\beta^2}{45\alpha^2 + \beta^2} \tag{4.33}$$

P can vary between 0 for a totally symmetric vibration that completely reverses the polarization direction (i.e., the vibration acts as a mirror reflecting the circular polarization) and 6 for an asymmetric vibration that partially scrambles the circularity of the scattered radiation.

4.5 A comparison between Raman and fluorescence spectroscopy

Fluorescence is introduced here as an effect that frequently confounds Raman spectroscopy. It turns out that fluorescence and Raman spectroscopy have similar vibronic origins and may occur together in an experiment, albeit at quite different time scales. Here, the theory of fluorescence spectroscopy is presented qualitatively, and from this discussion, it should become clear that the problem that confounds Raman spectroscopy often is not fluorescence of the sample, but fluorescence of sample impurities.

The following discussion is presented for a simple, diatomic molecule, although one could consider Figure 4.5 to hold for any vibrational coordinate, for example, a stretching coordinate in a polyatomic molecule. In the "FFR" discussion of the polarizability of non-resonant Raman spectroscopy in terms of its vibronic origin, it was assumed that the incident radiation has energy far from any real transition. In both resonance Raman and fluorescence spectroscopies this is not the case. In the latter, the incident photon excites a vibrationally excited state of an electronically excited state, as shown in Figure 4.5. Panel (a) depicts the electronic ground state potential (as mentioned above, for a diatomic molecule) with the corresponding anharmonic vibrational energy levels, as introduced in Section 1.4 and Figure 1.6. The upper energy diagram in panel (a) depicts the electronic potential surface for an electronically excited state. In general, the bond energy of the excited state (and therefore, the force constant acting between the atoms) is weaker than that of the ground state since an electron has been promoted into a non-bonding or anti-bonding orbital in the excitation process. Therefore, the potential curve is shallower, and displaced toward longer bond distance *r*, as shown in the figure.

A transition from the ground vibrational state of the electronic ground state to an excited state occurs according to the Frank–Condon principle that states that the transitions occur preferentially between states with the maximum overlap. This is shown by the gray arrows in Figure 4.5(a). Here, the square of the vibrational wavefunctions is shown, and the shaded areas indicate large overlap of these functions. Thus, according to

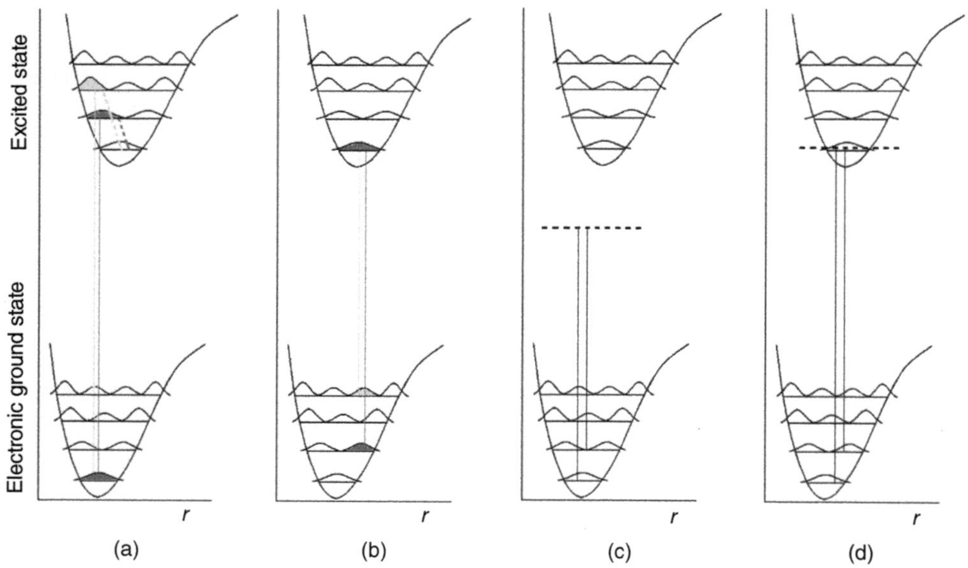

Figure 4.5 *Schematic diagrams representing fluorescence (a) and (b), spontaneous Raman (c), and resonance Raman (d) processes. See text for details*

the Frank–Condon principle, the most likely transitions are into the $n = 2$ and $n = 3$ vibrational states of the electronically excited state (n denotes, as before, the vibrational quantum number). Electronic (or more precisely, vibronic) transitions shown by the solid lines, occur very fast, on the sub-picosecond time scale. Re-emission from the excited vibronic state can occur very rapidly as well, as determined by the Einstein coefficients for spontaneous emission. However, the excited states can also partially deactivate non-radiatively, most likely by collisional deactivation. This can be visualized by the excited species undergoing collisions with the solvent molecules. Such deactivation does not obey the dipole selection rules discussed in Chapter 1 and can populate states that cannot be populated by regular dipole-allowed transitions. These collisional deactivation processes are shown by the dashed lines in Figure 4.5(a).

Fluorescence emission occurs from the ground vibrational state of the electronic excited state, and proceeds, again according to the Frank–Condon principle, into the vibrational states of the ground electronic states with the largest overlap – in the case shown in panel (b), the $n = 2$ and $n = 4$ vibrational states. Thus, the photons emitted in this fluorescence process are red-shifted (lower energy) from the exciting radiation, and often exhibit vibronic fine structure, that is, the fluorescent bands are split by the vibrational quantum.

In contrast to fluorescent processes, the spontaneous, non-resonant Raman process (FFR) is shown in Figure 4.5(c). Here, the dashed line represents the virtual state from which Raman scattering originates, and which is described as a sum of transitions into all real electronic excited states, weighted by an energy difference denominator, as discussed in Section 4.2 and shown in Figure 4.1. In the resonance Raman process, the virtual state coincides, or is energetically very close to a real state, as shown in Figure 4.5(d). In this case, the fluorescent photons and the Raman scattered photons, indeed, have the same energy and may overlap in the observed spectra. Interestingly, fluorescence and Raman spectra can be separated by the different time scales at which they occur. The emission of photons from the virtual state in Raman spectroscopy takes place in femtoseconds, whereas the fluorescent photons are emitted nanoseconds or even microseconds after the excitation. Using pulsed lasers and gated detectors (see Section 4.6), it is therefore possible to eliminate fluorescent contributions from the Raman spectra.

From the discussion of Raman and fluorescence processes, it is clear that the background emission that is often observed superimposed on Raman spectra and generally referred to as "fluorescent" background, cannot be from the sample in the case of "FFR" Raman scattering. This can be realized from an inspection of Figure 4.5(c), which shows that in this case, the real electronic states required for fluorescence are at far higher energy than the energy of the exciting laser light. Rather than attributing this background to fluorescence of the sample, it should be referred to as fluorescence of impurities or decomposition products of the sample. Thus, the best procedure to avoid background contribution to the spectra is sample purification. On the other hand, the high light intensity in the focal point of the exciting laser can cause the sample to photo-decompose, and the resulting decomposition products may fluoresce. This cause of spectral backgrounds can be avoided by rapidly flowing the sample through the laser focus, or utilizing a spinning sample cell. Since fluorescence quantum efficiency can be very high (many orders of magnitudes higher than that of Raman scattering), even nanomolar concentrations of impurities or decomposition products can cause substantial background intensities.

When investigating truly fluorescent samples such as dyes (see Figure 4.8), Raman and fluorescence spectra are truly superimposed. Here, fluorescence can be avoided by exciting the Raman emission with low energy photons that cannot access the electronic energy levels. This is the motivation to use near IR excitation for Raman spectroscopy, which led to the development of FT Raman spectroscopic data acquisition, to be discussed in Section 4.6. Biological samples, such as human tissue, also may exhibit true fluorescence since several amino acids and biological compounds fluoresce upon visible light excitation. In these and many other cases, it is advantageous to "burn off" the impurities or fluorescent molecules by exposing the sample to the laser light for many minutes. This often reduces the background emission to tolerable levels.

4.6 Instrumentation for Raman spectroscopy

The challenge in the design of Raman spectrometers lies in the fact that the number of Raman scattered photons is exceedingly small (only about 1 out of 10^{10} photons incident on the sample may undergo Raman scattering) and that this small number of frequency-shifted photons needs to be detected in the presence of much higher intensity of light that is scattered or reflected at the frequency of the incident laser. Thus, one focuses the exciting laser beam tightly to achieve a high photon flux in the excitation volume, and collects the Raman scattered photon with optical systems with as large apertures (low *f*-numbers) as possible to optimize the light collection. Next, the scattered light components are either diffracted by a monochromator or analyzed interferometrically. The detectors for Raman spectroscopy have high quantum yields and offer single photon sensitivity since the photons are much more energetic than those detected in IR spectroscopy.

4.6.1 Sources

Raman spectra are nearly exclusively excited by laser sources. Since Raman spectroscopy is such a weak effect, sources that produce highly monochromatic light at a flux of about 10^{16}–10^{18} photons/s are desirable. For mid-visible (500 nm, with a photon energy 4×10^{-19} J) radiation, this corresponds to a continuous wave (CW) laser power of between 4 and 400 mW. In the early days of laser Raman spectroscopy, gas lasers were used most frequently: the He—Ne laser emitting about 50 mW at 632.8 nm, the Ar ion laser emitting more than a watt at 514.5 and 488 nm (with a few other lines), and the Kr ion laser at 647.0 nm. For resonance Raman and many of the nonlinear Raman spectroscopic techniques, tunable dye lasers are commonly used. Early tunable dye lasers often consisted of a stream of free-flowing fluorescent dye pumped by a high power UV/visible laser. The fluorescent emission occurred in a laser cavity to provide the optical feedback and cause spontaneous emission.

The operation cost of gas lasers, in particular the Ar and Kr ion lasers, is rather high due to the limited lifetime of the plasma tubes. Consequently, they have recently been substituted by diode-pumped solid state lasers, in particular in the red end (at 785 and 1064 nm). These lasers do not require extensive water or forced air cooling, and convert a much higher fraction of the electric energy to light. In fact, a solid state laser emitting over 50 mW of power draws only about 100 W of electric power. Some of these lasers are based on the Nd:YAG or similar solid state lasers, and are frequency-doubled to produce a number of visible laser lines that are sufficiently monochromatic and stable for Raman spectroscopy.

For the nonlinear Raman experiments to be discussed in Chapter 6, high power, ultra-short pulse laser sources are utilized. These sources will be introduced later. Spontaneous Raman spectroscopy is generally observed *via* dispersive instruments (i.e., monochromators utilized in a spectrograph configuration) fitted with a multichannel detector, such as a charge-coupled device (CCD) camera. Such dispersive systems are discussed next, after a short foray into the history of dispersive Raman instrumentation. In addition, a short introduction into the CCD detector technology will be presented. Subsequently, interferometric FT-Raman instrumentation will be introduced.

4.6.2 Dispersive Raman instrumentation and multichannel detectors

In dispersive Raman instrumentation, the design requirements for the monochromator are more stringent than for dispersive IR instruments discussed earlier (Section 3.2). First, due to the presence of Rayleigh scattering, which may be orders of magnitude more intense than the Raman scattered intensities, the rejection or discrimination of stray light must be very high. Second, since the scattered light is still in the visible range, with relatively small frequency shift, the optical resolution of a monochromator system for Raman spectroscopy must be higher than that for IR monochromators. A simple example is given here to illustrate this point. In order to resolve two peaks at 2995 and 3000 cm^{-1} (3.3389 and 3.3333 μm, respectively) in IR spectroscopy, the resolving power needs to be about two parts in a thousand. Two Raman peaks at 2995 and 3000 cm^{-1} shift from the line of an exciting argon ion laser line at 514.5 nm (19,436 cm^{-1}) occur at 16,441 and 16,436 cm^{-1}, respectively (608.23 and 608.42 nm). To separate these lines, a resolving power of three parts in 10^4 is needed.

Thus, monochromators for Raman spectroscopy typically use longer focal lengths (50–100 cm), higher grating groove densities (1200–2400 lines/mm), and narrower slits (50–200 μm). The linear dispersion of such a monochromator is typically on the order of 0.5 nm mm^{-1}.

In the early 1970s, a number of instrument manufacturers began introducing Raman instrumentations that could compete with the IR instrumentation available at that time in terms of convenience and ease of data acquisition. These instruments incorporated gas lasers, 90° scattering geometry (see Section 1.6), double or triple monochromators, and photomultiplier detectors that had been optimized for the detection of low light level signals. The discussion of Raman instrumentation in the first edition of this book was still focused on this kind of Raman spectrometers. The samples, contained in melting point capillaries, cuvettes, spinning cells (to avoid sample heating), or even as free sample jets, were illuminated by the laser beam focused into the sample. For right angle scattering, the Raman scattered light was collected by a large aperture lens and focused into the monochromator, which then were nearly exclusively scanning double or triple monochromators with focal lengths of about 1 m, and grating groove densities of about 1200 mm^{-1}. Double or triple monochromators were necessary to discriminate against Rayleigh scattering. Photomultiplier detectors, consisting of a photo-sensitive cathode at −1800 V, amplifying dynodes and an anode typically at ground potential, amplified the primary photoelectrons emitted by the cathode after exposure to photons up to 10^6–10^8 fold such that a pulse of electrons could be detected and counted; yet the quantum efficiency of the photocathode generally was quite low (~10%). Compared to modern instrumentation, these early laser Raman systems were very slow, since only one spectral element was collected at a time and the grating was rotated to sweep all the different Raman-scattered frequencies by the detector consecutively.

Several technical advances of the past two decades have decreased the acquisition time for Raman spectra by about three orders of magnitude. One of these advances occurred in the design of holographic filters to remove the un-shifted frequencies of light collected from the sample. These may result from Rayleigh scattering or reflection of the laser beam, particularly in backscattering geometry (see Figure 4.4). These filters remove the requirement for double or triple monochromators since they can be configured as notch filters (cutting out a narrow frequency range centered at the laser wavelength) or as long pass filters, cutting out the frequency range typically 100 cm^{-1} or so lower than the laser wavelength for the observation of Stokes Raman spectra. The discrimination of these filters is about 10^8, allowing Raman scattering to be observed in back-scattering configuration with single monochromators, which, in turn, increases the light flux through the system.

A second major improvement is in the grating technology. With modern holographic gratings, stray light rejection is superior as compared to mechanically ruled gratings. Thus, single monochromators can be used in a spectrograph configuration, with a multichannel detector mounted in the exit plane. Furthermore, holographic gratings can be produced with higher groove density (1800–2400 lines/mm) allowing monochromators with shorter focal length and lower numerical aperture to be designed at the same linear dispersion (see Eq. 3.10). A 30 cm focal length monochromator, with a typical multichannel detector, can produce an optical resolution of about 2–5 cm^{-1} per detector element using a standard Czerny–Turner monochromator. Furthermore, curved holographic gratings also can be produced, thus eliminating the collimating mirrors shown for Czerny–Turner monochromators in Figure 4.6. The resulting monochromators often have no moveable parts and are extremely rugged. The third major improvement to Raman instrumentation was in detector technology, to be discussed below. These detectors consist of hundreds or thousands of individual solid state detector elements, each detecting the intensity at one wavelength, arranged in the focal plane of the monochromator.

When using a grating-based monochromator in a spectrograph configuration, that is, without an exit slit but a multichannel detector mounted in the exit plane, the linear dispersion between detector elements is not

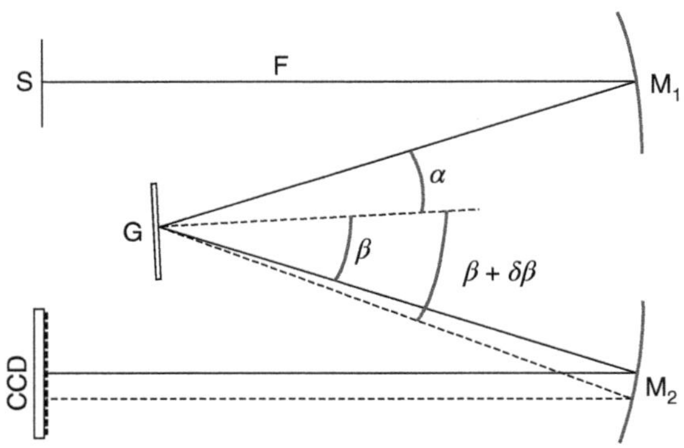

Figure 4.6 *Schematic of a spectrograph with a CCD (or diode array) detector. S: entrance slit, F: focal length, M_1 and M_2: collimating mirrors, G: grating, α: angle of incidence, measured with respect to the grating normal (long dashes), β: diffracted angle for center wavelength, and β + δβ: diffracted angle for different wavelength*

constant but varies with the cosine of the diffracted angle. Consequently, the detector elements in an array detector will not be registering in linear wavelength or linear wavenumber increments.

The actual wavenumber of the scattered radiation can be determined using the following procedure [6]; see Figure 4.6. Assuming a Czerny–Turner monochromator configuration, the included angle D is given by

$$D = \beta - \alpha \tag{4.34}$$

where α is the angle of incidence and β the diffracted angle for the center wavelength of the array detector element. Here, the negative sign of α results from the fact that both angles are defined with respect to the grating normal, and the angle of diffraction is assumed to be positive. The angle D is determined by the mechanical design of the monochromator, and corresponds to the angle described by angles α and β for the central element in the array detector.

The wavelength that impinges on detector element $\boldsymbol{j}$ with width w is given by the following expression:

$$\lambda_j = \frac{\left[\sin\alpha + \sin\left(\beta + \boldsymbol{j}\frac{w}{F}\right)\right]}{nd} \tag{4.35}$$

In this equation, n is the order of diffraction and d the spacing between adjacent lines of the grating. This equation follows from the definition of the linear dispersion of a monochromator,

$$\frac{d\lambda}{dx} = \frac{(d \cdot \cos\beta)}{nF} \tag{3.10}$$

The angle of incidence and the angle of diffraction need to be calculated for the central wavelength (solid ray in Figure 4.5) from the included angle according to

$$\alpha = \arcsin\left[\frac{dn\lambda}{2\cos\left(\frac{D}{2}\right)}\right] - \frac{D}{2} \quad \text{and}$$

$$\beta = \arcsin\left[\frac{dn\lambda}{2\cos\left(\frac{D}{2}\right)}\right] + \frac{D}{2} \tag{4.36}$$

Furthermore, the approximation

$$\delta\beta = \frac{\tan(jw)}{F} \approx \frac{jw}{F} \tag{4.37}$$

has been used; this approximation is good since the focal length of monochromators used in Raman spectroscopy is large compared to the width of a detector element. Thus, one needs to determine one wavelength or wavenumber accurately from a source with known emission wavelength. One can easily use the Rayleigh line from the laser for this purpose, since the low pass filter discussed above usually reduces the intensity of this line to a level that is not damaging to the detector. Once the detector element detecting this wavelength or any other well-known calibration line has been established, the wavenumber of all other detector elements can be calculated using the groove density, focal length, included angle in the monochromator, and the width of the detector elements. A detailed derivation of Eqs. 4.35–4.37 can be found in Ref. [6].

As pointed out above, the third and most important advance in modern Raman instrumentation occurred in the detector technology. While early instruments used a single detector (typically a photomultiplier tube), solid state multichannel detectors are now available that simultaneously sample the different wavelengths diffracted by the grating. The introduction of these detectors, in conjunction with the higher light levels due to the elimination of pre-monochromators, has reduced data acquisition times by two to three orders of magnitude, as compared to the scanning laser Raman spectrometers of the 1970s and 1980s.

Solid state multichannel detectors first appeared in the mid- and late 1970s. A multichannel or array detector consists of a large number (typically between 500 and a few thousand) of solid state detector elements that are placed into the focal plane of the monochromator without an exit slit. Each individual detector element simultaneously detects a different color of the diffracted light, and the resolution of the optical system is determined (among other factors) by the width of the individual detector elements.

Early array detectors included the so-called "diode arrays," which typically consisted of 512 or 1024 silicon photodiodes, each diode measuring about 0.025 mm wide and 2.5 mm high. Thus, the total light-sensitive area of the array was about $25 \times 2.5\ \text{mm}^2$. The rectangular shape of each diode made them ideally suited for slit- and grating-based spectrometers, since the entrance slit of a monochromator has a similar aspect ratio of 1:100 (typically 0.1 mm wide and 10 mm high). The image of the entrance slit was projected onto the exit plane of the spectrometer with suitable image demagnification. Thus, the image of the slit ideally filled each diode completely.

Silicon-based detectors are generally operated as shown in Figure 4.7. The photosensitive diode, together with a capacitor, forms an individual detector element. The capacitor is charged by an electronic circuit such that the charge cannot flow through the diode, which acts as a valve toward the flow of electricity. Photons impinging on the diode create electron hole conductors, thereby reducing the charge in the capacitor. After exposing the diode to light, the remaining charge is measured by special circuitry, which connects each diode consecutively to a charge-to-voltage converter/amplifier and a D/A converter, which measures the height of the pulse that corresponds to the remaining charge of the diode. The difference between original charge and the charge remaining after light exposure is proportional to the number of photons captured by the detector element. During the read cycle, the diode is also recharged.

Although the quantum yield is high in silicon (>80% for an anti-reflection-coated silicon photodiode), the S/N ratio is relatively poor in Si photodiode array detectors; that is, the relative change in charge in a diode

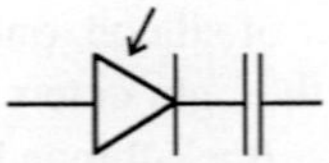

Figure 4.7 *Circuit diagram of a diode array or CCD detector element. The capacitor is charged at such a polarity that charge cannot flow through the photodiode in the absence of light-induced electron hole charge carriers*

brought on by the absorption of one photon is small compared to the readout noise. Thus, single-photon detection is not practical in photodiode arrays. To alleviate this problem, image intensifiers were used commonly with diode arrays. An image intensifier is a true photon multiplier: an incoming photon produces a photoelectron, which is amplified *via* a micro-channel plate electron amplifier. The resulting pulses of electrons are converted back to light by directing them toward a phosphorus screen, where they produce secondary photons. In this way, photon gains of between 10^3 and 10^4 could be achieved. With these amplified light levels, diode array detectors work satisfactorily.

Image intensifiers also offered the possibility of "gating" or rapidly turning the detector on or off. This can be accomplished by pulsing the high voltage on the electron amplifier in response to an external signal, such as a trigger from a laser pulse. Gating offers the ability to synchronize the detector to external events, which is particularly important when it is necessary to discriminate against fluorescent background scattering: since fluorescence is a much slower process than Raman scattering, it is possible to collect Raman photons in the first few hundred nanoseconds after an exciting laser pulse and gate out the fluorescent intensity, which occurs thereafter. The time-resolved Raman experiments discussed in Chapter 7 often employ gated detectors to discriminate Raman scattering created from the pump laser from that created by the probe pulse.

Since the mid-1980s, a new multichannel technology has emerged for use in Raman spectroscopy, and has reached total domination. This is the so-called CCD technology, originally developed for video cameras. The photosensitive area in the CCD is typically divided into 30 μm × 30 μm picture elements (pixels), composed of Si photodiodes. Compared to the detector elements in diode array detectors discussed earlier, the individual CCD detector elements are much smaller and therefore have a much higher specific detectivity, D^* (see Section 3.2). The pixel elements in a CCD camera used in Raman spectroscopy typically are arranged in a 1024 × 128 array. The diffracted wavelengths are projected along the long dimension of the camera, and the pixels in a column all detect the same wavelength. Electrically, each individual detector element is a photodiode/capacitor circuit, as described above for the diode array detector.

The main difference between CCDs and diode arrays is in the readout logic. The readout in a CCD camera proceeds as follows: the first element in the bottom row is read by converting its remaining charge, *via* a charge-to-voltage converter, to a voltage signal that is stored digitally. Next, the charges of all consecutive elements in the bottom row are shifted by one position by draining the remaining charge into the next pixel, and the first element is read again. Once the entire bottom row has been read, the charges in each pixel element in the next row up are drained into the pixel elements in the bottom row, and the readout process continues, until the entire array has been read. The efficiency of the charge transport (the "charge coupling") must be very high, on the order of 0.99999, since arrays often contain more than 100,000 pixels, and the charge in a pixel may have to be shifted over 1000 times until it is read. The read-out cycle of a CCD camera is very fast, and the sensitivity is excellent.

CCDs offer the advantage of a two-dimensional rectangular detector, with resolution in two perpendicular directions, whereas diode arrays were strictly one-dimensional arrays. This can be used advantageously in a standard spectrograph configuration, by "binning" (co-adding) the signal of pixels of the same column of the array. In the binning process, the pixels in a vertical column are co-added and read out together, which significantly speeds the reading process and increases the sensitivity of the measurement by the square root of the number of pixels binned together.

The sensitivity of a CCD chip can further be enhanced by a process known as back-thinning (also known as back-illumination). In this process, the layer of silicon, onto which the diodes and read-out circuitry have been deposited, is mechanically thinned, and the light enters the diode elements from the side that does not contain all the read-out electronic components. Back-thinned CCDs have quantum efficiencies of over 80% and single-photon detectivity. They have further increased the sensitivity of Raman spectroscopy and have allowed experiments to be carried out that were previously impossible.

4.6.3 Interferometric Raman instrumentation

Dispersive Raman spectrometers, utilizing CCD detectors, are actually quite simple and require only a grating and detector, aside from transfer optics and the pre-filters discussed above. However, Raman spectroscopy often is plagued by fluorescence effects that can be orders of magnitude more intense than the desired Raman signal. In order to reduce fluorescence, it is advantageous to excite far away from any possible visible or UV transitions of the sample (or any impurities contained in the sample), which implies excitation in the far red or near IR spectral region. The observation of Raman spectra with near IR radiation is hampered by the fact that Raman scattered intensities depend on the ν^4 factor (see Eq. 4.9): consequently, the use of long wavelength lasers to excite the Raman process seriously degrades the scattered intensities. A quick calculation shows a 20-fold decrease in the Raman scattering cross section between 488 nm excitation from an argon ion or solid state laser and a YAG laser at 1064 nm. Furthermore, the response of silicon CCD detectors drops off precipitously above about 900 nm; thus, standard dispersive Raman instruments using silicon CCD detectors have bad response in the C—H stretching region (3000 cm^{-1}) even when excited with a typical 785 nm near IR laser, and cannot be used with 1064 nm excitation.

Therefore, Raman spectroscopists have long been tempted to utilize the enormous advantages of interferometry in FT-Raman experiments, using near-IR detector technology. However, interferometers work best in the IR region of the spectrum for a number of reasons. First, the motion of the movable mirror must be parallel to the fixed mirror to better than the wavelength of light or the interference patterns will be distorted. Second, the wavelength of the reference beam (typically a He—Ne laser) should be much shorter than the wavelength of the light to be analyzed to ensure that the interferogram is sufficiently well defined. Thus, interferometry was commercialized in the IR spectral region with wavelength on the micrometer scale, but it became feasible in the 1990s to collect interferometric data in the near-IR (0.7–1.2 μm) range. It is now possible to collect interferometric data with the required precision for Raman spectroscopy in the red end of the visible and the near-IR portion of the visible spectrum. In addition, high-power lasers are now available in the near-IR region. Nd:YAG lasers, for example, are available that produce many watts of CW power output at 1.064 μm. The combination of near-IR lasers, improved interferometers, and solid state detectors for the near-IR (such as InGaAs) have made the observation of FT-Raman spectra possible.

As pointed out above, excitation in the near IR region can circumvent the fluorescence problem completely, since IR photons normally do not have sufficient energies to reach the vibronic states that cause fluorescence (see Figure 4.8). In fact, FT-Raman spectra of highly fluorescent dyes, such as rhodamine G, have been recorded free of fluorescence interference using IR laser excitation. Thus, FT Raman offers the major optical advantages discussed earlier in the chapter on FTIR spectroscopy, and the advantage of reduced fluorescence – while suffering from the disadvantage of a lower scattering cross section due to IR excitation. The advantages and disadvantages of FT-Raman methodology were elegantly demonstrated in the first report on FT-Raman spectroscopy by Chase [7].

FT-Raman, however, is by no means a trivial experiment. The main problems arise from laser shot noise during data acquisition. This noise is sampled both *via* the Rayleigh and the Raman scattering and is Fourier transformed into high-intensity noise throughout the entire spectral range. It can be so strong that it masks the Raman spectrum completely. This problem is sometimes referred to as the "multiplex disadvantage" [8]. Optical filters or a pre-monochromator can be used to remove the Rayleigh scattering and, with it, the source of this noise. Commercial FT-Raman instruments use mostly Nd:YAG lasers, notch filters at the laser wavelength to reduce the Rayleigh scatter, high-quality interferometers, and detectors such as InGaAs, which peak in the near-IR region. By switching a few optical elements, it is possible to use the instrument as an FTIR absorption spectrometer; thus, the possibility of utilizing one instrument to collect Raman and IR data has become a reality.

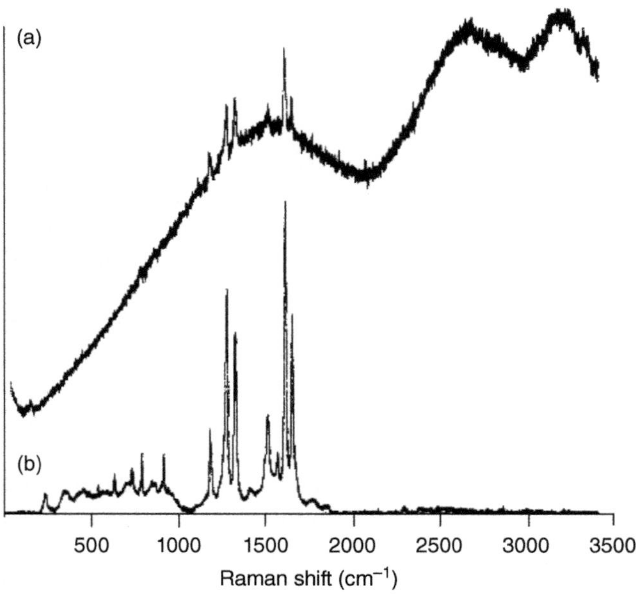

Figure 4.8 *Raman spectra of poly(p-phenylene)-terephthalamide. (a) Conventional dispersive Raman spectrum excited with 150 mW at 514.5 nm (backscatter geometry) and detected via a double monochromator and a photomultiplier. (b) FT-Raman spectrum, excited with 1.5 W at 1064 nm from a Nd:YAG laser. The FT-Raman spectrum is nearly completely devoid of fluorescent contributions. Chase, 1986, [7]. Reproduced with permission from American Chemical Society*

4.6.4 Raman microspectroscopy

Raman microspectroscopy (also referred to as Raman microscopy) is discussed in detail in Chapters 11 and 14.

References

1. Kauzmann, W. (1957) *Quantum chemistry*, Academic Press, New York, Chapter 16.
2. Nafie, L.A. (2001) Theory of Raman scattering, in *Handbook of Raman Spectroscopy* (eds I.R. Lewis and H. Edwards), Marcel Dekker, Inc., New York.
3. Long, D.A. (2002) *The Raman Effect: A Unified Treatment of the Theory of Raman Scattering by Molecules*, John Wiley & Sons, Ltd, Chichester.
4. Wilson, E.B., Decius, J.C., and Cross, P.C. (1955) *Molecular Vibrations*, McGraw-Hill.
5. Kiefer, W. and Topp, J.A. (1974) A method for automatic scanning of the depolarization ratio in Raman spectroscopy. *Appl. Spectrosc.*, **28** (1), 26–34.
6. Diem, M., Adar, F., and Grayzel, R. (1986) A simple algorithm to convert diode array spectral data to linear wavelength or wave number scales. *Comput. Enhanced Spectrosc.*, **3**, 29–33.
7. Chase, D.B. (1986) Fourier transform Raman spectroscopy. *J. Am. Chem. Soc.*, **108**, 7485–7488.
8. Schrader, B. (1990) Possibilities and limitations of FT-Raman spectroscopy, in *Practical Fourier Transform Infrared Spectroscopy* (eds J.R. Ferraro and K. Krishnan), Academic Press, San Diego, CA, pp. 168–199.

5

A Deeper Look at Details in Vibrational Spectroscopy

In this chapter, a refined discussion of certain aspects of vibrational spectroscopy is presented. In the previous chapters, various assumptions were made, for example, that normal modes of vibration do not interact or that the vibration and rotation of molecules are strictly separable. Here, more detailed theoretical arguments are presented that explain a number of spectral features that are not explained by the simplified earlier discussions.

The first of these effects to be discussed is Fermi resonance, which has been alluded to a number of times in Chapters 1 and 2. Fermi resonance is an effect that mixes the wavefunctions and energy eigenvalues of certain normal coordinates, and effects the transition moments into the mixed states such that normally weak bands can become very strong in the mixing process. Since this effect occurs quite frequently, a short review of the theory will be presented along with a number of commonly observed examples of Fermi resonance enhanced overtone and combination bands. Related to this effect is another mixing phenomenon, referred to as transition dipole coupling (TDC), which is observed frequently in polymers of repeating subunits, where the subunits exhibit identical or near-identical vibrational features. Here, the exciton-like coupling of the vibrational states of the subunits produce delocalized vibrational states, which explains the conformational sensitivity of vibrational spectroscopy.

Next, a theoretically trivial, but most useful concept is introduced: the concept of group frequencies. One of the advantages of vibrational spectroscopy is the fact that the same chemical moiety will exhibit similar vibrational frequencies, regardless of the chemical environment. Thus, vibrational spectroscopy can be used in a qualitative way to check for the presence or absence of specific groups, a fact that is promoted by all introductory texts on organic chemistry, for example, that introduce NMR and infrared (IR) spectroscopy as qualitative tools.

Subsequently, some aspects of rotational spectroscopy are presented to help understand the rotational fine structure of gas phase vibrational spectra. Here, the solution of the rotational Schrödinger equation and the derivation of the rotational eigenvalues are not presented, but the results important for vibrational spectroscopy are introduced.

Modern Vibrational Spectroscopy and Micro-Spectroscopy: Theory, Instrumentation and Biomedical Applications, First Edition.
Max Diem.

5.1 Fermi resonance

So far, it was assumed that the total molecular vibrational wavefunction ψ_{vib} is the product of the wavefunctions associated with each of the $3N-6$ normal coordinates, Q_k,

$$\psi_{\text{vib}} = \psi_1(Q_1) \cdot \psi_2(Q_2) \cdot \psi_3(Q_3) \cdot \cdots \tag{1.66}$$

and that the vibrational eigenfunctions form a vector space in which the vibrational Hamiltonian

$$\hat{H}\psi_{\text{vib}} = E\,\psi_{\text{vib}} \tag{5.1}$$

results in a diagonal matrix of eigenvalues:

$$\hat{H}\begin{bmatrix}\psi_1(Q_1)\\ \psi_2(Q_2)\\ \psi_3(Q_3)\\ \vdots\\ \psi_k(Q_k)\end{bmatrix} = \begin{bmatrix}E(Q_1) & 0 & 0 & 0 & 0\\ 0 & E(Q_2) & 0 & 0 & 0\\ 0 & 0 & E(Q_3) & 0 & 0\\ & & \cdots & & \\ 0 & 0 & 0 & 0 & E(Q_k)\end{bmatrix}\begin{bmatrix}\psi_1(Q_1)\\ \psi_2(Q_2)\\ \psi_3(Q_3)\\ \vdots\\ \psi_k(Q_k)\end{bmatrix} \tag{5.2}$$

In this case, the total vibrational energy is just the sum of all energies $E(Q_k)$, as discussed before (Eq. 1.70).

The picture discussed so far describes the vibrational states well if all energy levels are well separated and do not interact. However, under certain circumstances it is necessary to include the possibility that some vibrational coordinates interact. When such mixing occurs, the corresponding vibrational modes lose their identity, and the resulting modes must be described as linear combinations of the two interacting modes. Such an interaction may occur if there are two energy levels E_m and E_n in a molecule corresponding to the wavefunctions $\psi(Q)_m$ and $\psi(Q)_n$, which are accidentally degenerate or nearly degenerate, and if there is a perturbation operator H', which causes the mixing of the two states n and m. The perturbation may be due to dipolar coupling between degenerate transitions or the anharmonic potential energy terms in the case of interactions between a fundamental and the overtone of another vibrational transition:

$$E_{nm} = \langle \psi(Q)_m \,|H'|\psi(Q)_n \rangle \tag{5.3}$$

Under these conditions, the energy eigenvalues can no longer be written in diagonal form, but the energy matrix will appear as follows:

$$\begin{bmatrix}E(Q_1) & 0 & 0 & 0 & 0\\ 0 & E(Q_n) & E_{nm} & 0 & 0\\ 0 & E_{nm} & E(Q_m) & 0 & 0\\ & & \cdots & & \\ 0 & 0 & 0 & 0 & E(Q_k)\end{bmatrix} \tag{5.4}$$

The partial determinant containing the perturbation can be solved separately:

$$\begin{vmatrix}E_n & E_{nm}\\ E_{nm} & E_m\end{vmatrix} = 0 \tag{5.5}$$

where the notation $E(Q_n)$ was simplified to E_n. Equation 5.5 will be particularly important for the discussion in Section 5.2.

The resulting coupled energy eigenstates and eigenfunctions are referred to as E^+ and E^-, and ψ^+ and ψ^- respectively, where the positive and negative superscripts denote the symmetric and antisymmetric coupled states, respectively. The energy eigenstates are defined in terms of the non-perturbed energies as follows (see Ref. [1, Section II, 5c]):

$$E^{\pm} = \frac{1}{2}(E_n + E_m) \pm \sqrt{4E_{nm}^2 + \delta^2} \tag{5.6}$$

Here, $\delta = E_n - E_m$ is the energy difference of the unperturbed levels. The new wavefunctions for the mixed states can be written as follows:

$$\psi^+ = \left(\frac{1}{N}\right)(a\psi_m + b\psi_n) \quad \text{and} \quad \psi^- = \left(\frac{1}{N}\right)(a\psi_m - b\psi_n) \tag{5.7}$$

where a and b are the mixing coefficients, which can be expressed in terms of E_{nm} and δ according to

$$a^2 = \frac{\sqrt{4E_{nm}^2 + \delta^2} + \delta}{2\sqrt{4E_{nm}^2 + \delta^2}} \tag{5.8}$$

and

$$b^2 = \frac{\sqrt{4E_{nm}^2 + \delta^2} - \delta}{2\sqrt{4E_{nm}^2 + \delta^2}} \tag{5.9}$$

The effect of this interaction is to increase the splitting between the energy levels and to mix them. In the mixed state, it is not possible to assign one of the transitions to the state *n,* but it has to be assigned to the ψ^+ or ψ^- combination state. This splitting will be larger if the original energy difference is small and if the coupling energy is large. The interacting states must be of the same symmetry species (or contain the same irreducible representations, *cf.* Chapter 2) for the integral in Eq. 5.3 to be nonzero. The mixing of states is also accompanied by an equalizing effect of the vibrational intensities: the normally very weak overtone or combination band can obtain significant intensity from the fundamental with which it is in Fermi resonance.

A common example is discussed here to convey the order of magnitudes of the splitting and the interaction energy involved. In the C—H stretching region of a methyl group with C_{3v} symmetry (*cf.* Chapter 2), three peaks are observed between 2850 and 3050 cm^{-1}. A detailed assignment of these was presented in Section 2.6 for methyl chloride (see Table 2.3). The highest wavenumber peak, at 3039 cm^{-1}, is the antisymmetric stretching motion of E symmetry, and the two other peaks at about 2960 and 2880 cm^{-1} are a Fermi resonance enhanced doublet of the symmetric methyl stretching vibration of A_1 symmetry interacting with an overtone of the methyl antisymmetric deformation mode of E symmetry. This latter mode typically occurs at 1455 cm^{-1}; thus, its overtone would be expected at 2910 cm^{-1}, ignoring anharmonicity.

From an analysis of overtones and combination bands, the unperturbed symmetric stretching mode is expected to occur at 2930 cm^{-1}. Therefore, the unperturbed energy eigenvalues E_n and E_m (see Eq. 5.5) are 2930 and 2910 cm^{-1}, respectively. Thus,

$$\delta = E_n - E_m = 20 \text{ cm}^{-1} \tag{5.10}$$

and

$$\frac{1}{2}(E_n + E_m) = 2920 \text{ cm}^{-1}. \tag{5.11}$$

Using Eq. 5.6, one obtains for the energy of one of the mixed states, E^+ and E^-

$$E^+ = 2960 = \frac{1}{2}(E_n + E_m) + \sqrt{4E_{nm}^2 + \delta^2}$$

$$2960 = 2920 + \sqrt{4E_{nm}^2 + \delta^2} \tag{5.12}$$

$$E^- = 2880 = \frac{1}{2}(E_n + E_m) - \sqrt{4E_{nm}^2 + \delta^2}$$

$$2880 = 2920 - \sqrt{4E_{nm}^2 + \delta^2} \tag{5.13}$$

from which it follows that

$$(4\,E_{nm}^2 + \delta^2) = 1600$$

$$\text{and} \qquad E_{nm} = 17\ (\text{cm}^{-1}) \tag{5.14}$$

Substitution of this energy value into Eqs. 5.8 and 5.9 yields the values for the mixing coefficients a^2 and b^2 of 0.75 and 0.25, respectively. Thus, the wavefunction ψ^+ of the higher component of the Fermi doublet can be written as

$$\psi^+ = \psi_{2960} = 0.87\,\psi(v_s^{CH_3}) + 0.5\psi(2\delta_{as}^{CH_3}) \tag{5.15}$$

and that of the lower as

$$\psi^- = \psi_{2880} = 0.87\,\psi(v_s^{CH_3}) - 0.5\psi(2\delta_{as}^{CH_3}) \tag{5.16}$$

where the symbols $\psi(v_s^{CH_3})$ and $\psi(2\delta_{as}^{CH_3})$ are used to denote the symmetric stretching and the overtone of the antisymmetric deformation wavefunctions, respectively. The intensity of the Fermi-resonance-enhanced combination band at 2880 cm^{-1} is strongly enhanced and is nearly as strong as the fundamental with which it interacts. Using the procedures outlined in Chapter 2, it is easy to demonstrate that the overtone of an E symmetry vibration has a reducible representation of $[4\quad 1\quad 0]$, and that this reducible representation can be decomposed into $1A_1$, $1A_2$, and $1E$ reducible representations, and that it, indeed, contains the totally symmetric representation that is also the representation of the symmetric stretching coordinate. This is the group theoretical requirement that allows the interaction of the two energy levels.

This Fermi resonance example is observed basically for any molecule that contains methyl groups, as shown in Figure 5.1 for two quite different environments. Trace (a) depicts an IR absorption spectrum of the methyl stretching vibrations of 1,1,1-trichloroethane. Typically for IR spectra, the degenerate antisymmetric stretching mode at just below 3010 cm^{-1} is more intense than the symmetric stretching mode of A_1 symmetry, seen here at about 2950 cm^{-1}. This mode is in Fermi resonance with the overtone of the methyl deformation vibration at about 2870 cm^{-1}, as discussed above. Figure 5.1(b)–(d) shows the Raman spectra of aqueous solutions of various deuterated isotopomers of alanine, one of the simplest amino acids that exists at neutral pH in aqueous solution as a zwitterionic molecule, shown by the structure on the right of the figure [2]. Trace (c) shows the spectrum of alanine-d_1, in which the methyne hydrogen atom is substituted by a deuterium atom. In this molecule, the CH_3 group is attached to a carbon atom with three different substituents (D, CO_2^-, and NH_3^+). Although the overall symmetry group of alanine is C_1, which does not support any degenerate modes, the vibrational modes very much resemble those of a methyl group under C_{3v} symmetry: the symmetric stretching mode, again at about 2950 cm^{-1}, is in Fermi resonance with the deformation overtone, which appears strongly at about 2890 cm^{-1}. Quite typically for Raman spectra, the symmetric stretching mode is more intense than the antisymmetric stretching modes observed here at around 3000 cm^{-1}. The splitting in this latter band is attributed to the fact that no degeneracy is allowed in the C_1 point group, and that the formerly degenerate antisymmetric stretching modes now appear as two separate transitions.

When the methyl group in alanine is substituted by a CD_3 group, the only C—H stretching mode observed is the methyne C—H stretching transition, which appears as a broad band at about 2965 cm^{-1}, shown in trace (d). The native molecules, whose spectrum is shown in trace (b) is a superposition of traces (c) and (d), with minimal interactions between the vibrations of the methyl group and the methyne hydrogen. The deuteration

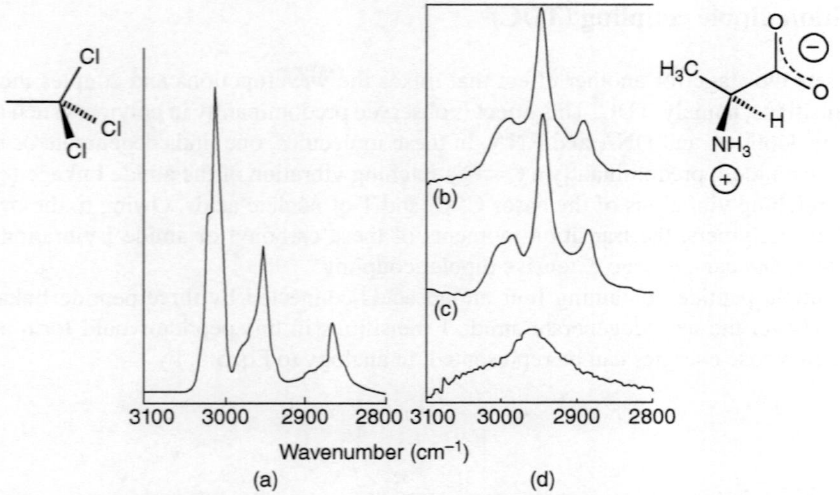

Figure 5.1 *Vibrational spectra features of methyl groups, including Fermi resonance. (a) Infrared absorption spectrum of liquid 1,1,1-trichloroethane. (b–d) Solution phase Raman spectra of the C—H stretching region of alanine and deuterated isotopomers. See text for detail*

studies reported here, carried out in the author's laboratory in the early 1980s [2], were necessary to sort out the contributions of the Fermi resonance overtone and of the methyne hydrogen; in previous papers, the methyne hydrogen stretching mode had been erroneously assigned to the 2890 cm^{-1} band.

The example above demonstrated the most common case of Fermi resonance, where one of the states is a combination band or an overtone. Overtones of non-degenerate fundamentals always contain the totally symmetric representation of a group, and those of degenerate states often do. Thus, overtones and combination bands often interact with totally symmetric vibrations. Many more examples exist in the vibrational spectra of simple molecules: in CCl_4, for example, which was discussed in Chapter 2, the antisymmetric C—Cl stretching vibration of T_1 symmetry at about 790 cm^{-1} interacts strongly with the combination mode $\nu_3(A_1) + \nu_2(T_2)$ (see Table 2.1), which is estimated to occur at 770 cm^{-1}. In this example, the fundamental and the combination band are equally strong in the Raman spectra (see Figure 2.13). Furthermore, the symmetric stretching mode $\nu_3(A_1)$ also is in Fermi resonance with an overtone of $2\nu_1(E)$, accounting for the low frequency, low intensity shoulder observed at about 435 cm^{-1}.

Another common example is that of CO_2, which is discussed next. This linear, triatomic molecule is expected to exhibit $3N - 5 = 4$ degrees of vibrational freedom, a symmetric stretching mode expected at about 1380 cm^{-1} (active in Raman only), a degenerate deformation (bending) mode at 667 cm^{-1}, and a very strong antisymmetric stretching mode at 2349 cm^{-1}. The latter two vibrations are both active in the IR spectrum only. Since the molecule possesses a center of inversion, the rule of mutual exclusion between Raman and IR activity holds. However, instead of the one band expected in the Raman spectrum, two bands are observed at 1285 and 1388 cm^{-1}. The former is the overtone of the deformation mode, in Fermi resonance with the symmetric stretching mode, and shifted substantially toward lower energy by this interaction.

Fermi-resonance-enhanced overtone and combination bands are common features in vibrational spectroscopy; an inspection of reported spectra of small molecules reveals hundreds of examples that have been established as early as the 1940s. The reader is referred to the comprehensive review of the early literature found in [1, Section III, 3].

5.2 Transition dipole coupling (TDC)

Equation 5.4 sets the stage for another effect that mixes the wavefunctions and couples the eigenvalues of vibrational transitions, namely, TDC. This effect is observed predominantly in polymers such as polypeptides, proteins, polynucleotides, and DNA and RNA. In these molecules, one finds degenerate or near-degenerate transitions of the amide I, predominantly a C=O stretching vibration of the amide linkage (*cf.* Section 10.2) or the C=O stretching vibrations of the bases C, G, and T of nucleic acids. Owing to the overall secondary structure of these polymers, the transition moments of these carbonyl or amide I vibrations are in a fixed geometric pattern, and can undergo extensive dipolar coupling.

Consider a simple peptide containing four amino acids connected by three peptide linkages. Following the discussion above, the near-degenerate amide I transitions in this peptide would form a set of coupled vibrational states whose energies can be represented, in analogy to Eq. 5.5, by

$$\begin{matrix} E_1 & V_{12} & V_{13} \\ V_{21} & E_2 & V_{23} \\ V_{31} & V_{32} & E_3 \end{matrix} \tag{5.17}$$

where the diagonal values $\mathbf{E_i}$ are the near-degenerate energy eigenvalues of unperturbed amide I transitions, and the off-diagonal elements V_{ij} are the interaction or coupling energies between the peptide transition moments, given to a first approximation by dipole–dipole interactions:

$$V_{ij} = \frac{\boldsymbol{\mu}_i \cdot \boldsymbol{\mu}_j}{|\mathbf{T}_{ij}|^3} - \frac{3(\boldsymbol{\mu}_i \cdot \mathbf{T}_{ij})(\boldsymbol{\mu}_j \cdot \mathbf{T}_{ij})}{|\mathbf{T}_{ij}|^5} \tag{5.18}$$

In Eq. 5.18, $\boldsymbol{\mu}_i$ and $\boldsymbol{\mu}_j$ are the transition dipole moments of oscillators i and j and $\mathbf{T}_{ij}$ is the vector connecting the centers of mass coordinates, X_i, of each individual transition:

$$\mathbf{T}_{ij} = X_j - X_i \tag{5.19}$$

In Eq. 5.18,

$$V_{ij} = V_{ji} \tag{5.20}$$

Notice that Eq. 5.18 is the equivalent of Eq. 5.5 for three interacting transitions, and can be extended to include n coupled oscillators.

For the case of n coupled transitions, Eq. 5.17 is diagonalized numerically to yield the energy eigenvalues ν_k of the kth "exciton" component and the eigenvector matrix **C,** from which the transition dipole strength D_k for each of the k coupled transitions can be computed according to

$$D_k = \sum_{j=1}^{n} \sum_{i=1}^{n} c_{ik} c_{jk} \left(\boldsymbol{\mu}_i \cdot \boldsymbol{\mu}_j\right) \tag{5.21}$$

where c_{ik} are the appropriate elements of the eigenvector matrix. This formalism was developed and published [3] for the computation of IR and IR vibrational circular dichroism (VCD, see Chapter 8) intensities, and followed the original work by Tinoco [4] aimed at computing electronic circular dichroism (ECD) using the same "coupled oscillator" approach. In these earlier efforts, the term "exciton" was used frequently, where exciton refers to a delocalized, one-photon excitation into one of many coupled states of interacting transitions.

The TDC approach for the prediction of the observed IR spectra in polymeric systems should be considered an idealized situation; in real application, significant contributions from other sources than mere dipolar coupling need to be included. In particular, through-bond (kinetic energy) coupling is quite pronounced in peptides where the carbonyl groups of the amide linkages are only three bonds apart. In nucleic acids, there are many more bonds separating the carbonyl groups, and through space, dipole–dipole coupling represents the major interactions between adjacent carbonyl groups.

Therefore, the TDC model worked well for nucleic acids in which the carbonyl groups are located in parallel planes, about 0.3 nm apart. It was shown that in cytosine–guanine containing double-stranded polymers, TDC calculations could distinguish a CG from a GC sequence, based on computed IR absorption features [5], and canonical B- and Z-forms of model oligonucleotides [6].

For polyamino acids and peptides, further coupling mechanisms had to be included to produce satisfactory agreement with experimental data. Some of these studies were undertaken to interpret experimental results obtained from ^{13}C substituted polyamino acids; this substitution reduces the C═O stretching frequency by about 35 cm^{-1} and effectively uncouples the ^{13}C═O groups from the rest of the carbonyl groups [7]. Brauner *et al.* empirically refined the coupling interactions to such a degree that excellent agreement between observed spectra, and spectra computed from structural data from the protein data bank [8] were obtained. In proteins, as compared to nucleic acids, the situation is further complicated by changes in hydrogen bonding patterns and strengths that accompany various protein secondary structures.

5.3 Group frequencies

IR and Raman spectroscopy offer the distinct advantage that spectral features of the same molecular group are quite similar, relatively unaffected by the chemical environment. Thus, the methyl stretching and deformation frequencies are similar for CH_3Cl and CH_3Br, as shown in Table 5.1, and can be referred to as "group frequencies" of the methyl group.

Thus, vibrational spectroscopy offers the advantage that simple inspection of a spectrum reveals the presence of chemical groups, similar to proton NMR spectroscopy, where the same chemical groups appear at similar chemical shifts. This is in contrast to rotational spectroscopy in which the moment of inertia is measured for the entire molecule, and a methyl group attached to a benzene ring or to a chlorine atom will produce entirely different spectral signatures.

Thus, IR and Raman spectroscopy allow the qualitative identification of small molecules, and are therefore used in organic chemistry to check for the presence of certain functional groups. Introductory books on organic chemistry typically list several dozen characteristic vibrational frequencies of aliphatic, vinylic and aromatic C—H stretching frequencies, carbonyl groups, aromatic ring vibrations, and the vibrations of groups containing heteroatoms. A listing of such group frequencies can be found in Section A.5. For a more complete listing of group frequencies and their use, the reader is referred to books by Colthup *et al.* [9] and Dollish *et al.* [10].

In view of the discussion of the normal vibrations of molecules (see Chapter 1), where it was pointed out that during normal mode all atoms undergo displacement, the use of group frequencies may appear naïve. Indeed,

Table 5.1 *Vibrational modes and assignments for CH_3Cl and CH_3Br*

Designation	Observed transition frequencies (cm^{-1})		Mode description
	CH_3Cl	CH_3Br	
ν_1	732	611	C—X stretching
ν_2	1017	955	CH_3 rocking
ν_3	1355	1306	CH_3 symmetric deformation
ν_4	1452	1442	CH_3 antisymmetric deformation
ν_5	2937	2935	CH_3 symmetric stretching
ν_6	3039	3056	CH_3 antisymmetric stretching

group frequencies have to be used with care, as the following discussion will show. A single C—Cl "group," or bond, will produce strong IR and Raman bands around 750 cm^{-1}, which are more or less independent of the rest of the molecule. However, if a molecule contains a Cl—C—Cl group, two C—Cl stretching vibrations are observed: a symmetric Cl—C—Cl stretching mode at about 715 and an antisymmetric stretching mode at about 755 cm^{-1}. In a –CCl_3 group, the symmetric stretching mode occurs at even lower frequency, 665 cm^{-1}, whereas the antisymmetric stretching mode (of local E symmetry) remains at 760 cm^{-1}. This example demonstrates how the coupling of individual C—Cl stretching modes can produce new group frequencies that should no longer be described as simple bond stretching motions, but more complex combinations of them.

5.4 Rot-vibrational spectroscopy

In Chapter 2 (Figure 2.14), it was pointed out that for gaseous molecules that freely rotate in space, transitions between vibrational and rotational energy levels are observed. Thus, this section deals with the theoretical background of rot-vibrational transitions, and – by necessity – with the rotational energy level of molecules. Pure rotational transitions for gaseous molecules are observed in the far IR (< 100 cm^{-1}) for light molecules, and in the microwave spectral range (< 5 cm^{-1}) for heavy molecules. Consequently, rotational spectroscopy is often referred to as microwave spectroscopy. For heavy molecules, it is customary to express the transition energies in gigahertz, rather than in wavenumber, where the transition energy of 1 cm^{-1} approximately corresponds to 30 GHz.

For vibrational spectroscopists, the interaction of vibrational and rotational energy levels is a mixed blessing. On the positive side, the observation of rotational fine structure of vibrational peaks (see Figure 2.14) can provide information on a molecule's moment of inertia and thereby, on bond distances and bond angles. This is particularly relevant for molecules devoid of a permanent dipole moment; these molecules do not produce a pure rotational spectrum, and the rot-vibrational spectrum is a way to obtain structural data (bond distances and angles) nevertheless. On the negative side, however, the rotational band envelopes can make the assignment of the vibrational spectrum a difficult task, because of the appearance of rot-vibrational envelopes as shown in Figure 2.14, and the overall complexity of the spectra. Furthermore, the rot-vibrational spectrum of water vapor frequently appears in IR spectra, since the rot-vibrational manifold of this molecular system is extremely strong in IR absorption and can even be detected in well-purged instruments. Furthermore, it covers a broad range of mid-IR frequencies (from about 1300 to 1800 cm^{-1}).

In Section 5.4.1, a short summary of the pure rotational energy levels is presented, followed by a discussion of the selection rules and simple rotational spectral features. Unfortunately, the rotational Schrödinger equation can only be solved exactly for linear, spherical, and "symmetric top" molecules. A symmetric top is a molecule that possesses at least a threefold axis of symmetry, such as ammonia (C_{3v} point group, see Chapter 2). For these molecules, and molecules of higher symmetry, at least two of the three moments of inertia are equal, and the rotational Schrödinger equation can be solved explicitly. Molecules such as water (C_{2v} symmetry) are known as "asymmetric top" rotors, that is, they have three distinct moments of inertia. For any molecules with three distinct moments of inertia, the rotational Schrödinger equation cannot be solved analytically, and approximate solutions need to be utilized.

5.4.1 Classical rotational energy

Classically, any rotatory motion can be described by the rotational kinetic energy [11]

$$E_{kin} = \frac{L^2}{2I} \tag{5.22}$$

where L is the angular momentum and I is the moment of inertia. This definition is in complete analogy to the definition of the (linear) kinetic energy

$$E_{\text{kin}} = \frac{p^2}{2\,m} \tag{1.41}$$

Thus, the kinetic energy, in both cases, is proportional to the square of the corresponding momentum, divided by a quantity measuring the inertia, or "resistance," to this motion. The angular momentum in Eq. 5.22 is defined as

$$\boldsymbol{L} = \boldsymbol{r} \times \boldsymbol{p} \tag{5.23}$$

that is, the vector product of the linear momentum and the radius of the circular motion. Thus, $\boldsymbol{L}$ is a vectorial quantity, indicated in bold face.

Multiplying the vector components $\boldsymbol{r} = \boldsymbol{i}x + \boldsymbol{j}y + \boldsymbol{k}z$ and $\boldsymbol{p} = \boldsymbol{i}p_x + \boldsymbol{j}p_y + \boldsymbol{k}p_z$ in Eq. 5.23 reveals that

$$\boldsymbol{L} = \boldsymbol{i}L_x + \boldsymbol{j}L_y + \boldsymbol{k}L_z \tag{5.24}$$

with

$$L_x = yp_z - zp_y \tag{5.25}$$

and similar expressions holding for the other components of $\boldsymbol{L}$.

The (scalar) moment of inertia, as pointed out above, depends very much on the shape (symmetry) of the molecule. In the simplest case of a linear, diatomic molecule, the moment of inertia is given by

$$I = m_1 r_1^2 + m_2 r_2^2 \tag{5.26}$$

where m_1 and m_2 are the masses of the two atoms and r_1 and r_2 their distances from the center of mass. The center of mass condition is given by $m_1 r_1 + m_2 r_2 = 0$. For a diatomic molecule, one moment of inertia is zero (the one along the bond axis, z) because it is assumed that the atoms are point masses, and rotation about the bond connecting the two atoms experiences no inertia. The other two components of the moment of inertia, I_x and I_y, are equal. The same argument holds true for linear molecules with more than two atoms.

For molecules of C_{3v} symmetry or higher, but not for molecules belonging to spherical point groups, two of the moments of inertia are equal, and different from the third one. These molecules are referred to as symmetric top rotors. For molecules belonging to spherical point groups (T_d, O_h), such as CH_4 and SF_6, all three moments of inertia are equal and nonzero. These molecules are referred to as spherical top rotors.

For a nonlinear molecule in an arbitrary coordinate system, the moment of inertia is a tensor defined as

$$I = \begin{bmatrix} I_{xx} & I_{xy} & I_{xz} \\ I_{yx} & I_{yy} & I_{yz} \\ I_{zx} & I_{zy} & I_{zz} \end{bmatrix} = \begin{bmatrix} \sum_i m_i (r_y^2 + r_z^2) & -2\sum_i m_i r_{xi} r_{yi} & -2\sum_i m_i r_{xi} r_{zi} \\ -2\sum_i m_i r_{yi} r_{xi} & \sum_i m_i (r_x^2 + r_z^2) & -2\sum_i m_i r_{yi} r_{zi} \\ -2\sum_i m_i r_{zi} r_{xi} & -2\sum_i m_i r_{zi} r_{yi} & \sum_i m_i (r_x^2 + r_y^2) \end{bmatrix} \tag{5.27}$$

In Eq. 5.27, the summation is over all atoms in the molecule, m_i are the masses of the individual atoms, and r_x, r_y, and r_z their Cartesian coordinates. If the inertial tensor is written in arbitrary coordinates x, y, and z, its values will depend on the choice of the origin of the coordinate system, that is, different numerical values for the tensor terms I_{xx}, and so on, are obtained when the origin is shifted. The rotational motion of a molecule, or for that matter, any freely rotating body, can be described such that the translational and rotational motions are separated. If the center of mass of the molecule does not translate, the molecular motion is a pure rotational motion. Under these conditions, the inertial tensor can be written in diagonal form:

$$I = \mathbf{T}^T \begin{bmatrix} I_{xx} & I_{xy} & I_{xz} \\ I_{yx} & I_{yy} & I_{yz} \\ I_{zx} & I_{zy} & I_{zz} \end{bmatrix} \mathbf{T} = \begin{bmatrix} I_A & 0 & 0 \\ 0 & I_B & 0 \\ 0 & 0 & I_C \end{bmatrix} \tag{5.28}$$

Here, **T** is a coordinate transform matrix between an arbitrary coordinate system and the principal axes of inertia; in this coordinate system, the inertial tensor is diagonal and the diagonal terms I_A, I_B, and I_C are known as the principal moments of inertia.

The square of the total angular momentum, $\boldsymbol{L}^2$, can be written as a sum of the individual principal components according to

$$\boldsymbol{L}^2 = L_A^2 + L_B^2 + L_C^2 \tag{5.29}$$

and thus, the rotational kinetic energy is

$$2T = \frac{L_A^2}{I_A} + \frac{L_B^2}{I_B} + \frac{L_C^2}{I_C} \tag{5.30}$$

Since the potential energy of a freely rotating body is zero, the total energy of the system is given by the kinetic energy, that is, $E = T$. The kinetic energy is generally expressed in terms of the *rotational constants A, B, and C,* rather than I_A, I_B, and I_C defined according to

$$A = \frac{h}{(8\pi^2 I_A)} \quad B = \frac{h}{(8\pi^2 I_B)} \quad C = \frac{h}{(8\pi^2 I_C)} \tag{5.31}$$

with the convention that $I_A \leq I_B \leq I_C$. Notice that in this convention, the rotational constants are expressed in units of 1/s = Hz. To convert to energy units, the constants would have to be multiplied by Planck's constant. In terms of the rotational constants, the rotational kinetic energy is written as

$$E = \frac{4\pi^2}{h}(AL_A^2 + BL_B^2 + CL_C^2) \tag{5.32}$$

The rotational constants are related to the quantum mechanical energy eigenvalues observed in rotational spectroscopy. They contain all the structural parameters of a molecule and allow the determination of molecular structures to within picometer and millidegree accuracy.

5.4.2 Quantum mechanics of rotational spectroscopy

In analogy to the discussion of vibrational spectroscopy, a classical model for energies was derived above. The classical models, for both the rotation and the vibration of molecules, predict the molecular energies properly in a steady state situation, that is, without interactions with light (photons). In order to predict these interactions and the possibility of observing a spectrum due to the molecular system undergoing transitions between stationary state energy levels, quantum mechanics needs to be invoked.

As before, the transition to quantum mechanics involves substitution of

$$p = \frac{\hbar}{i}\frac{\partial}{\partial x} \tag{1.42}$$

for the linear momentum in the equation for the angular momentum (Eq. 5.25)

$$L_x = yp_z - zp_y = \frac{\hbar}{i}\left(y\frac{\partial}{\partial z} - z\frac{\partial}{\partial y}\right), \tag{5.33}$$

with equivalent expressions holding for the other two Cartesian components of the angular momentum. Because of the differentiation required when operating with the L_x, L_y, or L_z operators on a function $f(x, y, z)$, it follows that the commutator of these operators, defined as

$$[L_x, L_y] = L_x(L_y f(x,y,z)) - L_y(L_x f(x,y,z)) \neq 0 \tag{5.34}$$

Equation 5.34 implies that applying the same two operators, but in reverse order, yields different results such that the difference between the two processes is not zero. In fact, it can easily be shown that

$$[L_x, L_y] = L_x(L_y f(x,y,z)) - L_y(L_x f(x,y,z)) = i\hbar L_z \tag{5.35}$$

The consequences of Eqs. 5.34 and 5.35 are quite far reaching. A mathematical theorem states that two operators, $\hat{Q}$ and $\hat{R}$, can only be diagonalized or solved for explicitly in the same vector space if they commutate – that is, if their commutator is zero. Since the components of the angular momentum operator do not commutate, it is impossible to experimentally determine the three principal moments of inertia in an asymmetric top rotor.

However, the total angular moment operator, $\boldsymbol{L}^2$, does commutate with all the components of the angular momentum operator and thus, it is possible to determine the total moment of inertia and one of the three components (normally assumed to be the L_z operator).

The rotational Schrödinger equation is set up for four distinct cases:

1. All rotational constants of the molecule are equal (spherical top rotor). The total rotational energy can be expressed in terms of $\boldsymbol{L}^2$ only. By convention, the rotational constant B was selected to represent the rotational energy (*cf.* Eq. 5.32) as

$$E = \frac{4\pi^2}{h}(B\,\boldsymbol{L}^2) \tag{5.36}$$

2. For linear molecules, one rotational constant is zero (the one along the bond axis), and the other two are equal. This leads to an energy equation for linear molecules identical to the one for spherical molecules.
3. For symmetric tops, either $I_A = I_B < I_C$ (oblate symmetric top rotor) or $I_A < I_B = I_C$ (prolate symmetric top rotor). For the former case,

$$E = \frac{L_A^2}{2I_A} + \frac{L_B^2}{2I_B} + \frac{L_C^2}{2I_C} = \frac{L_A^2 + L_B^2}{2I_B} + \frac{L_C^2}{2I_C} \tag{5.37}$$

Since

$$\boldsymbol{L}^2 = L_A^2 + L_B^2 + L_C^2 \tag{5.29}$$

$$L_A^2 + L_B^2 = \boldsymbol{L}^2 - L_C^2 \tag{5.38}$$

$$E = \frac{\boldsymbol{L}^2 - L_C^2}{2I_B} + \frac{L_C^2}{2I_C} = \frac{\boldsymbol{L}^2}{2I_B} + L_C^2\left(\frac{1}{2I_C} - \frac{1}{2I_B}\right) \tag{5.39}$$

$$= \frac{4\pi^2}{h}[B\boldsymbol{L}^2 + (C - B)\,L_C^2] \tag{5.40}$$

For the prolate top rotor, the corresponding equation is

$$E = \frac{4\pi^2}{h}[B\boldsymbol{L}^2 + (A - B)\,L_C^2] \tag{5.41}$$

In both equations, the two operators, $\boldsymbol{L}^2$ and L_C^2 commutate, and have eigenvalues in the same vector space. Thus, the problem can be solved.
4. For the asymmetric top rotor, the operators do not commutate, and an explicit solution is not achievable.

The rotational Schrödinger equation is defined in spherical polar coordinates r, θ, and ϕ, rather than in x, y, and z. In this coordinate system, the $\boldsymbol{L}^2$ operator is defined as

$$\boldsymbol{L}^2 = -\hbar^2\left[\frac{\partial^2}{\partial\theta^2} + \cot\theta\,\frac{\partial}{\partial\theta} + \frac{1}{\sin^2\theta}\,\frac{\partial^2}{\partial\phi^2}\right] \tag{5.42}$$

and the L_C operator as

$$L_C = i\hbar \frac{\partial}{\partial \phi} \tag{5.43}$$

The solution of Eq. 5.43 is quite simple, and yields a function $T(\phi)$ that obeys the boundary condition

$$T(\phi) = T(\phi + 2\pi) \tag{5.44}$$

and is given by

$$T(\phi) = (2\pi)^{-1/2}\, e^{iK\phi} \tag{5.45}$$

where $K = 0, \pm 1, \pm 2, \pm 3, \ldots$.

The procedure for solving Eq. 5.42 is similar to the way Hermite's differential equation was solved (see Appendix B), and is not presented here. The reader is referred to typical books on quantum mechanics [12, 13]. The solutions of both Eqs. 5.42 and 5.43 are referred to as the spherical harmonic functions $Y(\theta, \phi)$. Similar to the solutions of the vibrational Schrödinger equation, the spherical harmonics are polynomials in the variables θ and ϕ, and contain two indexes, J and K, that determine the power of the polynomials in the variables θ and ϕ (just as n defined the power of the variable in Hermite's polynomials, $H_n(z)$).

Again, quite similar to the solutions of the vibrational Schrödinger equation, the eigenvalues of the rotational Schrödinger equation are quite simple expressions:

$$\boldsymbol{L}^2 Y(\theta, \phi) = J(J+1)\,\hbar^2\; Y(\theta, \phi) \tag{5.46}$$

where J assumes integer values, $J = 0, 1, 2, 3, \ldots$, and

$$L_C Y(\theta, \phi) = K\,\hbar\; Y(\theta, \phi) \tag{5.47}$$

with K assuming values given in Eq. 5.45, up to a limit defined by the value of J: $K = 0, \pm 1, \pm 2, \pm 3, \ldots, \pm J$. Readers familiar with the quantum mechanics of the hydrogen atoms will recognize that the eigenfunctions and eigenvalues presented here are the same for the radial part of the hydrogen atom, and the limitations of the indexes of the spherical harmonics determine the splitting of the s, p, d, f orbitals into the sublevels determined by the magnetic quantum number.

Equations 5.36–5.47 lead to a surprisingly simple equation for the rotational energy of *spherical* and *linear* molecules:

$$E_{\text{rot}} = B\,J\,(J+1) \tag{5.48}$$

where B was defined above as

$$B = \frac{h}{(8\pi^2 I_B)} \tag{5.31}$$

A quick analysis of the units here again reveals that energy here is expressed in units of hertz (see Eq. 5.31):

$$B: \quad \frac{\text{J s}}{\text{kg}\ \ \text{m}^2} = \frac{\text{kg m}^2\ \text{s}^{-1}}{\text{kg m}^2} = \text{s}^{-1} \tag{5.49}$$

To express these energies in units of Joule, Eq. 5.31 would contain h^2, rather than h, in the numerator.

Since J can assume values of $0, 1, 2, 3, \ldots$, one finds the corresponding rotational energies to be $E_{J=0} = 0, E_{J=1} = 2B, E_{J=2} = 6B, E_{J=3} = 12B, \ldots$.

An energy level diagram for spherical or linear molecules is shown in Figure 5.2(a). The selection rules for rotational transitions are derived from the recursion formula of the Legendre polynomials that are part of the spherical harmonic functions, in analogy to the derivation of the vibrational selection rules (*cf.* Eqs. 1.98–1.102).

$$\Delta J = \pm 1 \tag{5.50}$$

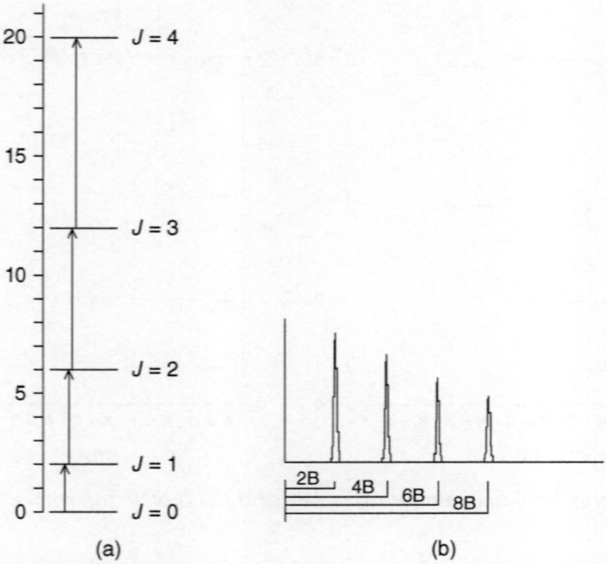

Figure 5.2 (a) Energy level diagram for symmetric top and linear rotors. (b) Schematic rotational spectrum for these two cases

Thus, transitions between adjacent energy levels are allowed, both up and down the energy ladder. The transition moments for rotational transitions are nearly independent of the energy level from which the transition originates; thus, the intensity of transitions is determined mainly by the Boltzmann distribution.

For the following discussion, it is assumed that a transition occurs from state J to a more energetic state $J' = J + 1$. This transition occurs at a transition energy (frequency) of

$$\begin{aligned}\Delta E &= BJ'(J' + 1) - BJ\,(J + 1) = BJ(J + 1)(J + 2) - BJ(J + 1)\\ &= 2B(J + 1)\end{aligned} \tag{5.51}$$

This is shown by the vertical arrows in Figure 5.2(a) depicting the energy from the $J = 0$ to $J = 1$ state to be 2B, from the $J = 1$ to $J = 2$ state to be 4B, and so on, where J is the lower energy state. Thus, a rotational spectrum for spherical and linear molecules consists of transitions at 2B, 4B, 6B, and so on, and consequently the spectrum appears as equally spaced transitions, the spacing being 2B. This is shown schematically in Figure 5.2(b).

For symmetric top rotors, the rotational energies are, according to Eqs. 5.40, 5.41, 5.46, and 5.47,

$$E_{\text{rot}} = B\,J\,(J + 1) + (C - B)K^2(\text{oblate}) \tag{5.52}$$

$$E_{\text{rot}} = B\,J\,(J + 1) + (A - B)K^2(\text{prolate}) \tag{5.53}$$

with selection rules

$$\Delta J = \pm 1 \quad \text{and} \quad \Delta K = 0 \tag{5.54}$$

An energy level diagram for the oblate symmetric top rotor is shown in Figure 5.3(a), and for the prolate top rotor in Figure 5.3(b). From the aforementioned convention $I_A \leq I_B \leq I_C$, it follows that $A \geq B \geq C$ and therefore, $(C - B)$ is always negative and $(A - B)$ is always positive. Thus, the energy levels for increasing K values are decreasing for the oblate, and increasing for the prolate top. This increase/decrease occurs with

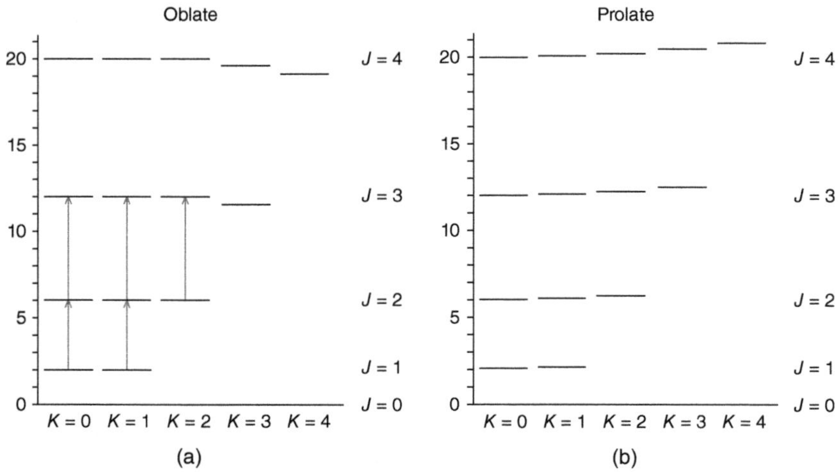

Figure 5.3 *Energy level diagram for (a) oblate and (b) prolate top rotors. See text for detail*

the square of K. However, since the selection rules require that transitions occur within a given K-level, the transition energies for the $J=1$ to $J=2$ transition for $K=0$ to $K=1$ are still the same, as indicated by the gray arrows in Figure 5.3(a). Thus, these transitions are, to a first order approximation, still degenerate, and the same spectra with equidistantly spaced lines (at $2B$ spacing) are observed for symmetric top rotors. Only under the influence of an external electric field (Stark splitting), or when centrifugal distortion is considered, will this degeneracy be lifted. Under these circumstances, the transitions from $J=1$ to $J=2$ will appear as a doublet, and the $J=1$ to $J=2$ transition as a triplet, and so on.

To a first approximation, the transition moment for the rotational transitions is equal; thus, the intensity of the observed spectral lines depends mostly on the population of each of the energy levels. Assuming that the rotational energy differences are very small, indeed, and the thermal energy is large, the Boltzmann distribution (Eq. 1.103) predicts that many rotational energy levels have similar populations. The probability P_J of finding a molecule in a given rotational level J, is given by

$$P_J = \frac{B(2J+1)}{kT} e^{-\frac{BJ(J+1)}{kT}} \tag{5.55}$$

In this equation, the exponential expression is the normal Boltzmann distribution. The factor $(2J+1)$ represents the degeneracy of each of the energy levels, and the factor B/kT results from the total number of occupied rotational states, given by

$$\int_0^{\infty} (2J+1)e^{-\frac{BJ(J+1)}{kT}}\, dJ = \frac{kT}{B} \tag{5.56}$$

Here, one assumes that the energy levels are so closely spaced that J becomes a continuous variable. The intensity distribution, as a function of the rotational transition, is shown schematically in Figure 5.4.

Asymmetric top rotors, as mentioned above, cannot be treated in a simple model described for the other tops above. Rather, approximate methods are used in which the molecule is described by its asymmetry parameter κ, which is defined as

$$\kappa = \frac{2B - A - C}{A - C} \tag{5.57}$$

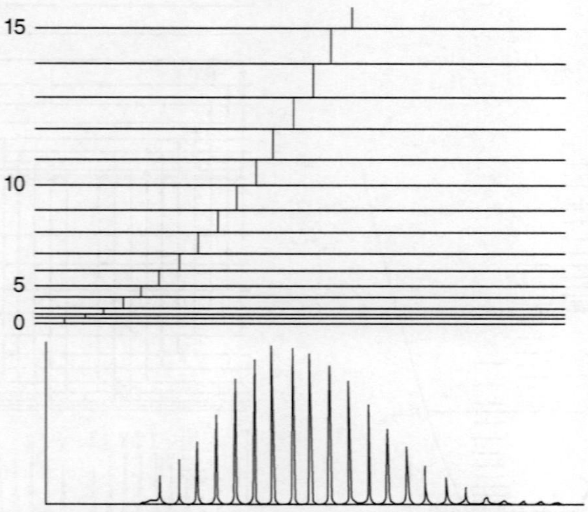

Figure 5.4 Transition energies for given J levels (top) and transition intensities (bottom), derived from Eq. 5.55

This parameter takes the value of -1 for the prolate top (with $B = C$) and $+1$ for the oblate top ($A = B$). Thus, transition energies $E(\kappa)$ are interpolated for an asymmetric top rotor between the oblate and prolate top boundaries, and the overall rotational energy can be approximated by

$$E_{\mathrm{rot}} = \frac{A+C}{2} J(J+1) + \frac{A-C}{2} E(\kappa) \tag{5.58}$$

Pure rotational spectra can also be observed *via* the Raman effect. For small molecules such as H_2, N_2, O_2, CO_2, and so on, equidistant lines on both sides of the Rayleigh line are due to rotational Raman spectra. Symmetric top rotors do not exhibit rotational Raman spectra, but linear rotors and symmetric top rotors undergo rotational transitions with the following selection rules:

$$\Delta J = \pm 2 \quad \text{for linear molecules} \tag{5.59}$$

and

$$\Delta J = \pm 1, \pm 2 \quad \text{and} \quad \Delta K = 0 \quad \text{for symmetric tops} \tag{5.60}$$

These transitions are quite strong; shining a focused laser beam through air easily allows collection of the rotational spectra of oxygen and nitrogen.

5.4.3 Rot-vibrational transitions

For gaseous molecules, the rotational fine structure due to the interaction of rotational and vibrational transitions is readily observed. This interaction can be investigated by writing rot-vibrational wavefunctions as products of the pure rotational and the pure vibrational wavefunctions:

$$\psi_{\mathrm{rotvib}} = \psi_{\mathrm{rot}} \psi_{\mathrm{vib}} \tag{5.61}$$

and the corresponding energies simply as the sum of rotational and vibrational energies:

$$E_{\mathrm{rotvib}} = E_{\mathrm{rot}} + E_{\mathrm{vib}} \tag{5.62}$$

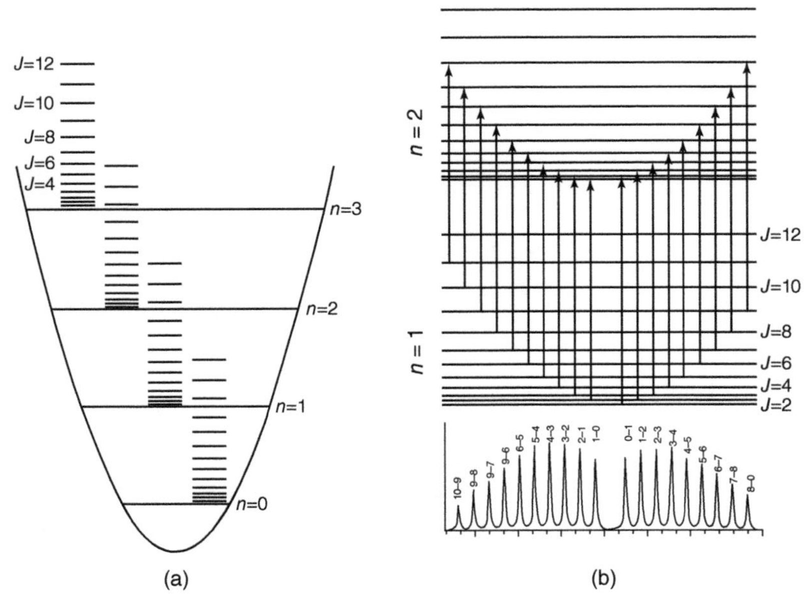

Figure 5.5 (a) Rot-vibrational energy level diagram for a diatomic molecule obeying the harmonic oscillator equations. (b) Rot-vibrational transitions for n = 1–2, with ΔJ = −1 ("P-branch," left half) and ΔJ = +1 ("R-branch," right half). The Q-branch, for which ΔJ = 0, may or may not be allowed according to the symmetry of the molecule. Adapted from Kauzmann, 1957, [13]

For a diatomic molecule, Eq. 5.62 would read as

$$E_{\text{rotvib}} = BJ(J+1) + h\nu\left(n + \frac{1}{2}\right) \tag{5.63}$$

In Eq. 5.63, both the vibrational and rotational energies are expressed in units of energy; that is, B is expressed in hertz and the vibrational energies in terms of the frequency (rather than the wavenumber) of the transition. The energy level diagram for a diatomic molecule is shown in Figure 5.5(a).

The selection rules for rot-vibrational transitions are

$$\Delta n = \pm 1; \quad \Delta J = \pm 1 \tag{5.64}$$

This leads to rot-vibrational transitions from the rotational sublevels of the ground vibrational state to rotational sublevels of the vibrationally excited state with either lower or higher rotational quantum numbers, as shown in Figure 5.5(b). The energy of the rotational sublevels will be added to the individual energies of the normal modes of the molecule.

Taking into account anharmonicity and centrifugal distortion, Eq. 5.63 needs to be augmented to read

$$E_{\text{rotvib}} = \left[BJ(J+1) + h\nu\left(n + \frac{1}{2}\right)\right] - \left[h\nu\chi\left(n + \frac{1}{2}\right)^2 + DJ^2(J+1)^2\right] \tag{5.65}$$

The anharmonicity terms in Eq. 5.65 were defined before (*cf.* Eq (1.62)), and the centrifugal distortion constant D accounts for the change of the moment of inertia at higher rotational energy.

By definition, the rot-vibrational transitions for which $\Delta J = -1$ (the low frequency progression of rot-vibrational peaks) are referred to as the P-branch, those with $\Delta J = 0$ are referred to as the Q-branch,

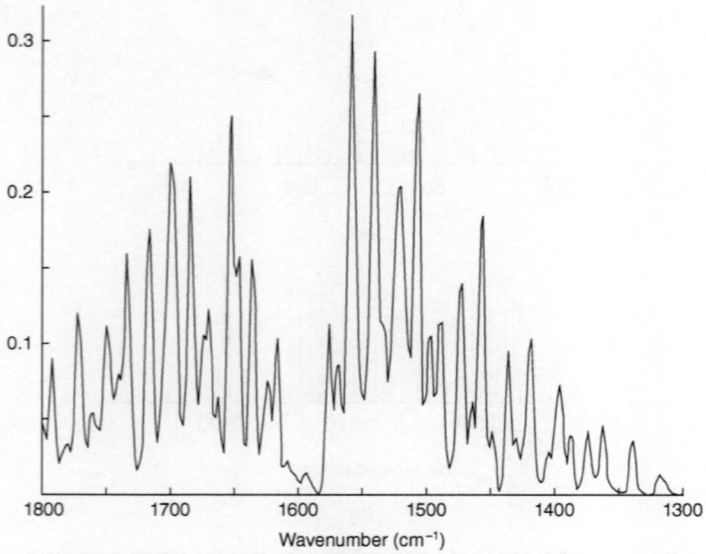

Figure 5.6 *Rot-vibrational absorption spectrum of water vapor at room temperature and ambient humidity, measured through an un-purged infrared microscope with a path length of about 1 m*

and those with $\Delta J = +1$ as the R-branch. Symmetry rules (to be discussed below) sometimes prohibit the occurrence of the Q-branch, and typical rot-vibrational spectra may appear as shown in the bottom of Figure 5.5(b). Notice that there are many transitions in the Q-branch, since the $\Delta n = 1$, $\Delta J = 0$ transitions can originate from $J=0$, $J=1$, $J=2$, etc.

As pointed out before, the rotational transitions for an asymmetric top rotor are very complicated; therefore the rot-vibrational envelopes also are quite complex and lack the simple progression of rotational lines, as shown in the next example.

Figure 5.6 shows the rot-vibrational spectrum of the symmetric deformation mode of an asymmetric top rotor, water(vapor). The intensity pattern is referred to as a B-type rot-vibrational band envelope where the vibrational transitions with $\Delta J = 0$ (the Q-branch) are forbidden. The spectrum shown in Figure 5.6 is the nemesis of vibrational spectroscopists working with biological systems in which the protein amide I and amide II vibrations are superimposed on the water vapor spectrum in this spectral region. Even spectra from a well-purged instrument will have some residual water vapor contributions that may give the appearance of protein features, particularly in second derivative spectra. This spectrum is reproduced here to warn scientists of the presence of this water background.

Finally, the rot-vibrational envelopes observed at low resolution, or for heavy molecules, are introduced. For heavy molecules, the moments of inertia are significantly larger than for small molecules; hence, the rotational constants are much smaller, and the rotational lines are spaced much more closely. For such molecules, rot-vibrational "envelopes" are observed, as shown in Figure 5.8. Such envelopes may have different appearance for linear and symmetric top rotors. For linear, hetero- and homonuclear diatomic molecules, one expects spectra exhibiting P and R branches, with the Q-branch missing, as discussed in Figures 5.5 and 5.6; however, the absence of a permanent dipole moment prohibits these transitions to be observed in absorption for the homonuclear diatomic molecules (however, they can be observed in Raman scattering).

In linear polyatomic molecules, vibrational transitions that do not change the dipole moment are forbidden in absorption; however, transitions that change the dipole moment, such as the antisymmetric stretching vibration

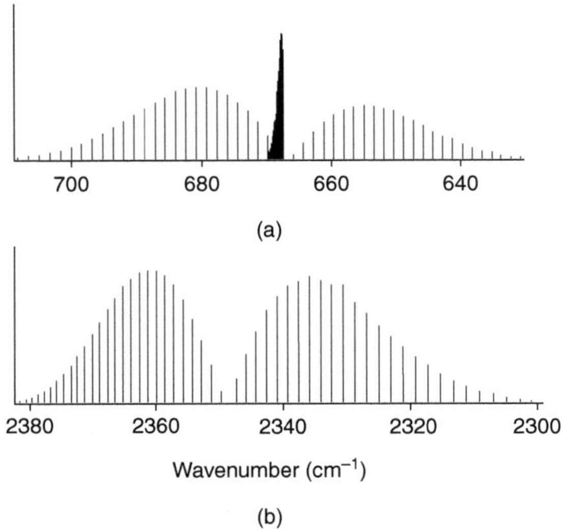

Figure 5.7 *Simulated rot-vibrational spectral band profiles for the deformation mode ν_1 (trace a) and the antisymmetric stretching mode ν_3 of CO_2 (trace b). From vpl.astro.washington.edu/spectra/co2.htm*

of CO_2 (*cf.* Chapter 2) do exhibit a rot-vibrational spectrum. For this vibration, the transition dipole lies along the molecular axis; consequently, the band envelope is referred to as a "parallel" envelope, characterized by the absence of a Q-branch, with an appearance similar to the rot-vibrational spectrum of a heteronuclear diatomic molecule. This is shown in Figure 5.7(b). In the CO_2 deformation mode, on the other hand, the dipole moment change is perpendicular to the molecular axis; consequently, the resulting band envelopes are referred to as "perpendicular" bands. Perpendicular bands have a pronounced Q-branch, broadened by the overlap of the $n=0$ to $n=1$ transitions for different J-values, *cf.* Figure 5.7(a).

The only bands of spherical top molecules (T_d and O_h symmetry) observed in absorption spectroscopy are the triply degenerate modes. These vibrations exhibit rot-vibrational spectra of the perpendicular type, with distinct Q-branches. In Figure 2.14, the rot-vibrational spectrum of the ν_4 mode (the antisymmetric stretching mode) of methane was presented. In this molecule, the "Q-branch" transitions are allowed and account for the central peak at about 3019 cm^{-1}. Inspection of Figure 2.14 also reveals that the spacing between the rotational transitions becomes smaller from lower to higher wavenumber. This is due to the centrifugal distortion effect: as the molecule rotates faster, centrifugal forces stretch the bond, thereby increasing the moment of inertia and decreasing the rotational constants, resulting in more closely spaced lines (see Eq. 5.65).

Symmetric top molecules again exhibit parallel or perpendicular rot-vibrational band profiles, depending on whether the dipole change occurs along, or perpendicular to, the preferred axis (the C_3 axis). Parallel bands, such as the C—Cl stretching mode in CH_3Cl, do not exhibit a Q-branch and appear similar to parallel bands in polyatomic linear molecules.

In even more complicated molecules, such as the prototypical asymmetric (chiral) molecule bromochlorofluoromethane (HCBrClF), the observed band envelopes are quite complex. For asymmetric top rotor molecules, the observed rot-vibrational band contours are classified as A, B, and C-type band envelopes. These band envelopes refer to the direction of the dipole change along the smallest, intermediate, or largest axis of inertia. Such band envelopes can be computed from structural data within the rigid rotor approximation. B-type band envelopes resemble the parallel type band envelopes of symmetric top rotors in that they do not exhibit a Q branch, but have intense P and R branches. Both A- and C-types exhibit P,

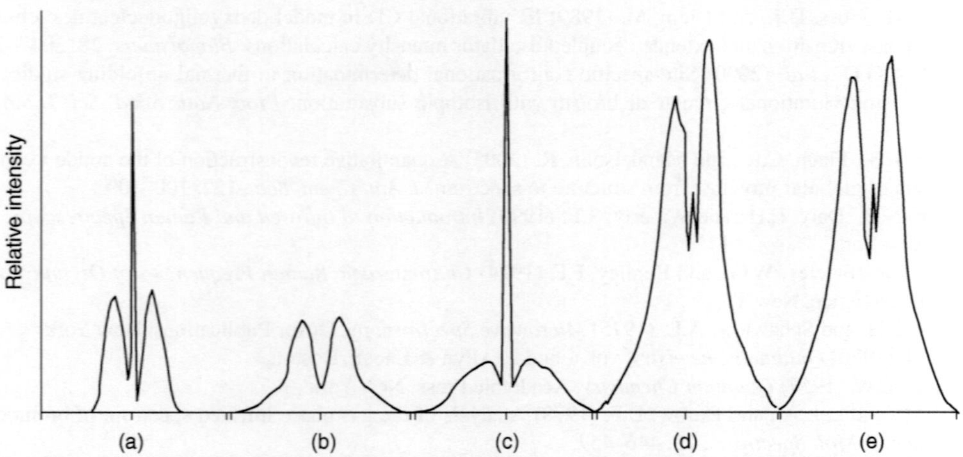

Figure 5.8 *Computed and observed band rot-vibrational envelopes for HCClFBr. Traces (a)–(c) represent computed A-, B-, and C-type rot-vibrational band envelopes, plotted from −30 to +30 cm*$^{-1}$ *from the band center. Panels (d) and (e) represent observed and computed band envelopes for the C—F stretching mode of HCClFBr. Diem* et al.*, 1978, [14]. Reproduced with permission from Elsevier*

Q, and R branches: the C-type envelopes have broader, less intense P and R-branches with very pronounced Q-branch intensity, whereas the A-type bands show more intense P- and R-branches. This is shown in Figure 5.8 that depicts in traces (a)–(c) the pure A-, B-, and C-type envelopes for an asymmetric top rotor with the rotational constants computed for HCBrClF.

In an asymmetric rotor without any symmetry, the transition dipole moments will most likely not coincide with an inertial axis; therefore, all observed band envelopes for HCBrClF are hybrids of pure A, B, and C-type band shapes. This is shown in Figure 5.8(d) and (e) for ν_4, assigned to be the C—F stretching mode. Trace (d) shows the observed and trace (e) the computed envelopes. The computations were based on a vibrational mode that involved all atomic displacements; this mode predicted about 80% B-type and 20% C-type band envelope contributions. Furthermore, the vibrational computation predicted a relatively small shift in the band center (0.5 cm^{-1}) due to ^{35}Cl/^{37}Cl isotopic effects. Obviously, these computed parameters reproduced the observed band shapes quite well; similar results were obtained for the other eight vibrational modes, as well as those of the deuterated isotopomer, DCBrClF. These studies were carried out [14] to ascertain that the empirical force field calculations used then properly predicted transition moment directions that agreed with the observed rot-vibrational envelopes that provide a reference frame for the direction of the dipole change in the inertial coordinate system.

References

1. Herzberg, G. (1945) *Molecular Spectra and Molecular Structure*, Van Nostrand Reinhold Co, New York.
2. Diem, M. *et al.* (1982) Vibrational circular dichroism in amino acids and peptides. 4. Vibrational analysis, assignment, and solution phase Raman spectra of deuterated isotopomers of alanine. *J. Am. Chem. Soc.*, **104**, 3329–3336.
3. Xiang, T., Goss, D.J., and Diem, M. (1993) Strategies for the computation of infrared CD and absorption spectra of biological molecules: ribonucleic acids. *Biophys. J.*, **65**, 1255–1261.
4. Tinoco, I. (1962) The exciton contribution to the optical rotation of polymers. *Radiat. Res.*, **20**, 133–139.
5. Birke, S.S. and Diem, M. (1995) Conformational studies of the smallest structural motifs of DNA detectable via vibrational circular dischroism: cytidylyl-(3′-5′)-guanosine and guanylyl-(3′-5′)-cytidine. *Biophys. J.*, **68**, 1045–1049.

6. Gulotta, M., Goss, D.J., and Diem, M. (1989) IR vibrational CD in model deoxyoligonucleotides: observation of the B-Z phase transition and extended coupled oscillator intensity calculations. *Biopolymers*, **28**, 2047–2058.
7. Silva, R.A.G.D. *et al.* (2000) Site-specific conformational determination in thermal unfolding studies of helical peptides using vibrational circular dichroism with isotopic substitution. *Proc. Natl. Acad. Sci. U.S.A.*, **97** (15), 8318–8323.
8. Brauner, J.W., Flach, C.R., and Mendelsohn, R. (2005) A quantitative reconstruction of the amide I contour in the IR spectra of globular proteins: from structure to spectrum. *J. Am. Chem. Soc.*, **127**, 100–109.
9. Colthup, N.B., Daly, L.H., and Wiberly, S.E. (1990) *Introduction to Infrared and Raman Spectroscopy*, Academic Press, New York.
10. Dollish, F.R., Fateley, W.G., and Bentley, F.F. (1974) *Characteristic Raman Frequencies of Organic Compounds*, Wiley-Interscience, New York.
11. Townes, C.H. and Schawlow, A.L. (1975) *Microwave Spectroscopy*, Dover Publications, New York.
12. Levine, I. (1970) *Quantum Chemistry*, vol. 1 and 2. Allyn & Bacon, Boston.
13. Kauzmann, W. (1957) *Quantum Chemistry*, Academic Press, New York.
14. Diem, M., Nafie, L.A., and Burow, D.F. (1978) Analysis of the gas phase infrared spectrum of bromochlorofluoromethane. *J. Mol. Spectrosc.*, **71**, 446–457.

6

Special Raman Methods: Resonance, Surface-Enhanced, and Nonlinear Raman Techniques

The advent of tunable, high-power, and ultra-short pulse lasers has spawned the development of new Raman techniques that have truly revolutionized the field of vibrational spectroscopy. Perhaps with the exception of FT-NMR techniques, using various pulse sequences to create multidimensional NMR processes, no other spectroscopic method has experienced such an explosive expansion during the past 30 years as has Raman spectroscopy.

In this chapter, several of these areas are introduced. The first of these techniques to be discussed, the resonance Raman effect, was experimentally verified in 1972 [1] and has had profound impact on the study of the structure and dynamics of biophysical systems. In resonance Raman spectroscopy, the Raman scattering cross section is enhanced by several orders of magnitude; the enhancement is due to interactions of vibrational and electronic excitation in the scattering process. Resonance Raman spectroscopy offers the enormous advantage that only groups on which the electronic transition is located will experience the resonance enhancement; thus, in addition to the enhanced intensities, this technique allows for spatial selection of diverse groups to be observed within macromolecules.

At about the same time, enormously enhanced Raman scattering from molecules adsorbed on electrochemically deposited silver surfaces was first reported. This effect, known as surface-enhanced Raman scattering (SERS) [2], has matured into a highly useful method in electrochemistry and nanotechnology, and has allowed the detection of single molecules *via* the "tip-enhanced" Raman scattering (TERS) [3] effect, which is a logical extension of SERS.

Several "nonlinear" Raman effects are discussed in this chapter, where the term "nonlinear" implies a dependence of the induced effects on the square or cube of the laser intensity. The first of the nonlinear Raman effects were reported even earlier than resonance Raman and SERS: both the hyper-Raman effect, a non-coherent three photon effect, and coherent anti-Stokes Raman scattering (CARS) (although not reported under this name) were observed at the Ford Motor Company Research laboratory in 1965 using a ruby laser for excitation [4]. This scientific achievement occurred a mere 5 years after the first experimental verification of a

Modern Vibrational Spectroscopy and Micro-Spectroscopy: Theory, Instrumentation and Biomedical Applications, First Edition.
Max Diem.

visible laser in 1960 (which was, incidentally, also a ruby laser). In the present times of financial hardship for any scientific endeavor, it may come as a surprise that this research was performed in the laboratory of private companies.

Subsequently, several other nonlinear Raman phenomena have been reported, some of which are introduced in this chapter. Finally, a short discussion on the laser technology that enables these nonlinear Raman techniques is presented.

6.1 Resonance Raman spectroscopy

It was pointed out before that resonance enhancement occurs in Raman spectroscopy if the photon energy of the exciting light is close to the energy of an electronic transition of the sample. In this case, one of the terms in the denominator of the polarizability expression (Eq. 4.10) becomes very small since $\omega_{0m}^2 \approx \omega^2$ and consequently, the corresponding transition component in the sum over all excited states becomes very large.

The basic theory of resonance Raman intensities is usually discussed in terms of the scattering tensor, rather than the polarizability, which has the form

$$(\alpha_{\alpha\beta})_{nm} = \frac{1}{\hbar}\sum\nolimits_r \left[\frac{\langle n|\,\mu_\alpha|r\rangle\langle r|\mu_\beta|m\rangle}{(\omega_{rm}-\omega+i\Gamma)} + \frac{\langle n|\mu_\alpha|r\rangle\langle r|\mu_\beta|m\rangle}{(\omega_{rm}+\omega+i\Gamma)}\right] \tag{6.1}$$

As before in Eq. 4.10, the subscripts α and β of the scattering tensor α refer to all permutations of the Cartesian coordinates x, y, and z. In Eq. 6.1, r is the intermediate (real or virtual) state, and each element of the scattering tensor is defined as the sum over *all* vibronic states of the molecule. The subscripts n and m denote the final and original states of the system.

The scattering tensor equation explicitly contains a damping term $i\Gamma$ in the denominator, which was neglected in the previous discussion of non-resonant Raman spectroscopy. This damping term physically is the lifetime of the intermediate state and prevents the denominator from becoming exactly zero at the resonance condition. In the "far-from-resonance" case [5], and if the molecule is initially in its ground state, Eq. 6.1 assumes the form of the polarizability tensor introduced in Eq. 4.10. Details on the discussion of the scattering tensor can be found in the literature (e.g., see [6, Sections 1.4 and 1.5]).

Within the Born–Oppenheimer approximation, the transition moments in the numerator of Eq. 6.1 can be separated into pure electronic transition moment M_e^α between states r and m, and the Franck–Condon overlap integrals between the vibrational wavefunctions:

$$\langle j|v\rangle\langle v|i\rangle \tag{6.2}$$

Here, states i and j represent the vibrational states of the ground electronic state and v is a vibrationally excited state of the resonant excited state [7]. Equation 6.2 thus represents how much the resonance excited state is displaced along the vibrational coordinate.

For the discussion of resonance enhancement, all states involved are written as the products of vibrational and electronic wavefunctions, and the dipole transition moments are evaluated separately for the purely electronic and vibrational wavefunctions. This allows the scattering tensor to be written as the sum of two terms, referred to as the A and B terms:

$$A = \frac{M_e^\alpha M_e^\beta}{\hbar}\sum\nolimits_v \frac{\langle j|v\rangle\langle v|i\rangle}{(\omega_{iv}-\omega+i\Gamma)} \tag{6.3}$$

$$B = \frac{M_e^\alpha}{\hbar}\left(\frac{\partial M_e^\alpha}{\partial Q}\right)\cdot\sum\nolimits_v \frac{\langle j|Q|v\rangle\langle v|i\rangle + \langle j|v\rangle\langle v|Q|i\rangle}{(\omega_{iv}-\omega+i\Gamma)} \tag{6.4}$$

The equation for the *A* term describes the resonance enhancement in totally symmetric modes, whereas the *B* term dominates when the vibrational modes mix the two excited electronic states. The resonance enhancement due to the *A* and *B* terms have different frequency responses, which determine the onset of resonance enhancement as the laser wavelength approaches, in energy, an electronic transition [8]. This aspect is particularly important when discussing the resonance Raman spectra of inherently symmetric moieties, such as the iron–porphyrin groups in heme proteins [9].

The enormous intensity enhancement in resonance Raman spectroscopy has revolutionized the vibrational spectroscopic field of biomolecules, such as proteins and nucleic acids, for two obvious reasons. First, molecules can be studied at concentrations of about 10^{-5} M, whereas in non-resonant standard Raman spectroscopy, the lower concentration limit is about 10 mM. This is highly important since the solubility of proteins in aqueous media is often in the micromolar range. Second, the ability to excite only transitions of interest provides the researcher with a method to focus on specific parts of the samples (*cf.* Figure 6.1).

Resonance Raman spectra were first reported [1] for proteins with chromophores that exhibit visible absorption spectra (hemoglobin, cytochrome C). In these proteins, excitation with visible laser sources provided an enormous Raman intensity enhancement in the vibrations of the heme group as shown in Figure 6.1. These vibrations were shown to exhibit sensitivity toward the oxidation state and coordination geometry of the central Fe atom, even allowing dynamic studies on the Fe–ligand association and dissociation kinetics [10].

In Figure 6.1, the strong band in the spectrum shown in panel (a), when excited in the Soret band (for carboxyhemoglobin, HbCO at about 420 nm), is a marker for the oxidation state of the iron atom, whereas the bands just below 1600 cm^{-1} are spin state markers. When the same molecule is excited with 230 nm laser light, at the wavelength in the UV–vis absorption spectrum (panel c) where the aromatic amino acids start to absorb, the observed resonance Raman changes drastically, as shown in panel (b) of Figure 6.1. This discussion demonstrates how the excitation wavelength affects the groups that are probed in resonance Raman spectroscopy.

Other resonance Raman studies focused on the intermediates and conversation dynamics in the light harvesting proteins such as bacteriorhodopsin [12] and similar molecules. Here again, the prosthetic group and its interaction with the protein bundles spanning the cellular membrane could be studied [13]. Since the light-produced intermediates have short life times, they most advantageously are studied by time-resolved techniques, which are introduced in Chapter 7.

When UV and deepUV laser sources became available, resonance Raman spectroscopy took another leap forward, since enhancement of aromatic amino acid residues as well as the protein backbone $\pi^* \leftarrow n$ and $\pi^* \leftarrow \pi$ electronic transitions became possible. This provided biophysicists with a new tool to study protein

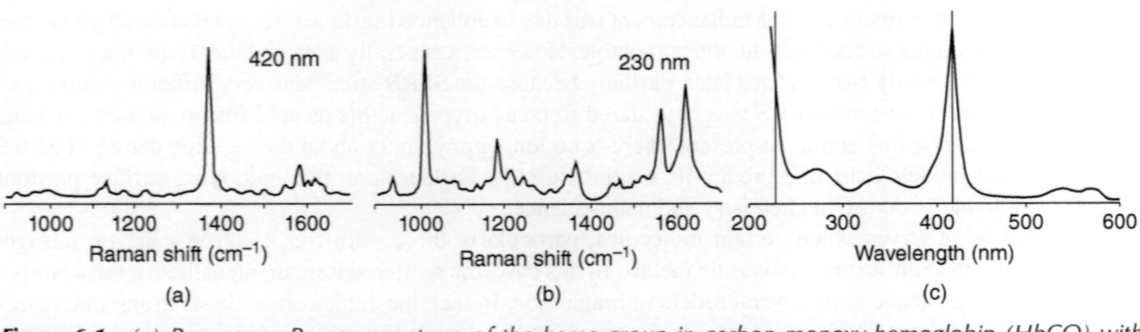

Figure 6.1 *(a) Resonance Raman spectrum of the heme group in carbon monoxy-hemoglobin (HbCO) with 420 nm excitation. (b) Resonance Raman spectrum of the HbCO aromatic amino acids (Tyr, Trp) with 230 nm excitation. (c) UV–vis absorption spectrum of HbCO. Adapted from Mukerji, 2012, [9] with permission from John Wiley & Sons*

conformation and dynamics [7, 14]. Similarly, the UV-absorptions of the bases in nucleic acids could also be excited with near UV-lasers, and resonance Raman studies of DNA and RNA were found to reveal many structural details (see Chapter 10). In a series of very elegant studies, Thomas [11] reported Raman spectra of viruses and bacteriophages excited by visible, near UV, and deep UV radiation, yielding the non-resonant Raman spectra of these biological entities, and information on the nucleic acid packing and conformation, and their protein features, all by just tuning the excitation wavelengths. The reader is referred to one of the many review articles and books on the biological applications of resonance Raman spectroscopy [8].

More recently, resonance Raman spectroscopy has been used to investigate the stacking interactions of hemoglobin in human erythrocytes (red blood cells). Since this experiment is carried out microscopically, it is discussed in more detail in Chapter 14. These results are particularly important for malaria research, since a cytotoxic form of hemoglobin (hemozoin) is secreted by the malaria parasite, and this form can be detected by its resonance Raman spectrum [15].

Resonance Raman spectra are observed *via* instrumentation very similar to those described in Chapter 4 for non-resonant Raman spectroscopy; however, laser sources that provide excitation wavelengths in the UV and deep UV spectral ranges are needed. Such laser sources add a level of instrumental complexity that is not encountered in non-resonant Raman spectroscopy, where commercial lasers are used routinely. Excitation wavelengths in the UV and deep UV spectral regions are accessible from near infrared (IR) and visible lasers by nonlinear processes such as frequency doubling, frequency quadrupling, and hydrogen shifting. Frequency doubling, the combination of two long wavelength photons into one photon of half the wavelength, or twice the frequency, occurs by a nonlinear process in certain crystalline materials; for example, frequency doubling of the 514.5 nm line of an Ar ion laser provides photons at 257.2 nm, whereas frequency quadrupling of the 1064 nm output of a Nd:YAG laser yields 266 nm photons. Hydrogen shifting is a nonlinear spectroscopic process (stimulated Raman spectroscopy, SRS, see Section 6.3.2) in which a high-power laser pulse, sent through a gas cell filled with molecular hydrogen, produces Raman-shifted light (shifted by 4157 cm^{-1}) that is emitted as a coherent laser signal. Using successive hydrogen shifters, a number of deep UV laser lines for the excitation of resonance Raman spectra can be obtained. These nonlinear methods to generate new excitation frequencies are introduced later in this chapter.

6.2 Surface-enhanced Raman scattering (SERS)

In the 1970s, a peculiar enhancement of Raman scattered intensities was observed when certain molecules were adsorbed at metal surfaces [2, 16]. Since the intensity enhancement could be as high as a million-fold, the early interpretation, namely, that the enhancement was due to enhanced surface area and increased population of the adsorbate on the surface, was no longer tenable. However, a generally accepted theory for the effect was not put forth until nearly two decades later, partially because the SERS effect was very difficult to reproduce and quantify. For some time, SERS was considered to be as irreproducible as cold fusion, and its existence was seriously doubted by some. At present, there is no longer any doubt about the genuine nature of SERS, and theoretical models exist that predict its magnitude [17]. Furthermore, the underlying surface plasmon effects have found wide use in chemistry and nano-science.

SERS can be observed when certain molecules, particularly those with free electron pairs on nitrogen atoms, are adsorbed on surfaces of certain metals. In this case, the scattered Raman intensities of the adsorbed molecules can be enhanced by several orders of magnitude. In fact, the enhancement is so strong that monolayers of molecules can be observed even in the presence of solvents. Typical examples for which SERS had first been reported were aqueous solutions of pyridine, or other amines, on either silver electrodes or silver sol particles. Other metals, notably noble metals, have shown similar surface enhancements, as have a number of metal oxides.

Since a number of different effects contribute to the surface enhancement, the actual mechanism of SERS was initially not fully understood. Now, it is widely accepted that there are two effects underlying enhancement: an increased electromagnetic field at the metal surface caused by "surface plasmons", and a "charge transfer"-type interaction between the metal orbitals and those of the adsorbate.

The physics behind the first of these effects can be visualized as follows: consider a metal sphere with a size much smaller than the wavelength of the light irradiating it. The electric field E_s at the surface of the particle, induced by the laser radiation field E_L, is given by

$$E_s = \frac{\varepsilon_1(\omega) - \varepsilon_2}{\varepsilon_1(\omega) + 2\varepsilon_2} E_L \tag{6.5}$$

where $\varepsilon_1(\omega)$ is the frequency-dependent dielectric function of the metal and ε_2 the dielectric permittivity of the surroundings (solvent). The dielectric function of a material is a complex quantity related to the refractive index and the absorption coefficient. ε_2, in general, is frequency dependent as well, but for the wavelength ranges that are relevant here (within the range of the plasmon resonance of the metal particle), it can be considered constant. Equation 6.5 becomes very large at the surface plasmon resonance condition

$$\varepsilon_1(\omega) = -2\varepsilon_2 \tag{6.6}$$

At this condition, the resulting surface plasmons may be visualized to be oscillations of the charges at the surface of the metal sphere, and can be viewed as collective electromagnetic oscillations obeying spherical harmonic equations. The surface plasmons contribute enormously to the local electric field a molecule experiences when bound to the surface of a metal particle, and is responsible for enhancement of the Raman signal by about four orders of magnitude. The roughness of the silver electrodes required for SERS to be observed can be visualized as contributing to the surface plasmons. The initially low reproducibility of SERS can be attributed directly to the difficulty of producing surfaces of equal roughness, but these problems have now been overcome by methods to be discussed below.

The "chemical" or "charge-transfer" mechanism for surface enhancement is discussed next. This mechanism was postulated since experimental results for N_2 and CO on silver surfaces showed very different SERS enhancement. Since these two molecules are isoelectronic and have very similar structural parameters, it was postulated that their binding to the metal surface is quite different. This binding can best be described as a charge transfer between the molecular orbitals of the adsorbate and the partially filled orbitals of the metal. This can lead to changes in the energies of non- or anti-bonding orbitals in the adsorbate, which, in turn, may change the polarizability of the molecule. For example, one may envision that lowering the orbital energies of the adsorbate could bring these energy levels into pre-resonance with the exciting radiation. This mechanism generally contributes less to the enhanced intensities than the plasmon enhancement, about two orders of magnitude. Both mechanisms can also contribute simultaneously, to produce enhancement factors on the order of 10^7 or larger.

Upon binding to the surface, the symmetry of the adsorbate is lowered, resulting in a different vibrational spectrum. Raman bands that are symmetry forbidden in the isolated adsorbate may be allowed in the complex with the surface. Different possible binding structures can often be distinguished from the SERS spectra, and it has been established that many nitrogen-containing molecules such as pyridine bind in such a manner that the ring is perpendicular to the surface. An example of the spectral changes occurring upon binding to a metal surface is shown in Figure 6.2, which depicts the spontaneous Raman and SERS of pyridine. Unfortunately, the spectra shown do not permit an assessment of the surface enhancement, but provide a view of the spectral changes due to the surface binding. This figure also demonstrates that silver nanoparticles, as found in a silver colloid, as well as electrochemically produced surfaces can be used for the observation of SERS. The surface enhancement also can be utilized to increase the Raman scattering efficiency enormously over the (generally

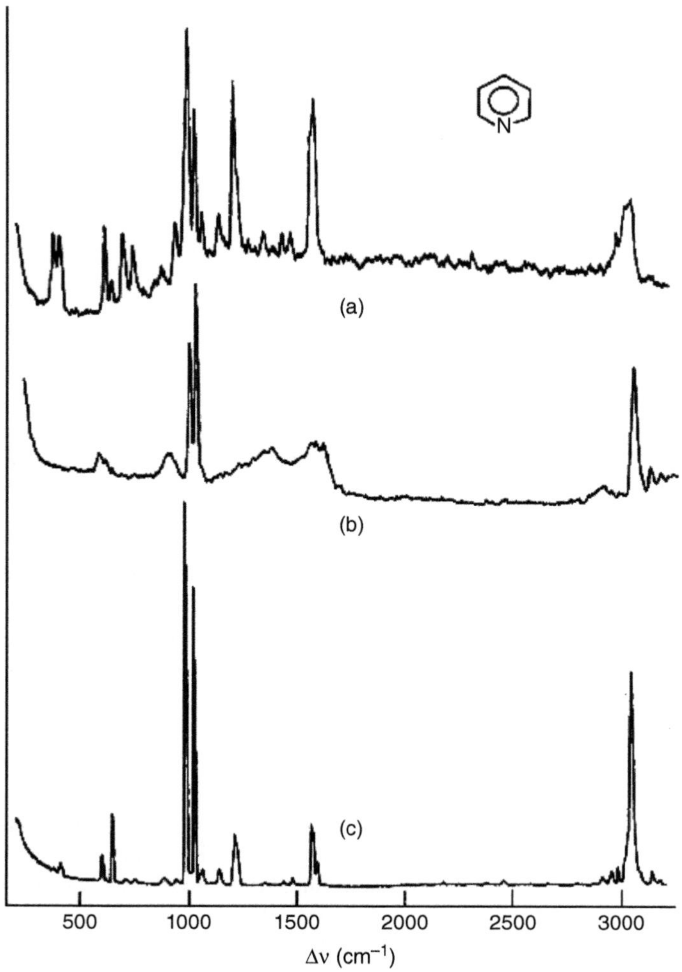

Figure 6.2 *(a) SERS spectrum of pyridine on a cold-deposited silver film and (b) on silver colloid in water. (c) Raman spectrum of liquid pyridine. Moskovits, 1985, [17]. Reproduced with permission from the American Physical Society*

unwanted) fluorescence process, which often accompanies Raman scattering. Highly fluorescent dyes have been studied by SERS on silver sol surfaces [18].

The surface plasmon-based theories of SERS enhancement were put forth upon observation of the dependence of the enhancement on particle size and metal optical properties. For example, it was well known that a silver sol actually appears colored due to the absorption of the light by the plasmon resonance effect, and that the wavelength of maximum absorbance depends on the solvent, the size of the metal spheres, and the nature of the metal. These studies paved the way for systematic studies of the Raman enhancement on particle size, morphology, and metal properties. This understanding, in turn, resulted in surfaces that produced reproducible SERS spectra. For example, nanostructures were prepared by nanolithography and subsequently coated by silver or other metals.

The fabrication of SERS-active surfaces was taken a step further in the efforts by Dluhy and coworkers [19] who deposited silver vapor at an oblique angle at specially primed surfaces to create silver nanorods with well-defined aspect ratios. These nanorods produced further enhancement of the Raman cross section in excess of eight orders of magnitude, allowing the detection of virus concentrations of as low as 100 PFU/mL, where PFU denotes "plague-forming unit," a measure used in virology to define active virus concentration.

Another potentially useful application of SERS is the design of "Raman nano-beacons." These consist of small metal spheres (20–200 nm in diameter), and a covalently attached molecular probe (dyes) whose Raman spectrum is enhanced by the proximity to the metal surface. The entire metal probe assembly may further be modified chemically (e.g., by enclosing it in a layer of silicate grown on the surface of the assembly) to render the probes chemically inert. Such a nanoassembly will exhibit a strong, SERS-enhanced signal of the probe. The nanoparticle can further be chemically modified to contain specific biomolecules that bind to specific receptor sites on cells. By designing different nanoparticles, for example, where one specific antibody is bound to nanoparticles with one dye, and another antibody is bound to nanoparticles with another dye, multiplexed binding assays can be carried out using SERS imaging as the detection method. Such results have been reported in the literature [20].

An ultimate SERS experiment, where the metal surface is the tip of a metal nanowire in contact with the sample, is known as TERS and is discussed in the Raman microscopic chapter (Chapter 11).

6.3 Nonlinear Raman effects

6.3.1 Spontaneous (incoherent) nonlinear Raman effects

With the advent of high-power pulsed lasers, several totally novel spectroscopic effects were discovered. These effects all have some principles in common with the classical Raman effect, but the scattered intensities of these effects depend nonlinearly on the intensity of the exciting electromagnetic radiation and are, therefore, referred to as "nonlinear Raman spectroscopies." The electric field in a pulsed laser can exceed 10^{10} V/m, which is about 100,000 times stronger than the field strength of a continuous wave (CW) laser commonly used to excite Raman spectra. At these high laser fields, the induced dipole moment can no longer be represented by Eq. 4.1

$$\boldsymbol{\mu}_{\text{ind}} = \overline{\alpha} \cdot \mathbf{E} \tag{4.1}$$

but needs to be rewritten as a series expansion to include higher order contributions to the induced electric dipole:

$$\boldsymbol{\mu}_{\text{ind}} = \overline{\alpha}\,\mathbf{E} + \frac{1}{2}\overline{\overline{\beta}}\mathbf{E}^2 + \cdots \tag{6.7}$$

where $\overline{\overline{\beta}}$ is known as the first hyper-polarizability tensor of rank 3 (a $3\times3\times3$ matrix). Equation 6.7 is often written in terms of macroscopic dielectric susceptibilities:

$$\mathbf{P} = \chi^{(1)}\mathbf{E} + \chi^{(2)}\mathbf{EE} + \chi^{(3)}\mathbf{EEE} + \cdots \tag{6.8}$$

where $\chi^{(n)}$ is a tensor of rank $n+1$, known as the dielectric susceptibility, which relates the induced electric macroscopic polarization **P** to the electric field **E** of the exciting radiation. In Eqs. 6.7 and 6.8, the first terms on the right-hand side describe the polarizability and macroscopic polarization, respectively, which are responsible for the effects discussed previously (Chapter 4). The second term in Eq. 6.7 explicitly contains the cause for hyper-Raman scattering, whereas the second and third terms in Eq. 6.8 account for nonlinear effects such as frequency doubling and nonlinear Raman effects such as CARS.

6.3.1.1 The Hyper-Raman effect

Hyper-Raman scattering results from the second term in Eq. 6.7, and is a nonlinear, non-coherent form of nonlinear Raman spectroscopy. Hyper-Raman and hyper-Rayleigh spectroscopies are three-photon processes depicted schematically in Figure 6.3(a). Two photons of frequency ω_1 (up arrows) create two virtual states, shown by the dashed lines. A hyper-Rayleigh photon at frequency $2\omega_1$ or a hyper-Raman photon at frequency $2\omega_1 - \omega_m$ is created from the upper virtual state, where ω_m is the frequency of a molecular vibrational quantum. This is shown by the downward arrows in Figure 6.3(a). The two laser photons need not necessarily have the same frequency: the hyper-polarizability can also mix photons of different frequencies, ω_1 and ω_2, resulting in what has been referred to as "non-degenerate" hyper-Rayleigh and hyper-Raman scattering [21] with frequencies $\omega_1 + \omega_2$ and $\omega_1 + \omega_2 - \omega_m$, respectively. This is shown in Figure 6.3(b). Just as the hyper-Rayleigh effect can be viewed as the incoherent form of frequency doubling, the non-degenerate hyper-Rayleigh effect is the non-coherent analog of sum-frequency generation (SFG).

In analogy of the Raman scattering tensor,

$$(\alpha_{\alpha\beta})_{nm} = \frac{1}{\hbar}\sum_r \left[\frac{\langle i|\,\mu_\alpha|r\rangle\,\langle r|\mu_\beta|f\rangle}{(\omega_r - \omega + i\Gamma)} + \frac{\langle i|\mu_\alpha|r\rangle\langle r|\mu_\beta|f\rangle}{(\omega_r + \omega + i\Gamma)}\right] \tag{6.9}$$

the hyper-polarizability for the initial and final, and intermediate states i, f, r, and s can be defined in terms of the transition moments M_{fr}, M_{rs}, and M_{si} (in the far-from resonance approximation) as

$$\beta_{\alpha\beta\gamma} = \frac{2}{\hbar^2}\sum_{r,s}\left[\frac{M_{fr}M_{rs}M_{si}}{(\omega_{si} - \omega - i\Gamma_{si})\,(\omega_{rf} + \omega - i\Gamma_{rf})} + \frac{M_{fr}M_{rs}M_{si}}{(\omega_{sf} + 2\omega - i\Gamma_{sf})(\omega_{rf} + \omega - i\Gamma_{rf})}\right.$$
$$\left. + \frac{M_{fs}M_{sr}M_{ri}}{(\omega_{si} - 2\,\omega - i\Gamma_{si})\,(\omega_{ri} - \omega - i\Gamma_{ri})}\right] \tag{6.10}$$

In Eq. 6.10, the subscripts f and i denote the final and initial states and s and r the two intermediate virtual states. Comparison of Eqs. 6.9 and 6.10 suggests that the selection rules should be different for Raman and

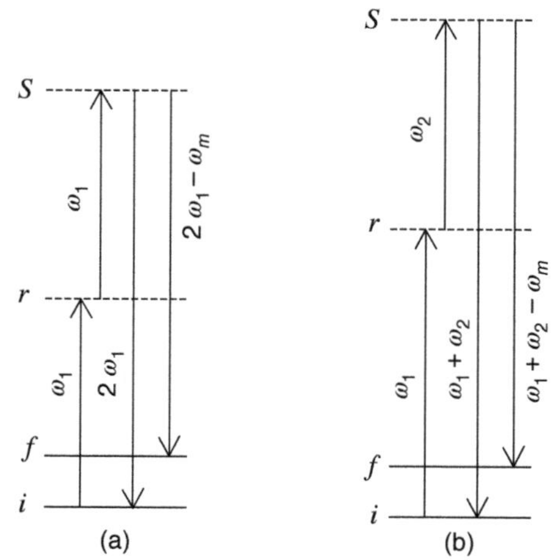

Figure 6.3 Schematic energy level diagram for degenerate (a) and non-degenerate and (b) hyper-Rayleigh and hyper-Raman scattering. See text for details

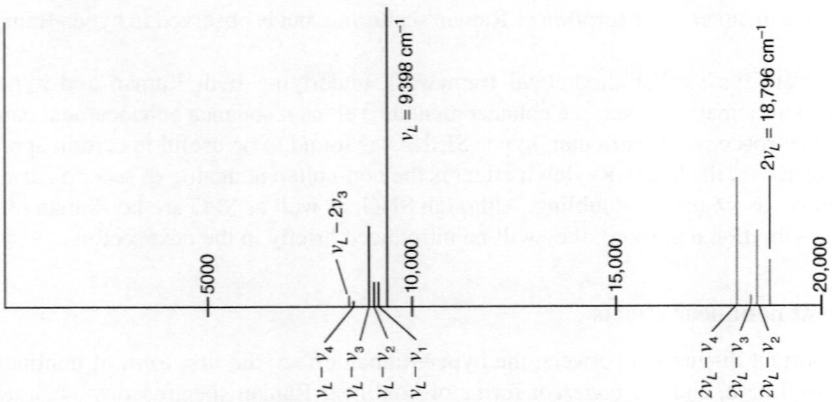

Figure 6.4 *Schematic Raman and hyper-Raman spectra of liquid CCl_4, excited at 1.064 μm. See text for details.* (See insert for color representation of this figure.)

hyper-Raman spectroscopy since two or three photons are involved in these processes, respectively. In fact, low symmetry vibrations often are allowed, and are more intense in hyper-Raman spectroscopy, whereas totally symmetric vibrations may be very weak or entirely forbidden. Owing to the different form of the hyper-polarizability tensor, as compared to the polarizability, the limiting values of the hyper-Raman depolarization ratios differ from those encountered in Raman spectroscopy and vary from 1/9 (polarized band) to 2/3 (depolarized band) for linearly polarized incident light and right angle scattering geometry.

A schematic hyper-Raman spectrum is shown in Figure 6.4. This figure emphasizes the different intensity patterns observed for Raman and hyper-Raman spectral features of the same molecule, CCl_4. This hypothetical spectrum is drawn for excitation with an Nd:YAG laser excitation at 1064 nm or 9398 cm^{-1}. The Raman peaks (see Figure 2.13) are drawn schematically (black traces), displaced from the Rayleigh line at 9398 cm^{-1} by the vibrational frequencies of 215, 314, 459, and 790 cm^{-1}, *cf.* Table 2.1, as discussed earlier. Although the wavenumber axis is drawn to scale, the intensity axis is not, since the Rayleigh line at 9398 cm^{-1}, in reality, may be a million-fold more intense than the (Stokes) Raman spectrum. The anti-Stokes Raman bands are not shown.

In order to emphasize the differences between overtones ($\nu_L - 2\nu_m$, also referred to as "harmonics"), and hyper-Raman bands, the position of the first overtone of the symmetric stretching mode (at about 900 cm^{-1}) is shown as the lowest wavenumber line in Figure 6.4.

The corresponding hyper-Rayleigh and hyper-Raman bands are shown in gray. Again, the intensities of the hyper-Raman spectra are not drawn to scale with respect to the Raman spectra. However, it is true that the hyper-Rayleigh line (the "harmonic" of the laser frequency) is far less intense in comparison with the hyper-Raman peaks, and, in fact, may be symmetry forbidden in certain molecules. Also as pointed out before, the selection rules and relative intensities for hyper-Raman spectra are quite different from those of Raman spectra: in hyper-Raman, as in the IR spectrum, the antisymmetric stretching mode of CCl_4, ν_4 at 790 cm^{-1} is by far the most prominent peak, whereas the symmetric stretching mode ν_3 at 459 cm^{-1} is very strong in Raman, but very weak in hyper-Raman spectroscopy. The doubly degenerate ν_1 mode is forbidden in hyper-Raman spectroscopy [21].

Since the selection rules are all different for Raman, hyper-Raman, and IR spectroscopy, the hyper-Raman effect complements the two other forms of vibrational techniques in the sense that it allows the observation of vibrational modes that cannot be observed in Raman or IR spectroscopy. The torsional vibration in tetrachloroethene, C_2Cl_4, for example, which transforms as the A_u irreducible representation of the D_{2h} point group, is not active in either IR absorption or Raman scattering, but is observed in hyper-Raman spectroscopy at $110\,\text{cm}^{-1}$.

As expected from the similar theoretical framework underlying both Raman and hyper-Raman spectroscopy, one can anticipate that surface enhancement as well as resonance enhancement can be observed in hyper-Raman spectroscopy. In particular, hyper-SERS was found to be useful in certain applications [22].

As pointed out above, the hyper-Rayleigh effect is the non-coherent analog of second harmonic generation (SHG), also known as frequency doubling. Although SHG, as well as SFG are no Raman effects (since they do not involve a vibrational photon), they will be introduced briefly in the next section.

6.3.2 Coherent nonlinear effects

There is an important distinction between the hyper-Raman effect, the first form of nonlinear Raman spectroscopy discussed here, and the *coherent* forms of nonlinear Raman spectroscopy, such as the stimulated Raman, inverse Raman, Raman gain, and coherent anti-Stokes Raman effects.

In the former technique, the hyper-Raman scattered photons are scattered non-coherently (spontaneously) into all 4π steradians, just as in the case of spontaneous Raman scattering. Also, in both Raman and hyper-Raman effects, all normal modes are excited simultaneously when a molecule is exposed to the exciting laser radiation. In the coherent, nonlinear Raman effects, on the other hand, the scattered light exits the sample as a coherent beam with the properties of laser light. Since some of the nonlinear Raman techniques require two input laser beams, only one normal mode, determined by the frequency difference between the two laser beams, is excited, leading to a much higher scattering efficiency. Thus, some of the nonlinear effects are very strong (albeit still not easily observed), whereas the incoherent hyper-Raman effect is so weak that it is still not a particularly practical technique.

In the remainder of this section, some of the theory underlying the coherent nonlinear techniques is introduced. The equation most commonly used as a starting point for the discussion of nonlinear effects is Eq. 6.8

$$\mathbf{P} = \chi^{(1)}\mathbf{E} + \chi^{(2)}\mathbf{EE} + \chi^{(3)}\mathbf{EEE} + \cdots \tag{6.8}$$

which was introduced before.

6.3.2.1 Second harmonic generation

A very commonly nonlinear optical effect, albeit not a nonlinear Raman process, namely coherent frequency doubling, is discussed first. This process, which is also known as second harmonic generation (SHG), is mediated by the second-order dielectric susceptibility $\chi^{(2)}$. Two photons of angular frequency ω are combined in a non-centrosymmetric crystal into one photon with angular frequency 2ω (where $\omega = 2\pi\nu$). Since the molecular system involved in this frequency-doubling process is left in the original state after the new photon is created, the momenta and energies of the photons involved must be conserved. The momentum conservation is indicated by writing the electromagnetic fields as

$$\mathbf{E}(z, t) = E_0 e^{i(kz-\omega t)} \tag{6.11}$$

where $\boldsymbol{k}$ is the wave (or momentum) vector of the photon, defined by

$$\boldsymbol{k} = \frac{2\pi}{\lambda} \tag{6.12}$$

and

$$p = \frac{h}{\lambda} = \hbar k \tag{6.13}$$

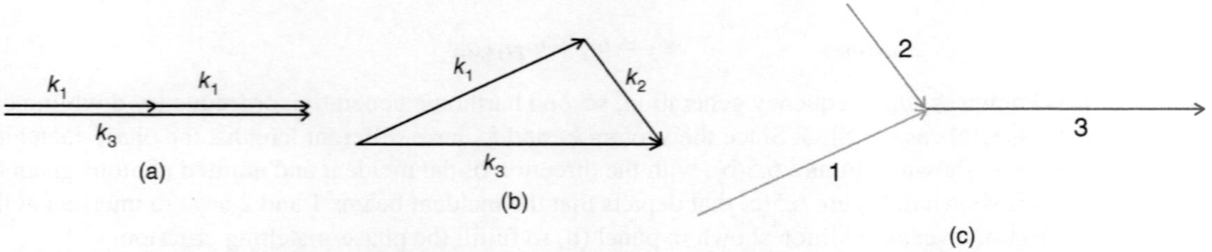

Figure 6.5 *(a) Phase-matching diagram for frequency doubling (SHG). (b) Phase-matching diagram for sum-frequency generation (SFG). (c) Directions of the incident beams 1 and 2 to create beam 3 in SFG. (*See insert for color representation of this figure.*)*

The conservation of momenta requires that $\mathbf{k}_3$, the momentum of the frequency-doubled photon, is given by

$$\mathbf{k}_3 = \mathbf{k}_1 + \mathbf{k}_1 \qquad \mathbf{k}_1 + \mathbf{k}_1 - \mathbf{k}_3 = 0 \tag{6.14}$$

where $\mathbf{k}_1$ and $\mathbf{k}_1$ are the momenta of the original photons. Equation 6.14 is often referred to as the phase-matching condition in nonlinear optics. Figure 6.5(a) depicts the phase-matching condition for frequency doubling. Here, Eq. 6.14 and Figure 6.5(a) predict that the frequency-doubled photon emerges from the (nonlinear) crystal material collinearly with the incident beam.

Typical nonlinear materials used for SHG are $LiIO_3$, $KNbO_3$, $LiNbO_3$, KH_2PO_4 (KHP), KD_2PO_4 (KDP), LiB_3O_5 (LBO), β-BaB_2O_4 (BBO), GaSe, $KTiOPO_4$ (KTP), and $(NH_4)H_2PO_4$ (ADP), where the commonly used (engineering) abbreviations, given in parentheses, strike horror into the hearts of chemists. Frequency doubling has become such a commonplace technique that green laser pointers, available for under $40, contain a frequency-doubled diode-pumped solid state laser, which gives a nice green spot on a reflective screen (see Figure 6.6).

When the photons incident on the nonlinear crystals do not have the same frequency, they still can interact and combine to form a new photon according to

$$\mathbf{k}_3 = \mathbf{k}_1 + \mathbf{k}_2 \qquad \mathbf{k}_1 + \mathbf{k}_2 - \mathbf{k}_3 = 0 \tag{6.15}$$

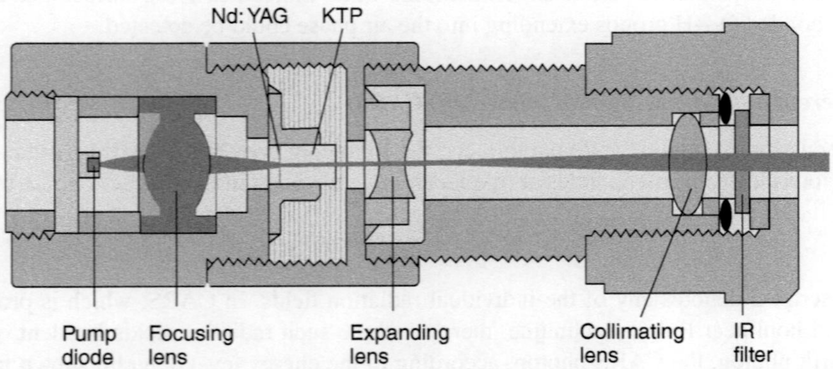

Figure 6.6 *Schematic of a (green) laser pointer, using frequency doubling. Light is provided by a high-power diode laser, which pumps a Nd:YAG laser (1064 nm). The output of this laser is frequency doubled in a potassium titanyl phosphate nonlinear crystal providing green light at 532 nm. Modified from en.wikipedia.org/wiki/Laser_pointer. Licenced under the terms of Creative Commons 3.0 CC-BY-SA licence. (See insert for color representation of this figure.)*

and

$$\omega_3 = \omega_1 + \omega_2 \tag{6.16}$$

This process is known as sum-frequency generation; second harmonic generation or frequency doubling can be considered a special case of SFG. Since the vectors $\boldsymbol{k}_1$ and $\boldsymbol{k}_2$ have different lengths, the phase-matching condition appears as shown in Figure 6.5(b), with the direction of the incident and emitted photons given by the arrows. This is shown in Figure 6.5(c) that depicts that the incident beams 1 and 2 have to intersect at the angle derived from the vector addition shown in panel (b) to fulfill the phase-matching criterion.

Since the directions of the incident and emitted photons are different and the refractive indices within the nonlinear crystal may differ along the different directions, Eq. 6.14 should be written as

$$k_3 = n_1 k_1 + n_2 k_2 \tag{6.17}$$

To account for the phase-matching conditions of nonlinear processes, the wave vector notation of the incident and emitted fields is included in the equations for the induced polarization. For SFG, for example, the second-order susceptibility term is written to include the phase-matching condition as

$$\chi^{(2)}\mathbf{E}(\omega_1)\mathbf{E}(\omega_1)\ e^{i(\boldsymbol{k}_1+\boldsymbol{k}_2-\boldsymbol{k}_3)z} \tag{6.18}$$

Here, the sign convention in the exponent indicates an emitted photon at ω_3 and incident photons at ω_1 and ω_2.

SFG of narrow visible laser pulses and broad IR pulses on interfaces was used in a set of elegant and innovative experiments to probe molecular properties at the boundary between water and air [23]. Although not a Raman experiment *per se*, this work exemplifies the power of nonlinear spectroscopy in general. Here, the SFG signal, I_{SFG}, is proportional to

$$I_{\mathrm{SFG}} \propto |\chi^{(2)}|^2 I_{\mathrm{vis}} I_{\mathrm{IR}} \tag{6.19}$$

where I_{vis} and I_{IR} are the intensities of the visible and IR laser pulses. If the frequency of the incident IR radiation is resonant with a vibrational mode, the SFG signal is strongly enhanced. In this way, the water molecules at an air–water interface could be probed without the interference of bulk water molecules, since the second-order nonlinear susceptibility is non-zero only for non-centrosymmetric systems. An interface is, of course, non-centrosymmetric. In this experiment, the water molecules at the surface that had "dangling," non-hydrogen-bonded O—H groups extending into the air phase could be detected.

6.3.2.2 Coherent anti-Stokes Raman scattering (CARS)

For further discussion of nonlinear Raman effects, the following conventions will be used. The medium is exposed to various electromagnetic fields at frequencies ω_a, traveling in the z-direction. Such a field is represented by

$$\mathbf{E}(\omega_a) = E_0 e^{i(k_a z - \omega_a t)} \tag{6.20}$$

where the subscript a denotes any of the individual radiation fields. In CARS, which is probably the most commonly used nonlinear Raman technique, there are three such radiation fields incident on the molecule to create a fourth photon, the CARS photon, according to the energy level diagram shown in Figure 6.7(a). In the following paragraphs, the four events necessary for the creation of a coherent anti-Stokes photon are described as if they occurred consecutively, although in reality, the four processes do not occur as separate events but are a four-wave mixing phenomenon mediated by the third-order nonlinear susceptibility $\chi_{\alpha\beta\gamma\delta}$.

In CARS, the sample is illuminated by two lasers, one of them with a fixed wavelength usually referred to as the pump laser ω_P or ω_1, and a second tunable laser, referred to as the Stokes frequency ω_S or ω_2.

A photon $\hbar\omega_P$ at the pump frequency promotes the system into a virtual state, shown by the lower dashed line in Figure 6.5(a). A photon $\hbar\omega_S$ from the laser at the (Stokes) Raman frequency causes the system to populate the vibrationally excited state, shown by the upper solid line. The vibrationally excited state in turn interacts with a second pump photon, $\hbar\omega_P$, to populate another virtual state that undergoes a transition back to the ground state. The energy released in this last step is carried off by a photon of frequency

$$\hbar\omega_{\text{AS}} = \hbar(2\omega_P - \omega_S) = \hbar(\omega_P + \omega_M) \tag{6.21}$$

where $\hbar\omega_M$ is the energy of one of the molecule's vibrational modes. Thus, the wavelength of the emitted photon is that of an anti-Stokes Raman process, and the emission of the anti-Stokes photon occurs only if the wavelength of the tunable Stokes laser fulfills the condition

$$\hbar\omega_S = \hbar(\omega_P - \omega_M) \tag{6.22}$$

For this process, in analogy to the discussion of SHG (Eq. 6.18), the term responsible for the CARS process can be written as

$$P_i^{(3)} = \sum\nolimits_{j,k,l} \chi_{i,j,k,l}^{(3)}\ \mathbf{E}(\omega_p)\mathbf{E}(\omega_p)\mathbf{E}(\omega_s)\ e^{i(k_p - k_s + k_p - k_{\text{AS}})z} \tag{6.23}$$

where the sign associated with each term in the exponent indicates whether a photon is annihilated or created. This exponential expression contains the four-wave vectors of the interacting electromagnetic fields. Since the molecular system is left in the original state after the creation of the CARS photon, the wave vectors need to add up to zero (see preceeding text). This leads to the phase-matching condition for CARS, which can be written as

$$\boldsymbol{k}_p - \boldsymbol{k}_s + \boldsymbol{k}_p - \boldsymbol{k}_{\text{AS}} = 0 \ \text{ or } \ \boldsymbol{k}_{\text{AS}} = \boldsymbol{k}_p - \boldsymbol{k}_s + \boldsymbol{k}_p \tag{6.24}$$

and visualized in Figure 6.7(b). It implies that the pump and Stokes beams must intersect at an angle given by the vector addition in Figure 6.7(b) for CARS photons to be generated.

The CARS intensity scattered at ω_{AS} is given by

$$I_{\text{AS}} = \left(\frac{4\pi^2\omega_{\text{AS}}}{c^2}\right)^2 I_P^2\, I_S\, |\chi|^2\, z^2 \tag{6.25}$$

where all symbols have their usual meaning and z is the distance over which phase matching is valid.

At this point, it is appropriate to investigate the form of the third-order susceptibility tensor of rank 4, used in Eqs. 6.23 and 6.25. This tensor has 81 ($=3^4$) elements, of which only 21 are non-zero in isotropic media.

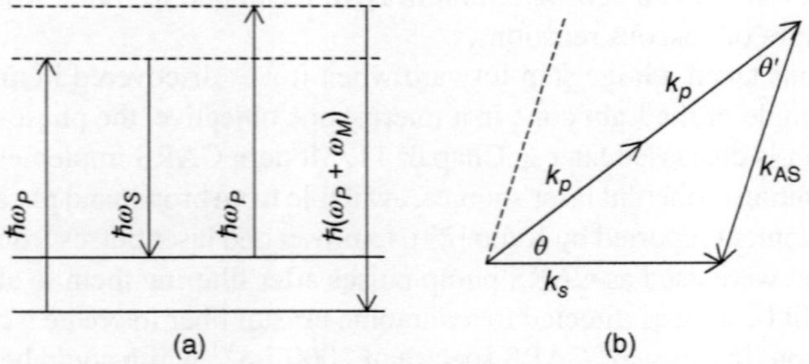

Figure 6.7 (a) Schematic energy level diagram for the CARS process. (b) Phase-matching condition. θ denotes the angle between pump and Stokes beams. θ′ is the angle between the pump and the CARS beam that is emitted along the dotted line

In fact, only tensor elements for which all four indices are the same (e.g., χ_{xxxx}) and those for which there are two pairs of identical indices (e.g., χ_{xxyy}, χ_{xyyx}, or χ_{xyxy}) are non-zero.

The 21 non-zero elements exhibit only four different numeric values, commonly referred to as χ_{1111}, χ_{1122}, χ_{1221}, or χ_{1212}. With that, Eq. 6.23 can be rewritten as

$$P_x^{(3)}(\omega_{AS}) = D[\chi_{1111} + \chi_{1122} + \chi_{1221} + \chi_{1212}]\ \mathbf{E}(\omega_p)\mathbf{E}(\omega_p)\mathbf{E}(\omega_s)\ e^{i(\boldsymbol{k}_p - \boldsymbol{k}_s + \boldsymbol{k}_p - \boldsymbol{k}_{AS})z} \tag{6.26}$$

where D is an integer factor between 1 and 6 that indicates how often each susceptibility term must be counted [24]. This equation assumes that the fields are polarized along the x-axis.

In the discussion so far, it was assumed that two photons of frequency ω_P from the pump laser interact with the sample (see Eq. 6.26). This need not necessarily be the case, and three different frequencies can interact within the sample. In this case, the anti-Stokes photon has a frequency of

$$\hbar\omega_{AS} = \hbar(\omega_1 + \omega_2 - \omega_S) \tag{6.27}$$

In both cases discussed so far (two identical or non-identical pump photons), the anti-Stokes photons leave the sample as a collimated, coherent laser beam, for which the spectral resolution is given by the line width of the exciting lasers. As pointed out before, spectral information on only one normal mode at a time is obtained if the Stokes laser is scanned to cover the spectral range.

Macroscopic CARS measurements are difficult to carry out, since they require careful matching of the wave vectors of two laser beams. Originally, CARS was carried out only at one selected wavelength; however, in order to observe broadband CARS signals, Carreira *et al.*, [25] developed a fully computerized CARS spectrometer, consisting of two nitrogen laser pumped tunable dye lasers, with a wavelength range from 260 to 740 nm. Both lasers, as well as the detector, were mounted on stages that could be rotated under computer control about the point where the pump and Stokes laser beams intersect, and from which the CARS beam emerges. During a CARS experiment, the wavelength of one of the dye lasers was maintained at a fixed value, whereas the other laser was scanned to fulfill Eq. 6.21. This instrument allowed CARS spectra of liquid samples to be collected over a range of many hundred wavenumbers. This turned out to be particularly difficult due to the dispersion of the refractive index, which makes phase matching more difficult (see Eq. 6.17).

CARS scattering from gases is somewhat easier to observe, since the refractive index does vary more subtly with wavelength in gases than in liquids. The enormous sensitivity of CARS makes it an ideal tool to investigate very low concentrations of gaseous samples. Rotational and rotational–vibrational spectra, with sub-wavenumber resolution, have been collected at sample pressures of a few millibars. Combustion processes, within the exhaust system of combustion and jet engines, or in flames, have successfully been studied *via* CARS spectroscopy, which is a sensitive, noninvasive technique to monitor products, reaction mechanisms, and the temperature of gaseous reactions.

CARS spectroscopy has taken a huge step forward when it was discovered in microscopic measurements [26]. In the large solid angle of the light cone in a microscope objective, the phase-matching angle is always fulfilled [27]. This aspect is discussed later in Chapter 11. Modern CARS implementations also have utilized broadband (super-continuum) coherent laser sources, available from broadband photonic crystal fibers [28]. In an elegant optical arrangement reported by Kano [29], femtosecond laser pulses from a Ti:Sapphire laser were split to obtain pulses that were used as CARS pump pulses after filtering them to about 20 cm^{-1} band width. The other part of the split beam was directed to a photonic crystal fiber to create a coherent super-continuum pulse for Stokes excitation. In this way, CARS spectra of 2000 cm^{-1} width could be collected simultaneously in a microscope set-up (*cf.* Figure 6.8).

However, the line shapes observed in CARS are usually distorted by dispersive line shape elements. This is because the third-order susceptibility tensor elements are composed of resonant and non-resonant terms.

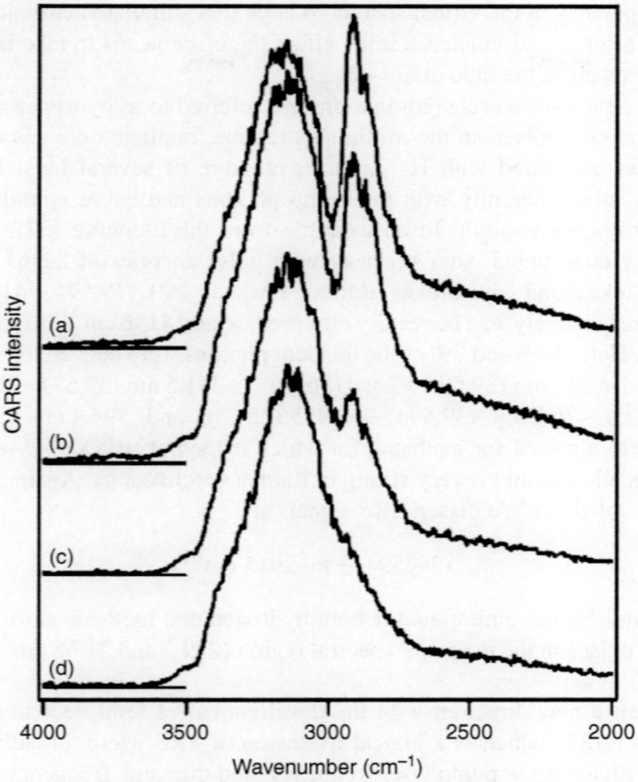

Figure 6.8 Broadband micro-CARS spectra of cellular components: (a) nucleolus, (b) chromosome, (c) cell membrane, and (d) background. Kano, 2008, [29]. Reproduced with permission from John Wiley & Sons

Depending on which of these terms contribute most, Lorentzian or dispersive line shapes may be obtained. The correction of these distorted lines has been reported in the literature [30] but is beyond the scope of this discussion. Furthermore, the signal in CARS is, in general, not proportional to the concentration of the sample, and mathematical procedures have been developed to extract a part of the CARS signal that is directly comparable to the spontaneous Raman effect and that is proportional to sample concentration. To this end, the imaginary part of the resonant $\chi^{(3)}$ contribution has to be separated from the non-resonant parts by determining the phase relationship between the two parts. Details have been discussed in the literature [31].

6.3.2.3 *Stimulated Raman scattering (SRS) and femtosecond stimulated Raman scattering (FSRS)*

Another nonlinear Raman effect due to the third-order susceptibility is SRS. In stimulated Raman scattering, the sample is illuminated with only one laser at the frequency ω_P. If the intensity of this pump laser increases past a threshold level, the intensity of the Stokes Raman scattering becomes sufficiently large that nonlinear mixing of the Raman radiation field with that of the pump laser causes coherent laser output at ω_S to occur, where $\hbar\omega_S = \hbar(\omega_P - \omega_M)$ (see Eq. 6.22). This effect can convert up to 50% of the incident photons into Raman scattering, compared to an efficiency of spontaneous Raman spectroscopy on the order of 10^{-10}–10^{-12}.

Since the effect that generates the Stokes laser beam is a Raman process, the stimulated Raman process is also known as a Raman laser. Conceptually, one may visualize the stimulated Raman effect as follows: the number of photons temporarily in the virtual state is so large that stimulated emission from this state occurs. The theoretical treatment of the stimulated Raman effect therefore needs to take into account the radiation field of the laser at ω_p as well as the field at $\omega_p - \omega_m$.

SRS is used to create new laser wavelength in a process referred to as hydrogen shifting (*cf.* Section 6.1). To this end, high-power laser pulses, in the millijoules regime, impinge on a gas cell, typically fitted with Brewster angle windows and filled with H_2 gas to a pressure of several bars. In this arrangement, the Raman shifted photons mix coherently with the pump photons and cause stimulated Raman emission at the Stokes and anti-Stokes wavelength. In an example from the literature [32], 266 nm (=39,594 cm^{-1}) pulses from a frequency-quadrupled Nd:YAG laser with pulse energies of 22 mJ were used to illuminate a hydrogen sample. Stokes and anti-Stokes Raman lines at 299 (39,594 − 4156 cm^{-1}) and 239.5 nm (39,594 + 4156 cm^{-1}), respectively, can be readily observed, where 4156 cm^{-1} is the vibrational frequency of the H_2 stretching mode. Here, 14% and 7% of the incident photons were converted by the SRS effect. Further SRS lines were observed at 217 nm (39,594 + 2 × 4156 cm^{-1}), 341.5 nm (39,594 − 2 × 4156 cm^{-1}), 397.9 nm (39,594 − 3 × 4156 cm^{-1}), 476.7 nm (39,594 − 4 × 4156 cm^{-1}), and 594.4 nm (39,594 − 5 × 4156 cm^{-1}). Similar experiments were reported for methane, for which the symmetric C—H stretching mode occurs at 2915 cm^{-1}; this mode is allowed and is very strong in Raman spectroscopy. Again, starting with laser pulses from a frequency-quadrupled Nd:YAG laser, SRS signals at

$$(39,594\ \pm n \cdot 2915\ \text{cm}^{-1}) \tag{6.28}$$

can be observed. The anti-Stokes emissions for both hydrogen and methane shifters are technically useful methods to create laser pulses in the deep UV spectral region (239.5 and 217.8 nm, and 246.9 and 230.3 nm, respectively).

SRS also has taken on a new direction with the development of femtosecond stimulated Raman spectroscopy, referred to as FSRS, which is a logical extension of SRS. Here, broadband femtosecond pulses are mixed coherently with a narrow pump laser frequency, and the same frequency mixing described before for SRS takes place. The major difference in FSRS is that all frequencies contained in the broadband pulse simultaneously can mix with the pump pulse; thus, the entire Raman spectrum can be probed at once. This is shown schematically in Figure 6.9, taken from [33]. This figure depicts the narrow Raman pump pulse (light grey) and the broadband femtosecond probe pulse (dark grey). Typically, this pulse is about 20 fs long and has a natural line width of about 1600 cm^{-1}. In the presence of a sample that exhibits allowed Raman transitions, some of the pump photons are transferred into the probe beam at the frequencies of the Raman modes. Ratioing the probe beam profile collected with and without the pump pulse yields the desired Raman spectrum, shown in the blue trace on top of Figure 6.9. This spectrum corresponds to a single laser pulse data acquisition (about 20 fs).

This time resolution is about the same as the low wavenumber vibrational frequency (a vibration at 500 cm^{-1} has a vibrational period of 60 fs); thus, FSRS can probe the time evolution of a vibrational mode and vibrational dephasing. This time scale is of significant interest, since the corresponding low frequency vibrations sample nuclear motion along reaction coordinates, which may be described in terms of the normal modes of the molecular systems during chemical reactions. Thus, FSRS offers very fast access to assess molecular reaction dynamics, and an entire branch of Raman spectroscopy has evolved around these concepts, and interested readers are referred to review articles by the Mathies group [33]. A number of items from these reviews are summarized below: first, the FSRS signal is phase matched according to

$$k_{\text{FSRS}} = -k_{\text{Raman}} + k_{\text{probe}} + k_{\text{Raman}} \tag{6.29}$$

which implies that the FSRS photons are emitted collinearly with the probe pulse. Second, the FSRS appears to violate the Heisenberg uncertainty principle in that the time–frequency product $\Delta\nu\ \Delta t$ is about an order of

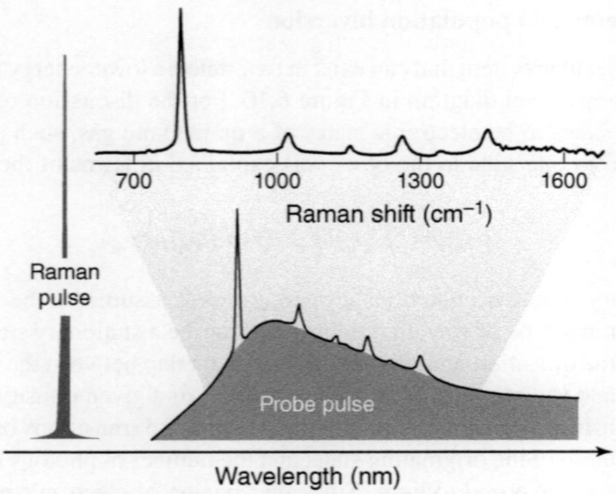

Figure 6.9 *Schematic diagram of FSRS. See text for details.* Lee et al., *2004, [33]. Reproduced with permission from AIP Publishing LLC*

magnitude better in FSRS than expected from the uncertainty principle that predicts [33]

$$\Delta\nu\,\Delta t \approx 5000\ \text{cm}^{-1}\ \text{fs} \tag{6.30}$$

This may be understood in terms of a disentanglement of energy and time resolution due to the fact that the broadband femtosecond pulse provides a molecular polarization with extremely high time resolution, whereas the non-time resolved detection of the FSRS photon provides independent and very high wavelength (frequency) resolution. Third, FSRS spectra appear identical, or at least very similar, to the spontaneous Raman spectra and are devoid of the line shape distortions observed in CARS. Furthermore, the spectra are linear in the concentration of the chemical to be analyzed.

The situation depicted in Figure 6.9 often is referred to as the "stimulated Raman gain" effect, since the (Stokes) Raman spectrum enhances the probe pulse at Raman resonance frequencies. Conversely, the anti-Stokes scattering causes a loss of intensity of the probe pulse and is consequently referred to as "stimulated Raman loss" spectroscopy. The research group of Xie at Harvard University has exploited microscopically observed FSRS for a number of pioneering spectroscopic studies of biological and biomedical interest [34, 35].

6.4 Continuous wave and pulsed lasers

In this section, a short introduction into the operation of continuous and pulsed lasers is presented, since laser technology is intimately linked to the progress in Raman and nonlinear Raman techniques. In fact, it is safe to say that without lasers, Raman spectroscopy would have never developed into the powerful spectroscopic method it is now, and nonlinear Raman effects probably could not have been detected at all. The discussion here starts with a very brief review of general laser operation, and then presents the ideas of pulsed laser operation.

Among the many laser types that now exist – gas lasers, solid state lasers, diode lasers, fiber lasers, and so on, gas lasers are used to elaborate upon the principles of laser operation since the principles here are most easily understood.

6.4.1 Einstein coefficients and population inversion

Consider an atomic or molecular system that can exist in two states, a lower energy ground state and an excited state, as shown by the energy level diagram in Figure 6.10. For the discussion of laser action, it is perhaps easiest to visualize these states to be electronic states of a monatomic gas, such as neon. In the discussions in Chapter 1, transition from one state to the other was explained in terms of the transition moment $\langle \boldsymbol{\mu} \rangle_{nm}$, defined by

$$\langle \boldsymbol{\mu} \rangle_{nm} = \int \psi_n \mu \psi_m \, d\tau = \langle \psi_n | \mu | \psi_m \rangle \tag{1.93}$$

In Chapter 1, the stationary state wavefunctions ψ_n and ψ_m were assumed to be vibrational wavefunctions, but the principle holds for any type of wavefunctions that describe a stationary state, for example, electronic energy levels. As before, the transition moment describes the overlap between the two wavefunctions and the dipole moment operator, and is a measure of the "allowedness" of a given transition.

However, the dipole transition moment does not express the rate of transitions between the two states. This rate depends on the population of the originating state, and the number of photons impinging onto the system, in the case here, the neon gas exposed to light. Since the spacing of electronic energy levels such as those found in gaseous neon are large as compared to room temperature, one may assume, for simplicity, that by far the most populated state is the ground state $|\psi_0\rangle$. The rate of absorption, that is, the number of photons being absorbed by the gaseous molecules undergoing a transition from the ground state to the excited state $|\psi_k\rangle$ per unit time, is given by the expression

$$n_0 B_{0k} \rho(\nu_{0k}) \tag{6.31}$$

Here, n_0 denotes the population of the ground state and B_{0k} is the Einstein coefficient for absorption, given by

$$B_{0k} = \frac{2\pi}{3\hbar^2} \, |\langle \psi_k | \mu | \psi_0 \rangle|^2 \tag{6.32}$$

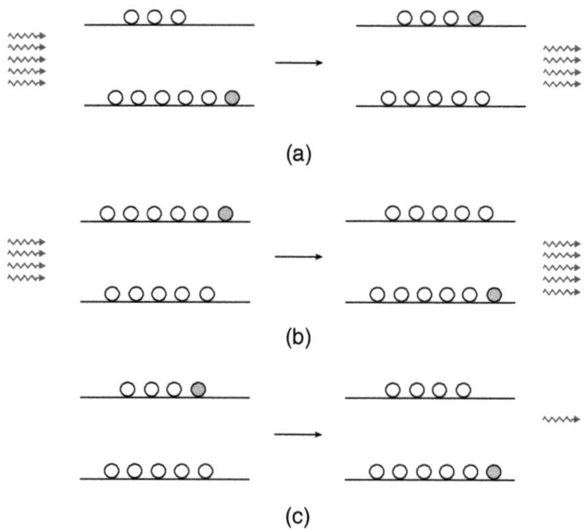

Figure 6.10 *(a) (Stimulated) absorption, (b) stimulated emission, and (c) spontaneous emission processes. (See insert for color representation of this figure.)*

and $\rho(\nu_{0k})$ is the radiation density, that is, the number of photons at frequency ν_{0k} that have the correct energy to cause the transition from the ground to the excited state. The population n_0 of the ground state is given by the Boltzmann equation (Eq. 1.103)

$$\frac{n_k}{n_0} = e^{-\frac{E_k - E_0}{kT}} \tag{1.103}$$

(For simplicity, the degeneracy factors in the original Eq. 1.103 were set to unity here.) Since the energy difference in the numerator of the exponent in Eq. 1.103 is positive, and the denominator is positive as well, the Boltzmann distribution ensures that n_0 is larger than n_k. Thus, the expression in Eq. 6.31 predicts that in the presence of a radiation field of the proper frequency, transitions from the lower to the upper state will occur. This is shown schematically in Figure 6.10(a), where photons (wavy arrows) impinge on the sample of gaseous atoms, and promote an atom (shown shaded) from the ground to the excited state. In this process, a photon is annihilated.

The corresponding process of "stimulated emission" is shown in panel (b) of Figure 6.10. Here, photons impinge on a set of atoms that exhibit "population inversion," a situation where there are more species in the excited state than in the ground state. In such a case, the incident photons stimulate one of the excited species to release a photon; thus, the photon density $\rho(\nu_{0k})$ has increased in the process. However, according to the Boltzmann relationship, the concept of population inversion cannot occur in a two-state system since it would require negative absolute temperatures. This aspect will be further elaborated upon shortly. The rate expression for this process is given by

$$n_k B_{k0} \rho(\nu_{0k}) \tag{6.33}$$

Here, n_k is the population of the excited state that must be larger than that of the ground state, and

$$B_{k0} = B_{0k} \tag{6.34}$$

The equality of the Einstein coefficients for absorption and stimulated emission results from the fact that the dipole transition moments are the same for both processes.

Another deactivation process, known as spontaneous emission, is shown in panel (c). Here, a system in thermal equilibrium (i.e., a system obeying the Boltzmann distribution) creates a photon spontaneously by one of the atomic species in the excited state dropping back into the ground state and emitting a photon. The rate of this step depends on the expression

$$n_k A_{k0} \tag{6.35}$$

where A_{k0} is the Einstein coefficient for spontaneous emission that also depends on the transition moment, but is about a million-fold smaller than B_{k0} for visible light.

Thus, the rate of light absorption (transition into the more highly excited state) is given by Eq. 6.31, whereas for the combined deactivation process, Eq. 6.36 holds:

$$n_k[A_{k0} + B_{k0}\rho(\nu_{0k})] \tag{6.36}$$

In a system that involves more than two states, population inversion can be achieved. Such a system is shown in Figure 6.11(a).

In the case shown in Figure 6.11(a), the system is exposed to a radiation field at ν_{31}. If the coefficient for spontaneous emission A_{21} is much larger than A_{32}, the population in state 2 is drained fast, and a population inversion between states 3 and 2 can occur. Laser action may happen between these states; this action is initiated when a photon at ν_{32} is created be spontaneous emission. This photon subsequently starts to stimulate species in state 3 and thus, amplification of the radiation field at ν_{32} occurs.

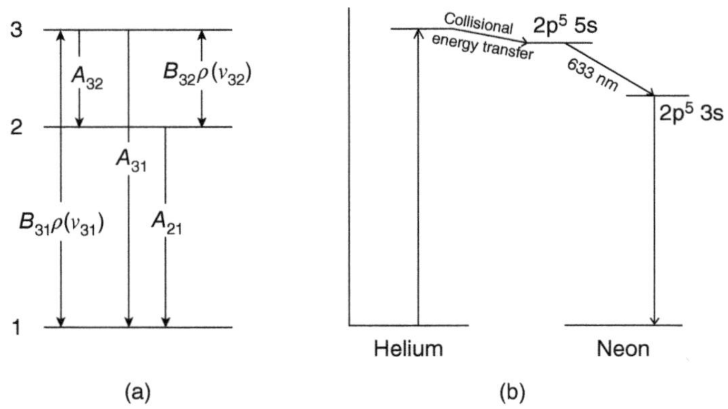

Figure 6.11 (a) Schematic energy level diagram of a three-level system in which population inversion can be achieved and (b) simplified energy level diagram of the He—Ne laser

6.4.2 Operation of a gas laser

One of the most commonly produced and used gas lasers is the He—Ne laser that emits a bright red laser line at 632.8 nm. The operation of lasers can be explained using the He—Ne laser for an example. A laser, in general, consists of a gain medium and a resonator structure, as shown in Figure 6.12. The resonator structure consists of the mirrors that reflect the light created in the gain medium back and forth to stimulate excited species in the gain medium to emit. The resonator structure often is a flat and a spherical mirror to focus the light into the gain medium. One of the mirrors may be partially transparent to allow some of the light, typically 1%, to escape from the resonator and create the laser beam.

The gain medium in a He—Ne laser is a mixture of helium and neon, at a ratio of about 10:1, contained in a gas cell (shown in Figure 6.12 fitted with Brewster angle windows to minimize reflection losses and to linearly polarize the laser output). A DC current of 10–20 mA at a few hundred volts is passed through the gas mixture. Collisions between electrons and He atoms cause transitions of the He atoms into a highly energetic state. These excited He atoms, in turn, collide with Ne atoms, promoting one of the 2p electrons into a 5s orbital; see Figure 6.11(b). This energy level corresponds to energy level 3 in Figure 6.11(a). Laser action occurs, among other transitions, between the $2p^5 5s^1$ and the $2p^5 3s^1$ states. This transition is responsible for the 632.8 nm laser line. The deactivation of the $2p^5 3s^1$ states is very fast, thus, allowing the population inversion between the two excited states to build. He—Ne lasers often are operated in continuous wave (cw) mode, which implies a continuous output, typically in the 0.1–50 mW regime. However, it is frequently desirable to operate a laser at the same average output power, but concentrate the output into pulses with high peak power, to be discussed next.

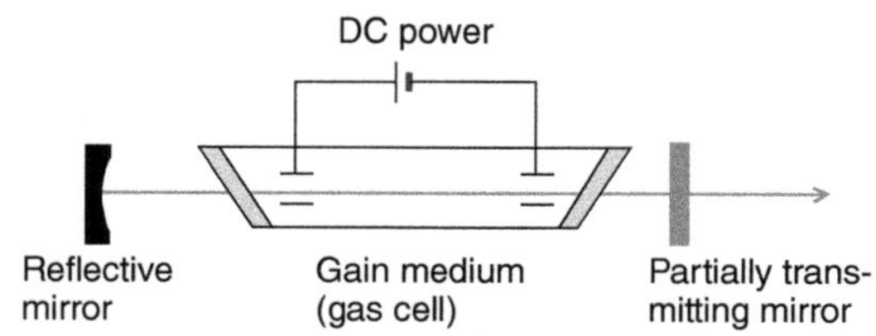

Figure 6.12 Schematic of a gas laser, consisting of the resonator structure, defined by two mirrors and the gain medium

6.4.3 Principles of pulsed lasers

The motivation of operating lasers in a pulsed, rather than CW mode is that at the same average laser power, significantly higher peak power values can be obtained that are required for the nonlinear optical effects. This is demonstrated next for a simple example of a laser operating at 500 nm with an average output power of 100 mW. Photons with this wavelength have an energy of

$$E = \frac{h\,c}{\lambda} = 4 \times 10^{-19}\ \text{(J)} \tag{6.37}$$

Thus, an average output power of 100 mW translates into a flux of

$$\frac{100}{4 \times 10^{-19}}\ \frac{\text{mW}}{\text{J/photon}} = 2.5\ \times 10^{17}(\text{photons/s}) \tag{6.38}$$

If this laser is operated in pulsed mode with a pulse length of 1 ns and a repetition rate of 1 kHz, the total number of photons is divided into 1000 pulses; thus, each pulse contains $2.5\ \times 10^{14}$ photons. Multiplying this number of photons with the energy of each photon (Eq. 6.37) gives the energy of each pulse as

$$E = 2.5\ \times 10^{14} \cdot\ 4 \times 10^{-19} \left(\text{photons}\ \frac{\text{J}}{\text{photon}}\right) = 0.1\ \text{mJ} \tag{6.39}$$

Since this energy is delivered in 1 ns, the peak power P delivered in each pulse is

$$P = \frac{E}{t} = \frac{10^{-4}}{10^{-9}}\left(\frac{\text{J}}{s}\right) = 10^{5}\ \text{(W)} \tag{6.40}$$

This power level can be increased relatively easily into the gigawatt regime. In addition to the increase in peak power of the laser pulses, another interesting phenomenon occurs when proceeding to shorter and shorter pulses. As this happens the frequency width of the pulses increases naturally, as predicted by Heisenberg's uncertainty principle (Eq. 1.38), written in the frequency–time domain

$$\Delta t\,\Delta \nu \geq \frac{\hbar}{2} \tag{6.41}$$

Thus, a 1 ps pulse has a band width of about 3 cm^{-1}, whereas a 10 fs pulse has a natural line width of about 300 cm^{-1}; in practice, the line width is larger and depends on the shape of the pulse. These pulses can be compressed or expanded in the frequency domain by a pair of gratings and appropriate filtering.

6.4.4 Operation of pulsed lasers

Creation of pulsed laser output may be achieved *via* two methods: by changing the resonator quality of the laser (Q-switching and cavity dumping) and mode locking. The term "Q-switching" implies a change in Q, the "quality factor" of the resonator, defined as the ratio of energy input and energy loss for one photon travel cycle in the cavity.

Q-switching is carried out by inserting a modulator into the laser cavity that inhibits the optical feedback. This could be, for example, a Pockels cell set to rotate the plane of the linearly polarized standing wave inside the laser cavity. Without the optical feedback, no stimulated emission occurs and the population inversion builds up to very high levels. By zeroing the potential over the Pockels cell for a short time, it will no longer rotate the plane of polarization; consequently, the stored energy can be emitted as a short pulse, before the potential over the Pockels cell is restored. Q-switching produces pulses of extremely high peak power and energy at relatively slow repetition rates.

While in the Q-switching experiment described above the laser pulses leave the resonator structure through a partially transparent resonator mirror (see Figure 6.12), for cavity dumping both mirrors are 100% reflective, and the modulator inside the cavity, when turned on, deflects the beam out of the cavity.

Mode-locking can produce pulses as short as a few femtoseconds, but operates on a different principle. In a laser cavity, there are several longitudinal modes being amplified at the same time. The longitudinal modes arise from the fact that only standing waves with integer number of nodes can exist in the cavity. These nodes are spaced, in frequency, by

$$\Delta\nu = \frac{c}{2L} \tag{6.42}$$

where c is, as usual, the velocity of light and L the length of the laser cavity. Assuming $L = 30$ cm for a small gas laser, the longitudinal modes of the laser are spaced by 0.5 GHz, which amounts to about 0.016 cm^{-1}. In a gas laser, the amplification width of the resonator, given mostly by Doppler broadening, is about 1.5 GHz, so only about three of the longitudinal modes are amplified and actively lasing. In a solid state laser (Ti:sapphire), the amplification band width may be as high as hundreds of terahertz; thus, there may be thousands of longitudinal modes lasing.

Normally, the phases of these longitudinal modes are not fixed, and they oscillate independently. In mode locking, one fixes the phases of the longitudinal modes with respect to each other. This is achieved by inserting a modulator, such as a Pockels cell into the cavity, and operating this modulator at a frequency that equals the cavity round trip time of a photon. Assuming again a cavity length of 30 cm, this round trip time is

$$t = 2 \times 10^{-9} \text{ (s)} \tag{6.43}$$

corresponding to a modulation frequency of 500 MHz. This modulator acts as a shutter, and only the mode that has a photon round trip time corresponding to the modulation frequency is transmitted. This method allows pulses as short as a few femtoseconds to be produced in lasers such as the Ti:sapphire. The width of such a short laser pulse is so high that only materials with a large amplification width can create these ultra-short pulses.

6.5 Epilogue

This chapter explored some of the techniques that make Raman spectroscopy one of the most versatile optical techniques to study molecular structure and dynamics. Several of these techniques – in particular, the nonlinear methods – have experienced explosive growth during the past decade, mostly due to the availability of pulsed lasers with extremely short pulses, high repetition rates, and high power. Results of novel techniques and improvements appear at every major conference on advanced spectroscopic methods, and at present, new developments seem unlimited.

In view of the originally slow growth and acceptance of Raman spectroscopy, these new developments are nothing short of amazing, and the different ways to probe molecular structure and dynamics have far surpassed the direct absorption-based approaches.

References

1. Spiro, T.G. and Strekas, T.C. (1972) Resonance Raman spectra of hemoglobin and cytochrome c: inverse polarization and vibronic scattering. *Proc. Natl. Acad. Sci. U.S.A.*, **69** (9), 2622–2626.
2. Fleischmann, M., Hendra, P.J., and McQuillan, A.J. (1974) Raman spectra of pyridine adsorbed at a silver electrode. *Chem. Phys. Lett.*, **26**, 163–166.

3. Rasmussen, A. and Deckert, V. (2006) Surface- and tip-enhanced Raman scattering of DNA components. *J. Raman Spectrosc.*, **37**, 311–317.
4. Maker, P.D. and Terhune, R.W. (1965) Study of optical effects due to an induced polarization third order in the electric field strength. *Phys. Rev.*, **137**, A801–A818.
5. Nafie, L.A. (2001) Theory of Raman scattering, in *Handbook of Raman Spectroscopy*, Chapter 1, 1–10 pp (eds I.R. Lewis and H. Edwards), Marcel Dekker, Inc., New York.
6. Koningstein, J.A. (1972) *Introduction to the Theory of the Raman Effect*, D.Reidel Publising Co., Dordrecht, Holland.
7. Asher, S.A. (1988) UV resonance Raman studies of molecular structure and dynamics: applications in physical and biophysical chemistry. *Annu. Rev. Phys. Chem.*, **39**, 537–588.
8. Spiro, T.G. and Stein, P. (1977) Resonance effects in vibrational scattering from complex molecules. *Annu. Rev. Phys. Chem.*, **28**, 501–521.
9. Mukerji, I. (2012) *Resonance Raman Spectroscopy*, eLS: Structural Biology, Wesleyan University, Middletown, CT.
10. Rousseau, D.L. and Ondria, M.R. (1985) Resonance Raman spectra of photodissociated hemoglobins. *Biophys. J.*, **47** (4), 537–545.
11. Thomas, G.J. Jr., (1999) Raman spectroscopy of protein and nucleic acid assemblies. *Annu. Rev. Biophys. Biomol. Struct.*, **28**, 1–27.
12. Lewis, A., Spoonhower, J., Bogomolni, R.A. *et al.* (1974) Tunable laser resonance Raman spectroscopy of bacteriorhodopsin. *Proc. Natl. Acad. Sci. U.S.A.*, **71** (11), 4462–4466.
13. Terner, J., Hsieh, C., Burns, A.R. *et al.* (1979) Time-resolved resonance Raman spectroscopy of intermediates of bacteriorhodopsin: the bK590 intermediate. *Proc. Natl. Acad. Sci. U.S.A.*, **76** (7), 3046–3050.
14. Thomas, G.J. Jr., (1975) Raman spectroscopy of biopolymers, in *Vibrational Spectra and Structure* (ed J. Durig), Marcel Dekker, Inc., New York, pp. 239–315.
15. Wood, B.R., Langforda, S.J., Cookeb, B.M. *et al.* (2003) Raman imaging of hemozoin within the food vacuole of Plasmodium falciparum trophozoites. *FEBS Lett.*, **554**, 247–252.
16. Jeanmaire, D.L. and Van Duyne, R.P. (1977) Surface Raman spectroelectrochemistry part I. Heterocyclic, aromatic, and aliphatic amines adsorbed on the anodized silver electrode. *J. Electroanal. Chem.*, **84**, 1–20.
17. Moskovits, M. (1985) Surface enhanced spectroscopy. *Rev. Mod. Phys.*, **57**, 783–828.
18. Lee, P.C. and Meisel, D. (1982) Absorption and surface-enhanced Raman of dyes on silver and gold sols. *J. Phys. Chem.*, **86**, 3391–3395.
19. Chaney, S.B., Shanmukh, S., Zhao, Y.-P., and Dluhy, R.A. (2005) Aligned silver nanorod arrays produce high sensitivity surface-enhanced Raman spectroscopy substrates. *Appl. Phys. Lett.*, **87**, 031908–031911.
20. Schluecker, S. (2009) SERS microscopy: nanoparticle probes and biomedical applications. *ChemPhysChem*, **10** (9-10), 1344–1354.
21. Ziegler, L.D. (1990) Hyper Raman spectroscopy. *J. Raman Spectrosc.*, **21**, 769–779.
22. Kneipp, J., Kneipp, H., Wittig, B., and Kneipp, K. (2007) One- and two-photon excited optical pH probing for cells using surface-enhanced Raman and hyper-Raman nanosensors. *Nano Lett.*, **7** (9), 2819–2823.
23. Sovago, M., Campen, R.K., Bakker, H.J. *et al.* (2009) Hydrogen bonding strength of interfacial water determined with surface sum-frequency generation. *Chem. Phys. Lett.*, **470**, 7–12.
24. Harvey, A.B. (ed) (1981) *Chemical Applications of Non-linear Raman Spectroscopy*, Academic Press, New York.
25. Carreira, L.A., Goss, L.P., and Malloy, T.B.J. (1981) in *Chemical Application of Nonlinear Raman Spectroscopy*, 321–376 (ed A.B. Harvey), Aademic Press, New York.
26. Cheng, J.-X., Volkmer, A., Book, L.D., and Xie, X.S. (2001) An epi-detected anti-Stokes Raman scattering (E-CARS) microscope with high spectral resolution and high sensitivity. *J. Phys. Chem. B*, **105**, 1277–1291.
27. Evans, C.L., Potma, E.O., Puoris'haag, M. *et al.* (2005) Chemical imaging of tissue in vivo with video-rate coherent anti-Stokes Raman scattering microscopy. *Proc. Natl. Acad. Sci. U.S.A.*, **102** (46), 16807–16812.
28. Krafft, C., Dietzek, B., and Popp, J. (2009) Raman and CARS microspectroscopy of cells and tissues. *Analyst*, **134**, 1046–1057.
29. Kano, H. (2008) Molecular vibrational imaging of a human cell by multiplex coherent anti-Stokes Raman scattering microspectroscopy using a supercontinuum light source. *J. Raman Spectrosc.*, **39**, 1649–1652.

30. Liu, Y., Lee, Y.J., and Cicerone, M.T. (2009) Fast extraction of resonant vibrational response from CARS spectra with arbitrary nonresonant background. *J. Raman Spectrosc.*, **40**, 726–731.
31. Day, J.P.R., Domke, K.F., Rago, G. *et al.* (2011) Quantitative Coherent Anti-Stokes Raman Scattering (CARS) microscopy. *J. Phys. Chem. B*, **115**, 7713–7725.
32. Gondal, M.A. and Dastageer, A. (2009) Relative intensity distribution of the Stimulated Raman Scattered (SRS) lines of methane and hydrogen excited with 266 nm UV radiation. *Appl. Math. Inf. Sci.*, **3** (3), 345–354.
33. Lee, S.-Y. and Zang, D. (2004) Theory of femtosecond stimulated Raman spectroscopy. *J. Chem. Phys.*, **121** (8), 3632–3642.
34. Freudiger, C.W., Min, W., Saar, B.G. *et al.* (2008) Label-free biomedical imaging with high sensitivity by stimulated Raman scattering microscopy. *Science*, **322**, 1857–1861.
35. Saar, B.G., Freudiger, C.W., Stanely, C.M. *et al.* (2010) Video-rate molecular imaging in vivo with stimulated Raman scattering. *Science*, **330**, 1368–1370.

7

Time-Resolved Methods in Vibrational Spectroscopy

7.1 General remarks

As compared to other spectroscopic techniques, methods employing vibrational spectroscopy have the advantage that vibrational transitions occur at very fast time scales (femtoseconds). In NMR spectroscopy, on the other hand, transitions occur on the millisecond to microsecond time scale. Thus, very fast conformational changes cannot be detected. Similar, fluorescence spectroscopy occurs on the nanosecond to microsecond time scale; only pure electronic transitions can occur as fast as the molecular vibrational transitions, which take on the order of 10^{-14} s. Therefore, vibrational spectroscopy – both Raman and infrared – can be used, in principle, to monitor elementary chemical and photochemical reactions in "real time." Since chemical reactions mostly occur along vibrational coordinates, for example, the stretching vibration of a leaving group, vibrational spectroscopy, indeed, provides direct insight into the dynamics of chemical reactions.

Of course, the standard Raman and infrared methods discussed so far do not permit the detection of "real time" spectroscopic effects in the picosecond realm (except for femtosecond stimulated Raman scattering; see Chapter 6, which can probe the coherence and dephasing of molecular vibrations). In infrared spectroscopy, rapid-scanning interferometers can acquire data in the low millisecond time regime. Thus, if interferograms are collected continuously and stored, performing the fast Fourier transform off-line at a later time, slow chemical reactions that occur within tens of milliseconds can be followed directly. For Raman spectroscopy, the acquisition times are even longer, and only very slow chemical reactions – those that occur in the seconds to minutes time scales, can be measured directly.

Thus, methods need to be employed that permit the acquisition of spectral data at nanosecond time scales. This is the subject of time-resolved vibrational spectroscopy. In general, time-resolved vibrational spectroscopy requires reactions that can be initiated photochemically, for example, by a short laser pulse, and that return to the initial state (i.e., the state before the laser excitation) in a cyclic chemical reaction. The time between the exciting pulse (the "pump" pulse) and the data acquisition defines the time resolution of the measurements.

Modern Vibrational Spectroscopy and Micro-Spectroscopy: Theory, Instrumentation and Biomedical Applications, First Edition.
Max Diem.

7.2 Time-resolved FT infrared (TR-FTIR) spectroscopy

7.2.1 Experimental aspects

As pointed out above, the time resolution achievable using fast scanning interferometers is in the low millisecond regime, basically determined by the mirror velocity and the speed at which the analog signal from the detector can be digitized. A quick calculation shows that for acquiring an interferogram consisting of 4096 data points (corresponding to a 4 cm^{-1} resolution spectrum) in 1 ms, intensity data points must be collected at least every 250 ns, corresponding to a sampling rate of 4 MHz or larger. Although faster A/D converters are available, the conversion step is in fact the limiting factor determining the time resolution using a fast scanning interferometer. The measurements based on fast scanning interferometers have the advantage that, in principle, a single transient event can be observed, as long as the signal-to-noise ratio is sufficient.

For faster measurements, step-scanning interferometers have been used for most of nanosecond time resolution Fourier transform infrared FTIR spectroscopy. Step-scanning interferometers allow the mirror to be moved in discrete steps, typically one half of the reference laser wavelength or about 316 nm (see Chapter 3) if a He–Ne laser is used. At each of the mirror positions, the time evolution of the signal at the detector is measured.

Fast photochemical reactions typically are studied in time-resolved Fourier transform infrared (TR- FTIR) measurements where the reaction can be initiated by a laser pulse and can be repeated thousands of times. These reactions return to its original state within milliseconds. The protocol for acquiring step-scanning, time-resolved measurements, depicted in Figure 7.1, is as follows: the interferometer is scanned to a given position of optical retardation, indicated by the point marked x_n along the optical retardation axis [1]. At each of these sampling positions (retardations), the time course of the reaction is measured: once the mirror has settled, the reaction is initiated by a laser flash, as indicated by the flash symbol. This time point is designated

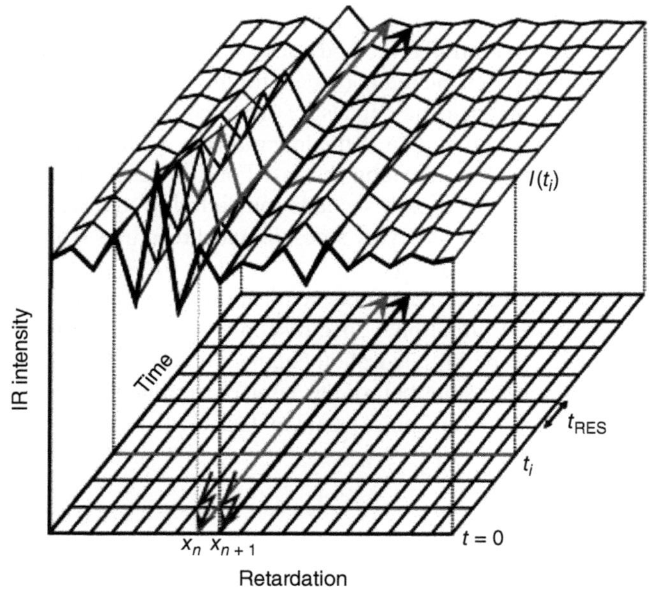

Figure 7.1 Schematic of data acquisition for step scanning time-resolved FTIR measurements of cyclic reactions. See text for details. Rammelsberg et al.*, 1997, [1]. Reproduced with permission from Society for Applied Spectroscopy*

as t_0. The time evolution at the detector signal is measured along the time axis in Figure 7.1 at time intervals t_{RES}. After the signal has decayed to a constant level and the chemical reaction has returned to its original state, the process is repeated at a new retardation position, x_{n+1}. When the appropriate number (2k, 4k, 8k, etc.) of retardation positions have been measured to produce an interferogram for the desired spectral resolution, the dataset is rearranged such that interferograms are constructed for different time points t_1, t_2, t_3, and so on. One of the interferograms, $I(t_i)$, created this way is shown as the gray interferogram at time t_i in Figure 7.1. Each interferogram is separated from the previous one by the time resolution t_{RES}, which is determined by the response time of the detector and the fastest achievable digitization of the time response of the detector signal. An excellent review of various time-resolved infrared measurements was recently published by Smith and Palmer [2].

For non-cyclic reactions, an elegant way was developed [3] to collect the spectral dataset by raster scanning the sample through the focus of a microscope, and initiating the reaction, as before, by a laser pulse. Rather than waiting for the reaction to return to its original state, the sample was scanned to the next pixel position, and the process was repeated.

Typically, high-power pulsed (1–10 ns) dye lasers are used to start the photochemical reaction. By acquiring an infrared spectrum before the reaction is initiated (i.e., before the laser pulse) one can obtain a reference, or zero-point spectrum, from which all subsequent spectra are subtracted [4]. This is shown in Figure 7.2 for a photochemical reaction to be discussed below. Difference spectroscopy, in general, offers the advantage that the regions of the spectrum that do not change during the chemical reaction are subtracted to zero, and only spectral changes appear in the difference spectra. For example, frequency shifts of a band appear as bisignate, sigmoidal patterns as shown for the amide I region in trace (b) of Figure 7.2, whereas intensity changes due to appearance or disappearance of a species appear as positive or negative bands in the difference spectra, for example, in the 1300–1450 cm^{-1} region in trace (b). By focusing on only the parts of the spectrum that undergo changes, enormous sensitivity enhancement can be achieved. An example of such time-resolved difference spectroscopy is discussed next. However, it should be stated that the dynamics of many chemical reactions have been elucidated by time-resolved infrared spectral measurements.

7.2.2 Applications of TR-FTIR spectroscopy

In the literature example [5] to be discussed, the hydrolysis reaction of Ras-bound GTP to Ras-bound GDP will be introduced (GTP: guanosine triphosphate, GDP: guanosine diphosphate). Here, the Ras protein is one of a family of guanine nucleotide-binding signal proteins found in cells that are involved in the regulation of a number of cellular processes. The Ras protein is one of the first molecules identified that can, upon mutation, become oncogenic, that is, promote the transition of a cell from normal to a cancerous state. Ras acts as a GTPase, and the oncogenic mutations of the Ras protein may affect its ability to catalyze the hydrolysis reaction. The detailed mechanism and dynamics of the Ras-mediated GTP hydrolysis reaction was the goal of this study. This example demonstrates impressively the details of reaction mechanism and dynamics revealed by TR-FTIR spectroscopy.

Figure 7.3 depicts the time evolution of the difference spectra of this hydrolysis reaction. Notice that the sign of the difference spectra between Figures 7.2 and 7.3 has been inverted. In addition to the changes in the di- and triphosphate vibrations between 1000 and 1300 cm^{-1}, the reaction is accompanied by changes in the amide I and amide II regions, indicating a significant protein conformational change during the hydrolysis reaction. When the Ras-mediated hydrolysis of GTP to GDP was followed by TR-FTIR spectroscopy over a time range spanning four orders of magnitude (Figure 7.3), a picture emerged that permitted a detailed assessment of the processes taking place during the hydrolysis reaction. The interpretation of the spectral results was aided by vibrational assignments based on isotopic labeling, with ^{18}O of the oxygen atoms at the γ-position (the terminal phosphate group that actually leaves during the hydrolysis reaction) of the GTP. Using

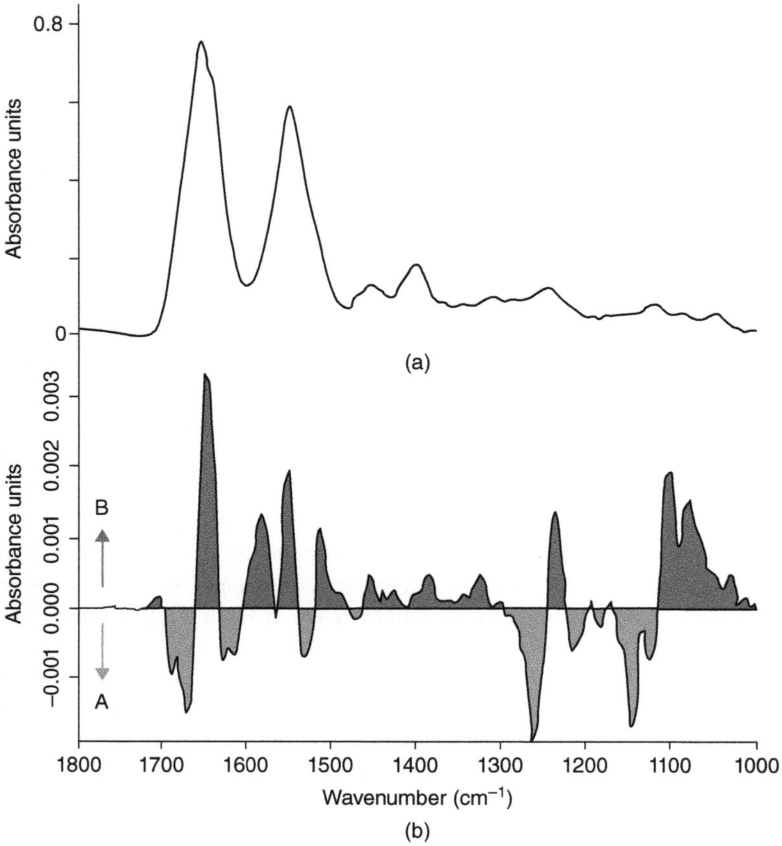

Figure 7.2 (a) FTIR spectrum of the Ras protein. (b) Difference FTIR spectrum between Ras-GTP and Ras-GDP. See text for details. Koetting and Gerwert, 2005, [4]. Reproduced with permission from John Wiley & Sons

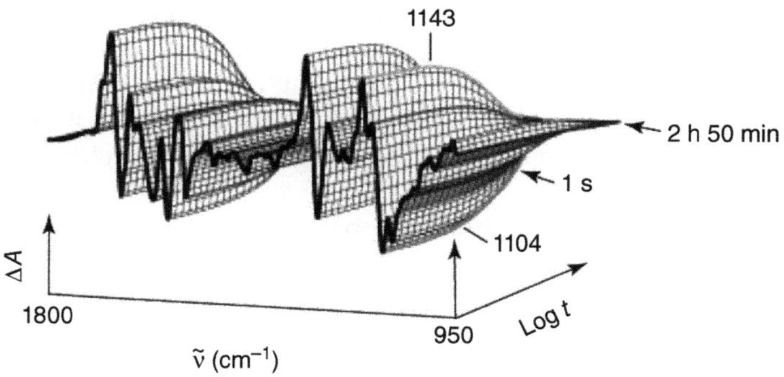

Figure 7.3 Time evolution of the Ras-GTP → Ras-GDP + P_i hydrolysis reaction. See text for details Gerwert, 2006, [5]. Reproduced with permission from John Wiley & Sons

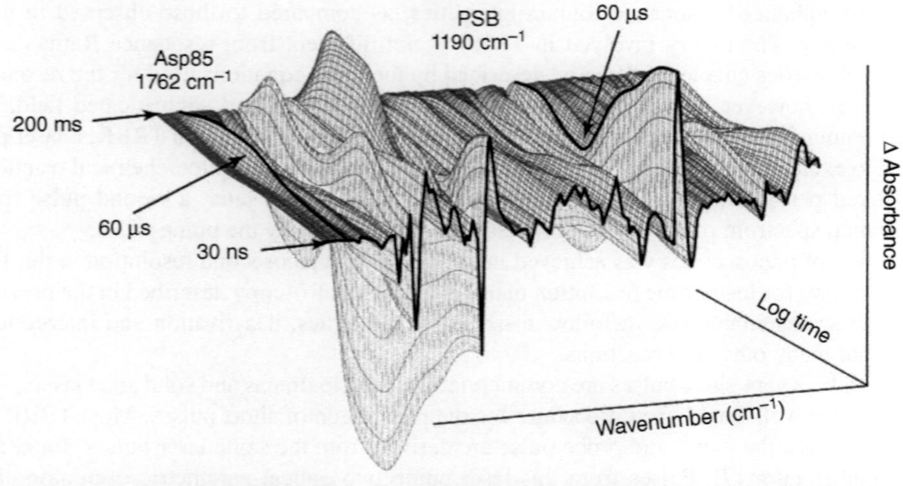

Figure 7.4 Time-resolved FTIR spectroscopy of bacteriorhodopsin allows observation of the deprotonation of the Schiff-base (PSB) and the concomitant protonation of Asp85. See text for details. Gerwert and Kötting, 2010, [6]. Reproduced with permission from John Wiley & Sons

the bond-specific information provided by the isotopically labeled γ-phosphate group, the events leading to the hydrolysis (bond breakage) could be elucidated, including a charge redistribution of the O—P bond of Ras-bound GTP, leading toward a dissociative transition state.

Another example of a reaction that has been studied extensively by time-resolved methods of vibrational spectroscopy is the photo cycle of bacteriorhodopsin (bR). This membrane-spanning protein consists of seven helical protein sections within which is embedded the photoactive prosthetic group, all-*trans* retinal. Upon exposure to light, this moiety isomerizes to 13-*cis* retinal in less than a picosecond (see Section 7.3). After a few picoseconds, the protein rearranges into a metastable state referred to as the "K" intermediate. After about 80 ms, the proton from the protonated Schiff base (PSB) is transferred to an amino acid residue of the protein helix (aspartic acid 85), and a proton is released to the outside medium through the helix bundle. This prototypical proton-pump returns to its original state within a few hundred milliseconds [6].

The TR-FTIR difference spectra for this photosystem are shown in Figure 7.4. In this graph, one can see the disappearance of the 1190 cm^{-1} signal of the PSB within the first 60 µs after the reaction was initiated, and the concomitant protonation of Asp85 that shows a signal at 1762 cm^{-1}. These most obvious signals are accompanied by changes in the protein envelope that indicate very subtle changes in the orientation of the membrane-spanning protein helices during the process. In conjunction with X-ray crystallographic results that reveal the overall composition of the protein helices and their orientation, TR-FTIR gives highly sensitive and specific information on the dynamic processes that occur in the photoinduced proton transfer of bR.

Time-resolved Raman (TRR) spectroscopy, to be presented in Section 7.3, also has contributed enormously to the present understanding of the detailed dynamics of these systems.

7.3 Time-resolved Raman and resonance Raman (TRRR) spectroscopy

7.3.1 Instrumental aspects

The ability to create short laser pulses (see Section 6.4) has given rise to a field known as TRR and time-resolved resonance Raman (TRRR) spectroscopy. Of these, the latter has been applied more frequently due to the vastly enhanced resonance Raman intensities, as compared to those observed in non-resonant Raman spectroscopy. The theory involved in TRRR is not different from resonance Raman spectroscopy discussed in the previous chapter and can be described by the same equations used for the resonance Raman effect. In practice, however, TRRR has evolved into a very different and sophisticated field, particularly from the viewpoint of the experimental aspects required for observing TRRR. In TRRR, a laser pulse (pump pulse) is used to excite a molecule into an excited electronic state, initiate a photochemical reaction, or create any other desired perturbation within a species. A short time interval later, a second pulse (probe pulse) excites the Raman spectrum of the transient species that was created by the pump pulse.

Time resolution of nanoseconds was achieved in the 1970s, and picosecond resolution in the 1980s. Thus, this technique allows for faster time resolution than TR-FTIR spectroscopy described in the previous section. Such short time scales enable one to follow dissociation dynamics, deactivation and intersystem crossing mechanisms, and many other fast reactions.

Lasers that produce very short pulses are commercially available. In gas and solid state lasers, for example, cavity dumping is a well-established procedure for the production of short pulses. Most TRRR approaches use instruments where the pump and probe pulse are derived from the same laser pulses, for example, from a titanium sapphire laser [7]. Pulses from this laser pump two optical parametric oscillators that produce separately tunable output from the near UV to the near IR. The time delay between the pump and the probe pulse can be achieved by varying the path the probe and pump pulses travel, *via* an optical delay line, before impinging on the sample. Optical delay lines are simple to implement at the time scales desired in TRRR, since a laser pulse, traveling at the speed of light, requires 1 ns for 30 cm. Thus, a path difference of 3 mm will introduce a time delay between pulses of 10 ps. The Raman scattered light, indicated by the dashed line in Figure 7.5, can be cleaned of fluorescent contribution by the Kerr gate, as shown in the figure. Since the fluorescence emission occurs at a time scale that is much slower, it is desirable to collect the Raman scattered light before the fluorescence emerges, and use a Kerr gate as a temporal filter. The Kerr effect describes the (nonlinear) change in refractive index of a medium when exposed to a high power laser pulse. In the Kerr gate used here, a laser pulse derived from the main laser is used to trigger the Kerr gate, consisting of two crossed polarizers and a nonlinear medium, to transmit light only during the short duration of the laser pulse.

Raman spectra are usually collected with standard spectrographs fitted with intensified charge-coupled device (ICCD) detectors. ICCDs consist of standard charge-coupled device (CCD) detectors, typically operating at liquid nitrogen temperature, to which a microchannel plate image intensifier has been added. These detectors also can be gated electronically for further fluorescence reduction.

Numerous laser pump/probe pulse cycles are integrated on the detector elements before readout to improve the signal-to-noise ratio of the Raman spectra. Although integration and readout of the spectra may take many

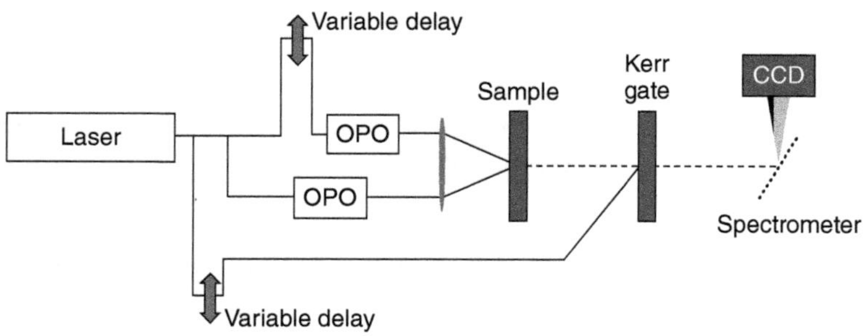

Figure 7.5 *Schematic diagram of a modern TRRR instrument. See text for details. Science and Technology Council, UK Central Laser Facility, 2013, [7]*

seconds, the time resolution achieved is determined by the time delay between pump and probe pulses, and the actual Raman acquisition time is determined by the total lengths of the probe pulses.

7.3.2 Applications of TRRR

First, some representative results of TRRR on a non-biological system are presented to demonstrate the relevance of this field to research in molecular dynamics and structure studies. A relatively slow time scale study of the photoinduced creation of a free radical anion in a substituted anthraquinone was reported by Moore *et al.* [8]. Here, anthraquinone-2,6-disulfonate was illuminated by pump laser pulses at 337 nm. These pulses create a free radical anion, for which the Raman spectrum was probed with delayed laser pulses at 480 nm. Delay times between pump and probe pulses were varied between 20 ns and 10 μs. The resulting Raman spectra are shown in Figure 7.6. All Raman bands exhibit the same time dependence and therefore are due to the same species that was identified to be the radical anion by its vibrational spectrum. After 10 μs, the system returned to its original state.

Most of TRRR applications, however, were in the field of biophysics. In particular, two photosystems have been extensively studied by TRRR: the binding and dissociation dynamics of oxygen, carbon monoxide, and carbon dioxide to metal porphyrin model systems, as well as intact hemoglobin (Hb), myoglobin, and cytochromes. The other photosystem of intense interest is rhodopsin, the visual pigment and primary light harvesting protein that was introduced in Section 7.2.

7.3.3 Heme group dynamic studies

Numerous TRRR have been carried out to study both the photoinduced dissociation kinetics of iron-bound ligands in Hb, myoglobin, and cytochromes, and the very fast changes in the tertiary structure of the protein in

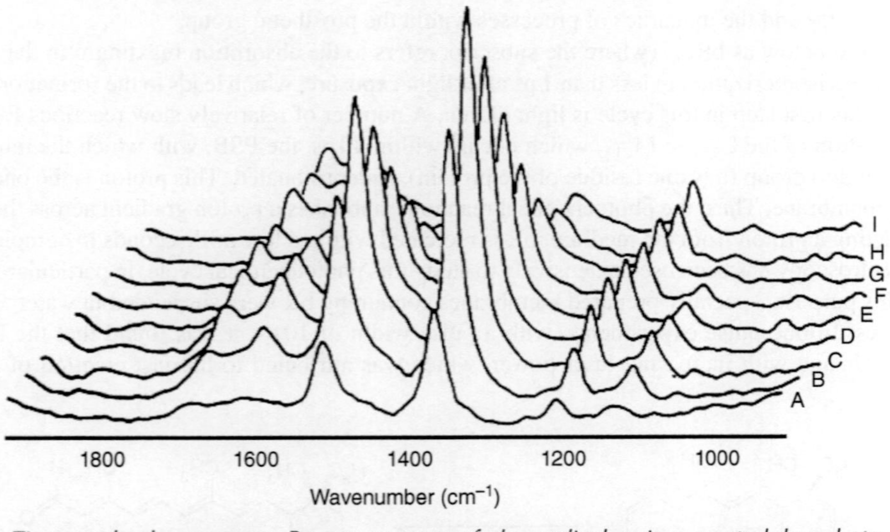

Figure 7.6 *Time-resolved resonance Raman spectra of the radical anion created by photo-excitation of anthraquinone-2,6-disulfonate. Pump wavelength 337 nm, probe wavelength, 480 nm. Time delays: A, 20 ns; B, 50 ns; and C, 100 ns. D, 500 ns; E, 1 μs; F, 2 μs; G, 5 μs; H, 10 μs; and I, probe laser only. Moore* et al.*, 1987, [8]. Reproduced with permission from Taylor & Francis Books (UK)*

the vicinity of the binding pocket [9]. The actual changes in structure and electronic state of the porphyrin core after the laser-pulse-induced dissociation of HbO_2 or HbCO can be interpreted by comparing the resonance Raman features of the products with those of deoxy-Hb. Marker bands for spin and oxidation states, as well as core size, are well known; thus, the observation of shifts in these bands within 30 ps after photodissociation indicates the very fast rearrangement of the Fe–porphyrin core. However, there are still slight frequency differences that persist for tens of nanoseconds between the photo-dissociated intermediate and the fully relaxed state. During this time, the protein chains around the porphyrin core will undergo significant structural changes, in part induced by the motion of the central iron atom out of the plane of the surrounding four ligated N-atoms. This motion is accompanied by a concomitant motion of the histidine residue in the fifth coordination position, which, in turn, moves one of the peptide chains of hemoglobin.

7.3.4 Rhodopsin studies

Both the visual pigment, rhodopsin, and the light-harvesting photosystem bR have been studies extensively by TRRR [10, 11]. The former is found in the retina of all mammals and many other living beings and is responsible for the primary process of vision, namely, the capture of a visible photon and its conversion to chemical energy. The second of the proteins above occurs in the purple membranes of algae, which are found in very saline water. These algae use bR as an antenna for visible light and utilize the energy captured to transport protons through the bacterium's purple membrane.

Both of these proteins contain retinal as their prosthetic groups. Retinal is a polyene, the structure of which is shown in Figure 7.7. The mechanism of the conversion of light to chemical energy involves in both cases a conformational change induced by absorption of a photon into the conjugated polyene structure, which weakens the double bond and allows isomerization at the 13 position.

The TRRR data offer time resolution in the 100 femtosecond regime; that is, this technique allows for time resolution that is orders of magnitude faster than TR-FTIR introduced above. However, resonance Raman spectroscopy is much less sensitive to the changes in protein structure; instead, it offers a detailed view into the photochemistry and the dynamics of processes within the prosthetic group.

bR, referred to below as bR_{570} (where the subscript refers to the absorption maximum in the visible spectrum), undergoes isomerization in less than 1 ps after light exposure, which leads to the formation of "K" state (K_{590}). Only this first step in this cycle is light driven. A number of relatively slow reactions follow. During the transformation of the L_{550} to M_{412}, which occurs within 40 ps, the PSB, with which the retinal is bound to the (side) amino group (a lysine residue of the protein), is deprotonated. This proton is the one transported through the membrane. Thus, the photochemical reaction establishes a proton gradient across the membrane. After abstracting a proton from the medium, bR is recreated within a few milliseconds to complete the cycle.

TRRR spectroscopy has been used extensively to study this photochemical cycle, in particular its very early steps. For this purpose, specially prepared membranes containing bR were suspended in water and subjected to laser pulses. In one-pulse experiments, with a pulse width of 10 ps, it was found that the Raman spectra of bR_{570} change with increasing laser power, which was attributed to the fast creation of a mixture of

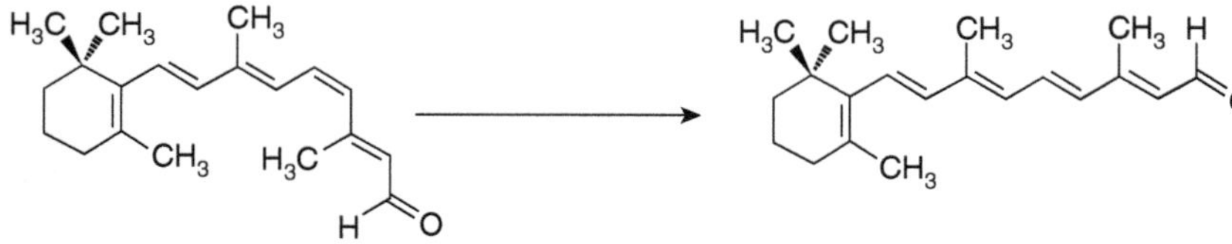

Figure 7.7 *Structure of retinal and its cis → trans isomerization*

a photoexcitation product and the original species within the length of the laser pulse. This product could be identified further by pump–probe experiments, with delay times of between 0 and 40 ps. It was found *via* the analysis of a number of marker bands that the first product was most likely an intermediate, twisted stage between all *trans*-retinal and 13-*cis*-retinal. The marker bands had been defined *via* detailed deuterium substitutions along the olefinic backbone of retinal. As experimental techniques are being refined, more details of these reaction pathways, and their dynamics, will undoubtedly be found, and other techniques, such as time-resolved infrared and UV visible absorption spectroscopies, have been adapted to study such fast processes. However, to date, time-resolved and resonance Raman techniques have proved invaluable in the determination of the processes involved in many biochemical reactions. In addition to the prosthetic groups introduced so far, a large number of different reaction centers of biological molecules have been studied, among them chlorophyll, iron–sulfur centers, and the active site in vitamin B.

References

1. Rammelsberg, R., Hessling, B., Chorongiewski, H., and Gerwert, K. (1997) Molecular reaction mechanisms of proteins monitored by nanosecond step-scan FT-IR difference spectroscopy. *Appl. Spectrosc.*, **51** (4), 558–562.
2. Smith, G.D. and Palmer, R.A. (2006) Fast time-resolved mid-infrared spectroscopy using an interferometer, in *Handbook of Vibrational Spectroscopy*, 625–640 (eds J.M. Chalmers and P.R. Griffiths), John Wiley & Sons, Ltd, Chichester.
3. Rammelsberg, R., Boulas, S., Chorongiewski, H., and Gerwert, K. (1999) Set-up for time-resolved step-scan FTIR spectroscopy of noncyclic reactions. *Vib. Spectrosc.*, **19**, 143–149.
4. Koetting, C. and Gerwert, K. (2005) Proteins in action monitored by time-resolved FTIR spectroscopy. *Chem. Phys. Chem.*, **6**, 881–888.
5. Gerwert, K. (2006) Molecular reaction mechanisms of proteins monitored by time-resolved FT-IR difference spectroscopy, in *Handbook of Vibrational Spectroscopy*, 1–20 (eds J.M. Chalmers and P.R. Griffiths), John Wiley & Sons, Ltd, Chichester.
6. Gerwert, K. and Kötting, C. (2010) Fourier transform infrared (FTIR) spectroscopy, in *Encyclopedia of Life Sciences*, 1–8 Wiley-Blackwell.
7. Science and Technology Council, UK Central Laser Facility (2013) http://www.stfc.ac.uk/CLF/Facilities/Ultra/Techniques/14368.aspx (accessed 12 January 2015).
8. Moore, J.N., Phillips, D., Hester, R.E. *et al.* (1987) Time-resolved resonance Raman spectroscopy of electronically excited species, in *Time Resolved Laser Raman Spectroscopy*, 23–56 (eds D. Phillips and G.H. Atkinson), Academic Press, London.
9. Friedman, J.M., Rousseau, D.L., and Ondrias, M.R. (1982) Time-resolved resonance Raman studies of hemoglobin. *Annu. Rev. Phys. Chem.*, **33**, 471–491.
10. Atkinson, G.H., Brack, T.L., Blanchard, D., and Rumbles, G. (1989) Picosecond time-resolved resonance Raman spectroscopy of the initial trans to cis isomerization in the bacteriorhodopsin photocycle. *Chem. Phys.*, **131**, 1–15.
11. Kim, J.E., McCamant, D.W., Zhu, L., and Mathies, R.A. (2001) Resonance Raman structural evidence that the cis-to-trans isomerization in rhodopsin occurs in femtoseconds. *J. Phys. Chem. B*, **105**, 1240–1249.

8

Vibrational Optical Activity

8.1 Introduction to optical activity and chirality

Vibrational optical activity (VOA) is a collective term applied to two spectroscopic techniques discovered during the early 1970s. The two techniques, which had been predicted on theoretical grounds, are infrared (IR) or vibrational circular dichroism (VCD), and Raman optical activity (ROA). Both these techniques are manifestations of natural optical activity, a phenomenon exhibited by chiral substances. The discovery of optical activity dates back to the mid-19th century when L. Pasteur physically separated mirror image crystals of ammonium sodium tartrate, a salt of tartaric acid. When these mirror image crystals were separately dissolved in water, the resulting solutions rotated the plane of incident linearly polarized light in opposite directions. The "optical rotation," or more generally, "optical rotatory dispersion" (ORD) is one of several manifestations of electronic optical activity, that is, an unequal response of the medium toward left or right circularly polarized light.

Linearly polarized light can be described as being composed of two co-propagating circularly (one left and one right) polarized light waves. If the refractive index of the sample toward right circularly polarized light, n_R, differs from that for left circularly polarized light, n_L, the two circularly polarized waves propagate with different velocity through the medium; consequently, the polarization plane of linearly polarized incident light will be rotated upon exiting the sample solution. Thus, ORD is described by the difference in refractive indices toward circular polarization as

$$\Delta n = n_L - n_R \tag{8.1}$$

Optical rotation was first observed in the visible part of the spectrum and was found to generally increase in magnitude with decreasing wavelength of light. Furthermore, it was found that at absorption maxima (typically in the visible or ultraviolet (UV) spectral region), the sign of the optical rotation underwent anomalous dispersion as shown by the top trace in Figure 8.1. In addition, it was found that within an absorption peak, there exists a non-zero difference between extinction coefficients toward left and right circularly polarized light:

$$\Delta \varepsilon = \varepsilon_L - \varepsilon_R \tag{8.2}$$

Modern Vibrational Spectroscopy and Micro-Spectroscopy: Theory, Instrumentation and Biomedical Applications, First Edition.
Max Diem.

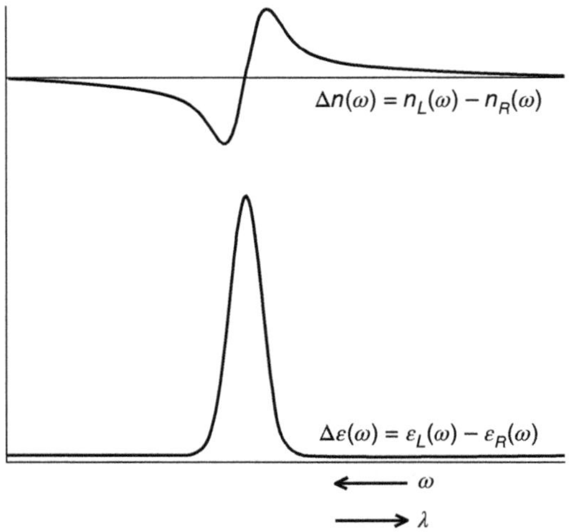

Figure 8.1 *Relationship between optical rotatory dispersion (ORD) and circular dichroism (CD). The particular sign pattern of the CD and ORD curves is called a "positive Cotton effect." For the other enantiomer, the CD curve would be negative, and the ORD curve a mirror image about the wavelength axis (negative Cotton effect)*

This effect is known as circular dichroism (CD). Since the transitions observed in the visible/UV regions are electronic transitions, this form of CD is often referred to as electronic circular dichroism (ECD). The relationship between ORD and CD is shown in Figure 8.1, which is analogous to Figure 3.1. Indeed, both figures depict coupled physical phenomena, related by the Kramers–Kronig transform.

ORD and CD are often introduced during the discussion of optical activity and stereochemistry in introductory courses in organic chemistry. Optical activity is a property exhibited by enantiomerically pure or enriched, chiral molecules. Chirality or handedness describes the relationship between two molecules (or items) that are related to each other just as the left hand is related to the right hand, namely, by being mirror images of each other. A molecule is said to be chiral when it is non-superimposable on its mirror image.

Chiroptical techniques, such as CD, are among the most fascinating spectroscopic phenomena, since they can distinguish molecules that are identical except for their configuration, that is, their handedness. Chiroptical methods use circularly polarized light to distinguish the two forms of molecules. Since circularly polarized light itself has handedness (it can be either left or right circularly polarized), different interactions of the two forms of light occur with chiral molecules and give rise to the observable effects enumerated above.

In VOA, the effects of the chirality on vibrational and not electronic transitions are observed. Thus both ROA and VCD are truly forms of vibrational spectroscopy. VCD is conceptually a straightforward experiment in which the differential absorption of left and right circularly polarized IR radiation by a vibrational transition of a chiral molecule is observed. As such, it is a direct extension of the principles of ECD toward a different spectral range, namely, that involving vibrational transitions [1].

ROA, similarly to Raman spectroscopy, is a scattering, rather than an absorption, phenomenon. It is observed by exciting the sample alternatively with left and right circularly polarized laser radiation and measuring the differential Raman scattering cross section as an intensity differential. In addition, there are other methods to observe ROA [1].

The discussion of these subjects in the following sections is at a rather level, since a number of review articles and books have been produced [2–4] by several researchers in the field, and the reader is referred to these works for more detailed discussions.

8.2 Infrared vibrational circular dichroism (VCD)

8.2.1 Basic theory

Conceptually, one may view VCD as an extension of the principles of ECD, normally observed in the UV spectral region, into the domain of vibrational transitions in the IR spectral region. In IR VCD, the experimental result is the differential absorption between left and right circularly polarized IR radiation transmitted by a chiral sample, defined as

$$\Delta A = A_L - A_R \tag{8.3}$$

Here, as in the remainder of this chapter, capital subscripts R and L are used to denote right and left circularly polarized radiation. The differential absorption ΔA for a transition from the vibrational ground state $|0\rangle$ to the vibrationally excited state $|1\rangle$ can be related to the quantum mechanical observable known as the *rotatory strength* R_{01} given by

$$R_{01} = \mathrm{Im}[\langle 0|\boldsymbol{\mu}|1\rangle \cdot \langle 1|\mathbf{m}|0\rangle] \tag{8.4}$$

where $\boldsymbol{\mu}$ and $\mathbf{m}$ are the electric and magnetic dipole operators. The first one was introduced in Eq. 1.78:

$$\boldsymbol{\mu} = \sum_i e_i x_i \tag{1.78}$$

whereas the latter is defined by

$$\mathbf{m} = \sum_i \frac{e_i}{2m_i}(\mathbf{r}_i \times \mathbf{p}_i) \tag{8.5}$$

where e, m, $\mathbf{r}$, and $\mathbf{p}$ are the charges, masses, positions, and momenta of particle i, and the summation is over all particles.

Just as the dipole strength D_{01} was defined before as

$$D_{01} \propto \langle 0|\boldsymbol{\mu}|1\rangle^2 = \frac{1}{\tilde{\nu}_0}\int \varepsilon(\tilde{\nu})\, d\tilde{\nu} \tag{1.113}$$

$$R_{01} = \mathrm{Im}[\langle 0|\boldsymbol{\mu}|1\rangle \cdot \langle 1|\mathbf{m}|0\rangle] = \frac{1}{\tilde{\nu}_0}\int \Delta\varepsilon(\tilde{\nu})\, d\tilde{\nu} \tag{1.8.6}$$

one can convert the differential absorption ΔA to $\Delta\varepsilon$ units, using the Lambert–Beer law (Eq. 1.112) and integrating over the VCD peak:

$$R_{01} = \mathrm{Im}[\langle 0|\boldsymbol{\mu}|1\rangle \cdot \langle 1|\mathbf{m}|0\rangle] = \int \frac{\Delta\varepsilon}{\tilde{\nu}}\, d\tilde{\nu} \tag{8.6}$$

As in all manifestations of natural optical activity, the desired observables – ΔA or R_{01} – arise through the interference of electric and magnetic dipole transition moments. The sensitivity of chiroptical techniques toward the handedness of the molecule results directly from the form of the magnetic momentum operator: it contains the vector product of momentum and position vectors, the result of which is another vector. The sign of this vector is determined by the handedness of the coordinate system, and it is well known that a vector product will change sign upon converting from a left- to a right-handed coordinate system. By changing the configuration of the molecule from one to the other enantiomer, the sign of the magnetic transition moment changes. Similarly, by keeping the configuration of the molecule fixed and changing from left to right circularly polarized light, the sign of the magnetic transition moment is reversed. The equations presented here are analogous to the expressions governing ECD in the visible/UV spectral range.

The ratio of the integrated VCD and absorption peaks can be defined as

$$g = \frac{\Delta A}{A} = \frac{4R_{01}}{D_{01}} \tag{8.7}$$

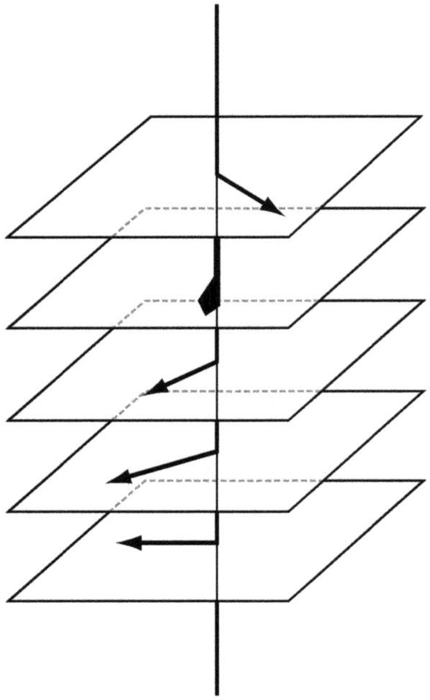

Figure 8.2 *"Spiral staircase" model of a dissymmetric arrangement of achiral transitions*

and is typically on the order of 5×10^{-4}–5×10^{-5}. Since VCD is a very small effect, its observation is rather difficult. However, through the diligent efforts of a few research groups in the field, VCD has now become a generally applicable spectroscopic technique.

8.2.2 Exciton theory of optical activity

The discussion of VCD, so far, has emphasized asymmetric species (those belonging to point group C_1, which has only the identity symmetry element). However, optical activity is also exhibited by dissymmetric species of pure rotational point groups, such as C_2. In these molecules that may be represented by a twisted allene structure, or a helical structure, inherently achiral chromophores or groups can produce optical activity that was first described by the "coupled oscillator" model by Tinoco [5] and Bayley *et al.* [6] for CD of electronic transitions.

As discussed in Chapter 5, transition dipole coupling is one way to describe the transitions of identical or near-identical groups in a dimeric or polymeric molecular system. For the following discussion, a hypothetical "*n*-mer" of identical transitions is introduced, where the achiral transitions, for example, C=O stretching vibrations, are arranged in what was referred to previously as a "spiral staircase" geometry, shown in Figure 8.2. In this hypothetical system, the transitions are in parallel planes and displaced from each other by a constant angle (pitch) and constant tread. This model can be used, to a first approximation, to describe the (hydrogen-bonded) carbonyl groups in a canonical DNA model.

As pointed out in Chapter 5, a one-photon excitation into the manifold of coupled carbonyl groups results in an "excitonic" state in which the excitation is highly delocalized between the coupled excited states. The energies of these excited states are defined, as pointed out before, by the eigenvalues of the interaction energies

$$\begin{matrix} E_1 & V_{12} & V_{13} \\ V_{21} & E_2 & V_{23} \\ V_{31} & V_{32} & E_3 \end{matrix} \tag{5.17}$$

where the diagonal values are the degenerate or near-degenerate energy eigenvalues of unperturbed transitions, and the off-diagonal elements V_{ij} are the interaction or coupling energies between the transition moments, given to a first approximation by dipole–dipole interactions:

$$V_{ij} = \frac{\boldsymbol{\mu}_i \cdot \boldsymbol{\mu}_j}{|\mathbf{T}_{ij}|^3} - \frac{3(\boldsymbol{\mu}_i \cdot \mathbf{T}_{ij})(\boldsymbol{\mu}_j \cdot \mathbf{T}_{ij})}{|\mathbf{T}_{ij}|^5} \tag{5.18}$$

In Eq. 5.18, $\boldsymbol{\mu}_i$ and $\boldsymbol{\mu}_j$ are the transition dipole moments of oscillators i and j and $\mathbf{T}_{ij}$ is the vector connecting the centers of mass coordinates, X_i, of each individual transition:

$$\mathbf{T}_{ij} = X_j - X_i \tag{5.19}$$

For the case of n coupled transitions, Eq. 5.17 is diagonalized numerically to yield the energy eigenvalues v_k of the kth "exciton" component and the eigenvector matrix **C,** from which the transition dipole strength D_k for each of the k coupled transitions can be computed according to

$$D_k = \sum\nolimits_{i=1}^{n} \sum\nolimits_{j=1}^{n} c_{ik} c_{jk} (\boldsymbol{\mu}_i \cdot \boldsymbol{\mu}_j) \tag{5.21}$$

where c_{ik} are the appropriate eigenvector matrix elements.

The corresponding rotational strengths for the kth exciton component are given by

$$R_k = -\frac{\pi \tilde{v}_0}{c} \sum\nolimits_{i=1}^{n} \sum\nolimits_{j>i}^{n} c_{ik} c_{jk} \, [T_{ij} \cdot \boldsymbol{\mu}_i \times \boldsymbol{\mu}_j] \tag{8.8}$$

where c is the velocity of light. For the dimeric case ($n = 2$), Eq. 8.8 simplifies to the well-known "coupled oscillator" equation [5]

$$R^{\pm} = \mp \frac{\pi \tilde{v}_0}{2c} [T_{12} \cdot \boldsymbol{\mu}_1 \times \boldsymbol{\mu}_2] \tag{8.9}$$

in which $R^{\pm}$ denotes the rotational strengths of the symmetric $|+\rangle$ and antisymmetric $|-\rangle$ combination states of the two transitions. Owing to superior computational methods, the equations introduced here are no longer used for the actual computation of VCD intensity, and *ab initio* calculations are generally used that explicitly account for the concepts introduced here. Nevertheless, the exciton formalism conveys a simple model for the optical activity produced by the dissymmetric interaction of achiral transitions.

8.3 Observation of VCD

VCD can be observed *via* dispersive or Fourier transform (FT) instrumentation. Until the introduction of optimized commercial FT-VCD spectrometers in the early 2000s, the performance of dispersive and FT-VCD spectrometers was comparable in signal-to-noise and data acquisition times. The same cannot be said for standard dispersive and FT IR spectrometers, where the performance of the latter is certainly orders of magnitude better than that of their dispersive counterparts.

Since VCD is such small an effect, it requires modulation techniques for its observation: VCD is observed by switching the IR radiation incident on the sample at high frequency between left and right circular polarization *via* a photoelastic modulator (PEM). The desired VCD information is then contained in an AC signal at the frequency of the PEM.

The central piece of all VCD instruments, therefore, is the PEM that uses the phenomenon of stress birefringence to introduce a phase difference between two orthogonally polarized components of light incident on the modulator. The PEM crystal is manufactured of a uniaxial piece of material that is transparent in the IR spectral range and is aligned with its unique axis (the z-axis) along the direction of propagation of the light. Under the influence of a mechanical stress or strain, applied perpendicularly to the direction of propagation of the light and along the crystal's x- or y-axis, the refractive indices along these axes, n_x and n_y, become unequal, causing light waves polarized along the x- and y-directions to travel with different velocities through the crystal. At the exit face of the crystal, circularly polarized light is produced if the retardation between the two orthogonal components of light is $\pm\lambda/4$. The alternating strain/stress is produced by squeezing the modulator crystal between two piezoelectric drivers. An AC voltage is applied to the piezoelectric crystals, and the amplitude of this voltage determines the stress/strain and therewith the retardation. Commercial modulators are available for the IR region, using CaF_2 above 1200 cm^{-1} and ZnSe down to about 600 cm^{-1} as stress optical materials.

In the measurement of VCD, the actual differential signal, at constant CD of the sample, will vary with the light level transmitted at a given wavelength. Thus two intensities must be monitored and their ratio determined in order to obtain ΔA. Denoting the signal at the modulator frequency as $I_{AC}(\tilde{\nu})$ and the overall transmission of the sample as $I_{DC}(\tilde{\nu})$, it can be shown from an analysis of the radiant energy at the detector that

$$\frac{I_{AC}(\tilde{\nu})}{I_{DC}(\tilde{\nu})} = \tanh(1.15\ \Delta A)\ \sin\ \alpha \tag{8.10}$$

Here, α denotes the retardation between the two linearly polarized components of light used to produce circularly polarized light. Equation 8.10 demonstrates that the ratio $I_{AC}(\tilde{\nu})/I_{DC}(\tilde{\nu})$ is proportional to ΔA. The term sin α itself varies sinusoidally with time, since the retardation α may be written as

$$\alpha = \alpha_0\ \sin(\omega_M t) \tag{8.11}$$

where ω_M is the oscillation frequency of the modulator and α_0 is the amplitude of the retardation. Using a number of simplifications, an approximation for the observed signal can be obtained:

$$\frac{I_{AC}(\tilde{\nu})}{I_{DC}(\tilde{\nu})} = (1.15\ \Delta A)\ J_1(\alpha_0\ )\sin(\omega_M t) \tag{8.12}$$

Here, J_1 is the first-order Bessel function. Inspection of Eq. 8.12 reveals that for the observation of ΔA, a signal in phase with ω_M must be monitored *via* a lock-in amplifier and divided by the DC signal. Details of the derivation have been reviewed in a number of publications [7].

The schematic of a typical dispersive, double modulation VCD spectrometer is shown in Figure 8.3 [7]. Light from an IR source was focused *via* gold-coated optical components into a monochromator optimized for a relatively narrow region in the mid-IR (800–1800 cm^{-1}). The light was chopped mechanically at low frequency to provide the "DC" signal, which is a measure of the overall light level transmitted by the optical components and the sample. Subsequently, the light was converted to alternating left and right circularly polarized light at the PEM. The PEM used in this instrument was a ZnSe modulator operating at 31.2 kHz.

After passing the sample, the light was detected *via* a HgCdTe detector operating at liquid nitrogen temperature. Signal analysis was carried out using two lock-in amplifiers in tandem, first demodulating the signal at the modulator frequency ω_M and subsequently at the chopper frequency ω_C to yield $I_{AC}(\tilde{\nu})$. $I_{DC}(\tilde{\nu})$ was measured separately using a voltage-to-frequency converter with a variable gain preamplifier with a dynamic range of better than four orders of magnitude. The ratio $I_{AC}(\tilde{\nu})/I_{DC}(\tilde{\nu})$ was computed by a dedicated personal computer [8]. The instrument operated in a "step-scan" mode where the monochromator was scanned at constant wavenumber increments, typically in 1 or 2 cm^{-1} steps, that were converted to wavelength increments, using basic monochromator optical design parameters (see Eq. 3.10). Subsequently, the modulation level (the voltage applied to the PEM) was adjusted for quarter-wave retardation, the sensitivity setting of the lock-in amplifier was adjusted under computer control to give at least three significant figures of digitized output, and the measurements were stored in computer memory. Scan speeds of 60 cm^{-1}/minute could easily

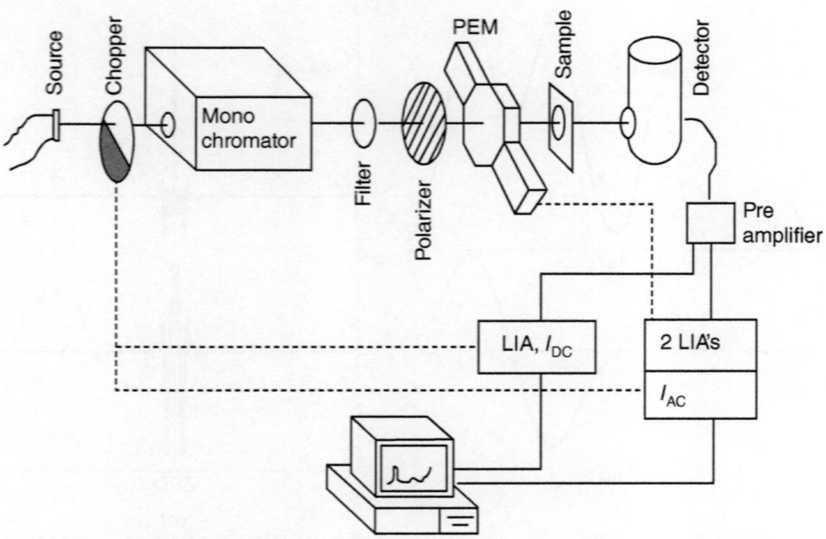

Figure 8.3 *Schematic of a dispersive VCD spectrometer. See text for details*

be realized by this spectrometer that allowed for repetitive scan signal averaging, and produced VCD spectra with excellent signal-to-noise ratio, comparable to that of FT-VCD instruments. The residual artifact in this instrument was reduced to a level of $\Delta A/A \approx 10^{-7}$ [9].

FT-VCD instruments also use a double modulation approach. One modulation occurs, as in the case of the dispersive method described above, *via* the PEM at a frequency determined by the size and hardness of the modulator crystal material, typically between 30 and 60 kHz. The second modulation is provided by the Fourier frequencies, and is dependent on the interferometer scan speed. As in the case of dispersive VCD, the signal at the detector is split into two paths: in one path, the signal is demodulated *via* a lock-in amplifier tuned to the PEM frequency. The output of this amplifier is proportional to the FT of the $I_{AC}(\tilde{\nu})$ component. The second signal path measures the total signal at the detector that corresponds to the FT of the $I_{DC}(\tilde{\nu})$ signal. Separate storage, FT, and division of the two signals result in the VCD spectra. Early investigations in FT-VCD [10, 11] were confounded by software problems resulting from the fact that the $I_{AC}(\tilde{\nu})$ interferogram does not exhibit a center burst *per se,* but two symmetrical bursts displaced about the zero path difference (ZPD). This is shown schematically in Figure 8.4. A comparison of the performance of dispersive and FT-VCD spectrometers, as of the mid-1990s, is shown in Figure 8.5.

Modern commercial FT-VCD utilize digital signal processors instead of lock-in amplifiers and have largely overcome the signal dynamic range problems inherent in FT-VCD, as compared to dispersive VCD where the amplification of the signal could be adjusted during a scan. Furthermore, the residual artifact has been reduced by the use of a second PEM to scramble any remaining polarization of the light after it has passed the sample. This approach was pioneered by Cheng *et al.* [12], and incorporated into the design of the most recent commercially available FT-VCD spectrometers. The residual polarization, due to birefringence in some of the optical elements, has been reduced to a level comparable to that observed for the dispersive instrument described above.

Reduction of the residual polarization artifact is essential for applications of VCD in real laboratory settings when only one enantiomer of the sample is available. If both enantiomers, or one enantiomer and the racemic modification are available, the artifact can simply be subtracted out.

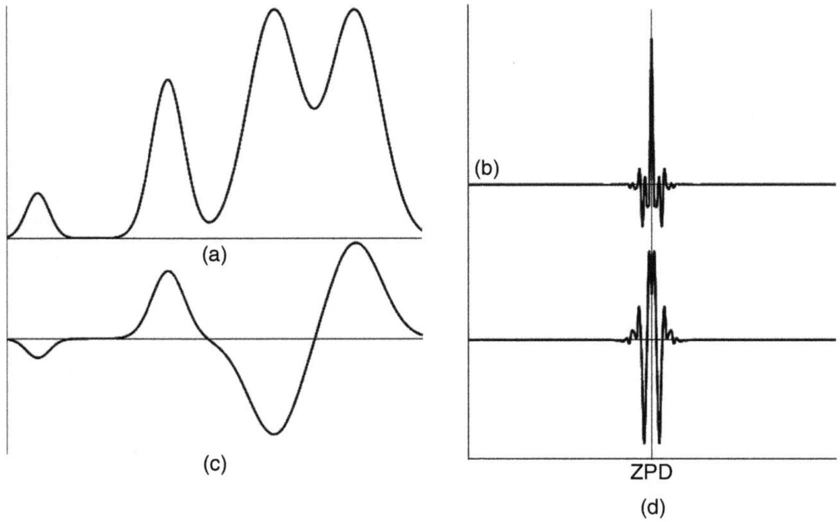

Figure 8.4 (a) Schematic of an all-positive spectrum and (b) its unfolded interferogram, exhibiting a sharp center burst at the ZPD. (c) Bisignate spectrum (e.g., a VCD spectrum) and (d) its interferogram devoid of a center burst

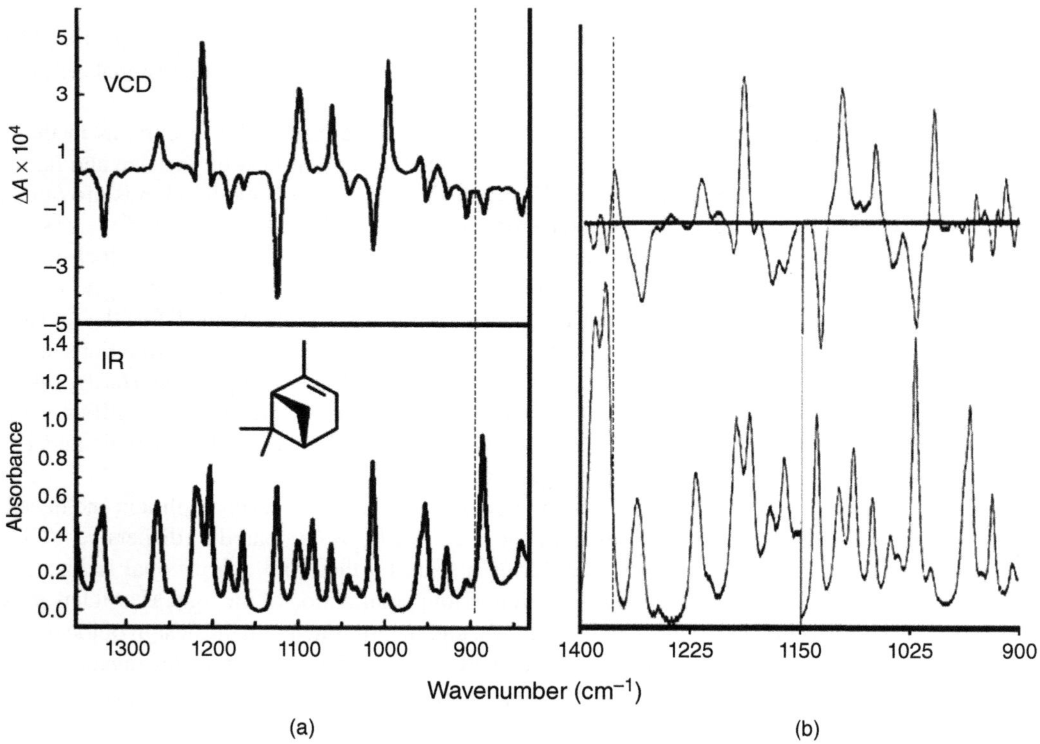

Figure 8.5 Comparison of (a) FT-VCD (top) and absorption (bottom) spectra of (−)-α-pinene and (b) dispersive VCD and absorption spectra. The dashed vertical lines denote the different spectral regions scanned. The vertical gray line in the dispersive spectra at 1150 cm^{-1} indicates a change in collection f-number and filter. Adapted from Nafie, 1996, [1]. Reproduced with permission from Society for Applied Spectroscopy

8.4 Applications of VCD

One of the most fascinating aspects of VCD is its sensitivity toward molecular conformation and conformational changes, where "conformational" changes may be defined as structural changes of a molecule resulting from internal rotation about single bonds without breakage and reformation of chemical bonds. A typical conformational transition in macromolecules is the one between α-helical and coil conformations in peptides. Such conformational changes alter the dihedral angles between certain probe groups, such as the amide I vibrations (*cf.* Chapter 10) in peptides. VCD intensity can be produced by the dipolar coupling of (virtually achiral) vibrational transitions, which are in a fixed, dissymmetric geometric pattern, such as a helix. This mechanism for the creation of VCD intensity, which was introduced in Section 8.2, is one of the dominant mechanisms giving rise to the conformational sensitivity of VCD in biomolecules.

8.4.1 VCD of biological molecules

Probably more than 80% of all VCD publications have dealt with studies on biomolecular systems of general interest to biophysicists and structural biologists, in particular amino acids, peptides, proteins, oligonucleotides, and nucleic acids. Some of these studies are reviewed in Chapter 10, where the structural sensitivity of both Raman and IR spectroscopy has been enormously enhanced by VOA. Here, results of VCD studies to elucidate the absolute configuration of small molecules, in particular of small, chiral drug molecules, are emphasized.

8.4.2 Small molecule VCD

VCD has had enormous impact for the determination of absolute configuration of small molecules, in particular of small molecules with more than one chiral center used in the pharmaceutical industry. For such molecules, ECD in the UV–visible spectra often is of little use since the CD of chromophores at different chiral centers and of different conformers overlap and may cancel. Thus, techniques that sample local configuration *via* vibrational transitions are of major importance. Consequently, VCD has been embraced by the pharmaceutical industry.

The procedure utilized for absolute configuration determination can be summarized as follows. After acquisition of the VCD spectrum in the vibrational fingerprint (typically about 800–2000 cm^{-1}) and the C—H stretching regions, these spectra are analyzed *via* quantum mechanical computations: first, the geometry of the molecule is optimized and the vibrational normal coordinates are calculated. This particular step has become routine during the past decade, and is summarized in Chapter 9. The geometry optimization for a molecule with one chiral center will result in the same energy for either the *R* or the *S* absolute configuration. Thus, one starts with either one of them, and proceeds with the VCD computations.

IR and VCD intensity computations are generally carried out using the Gaussian program suite. This program has been first released in the 1980s. In the release of the late 1990s, the code to compute VCD intensities was incorporated in the Gaussian 98 programs [13]. In general, density functional theory at a fairly high level of parameterization actually produces very good agreement between observed and computed VCD and IR intensities. For a rigid molecule with one chiral center, the observed VCD patterns are compared to the computed ones: if the intensity patterns agree, one may assume that the absolute configuration used to set up the molecule is correct; if computed and observed VCD intensities are mirror images of each other, the incorrect configuration was assumed. This is shown in Figure 8.6 [14].

If a molecule can exist in several conformational forms, the energy difference between the conformers is computed *via* the Gaussian program, and VCD and absorption spectra are computed for all conformers. The resulting VCD and absorption spectra are obtained by adding the spectra of all conformers, scaled by their

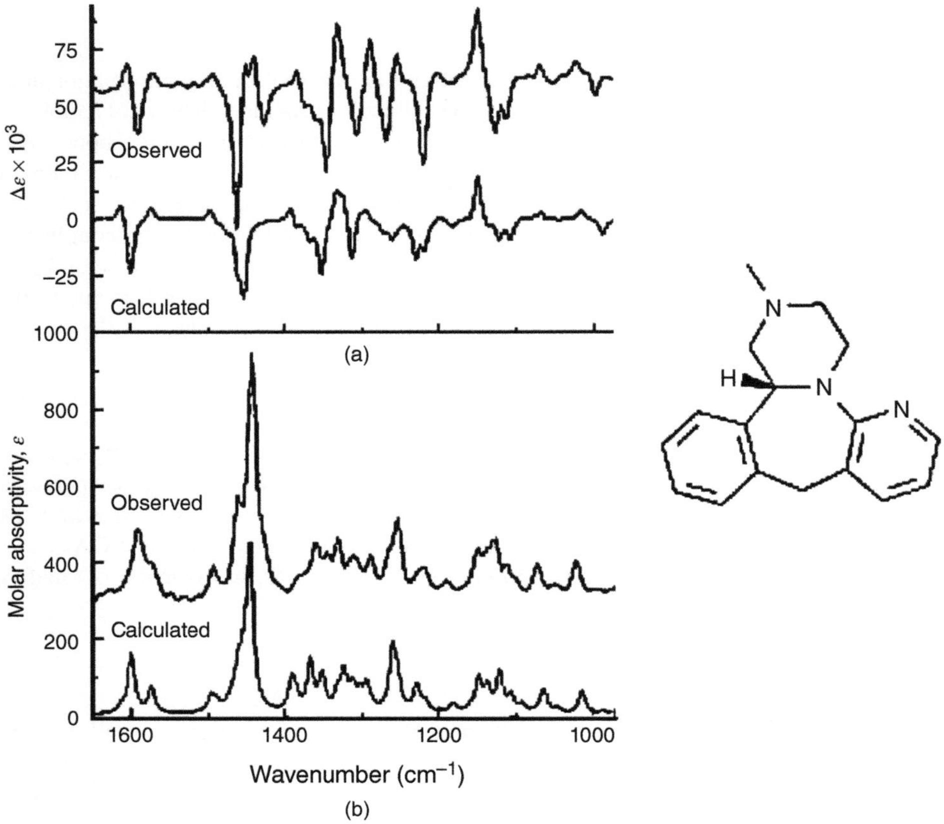

Figure 8.6 *Determination of the absolute configuration of the pharmaceutical compound mirtazapine. (a) VCD and (b) infrared absorption spectra. Observed spectra: (−)-mirtazapine; Calculated spectra:* R*-mirtazapine. The agreement between observed and computed spectra established the absolute configuration as (−)-*R*-mirtazapine. Freedman* et al.*, 2003, [14]. Reproduced with permission from John Wiley & Sons*

abundance at the temperature at which the measurement was carried out, using the Boltzmann relationship (see Eq. 1.102).

Similarly, for molecules with more than one chiral center, VCD and absorption spectra are computed for all possible combinations of absolute configurations at the chiral centers. This can be carried out for the entire molecule if it is not too large, or in sections containing the chiral centers. This strategy was applied successfully to study the absolute configuration and conformation of a number of antifungal drugs by Dunmire *et al.* [15].

8.5 Raman optical activity

ROA is the differential inelastic scattering of circularly polarized light from chiral molecules. ROA can be observed in several different forms. The easiest to visualize (but very difficult to observe) form of ROA is a straightforward extension of the principles of optical activity measurements, in general: the sample is excited alternatingly by left and right circularly polarized photons, and the difference in scattered intensity is measured. This form of ROA is referred to as incident circular polarization Raman optical activity (ICP-ROA). In scattered circular polarization ROA, the chiral sample is illuminated with linearly polarized light, and the circularity (ellipticity) of the scattered photon is measured. In addition, there are two more forms of ROA where both the incident light and the scattered photons are modulated. These methods are referred to as dual circular polarization Raman optical activity (DCP-ROA). Depending on whether the incident and scattered lights are modulated in phase or out of phase, these techniques are referred to as DCP_I and DCP_{II}.

The discussion in this book concentrates on only one of the techniques, the ICP-ROA observed in backscattering (180° scattering) and without a polarizer in the scattered beam. The ROA equivalent of the g-factor defined in Eq. 8.7 for VCD

$$g = \frac{\Delta A}{A} = \frac{4\,R_{01}}{D_{01}} \tag{8.7}$$

would be the dimensionless circular intensity differential (CID) Δ, defined as

$$\Delta = \frac{I_R - I_L}{I_R + I_L} \tag{8.13}$$

Here, I_R and I_L denote Raman intensities scattered for right and left circularly polarized light, respectively. In Eq. 8.13, the differential intensity is defined, for historic reasons, with the opposite sign convention than the one used in the definition of ΔA, which was defined above as the absorbance toward left minus the absorbance toward right circularly polarized light. For backscattering from resonance, the CID can be written in terms of the polarizability and optical activity tensor invariants (see below) as

$$\Delta = \frac{24\,[3\beta(G)^2 + \beta(A)^2]}{2c\,[45\alpha^2 + 7\beta(\alpha)^2]} \tag{8.14}$$

In Eq. 8.12, the denominator is the total Raman scattered intensity, as defined in Section 4.4, which was written then as $[45\,\alpha^2 + 4\,\beta^2]$ for linearly polarized and as $[45\alpha^2 + 7\beta^2]$ for unpolarized or circularly polarized light in Eq. 4.28. There, the first anisotropy β^2 was defined as

$$\beta^2 = (\alpha_{xx} - \alpha_{xx})^2 + (\alpha_{xx} - \alpha_{zz})^2 + (\alpha_{yy} - \alpha_{zz})^2 + 6(\alpha_{xy}^2 + \alpha_{xz}^2 + \alpha_{yz}^2) \tag{4.30}$$

In the formula given by Eq. 8.14, $\beta(\alpha)^2$ is used as an abbreviation for the anisotropy of the polarizability [4] in order to provide similar expressions for the anisotropies containing G and A:

$$\begin{aligned}\beta(\alpha)^2 &= \frac{1}{2}\,(3\,\alpha_{\alpha\beta}\alpha_{\alpha\beta} - \alpha_{\alpha\alpha}\alpha_{\alpha\alpha})\\ &= (\alpha_{xx} - \alpha_{xx})^2 + (\alpha_{xx} - \alpha_{zz})^2 + (\alpha_{yy} - \alpha_{zz})^2 + 6(\alpha_{xy}^2 + \alpha_{xz}^2 + \alpha_{yz}^2)\end{aligned} \tag{8.15}$$

Similarly, the anisotropic parts of the magnetic dipole and the electric quadrupole polarizability tensors are [4]

$$\beta(G)^2 = \frac{1}{2}\,(3\,\alpha_{\alpha\beta}G_{\alpha\beta} - \alpha_{\alpha\alpha}G_{\beta\beta}) \tag{8.16}$$

$$\beta(A)^2 = \frac{1}{2}\,\omega_0\alpha_{\alpha\beta}\varepsilon_{\alpha\gamma\delta}\,A_{\gamma\delta\beta} \tag{8.17}$$

The magnetic polarizability $G_{\alpha\beta}$ is defined in analogy to the electric polarizability (Eq. 4.10) as

$$\alpha_{\alpha\beta} = \frac{2}{\hbar}\sum\nolimits_m \frac{\omega_{0m}}{\omega_{0m}^2 - \omega^2}\langle\psi_{ev}|\mu_\alpha\,|\psi_{e'v'}\rangle\langle\psi_{e'v'}|\,\mu_\beta|\psi_{ev'}\rangle \tag{4.10}$$

$$G_{\alpha\beta} = -\frac{2}{\hbar}\sum\nolimits_m \frac{\omega_{0m}}{\omega_{0m}^2 - \omega^2}\,\langle\psi_{ev}|\mu_\alpha|\psi_{e'v'}\rangle\langle\psi_{e'v'}|m_\beta|\psi_{ev'}\rangle \tag{8.18}$$

Here, m is the magnetic dipole operator, defined in Eq. 8.5; thus, it can be seen that ROA, like all optical activity, is due to the interaction of electric and magnetic transition moments.

In addition to rotatory polarizability terms (also known as electric dipole magnetic dipole polarizability) described by Eq. 8.18, the interference between electric dipole and electric quadrupole transition moments also gives rise to circular differential scattering. The electric dipole–electric quadrupole polarizability is given by an expression similar to that of the regular polarizability, except that the second transition moment is substituted by terms of the form

$$\langle \psi_{e'v'} | \Theta_{\beta\gamma} | \psi_{ev'} \rangle \tag{8.19}$$

where the electric quadrupole operator is defined as

$$\Theta_{\alpha\beta} = \frac{1}{2}\sum_i e_i (3r_{i\alpha} r_{i\beta} - r_i^2 \delta_{\alpha\beta}) \tag{8.20}$$

In Eq. 8.20, r_i are the position vectors of particle i, δ is the Kronecker delta, and ω_0 in Eq. 8.17 is the frequency of the exciting radiation.

Typical differential scattering intensities are about 10^{-3}–10^{-4} of the Raman intensities. Although this effect is somewhat larger than the VCD differential intensities discussed before, its observation is significantly more complicated than that of VCD, and much more susceptible to artifacts. Furthermore, Raman scattered intensities are much lower than the light levels in IR spectroscopy. Therefore, ROA only recently has become a method with broad applicability.

8.6 Observation of ROA

The classical ROA experiment (as opposed to the one where the polarization of the scattered photons is measured) is carried out using a standard Raman instrument to which light modulation optics has been added. Since Raman scattering is a weak effect, detection of the scattered light *via* charge-coupled device (CCD) detectors is essential. In fact, the first practical routine ROA instrumentation was not constructed until CCD detectors were readily available.

Originally, modulation between left and right circularly polarized light was accomplished using electro-optic modulators (EOMs). In an EOM, the birefringence needed to produce circularly polarized light is induced by a high voltage applied across a uniaxial crystal, such as KD_2PO_4, along the propagation direction of the laser. By reversing the direction of the electric field applied to the crystal, both left and right circularly polarized light can be produced. The read-out of the CCD is synchronized to the (slow) modulation between the two circular polarization states, and Raman intensities for these polarization states are read out and co-added separately. A detailed account of the design of state-of-the-art ROA instrumentation was recently published [1].

EOMs are notoriously difficult to align, and are sensitive to slight temperature fluctuations. Both these effects can cause slightly elliptically polarized light to be produced, and the ellipticity of the light, in an optically active medium, can produce confounding effects that mimic ROA features. Several methods appeared in the early literature to circumvent these confounding effects [16]. However, truly routine ROA operation by non-spectroscopists was not possible until recently when the first commercial ROA spectrometers were introduced. The optical design of these avoids the use of EOMs, but rather uses a single (zero-order) quarter wave plate that is rotated mechanically to produce left and right circularly polarized light. This is shown in the schematic for a backscatter ROA instrument in Figure 8.7. An optional second wave plate/polarizer combination in the beam of the scattered light can be used to perform DCP_I and DCP_{II} measurements.

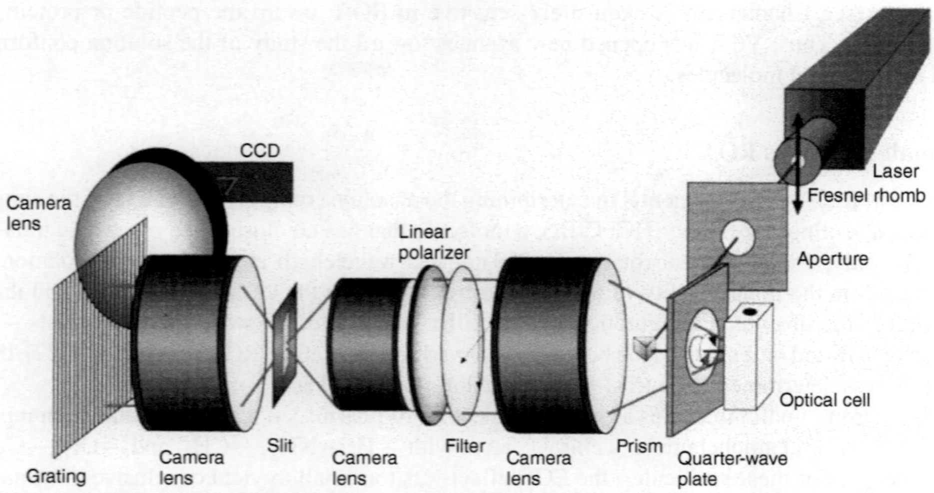

Figure 8.7 *Schematic of a commercial ROA instrument for the observation of ICP and DCP ROA. Nafie, 1996, [1]. Reproduced with permission from Society for Applied Spectroscopy*

8.7 Applications of ROA

The discussion presented in Section 8.4 above holds equally true for VCD and ROA in that both absolute configuration at selected chiral centers, and molecular conformational information can be obtained. Thus, similar to the discussion on the applications of VCD, the applications of ROA can be divided into configurational studies of small molecules, as well as conformational studies of large (bio)-molecular molecules such as peptides, proteins, and nucleic acids.

8.7.1 ROA of biological molecules

The first reported ROA of a biological molecule was that of the resonance-enhanced magnetic ROA spectrum of cytochrome c [17], at a time when the observation of ROA was still in its infancy. It was found that resonance ROA, in general, may be a very difficult experiment to carry out, due to the CD accompanying the visible absorption of a chiral molecule. However, in the case of cytochrome c, it turned out that at the excitation wavelength used (514.5 nm), the resonance enhancement is strong, but the CD is virtually zero. Thus, resonance ROA and magnetic resonance ROA were obtained with minimal interference from the ECD of the sample.

The first non-resonant ROA of a biological molecule, alanine, as an aqueous solution was reported by Barron *et al.* [18], followed by ROA studies of peptides and proteins by the same laboratory. A review of these efforts can be found in Ref. [19]. Most of the biomedical applications of ROA are reviewed in Chapter 10; yet, at this point, a few summary comments about ROA and a comparison with VCD results are appropriate.

In VCD, the most predominant features are those of the highly polar amide I vibrations (see Chapter 10) that interact, *via* transition dipole coupling and similar mechanisms, discussed in Section 8.2 to produce VCD spectral features that depend very sensitively on the relative orientation of adjacent peptide linkages. In ROA, such a dipolar coupling mechanism does not contribute to the observed spectral features; consequently, left-handed and right-handed protein helices show nearly constant amide I signals. However, the amide III spectral region (see Chapter 10) is exquisitely sensitive in ROA toward the peptide or protein secondary structure [20, 21]. Thus, VOA has opened new avenues toward the study of the solution conformation and dynamics of biological molecules.

8.7.2 Small molecules ROA

In this category, ROA was instrumental in determining the absolute configuration of one of the simplest chiral mole, bromochlorofluoro methane (HCFClBr), a molecule that has no close-lying electronic transitions and, consequently, only a small optical rotation at 589 nm, the wavelength at which optical rotation is usually reported. Based on the polarizability of the four atoms attached to the chiral carbon atom and the observed optical rotation, the absolute configuration of HCFClBr was deduced – with opposite results – by several research groups. Based on a comparison between observed and theoretical ROA experiments [22], the absolute configuration was determined to be (*S*)-(+)-bromochlorofluoromethane.

ROA of numerous small molecules as well as molecules of pharmaceutical interest has been reported [23], for example, for several amphetamine (a chiral C atom with —H, —NH_2, —CH_3, and —CH_2—C_5H_6 groups attached) analogs. For these molecules, the ECD effect was too small to yield conclusive information on the absolute configuration; however, ROA data provided this information. As in the discussion of VCD results for small molecules, ROA spectra of several conformeric forms of amphetamine needed to be computed and co-added, after scaling by their Boltzmann factors. For a detailed review of ROA results for a number of non-biological molecules, the reader is referred to the literature [3, 24].

References

1. Nafie, L.A. (1996) Vibrational optical activity. *Appl. Spectrosc.*, **50** (5), 14A–25A.
2. Nafie, L.A. (1997) Infrared and Raman vibrational optical activity: theoretical and experimental aspects. *Annu. Rev. Phys. Chem.*, **48**, 357–386.
3. Nafie, L.A. (2011) *Vibrational Optical Activity: Principles and Applications*, John Wiley & Sons, Ltd, Chichester.
4. Barron, L.D. (1982) *Molecular Light Scattering and Optical Activity*, Cambridge University Press, Cambridge.
5. Tinoco, I. (1962) The exciton contribution to the optical rotation of polymers. *Radiat. Res.*, **20**, 133–139.
6. Bayley, P.M., Nielsen, E.B., and Schellman, J.A. (1969) The rotatory properties of molecules containing two peptides groups: theory. *J. Phys. Chem.*, **73** (1), 228–243.
7. Diem, M., Roberts, G.M., Lee, O., and Barlow, A. (1988) Design and performance of an optimized dispersive infrared dichrograph. *Appl. Spectrosc.*, **42**, 20–28.
8. Lee, O. and Diem, M. (1992) Infrared CD in the 6 μm spectral region: design of a dispersive spectrograph. *Anal. Instrum.*, **20** (1), 23–43.
9. Xie, P. and Diem, M. (1996) Measurement of dispersive vibrational CD: signal optimization and artifact reduction. *Appl. Spectrosc.*, **50**, 675–680.
10. Nafie, L.A. and Diem, M. (1979) Theory of high frequency differential interferometry: application to the measurement of infrared circular and linear dichroism using Fourier transform spectroscopy. *Appl. Spectrosc.*, **33**, 130–139.
11. Nafie, L.A., Diem, M., and Vidrine, D.W. (1979) Fourier transform infrared vibrational circular dichroism. *J. Am. Chem. Soc.*, **101**, 496–503.
12. Cheng, J.C., Nafie, L.A., and Stephens, P.J. (1975) Polarization scrambling using a photoeleastic modulator: application to circular dichroism measurement. *J. Opt. Soc. Am.*, **65** (9), 1031–1035.
13. Frisch, M.J., Trucks, G.W., Schlegel, H.B., *et al.* (1998) *Gaussian 98*, Gaussian, Inc., Pittsburgh.
14. Freedman, T.B., Cao, X., Dukor, R.K., and Nafie, L.A. (2003) Absolute configuration determination of chiral molecules in the solution state using vibrational circular dichroism. *Chirality*, **15**, 743–758.
15. Dunmire, D., Freedman, T.B., Nafie, L.A. *et al.* (2005) Determination of the absolute configuration and solution conformation of the antifungal agents ketoconazole, itraconazole, and miconazole with vibrational circular dichroism. *Chirality*, **17**, S101–S108.

16. Hug, W. (1981) Optical artifacts and their control in Raman circular difference scattering measurements. *Appl. Spectrosc.*, **35** (1), 115–124.
17. Barron, L.D. (1975) Magnetic vibrational optical activity in the resonance Raman spectrum of ferrocytochrome c. *Nature*, **257** (5525), 372–374.
18. Barron, L.D., Gargaro, A.R., Hecht, L., and Polavarapu, P.L. (1991) Experimental and ab initio theoretical vibrational Raman optical activity of alanine. *Spectrochim. Acta*, **47A** (8), 1001–1016.
19. Barron, L.D. and Hecht, L. (1994) Vibrational Raman optical activity: from fundamentals to biochemical applications, in *Circular Dichroism: Interpretation and Applications* (eds K. Nakanishi, N.D. Berova, and R.W. Woody), VCH Publishers, New York, 179–216.
20. McColl, I.H., Blanch, E.W., Gill, A.C. *et al.* (2003) A new perspective on beta-sheet structures using vibrational Raman optical activity: from poly(L-lysine) to the prion protein. *J. Am. Chem. Soc.*, **125** (33), 10019–10026.
21. McColl, I.H., Blanch, E.W., Hecht, L. *et al.* (2003) A study of alpha-helix hydration in polypeptides, proteins, and viruses using vibrational Raman optical activity. *J. Am. Chem. Soc.*, **126** (26), 8181–8188.
22. Costante, J., Hecht, L., Polavarapu, P.L. *et al.* (1997) Absolute configuration of bromochlorofluoromethane from experimental and ab initio theoretical vibrational Raman optical activity. *Angew. Chem., Int. Ed. Engl.*, **36** (8), 885–888.
23. Berg, R.W., Shim, I., White, P.C. *et al.* (2012) Raman optical activity and Raman spectra of amphetamine species—quantum chemical model calculations and experiments. *Am. J. Anal. Chem.*, **3**, 410–421.
24. Polavarapu, P.L. (1988) *Vibrational Spectra: Principles and Applications with Emphasis on Optical Activity*, Elsevier, New York.

9

Computation of Vibrational Frequencies and Intensities

9.1 Historical approaches to the computation of vibrational frequencies

The origins of observing and qualitatively interpreting infrared and Raman spectra date back to the middle of last century. The qualitative interpretation, in terms of group frequencies and their dependence on factors such as isotopic substitution, solvation, and other structural effects, could be understood from basic physical principals derived from the vibration of very simple systems, such as diatomic molecules. However, even the transition from a diatomic to a triatomic molecule changes the vibrational spectral features to an extent that made a detailed interpretation difficult and the computations virtually impossible at that time.

However, the observation of group frequencies in vibrational spectroscopy suggested that the forces (expressed as the "force field") acting between atoms in similar moieties were transferable; thus, the concept of empirical force field calculation was born. This concept dominated the interpretation of vibrational spectra until the 1980s. Force-field-based calculations to reproduce vibrational frequencies, however, are only one part of the equation. The other part, the computation of vibrational intensities, posed an even more difficult problem since the classical descriptions of fixed charges moving along normal coordinates were hopelessly inadequate to predict vibrational intensities. In early attempts, estimated partial charges were attached to each of the oscillating atoms and changes in the dipole moments were computed from these charges and the atomic displacements of each atom during a vibrational mode. This approach was referred to as the "fixed partial charge" model [1]. A slight improvement was achieved using a "polar tensor" model [2, 3] in which the effect of charge displacement was assumed to be direction dependent. However, these polar tensors did not provide atomic properties transferable from one to another molecule, and the field of vibrational intensity calculations was relatively stagnant.

From a quantum mechanical viewpoint, the problem of computing vibrational frequencies is more straightforward, albeit computationally very involved: quantum mechanics reveals the overall energy of a molecular system quite accurately, after the geometry of a molecule is optimized. Subsequently, the second derivatives of the molecular energy with respect to Cartesian displacement coordinates (the so-called Hessian matrix) yields the force field, and after diagonalization of the potential energy matrix, the vibrational energy eigenvalues, related to the vibrational frequencies. The accuracy of the computed frequencies has steadily increased from the semi-empirical quantum mechanical methods (such as complete neglect of differential overlap (CNDO)) to

Modern Vibrational Spectroscopy and Micro-Spectroscopy: Theory, Instrumentation and Biomedical Applications, First Edition.
Max Diem.

large basis-set density functional theory (DFT) calculations. The quality of DFT-based frequency calculations is sufficiently good such that they can be used to assign vibrational spectra in cases of ambiguities.

Quantum mechanical calculations are based on optimized positions of the nuclei and electronic orbital centroids. Distortion of the nuclear geometry by vibrational coordinates and the ensuing redistribution of electron densities reveal the dipole moments, and thus, can predict vibrational intensities quite accurately. Many of the algorithms used to compute vibrational intensities are based on theoretical work by researchers in vibrational optical activity (see Chapter 8) who have advanced the theory of vibrational intensities (Raman, infrared, vibrational circular dichroism (VCD), and Raman optical activity(ROA)) to the level that these algorithms are now integrated into molecular orbital computational platforms such as Gaussian'09 or Hyperchem.

9.2 Vibrational energy calculations

9.2.1 Classical approaches: the Wilson GF matrix method

In the days before digital computers, the computation of vibrational frequencies was being carried out by electric calculators that could not perform much more than simple arithmetic operations. Thus, it was necessary to reduce the computational steps to a minimum because the computation using the secular equation defined earlier in mass-weighted Cartesian coordinates,

$$\begin{aligned} f_{11} - \lambda + f_{12} + f_{13} + \cdots + f_{1,3N} &= 0 \\ f_{21} + f_{22} - \lambda + f_{23} + \cdots + f_{2,3N} &= 0 \\ f_{31} + f_{32} + f_{33} - \lambda + \cdots + f_{3,3N} &= 0 \\ \cdots \end{aligned} \tag{1.21}$$

proved to be too complicated even for a triatomic molecule, requiring the solution of nine simultaneous equations. Furthermore, as pointed out before, the force constants defined in an arbitrary Cartesian coordinate system are not related to "chemically" significant force constants, such as bond stretching or angle deformation forces. Furthermore, even for a triatomic molecule, an 81-element force constant matrix would have been required (see Eq. 1.21). Such force constants could only be transferred from diatomic molecules, which were then the only molecular systems that yielded numerical values for force constants (see Eq. 1.28).

Thus, in the early to mid-1950s, an approach was developed that used "internal coordinates," namely, bond stretching (between two atoms), angle deformations (three atoms), torsional motions (four atoms), and out-of-plane (four atoms) coordinates to describe the motions of atoms during a normal mode of vibration. These internal coordinates, $\boldsymbol{R}$, are related to Cartesian displacement coordinates, $\boldsymbol{x}$, by the relation

$$\boldsymbol{R} = \boldsymbol{B}\boldsymbol{x} \tag{9.1}$$

The $\boldsymbol{B}$-matrix is a rectangular matrix of dimension $3N \times (3N - 6)$. The number of internal coordinates defined was about $3N - 6$ where, as before, N denotes the number of atoms in a molecule. The reason the above sentence contains the word "about" is that sometimes "redundant" coordinates were defined: for example, only five of the six angle deformation coordinates around a central carbon atom in a tetrahedral molecule are linearly independent; however, one sometimes defined all six coordinates, one of which was redundant. In this case, there would have been $3N - 5$ internal coordinates.

For a triatomic, nonlinear molecule such as water, the transition from Cartesian displacement coordinates (see Section 1.3) to internal coordinates significantly reduced the computational complexity: instead of 9 simultaneous equations with 9 eigenvalues (with the 6 translational and rotational equations with 0 eigenvalues) and 81 force constants, the vibrational computations in internal coordinates consisted of 3 equations with

9 force constants, some of which are related by symmetry; furthermore, these force constants were assumed to be transferable between similar molecules (or isotopic species, such as D_2O). The force constants, collectively referred to as the "force field," are discussed in more detail later.

The kinetic energy, expressed in internal coordinates, in analogy to Eqs. 1.8 and 1.34, is given by

$$2T = \dot{R}^T G^{-1} \dot{R} \tag{9.2}$$

where the dot refers to the time derivative of the coordinates, and the G-matrix to a kinetic-energy-related quantity, to be discussed below, that gave the entire computational approach the name "Wilson G-matrix" or "GF-matrix" method. The G-matrix was defined [4] as

$$G = R^T \left(\frac{1}{M}\right) R \tag{9.3}$$

Here, $\left(\frac{1}{M}\right)$ is a diagonal matrix containing the inverse of the masses of the individual atoms. In the space of internal coordinates, the secular equation could be written as

$$|F - G^{-1}\,\lambda| = 0 \tag{9.4}$$

$$|GF - E\lambda| = 0 \tag{9.5}$$

where E is the unit matrix.

The advantage of the "Wilson G-matrix method" was that the G-matrix elements could be tabulated for given internal coordinate geometries [4]. There were 11 different internal coordinates listed, 3 of which are shown in Figure 9.1.

Here, the nomenclature implies a simple bond stretching G-matrix element (G_{rr}^2), with atoms 1 and 2 involved. (G_{rr}^1) denotes a bond stretch–bond stretch interaction, with atom 1 shared in both motions. $G_{\varphi\varphi}^3$ denotes an angle deformation involving atoms 1–2–3. The corresponding G-matrix elements, for tetrahedral geometry with all bond angles equal ($\varphi = 109.47°$), can be found to be

$$G_{rr}^2 = \mu_1 + \mu_2 \tag{9.6}$$

$$G_{rr}^1 = -\frac{\mu_1}{3} \tag{9.7}$$

$$G_{\varphi\varphi}^3 = \left\{ \rho_{21}^2\mu_1 + \rho_{21}^2\mu_1 + \frac{1}{3}\,\left[3\rho_{21}^2 + 3\rho_{23}^2 + 2\,\rho_{12}\,\rho_{23}\right]\,\mu_2 \right\} \tag{9.8}$$

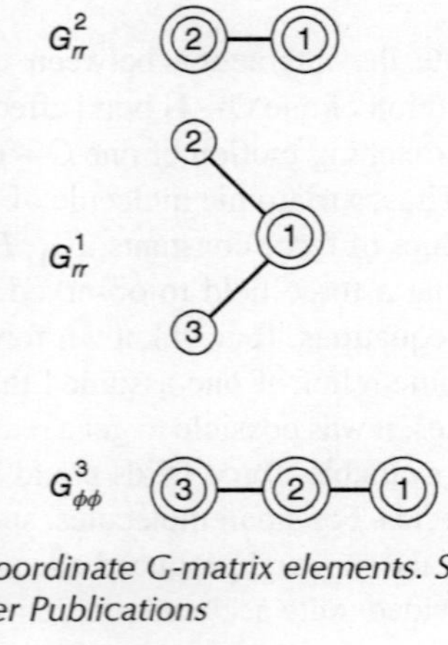

Figure 9.1 *Partial display of internal coordinate G-matrix elements. See text for details. Wilson* et al., *1955, [4]. Reproduced with permission from Dover Publications*

Here, μ denotes the reciprocal of the mass of the atom in question and ρ the reciprocal of the interatomic distance.

Parallel to the kinetic energy matrix elements, expressed in terms of the internal coordinates, the potential energy for the same internal coordinates was defined. The simplest force field used only diagonal terms, and was referred to as "valence force field" (VFF). For a symmetric triatomic molecule such as water, the determinant of the force constant matrix could be written as

$$|F| = \begin{vmatrix} F_{11} & & \\ & F_{11} & \\ & & F_{33} \end{vmatrix} \tag{9.9}$$

where the terms F_{11} denote the O—H stretching and F_{33} the H—O—H deformation force constants. In conjunction with the tabulated G-matrix elements, the problem of solving the vibrational secular equation was a manageable task. The normal coordinates, expressed in contributions of internal coordinates, could be obtained by the eigenvectors of $\boldsymbol{GF} - \boldsymbol{E}\lambda$ (Eq. 9.5).

It was found that the diagonal force constants for stretching modes were reasonably transferable for molecules with similar chemical structures, and were originally derived from force constants of diatomic molecules. Diagonal angle deformation, torsional, and out-of-plane force constants were obtained by numerically fitting force constants to reproduce observed vibrational frequencies. However, the VFF was not very reliable in predicting, for example, the vibrational frequencies of isotopically labeled molecules. This was a serious problem since – to a first approximation – the molecular force field should be independent of mass effects, and vibrational frequencies of isotopic species should be computable by merely changing the masses (kinetic energy).

In order to achieve transferability between different molecules, and to account for isotopic species, the VFF was augmented with off-diagonal force constants that needed to be fitted and adjusted to reproduce isotopic frequencies. This force field was generally referred to as the "generalized valence force field" (GVFF). For water, the force constant matrix would be

$$\begin{matrix} F_{11} & F_{11}^{11} & F_{33}^{11} \\ F_{11}^{11} & F_{11} & F_{33}^{11} \\ F_{33}^{11} & F_{33}^{11} & F_{33} \end{matrix} \tag{9.10}$$

Here, the off-diagonal elements indicate the interaction between different internal coordinates. F_{11}^{11}, for example, indicates how the stretching motion of one O—H bond affects the force constant of the other O—H bond. Similarly, F_{33}^{11} indicates how the stretching motion of one O—H bond affects the force constant of the H—O—H deformation force constant. Thus, a triatomic molecule of C_{2v} symmetry (see Chapter 2) requires, in the GVFF approach, four numeric values of force constants, F_{11}, F_{33}, F_{11}^{11}, and F_{33}^{11}.

For the reverse problem, namely, fitting a force field to observed frequencies, it is obvious the problem is indeterminate to some degree (three equations, four unknown force constants). Here, the assumption of transferable force constants provided some relief: if one assumed that the O—H stretching force constants were similar for chemically related species, it was possible to get a reasonable estimate for the rest of the force field. Even better, if isotopic data were available, force fields could be obtained that quite well reproduced the vibrational frequencies of related species. For small molecules, such as $^{35}ClCH_3$, $^{37}ClCH_3$, $^{35}ClCD_3$, and $^{37}ClCD_3$, such procedures worked satisfactorily, and produced adequate descriptions of the atomic motions for a normal mode of vibration, and provided quite accurate prediction of molecular vibrational frequencies.

9.2.2 Early computer-based vibrational analysis

Section 9.2.1 provided a historical insight into original methodologies for the computation of vibrational frequencies from structural parameters and a transferable force field such as the GVFF or the Urey–Bradley Force Field (UBFF, see, e.g., Ref. [5, Chapter 3]). The field of computational vibrational spectroscopy took an enormous leap forward when Schachtschneider from the Shell Development Company [6] published a set of Fortran programs that allowed rapid computation of vibrational frequencies using the Wilson G-matrix approach in 1965. These programs typically ran on IBM or similar main frame computers. Most notable was a program routine that allowed for the least square refinement of selected force constants. Suddenly, molecules much larger than those accessible previously could be analyzed; the interests at the Shell Development Company, for example, were in the areas of vibrational spectroscopy of alkanes.

By the late 1980s, it became practical to carry out similar computations on personal computers [7], at least for small molecules (<30 atoms). At this time, it become more practical to carry out the entire computation in Cartesian displacement coordinates, since the time savings due to looking up kinetic energy matrix elements in a table is insignificant compared to diagonalizing the total energy matrix in Cartesian coordinates. Furthermore, in mass-weighted Cartesian coordinates, the kinetic energy is already diagonal (see Eq. 1.8) whereas the GF matrix is neither diagonal nor symmetric. Thus, the entire problem would be reduced to inputting the force constants in the form of internal coordinates (such as a GVFF), but immediately transformed into Cartesian displacement coordinates according to

$$\boldsymbol{F}_q = \boldsymbol{W}\boldsymbol{B}^T \boldsymbol{F}_R \boldsymbol{B}\boldsymbol{W} \tag{9.11}$$

where $\boldsymbol{W}$ is a diagonal matrix of the inverse of the square root of atomic masses, $\boldsymbol{F}_q$ is the force constant matrix expressed in mass-weighted Cartesian coordinates, and $\boldsymbol{F}_R$ is the GVFF matrix in internal coordinates, with $\boldsymbol{B}$ defined by Eq. 9.1.

Diagonalization of the resulting total energy matrix revealed the vibrational eigenvalues as well as the eigenvectors that were required to define the normal modes of vibration. The eigenvector matrix also was needed in early attempts to compute infrared and VCD intensities, to be discussed in Section 9.4.

9.3 *Ab initio* quantum-mechanical normal coordinate computations

With the availability of quantum chemical software, such as Hyperchem, Gaussian, Turbomol, and others, the *ab initio* computation of molecular energies and molecular force fields became practical, and the empirical force field computations described in the previous section have become obsolete. Using extended basis sets and DFT, these calculations produce force fields by calculating the Cartesian electronic energy gradients with better accuracy than any empirical force field in the past.

This approach is based on finding for the eigenvalues of [8]

$$\sum\nolimits_{k=1}^{3N} (H_{jk} - \lambda_i M_{jk})\, X_{ki} = 0 \tag{9.12}$$

In Eq. 9.12, both indices j and k run from 1 to $3N$, where N, as before, refers to the number of atoms in the molecules. The eigenvalues λ_i are related to the vibrational frequencies ν_i by

$$\lambda_i = (2\pi\nu_i)^2 \tag{1.36}$$

M_{jk} is the diagonal matrix of the atomic masses and X_{ki} are the eigenvectors that transform from Cartesian displacement coordinates to normal coordinates (see Eq. 1.36). In Eq. 9.12, H_{jk} is the Hessian matrix of the partial derivatives of the total energy E of a molecule with respect to the Cartesian displacement coordinates:

$$H = \begin{array}{cccc} \dfrac{\partial^2 E}{\partial x_1^2} & \dfrac{\partial^2 E}{\partial x_1 \partial x_2} & \cdots & \dfrac{\partial^2 E}{\partial x_1 \partial x_{3N}} \\ \dfrac{\partial^2 E}{\partial x_2 \partial x_1} & \dfrac{\partial^2 E}{\partial x_2^2} & \cdots & \dfrac{\partial^2 E}{\partial x_2 \partial x_{3N}} \\ \vdots & \vdots & \vdots & \vdots \\ \dfrac{\partial^2 E}{\partial x_{3N} \partial x_1} & \dfrac{\partial^2 E}{\partial x_{3N} \partial x_2} & \cdots & \dfrac{\partial^2 E}{\partial x_{3N} \partial x_{3N}} \end{array} \tag{9.13}$$

These computations are still quite time consuming, and may take a few minutes for a molecule of 20 or 30 atoms. Depending on the total energy calculation used (basis set, etc.) all force constants may be found to be higher than those from empirical calculations and, therefore, yield vibrational frequencies that are higher than observed frequencies. In these cases, the entire force field is scaled by a constant factor, or as detailed by Williams *et al.* [9], a fitting procedure can be implemented that refines the scale factors appropriately. The recent version of the Gaussian suite of programs (Gaussian'09) allows the computation of anharmonic frequencies and force fields that should also improve the overall quality of observed frequencies.

9.4 Vibrational intensity calculations

In the previous section, the computation of the vibrational frequencies (the eigenvalues of the vibrational energy matrix) was discussed, both for early empirical approaches using transferable force constants and for an *ab initio* quantum mechanical approach where the forces acting between atoms are computed as the second derivatives of the total energy with respect to the normal coordinates of the molecule. The second part of the efforts required to compute or predict a vibrational spectrum deals with the problem of vibrational intensities. This is a good deal more complicated, since the exact atomic displacements as well the change in dipole moment accompanying the atomic displacement motion needs to be known. Classically, one may visualize the dipole moment change in terms of a partial charge attached to an atom; as the atoms move, a change in dipole moment will result. This approach was taken by researchers early on in efforts to compute vibrational intensities, but this approach was similarly limited as the approach of using transferable force fields for the computation of vibrational frequencies. The reason for this is that electronic motion is so much faster than nuclear motion; therefore, electrons redistribute during a molecular vibration, and the concept of "fixed partial charges" does not hold for molecular vibrations. Nevertheless, the discussion of computational vibrational intensities starts with a short, historical review of the fixed charge model.

9.4.1 Fixed partial charge method for infrared intensities

The infrared absorption intensity, or dipole strength D_{01}^{j} for the jth normal mode of vibration, for the transition from the vibrational ground state $|0\rangle$ to the first excited state $|1\rangle$is given by (see Eq. 1.114)

$$D_{01}^{j} = |\boldsymbol{\mu}_{01}|^2 = \frac{\hbar}{4\pi v_j} \left| \frac{\partial \boldsymbol{\mu}}{\partial Q_j} \right|^2 \tag{9.14}$$

In the case of a diatomic molecules with just one normal mode of vibration, the observed *intensity* can be directly related to the change in dipole moment. This is analogous to the determination of the force constant for the vibration of a diatomic molecule, where the observed *frequency* can be directly converted to a numerical value of the force constant.

To transition from diatomic to polyatomic molecules, it was hoped that certain parameters would be transferable and allow the empirical calculation of vibrational intensities. Thus, the first attempts at these calculations were made using the "fixed partial charge" model, where the charges would be transferable from one molecule to the next.

Early attempts to predict infrared absorption intensities were in part carried out by researchers in vibrational optical activity (in particular, VCD; see Chapter 8) in their quest to establish the sensitivity required for a VCD instrument to be able to detect this effect. Thus, the fixed partial charge model was developed, before the experimental verification of VCD to predict the magnitude of the effect and was published by Schellman [1]. In the fixed partial charge approach, the dipole operator is given by

$$\boldsymbol{\mu} = \sum_i \xi_i \boldsymbol{s}_i \tag{9.15}$$

Here, the summation is over all atoms i in a molecule and ξ_i are their fixed partial or shielded charges. The vector $\boldsymbol{s}_i$ represents the derivative of the position of atom i with respect to the normal coordinate Q_j.

For a given normal mode j, the dipole strength D_j for the ground to first excited vibrational state in the fixed partial charge approach is given by

$$D_{01}^{j} = |\boldsymbol{\mu}_{01}^{j}|^2 = \frac{\hbar}{4\pi \nu_j} \left| \left(\sum_i \xi_i \, s_i^j \right) \right|^2 \tag{9.16}$$

The change in dipole moment of a molecule in the fixed partial charge formalism can also be written as

$$\left(\frac{\partial \boldsymbol{\mu}}{\partial Q_j} \right) = \sum_i \xi_i \left(\frac{\partial \mathbf{R}_i}{\partial Q_j} \right) \tag{9.17}$$

where $\mathbf{R}$ is its equilibrium position of the ith nucleus and the term $(\partial \mathbf{R}_i/\partial Q_j)$ is the trajectory of the ith nucleus for the jth normal mode [10]. Unfortunately, this method is accurate in predicting vibrational intensities for selected vibrational modes only. For example, CH stretching intensities may be predicted reasonably well from Eq. 9.17 since the very nonpolar character of the C—H bond does not change appreciably upon bond elongation.

A more sophisticated approach, known as the "polar tensor" model, improves the predictions somewhat. In the polar tensor approach, the dipole change is expressed as

$$\left(\frac{\partial \boldsymbol{\mu}}{\partial Q_j} \right) = \sum_i \left(\frac{\partial \boldsymbol{\mu}}{\partial R_i} \right) \left(\frac{\partial \mathbf{R}_i}{\partial Q_j} \right) \tag{9.18}$$

Here, the terms $(\partial \boldsymbol{\mu}/\partial R_i)$ are the derivatives of the total molecular dipole moment with respect to the Cartesian displacements of nucleus i. This derivative matrix is known as the polar tensor. For symmetric molecules, Newton and Person [3] were successful in reducing this molecular polar tensor to an atomic polar tensor format. It was found that these experimentally derived atomic polar tensors were transferable to some degree from one molecule to another. However, even this model was far from a generally applicable approach to predict infrared intensities. The reason for the shortcomings of these models lies in that the "partial charges" of the atoms vary enormously during vibrations. In fact, they vary so much that vibronic methods need to be invoked to predict the dipole changes during a vibration properly. The charge redistributions, or charge flows, are so significant that only quantum mechanical calculations are able to predict the dipole changes properly.

9.4.2 Quantum mechanical infrared and Raman intensities: localized molecular orbitals

In order to explicitly account for charge redistribution during molecular vibrational motions, Nafie and coworkers [11–13] divided the expression for the change in dipole moment for the *j*th normal coordinate (Eq. 9.16) into nuclear and electronic contributions:

$$\left(\frac{\partial \boldsymbol{\mu}}{\partial Q_j}\right) = \sum\nolimits_n \left(\frac{\partial \boldsymbol{\mu}_n}{\partial Q_j}\right) + \sum\nolimits_i \left(\frac{\partial \boldsymbol{\mu}_i}{\partial Q_j}\right) \tag{9.19}$$

$$= \sum\nolimits_n e\, Z_n \mathbf{s}_{n,j} - \sum\nolimits_i e \boldsymbol{\sigma}_{i,j} \tag{9.20}$$

In Eqs. 9.19 and 9.20, the summations are over all nuclei n, which have nuclear charges $e\, Z_n$, and over all molecular orbitals i. The vector $\mathbf{s}_{n,j}$ has a similar meaning as defined above (Eq. 9.16), except that it describes the derivative of the *nuclear* coordinate with respect to the normal mode. Similarly, the vector $\boldsymbol{\sigma}_{i,j}$ describes the derivative of the centroid of the orbital position with respect to the normal coordinate. All derivatives in Eq. 9.19 are evaluated at the equilibrium positions of the atoms.

While the expressions for the change in dipole moment described by fixed partial charge model (Eq. 9.16) and for the "localized molecular orbital" (LMO) model in Eq. 9.20 appear similar, the complexity of Eq. 9.20 is orders of magnitude larger than that in Eq. 9.16. This is because the complete electronic energies and orbital centroids need to be computed for several displacement positions for each normal mode. At the time when these models were developed, most computations were not yet carried out – due to computational restrictions – at the *ab initio* level but used approximate methods such as CNDO.

In analogy to Eq. 9.19, Raman intensities can be computed according to

$$\left(\frac{\partial \boldsymbol{\alpha}_{\alpha\beta}}{\partial Q_j}\right) = \sum\nolimits_i \left(\frac{\partial \boldsymbol{\alpha}_{i}, \alpha\beta}{\partial Q_j}\right) \tag{9.21}$$

Here, the subscripts $\alpha\beta$ denote permutations of all Cartesian indexes x, y, z, and the summation is again over all orbitals. There are no nuclear contributions such as those in Eq. 9.19, since the nuclei do not contribute to the polarizability.

Thus, both infrared and Raman intensities can be computed for a molecule from the displacement vectors of the nuclei during a normal mode of vibration (see Eq. 9.12) and the displacement vectors of the centroids of the molecular orbitals. These can be viewed as shielding factors of the intensities one would obtain for the motions of the naked nuclei. Difficulty arises, however, from the fact that these shielding factors vary enormously during a normal vibration.

9.4.3 The finite perturbation method

Originally, the nuclear and orbital positions required in Eq. 9.20 were calculated for the equilibrium position, and subsequently, for the molecules distorted by the vibrational coordinates, at several amplitude values. These "brute force calculations" were impractical for most molecular systems of interest due to prohibitive computation times. In 1979, an elegant way to circumvent the computational burden was developed by Kormornicki and McIver [14]. Rather than computing the derivatives required in Eqs. 9.19 and 9.21 for different molecular geometries distorted by the vibrational coordinates, the authors proposed to evaluate these derivatives in the presence of a small electric field perturbation according to

$$\left(\frac{\partial \boldsymbol{\mu}_{\alpha}}{\partial Q_j}\right) = -\frac{\partial}{\partial Q_j}\left(\frac{\partial \varepsilon}{\partial F_\alpha}\right) = -\frac{\partial}{\partial F_\alpha}\left(\frac{\partial \varepsilon}{\partial Q_j}\right) = -\left(\frac{\partial g_j}{\partial F_\alpha}\right) \tag{9.22}$$

In Eq. 9.22, ε is the potential energy matrix of the electrons in the molecules and ∂F_α are the Cartesian components of the applied perturbing electric field. The term g_j is the gradient of the electronic potential energy along the normal coordinate Q_j. Equation 9.22 now needs to be written in terms of the sum of the contributions of the individual LMOs:

$$\left(\frac{\partial \boldsymbol{\mu}_\alpha}{\partial Q_j}\right) = -\sum_i \left(\frac{\partial g_{i,j}}{\partial F_\alpha}\right) + \sum_n \left(\frac{\partial \mu_{n,\alpha}}{\partial Q_j}\right) \tag{9.23}$$

where the second term corresponds to the nuclear contribution in Eq. 9.19. A comparison of Eqs. 9.23 and 9.19 yields the contributions of each localized orbital to the dipole moment derivatives:

$$\left(\frac{\partial \boldsymbol{\mu}_{i,\alpha}}{\partial Q_j}\right) = -\left(\frac{\partial g_{i,\alpha}}{\partial F_\alpha}\right) \tag{9.24}$$

The corresponding equations for Raman scattered intensities are

$$\left(\frac{\partial \boldsymbol{\alpha}_{\alpha\beta}}{\partial Q_j}\right) = -\frac{\partial}{\partial Q_j}\left(\frac{\partial \mu_\alpha}{\partial F_\beta}\right) = -\frac{\partial}{\partial F_\beta}\left(\frac{\partial \mu_\alpha}{\partial Q_j}\right) = -\left(\frac{\partial^2 g_j}{\partial F_\beta \partial F_\alpha}\right) \tag{9.25}$$

and

$$\left(\frac{\partial \boldsymbol{\alpha}_{\alpha\beta}}{\partial Q_j}\right) = -\sum_i \left(\frac{\partial^2 g_{i,j}}{\partial F_\beta \partial F_\alpha}\right) \tag{9.26}$$

Equations 9.23 and 9.26 explicitly express how to compute infrared absorption and Raman scattered intensities *via* the finite perturbation procedure, as applied to the localized orbital model. The computational savings in the finite perturbation approach, as compared to the standard method, is enormous: in the latter, at least one MO calculation needs to be carried out for each of the $3N-6$ normal modes for nuclei displaced along the normal coordinates. In the finite perturbation approach, on the other hand, the dependence of the intensity parameters on the normal modes is computed analytically from the total electronic energy in the equilibrium position. Thus, this method is amenable to much higher levels of MO calculations, and is now contained in MO programs such as Gaussian'09 or Hyperchem for *ab initio* calculations using standard Gaussian as well as density functional approaches.

An example demonstrating the excellent agreement between observed and computed infrared spectra for a small (six heavy atoms) cyclic molecule is shown in Figure 9.2. Here, both the force field and intensities were obtained by high level DFT methodology [15].

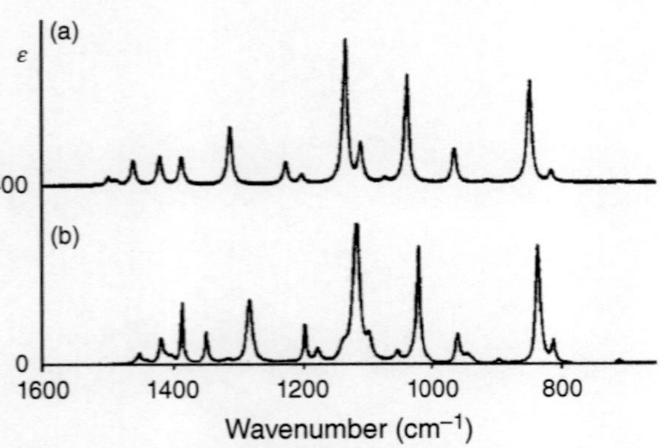

Figure 9.2 *Example of* ab initio *calculation of infrared absorption spectra. (a) Observed spectrum for 4-methyl-2-oxetanone. (b) Computed spectrum using* ab initio *density functional theory. Stephens, Devlin, Chabalowski, et al., [15]*

References

1. Schellman, J.A. (1973) Vibrational optical activity. *J. Chem. Phys.*, **58** (7), 2882–2886.
2. Person, W.B. and Newton, J.H. (1974) Dipole moment derivatives and infrared intensities. I. Polar tensors. *J. Chem. Phys.*, **61** (3), 1040–1049.
3. Newton, J.H. and Person, W.B. (1976) Dipole moment derivatives and infrared intensities II. Polar tensors in methyl halide molecules. *J. Chem. Phys.*, **64** (7), 3036–3049.
4. Wilson, E.B., Decius, J.C., and Cross, P.C. (1955) *Molecular Vibrations*, McGraw-Hill.
5. Diem, M. (1993) *Introduction to Modern Vibrational Spectroscopy*, Wiley-Interscience, New York.
6. Schachtschneider, J.H. (1965) *Vibrational Analysis of Polyatomic Molecules: FORTRAN IV Programs for Solving the Vibrational Secular Equation and Least-Squares Refinement of Force Constants*, Shell Development Company, Emeryville, CA.
7. Barlow, A. and Diem, M. (1991) Normal coordinate calculations as a classroom computer project. *J. Chem. Educ.*, **68** (1), 35–39.
8. Porezag, D. and Pederson, M.R. (1996) Infrared intensities and Raman-scattering activities within density-functional theory. *Phys. Rev. B*, **54** (11), 7830–7836.
9. Williams, R.W., Bone, I., and Weir, A.F. (2004) A scaled quantum mechanical force field for the alanine–alanine zwitterions in water based on scale factors for alanine, glycine, N-methylacetamide, acetate, formate, and methylamine. *J. Mol. Struct.*, **697**, 81–95.
10. Freedman, T.B. and Nafie, L.A. (1987) Stereochemical aspects of vibrational optical activity, in *Topics in Stereochemistry* (eds E.L. Eliel and S.H. Wilen), John Wiley & Sons, Inc., New York, pp. 114–202.
11. Nafie, L.A. and Polavarapu, P.L. (1981) Localized molecular orbital calculations of vibrational circular dichroism. I. General theoretical formalism and CNDO results for the carbon-deuterium stretching vibration in neopentyl-1-d-chloride. *J. Chem. Phys.*, **75** (6), 2935–2944.
12. Polavarapu, P.L. and Nafie, L.A. (1981) Localized molecular orbital calculations of vibrational circular dichroism. II. CNDO formulation and results for the hydrogen stretching vibrations of (+)-(3R)methylcyclohexanone. *J. Chem. Phys.*, **75** (6), 2945–2951.
13. Nafie, L.A. and Freedman, T.B. (1981) A unified approach to the determination of infrared and Raman vibrational optical activity intensities using localized molecular orbitals. *J. Chem. Phys.*, **75** (10), 4847–4851.
14. Kormornicki, A. and McIver, J.W. Jr., (1979) An efficient ab initio method for computing infrared and Raman intensities: application to ethylene. *J. Chem. Phys.*, **70** (4), 2014–2016.
15. Stephens, P.J., Devlin, F.J., Chabalowski, C.F., and Frisch, M.J. (1995) *Ab Initio Calculation of Vibrational Absorption and Circular Dichroism Spectra Using Density Functional Force Fields*, Army Research Laboratory.

Part II

Biophysical and Medical Applications of Vibrational Spectroscopy and Microspectroscopy

Introduction to Part II

In Part I, the basic principles of vibrational spectroscopy were introduced and elaborated upon using the spectroscopic features of small molecules as examples. Part II is devoted to biophysical and biomedical applications of vibrational spectroscopy. It starts with the macroscopic spectroscopy of the major biological components, a subject that has been explored extensively and includes studies on peptides, proteins, nucleic acids, and lipids, as well as specific biophysical systems such as the photosystems, heme proteins, and so on. These studies were carried out as early as the 1970s, when infrared and Raman spectroscopy became sufficiently sensitive to carry out biophysical studies in aqueous solution, and yielded a plethora of information on structure, interactions, hydration, and dynamics of biological systems. These studies could be carried out at concentrations much lower than those required for NMR studies, and gave a better indication of interactions and structural changes of biological systems under near-biological conditions. Furthermore, the extremely fast timescale of vibrational transitions, in general, allows the detection of conformational structures that appear only as broad averages in slow techniques such as NMR spectroscopy.

The other methods introduced in Part II deal with the spectroscopy of even larger sample systems, namely entire cells or sections of tissue. Although it is certainly possible to analyze such samples macroscopically, that is, by inserting such samples on a suitable substrate into the sample position in a standard Fourier transform infrared (FTIR) spectrometer, where the beam measures a few millimeters in diameter, such measurements are not very useful since they represent an average over many cells or tissue types. In order to correlate infrared or Raman spectra to any relevant medical features, it is necessary to carry out these measurements microscopically, that is, through a Raman or IR (infrared) microscope, and correlate spectral changes on the cellular level. An exception to this statement is the analysis of dried body fluids where no spatial resolution is needed, and which, therefore, can be carried out macroscopically.

Such infrared and Raman microspectral (or microscopic) techniques are introduced in Chapter 11. This instrumental chapter is followed by a chapter on the data analysis utilized for the interpretation of vibrational microspectral data sets. The methodology introduced in this chapter is of particular significance for two reasons. First, in infrared and Raman microspectral mapping (or imaging) experiments, data sets of 10 000–1 000 000 spectra are routinely collected; with spatial dimensions for each individually collected spectrum determined by the diffraction limit (see Chapter 11). Manual evaluation and interpretation of data sets of that size is, of course, impossible. Second, there is a natural variance in the spectra of (near-identical) cells and tissue types that require a careful statistical analysis of the spectral features and their changes that may be brought on by a variety of biological or medical conditions.

Finally, Chapters 13 and 14 deal with the biomedical applications of infrared and Raman microspectroscopy. This field has expanded, from a few papers per year in the 1990s to well in the hundreds of papers a year at the time of writing of this book. In fact, the RSC journal *Analyst* has published several special issues devoted to biomedical applications, and so has the *Journal of Biophotonics*. These two journals have contributed much to the dissemination of ideas in this field, and have become major venues of publishing research papers related to biomedically oriented research in vibrational spectroscopy. Since vibrational spectroscopy monitors the biochemical composition of a sample pixel, and changes in the composition due to a variety of conditions, these microspectral techniques allow novel diagnostic approaches that do not depend on cell morphology or tissue structure, but truly on changes in chemical composition. Thus, vibrational spectroscopic methods bridge classical cytological and histological approaches with a combination of genomic, proteomic, and metabolomics approaches. Although vibrational microspectral techniques do not have the pinpoint sensitivity to detect all up- or downregulated proteins in cancer, for example, as do some proteomic methods, they do present the ability to obtain a "bird's-eye" view of changes in the proteome, while simultaneously monitoring other compositional changes such as the metabolome.

10

Biophysical Applications of Vibrational Spectroscopy

10.1 Introduction

Even a cursory literature search of key words related to vibrational spectroscopy of peptides, proteins, nucleic acids, and lipids reveals that there are well over 10,000 papers published in these fields. Along with solution phase NMR spectroscopy, infrared (IR) and Raman spectroscopies have revealed a plethora of information on solution structure and dynamics of biomolecules. The main advantages of vibrational spectroscopy over other techniques used in structural biophysics are the low sample requirement (micro- to nanogram quantities), the ease of the measurement itself, the fast timescale of vibrational transitions, the ability to focus on particular regions of the spectrum using difference or resonance enhancement methods, the ability to operate in solution phases, and many more.

Since the early 1980s, the sensitivity of structural studies of nucleic acids, peptides, and proteins have seen a vast increase with the introduction of specialized techniques such as vibrational circular dichroism (VCD) and Raman optical activity (ROA, see Chapter 8), in addition to "classical" IR and Raman spectroscopies. In fact, VCD studies have revealed more direct structural information on peptides and proteins than all other vibrational techniques combined. Consequently, this chapter emphasizes results from all vibrational techniques, in particular those from VCD measurements.

This chapter reviews the basic results for peptides, proteins, nucleic acids, and lipids, and also explores some results that demonstrate how imaginative application of these techniques can reveal information that is not available from any other techniques. Unfortunately, the biophysical information provided by techniques of vibrational spectroscopy are often overshadowed by the "direct" structural methods, such as X-ray crystallography or solution phase NMR spectroscopy that can provide a highly accurate but rather static picture of the shape and interaction of biomolecules.

Modern Vibrational Spectroscopy and Micro-Spectroscopy: Theory, Instrumentation and Biomedical Applications, First Edition.
Max Diem.

10.2 Vibrations of the peptide linkage and of peptide models

10.2.1 Amino acids and the peptide linkage

Owing to their formidable size, low solubility, and structural complexity, proteins were realized early on to be poor subjects for initial studies using vibrational spectroscopy. Thus, small yet structurally similar model systems were used initially to help understand the vibrational features of proteins. To this end, peptide models and small peptides were studied in both IR and Raman spectroscopies.

A peptide (as well as a protein) consists of any of 20 naturally occurring amino acids, linked by peptide moieties, as shown in Figure 10.1(a). The 20 amino acids differ in the nature of the "R" group that may be simply an H atom, a methyl, or one of the several aliphatic groups, or aromatic, or charged groups. The —CO_2H group of one amino acid and the NH_2—group of another amino acid form a planar peptide linkage, as depicted in Figure 10.1(b). The vibrations of this peptide or amide linkage dominate the IR spectra of peptides and proteins, and are discussed next.

10.2.2 The vibrational modes of the peptide linkage

The original vibrational assignment of the peptide linkage can be traced back to the work of Miyazawa *et al.* [1] who used the molecule *N*-methylacetamide (NMAA, the structure shown in Figure 10.1(b) with methyl groups at the two terminal C positions) as a model for the peptide linkage. The vibrational modes for NMAA are listed in Table 10.1. Here, the amide A vibration at 3250–3300 cm^{-1} corresponds to an N—H stretching vibration that is, as expected, sensitive to solvent interaction (hydrogen bonding) and deuteration. The amide I vibration is mostly the C═O stretching mode at 1653 cm^{-1} (for NMAA) with minor contribution from the C—N stretching coordinate. As this mode is nearly insensitive to deuteration of the amide proton, a large

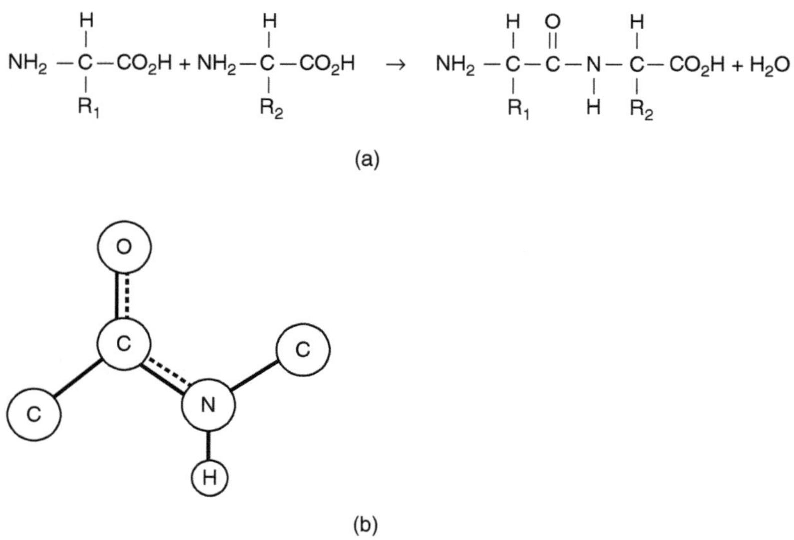

Figure 10.1 (a) Schematic structure of amino acids and the formation of a peptide linkage. R_1 and R_2 denote any of the 20 natural amino acid side chains. (b) Details of the peptide linkage. The carbonyl carbon atom and the nitrogen atom are both sp^2 hybridized; hence, the six atoms shown lie in one plane, and the carbon–nitrogen bond has multiple bond character, as indicated by the dotted line that represents a delocalized bond

Table 10.1 Observed frequencies (cm^{-1}) and vibrational assignment for the amide vibrations in NMAA

Name	Frequency	Approximate description[a]
Amide A	3250–3300	N—H stretch
Amide I	1620–1690	C═O stretch (C—N stretch)
Amide II	1540–1560	N—H deformation (C—N stretch)
Amide III	1230–1330	C—N stretch/N—H deformation[b]
Amide IV	630–750	O═C—N deformation
Amide V	700–750	N—H out-of-plane deformation
Amide VI	~600	C═O out-of-plane deformation

[a]Minor contributions in parentheses.
[b]Note that for real peptides, the amide III vibration is assigned to different vibrational coordinates.

contribution from the N—H deformation coordinate is unlikely. The amide I vibration has a very large dipole transition moment of

$$\mu \approx 3 \times 10^{-19} \text{ (esu cm)} \tag{10.1}$$

corresponding to an extinction coefficient of about 950 (L/(mol cm)), and is therefore the strongest band in IR spectra of peptides and proteins [2]. The transition moment of this vibration lies about 20° from the C═O bond direction (toward the adjacent C atom) [3]. The amide I vibration is exquisitely sensitive to hydrogen bonding and peptide conformation. This latter aspect is due in part to the transition dipole coupling discussed in Chapter 5. This coupling is strongly geometry dependent and, therefore, readily permits assessment of peptide and protein secondary structure.

At first glance, the amide I appears to be a typical "group frequency" as discussed in Chapter 5; however, owing to the coupling interaction of adjacent peptide linkages, the simple group frequency view needs to be reevaluated in terms of delocalized vibrational states. The mechanism of this interaction and the conformational sensitivity of the amide I vibration are discussed in Section 10.3.

As the N—H bond has primary amide character, the H atom readily exchanges with deuterium when a peptide is dissolved in D_2O, unless it is in the hydrophobic interior of a protein, or strongly hydrogen bonded. Upon deuteration, the amide I band is shifted by <5 cm^{-1}. In prolyl peptides, the amide I vibration occurs at significantly lower frequencies (about 1623 cm^{-1}) because a tertiary amide peptide linkage is formed [4].

The amide II vibration is a coupled C—N stretching and N—H deformation; this band shifts to about 1450 cm^{-1} upon deuteration; however, it shows very little conformational sensitivity and is not widely studied.

The other conformationally sensitive band of the peptide linkage is the so-called amide III vibration, observed between 1250 and 1350 cm^{-1}. Based on early studies on *N*-methylacetamide [1], this band was assigned to a C—N stretching/in-plane N—H bending mode; however, in real peptides, rather than the model system *N*-methylacetamide, the N—H deformation group strongly couples with the C—H deformations of adjacent C_α atom. This was demonstrated by the author and coworkers in Raman, IR, and VCD studies of

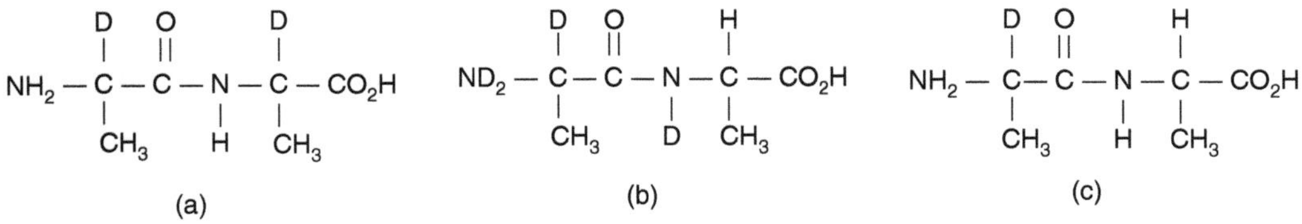

Figure 10.2 Selectively deuterated alanyl alanine isotopomers used for the following discussion. See text for details

selectively deuterated alanyl di- and tripeptides in aqueous solution [5–7]. The reassignment of this spectral region is discussed next in detail.

Figure 10.2 shows a series of alanyl alanine molecules indicating the positions that were deuterated. The spectrum of the dipeptide, deuterated at both C_α positions (Figure 10.2(a)), only shows a broad feature at $1336\,cm^{-1}$ that was assigned to the N—H in plane deformation mode. In species (b), the "unperturbed" C_α—H deformations of the C-terminal alanine residue are observed at 1330 and $1279\,cm^{-1}$, in analogy to alanine itself where they occur at 1351 and $1301\,cm^{-1}$.

In species (c), the deformation vibration of the C_α—H hydrogen adjacent to the peptide linkage couples strongly with the N—H deformation. This coupling of the two deformation coordinates, the $1330\,cm^{-1}$ C_α—H and the $1336\,cm^{-1}$ N—H deformation, produces increased splitting of the states as shown in Figure 10.3, and mixing of their wavefunctions, in analogy to the mixing of coordinates described under Fermi resonance in Chapter 5. Two new modes originate from this mixing, a symmetric combination at $1311\,cm^{-1}$ (polarized in the Raman spectrum) and an antisymmetric one at $1346\,cm^{-1}$ (depolarized).

This interaction between the C_α—H and the N—H vibrational coordinates appears to be the cause of the conformational sensitivity [8] of what is commonly assigned the amide III vibration, as it is well known that these interactions depend on the dihedral angle between the coupling groups. However, the study reported here better explained the reason for this conformational sensitivity of the amide III vibration and properly assigned this vibration as a coupled C_α—H/N—H coordinates.

The discussion presented so far was based on the vibrational assignment of a number of selectively deuterated alanyl dipeptides; the inclusion of diastereomeric species led to further insight into the nature of the amide III vibration, to be discussed next.

The solution phase IR and Raman spectra of the diastereomeric species L-alanyl-L-alanine and D-alanyl-L-alanine in the 1200–1500 cm^{-1} range are shown in Figure 10.4 [6, 9]. Here, the peaks at 1460, 1407, and $1370\,cm^{-1}$ are due to the methyl antisymmetric deformation, the symmetric stretching mode of the (deprotonated) carboxylate group, and the symmetric methyl deformation, respectively. These peaks are the same for both diastereomeric forms. Three peaks remain in the spectral region between 1270 and $1350\,cm^{-1}$: two unresolved peaks at $1325/1340\,cm^{-1}$ and a broad strong peak at $1279\,cm^{-1}$. The former pair exhibits geometric sensitivity, that is, the peaks split apart and change relative intensities and depolarization ratios when the

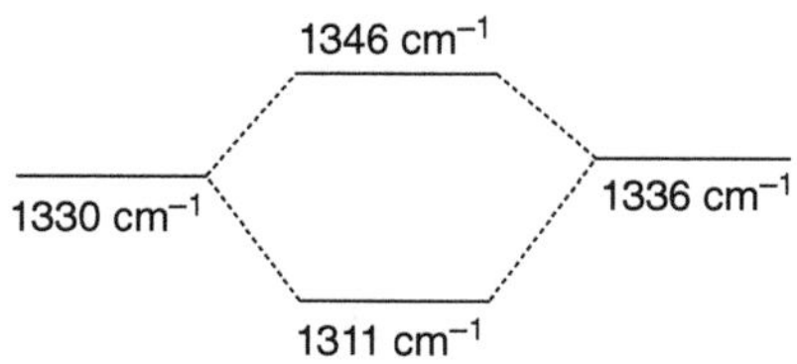

Figure 10.3 Energy level diagram for the coupling of the C_α—H deformation at $1330\,cm^{-1}$ and the N—H deformation at $1336\,cm^{-1}$ to result in coupled states

N-terminal L-alanine residue was substituted by a D-alanine [5]. The latter band at 1279 cm^{-1} (see Figure 10.4) was designated as the amide III band by previous workers. However, this peak disappeared when either of the C_α—H (at the C or N terminal alanine) is deuterated; thus, this peak cannot be the amide III band. Rather, the following reassignment was postulated: the 1270–1350 cm^{-1} region is a mixture of three hydrogen deformation modes, the C_N—H deformation of the N-terminal alanine residue, the N—H deformation, and the C_C—H deformation of the C-terminal residue. Consequently, this region (formerly the "amide III region") is now referred to as the "extended amide III" mode, and can be described by three amide III bands referred to as the amide III^1, amide III^2, and amide III^3 modes, all of which contain significant N—H deformation contributions. These three amide III modes are shown in Figure 10.5, based on empirical force field calculations. This force field was based on vibrational frequencies of six Ala-Ala isotopomers and two diastereomeric forms, and used IR, Raman, and VCD intensity arguments for the assignments [6, 9] and was recently verified by *ab initio* quantum mechanical calculations [10].

This reinterpretation of the amide III manifold described earlier and published between 1984 and 1992 by the author's group [6, 9] remains largely unknown in the scientific literature, and many researchers still describe the amide III band in terms of the original assignment in NMAA by Miyazawa [1], see Table 10.1. However, the revised assignment presented here was recently verified computationally by Weymuth and Reiher [10]. Furthermore, Raman and ROA studies on peptides and proteins by the research group of Barron

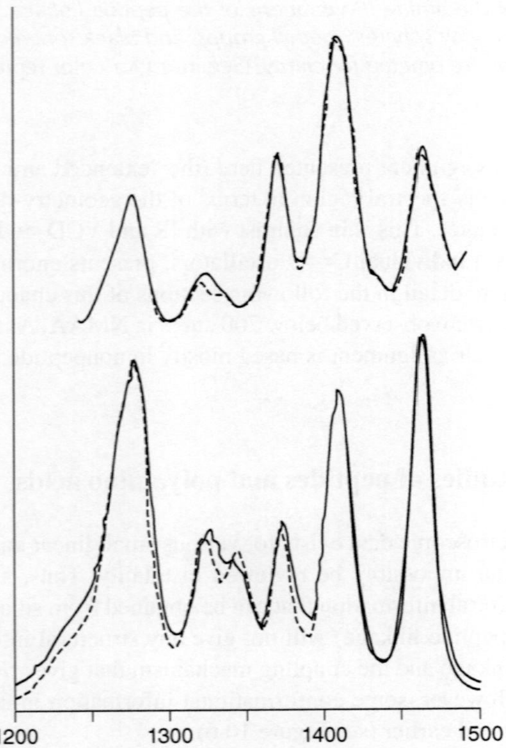

Figure 10.4 *Infrared absorption (top) and Raman (bottom) spectra of L-alanyl-L-alanine (solid trace) and D-alanyl-L-alanine (dotted trace)*

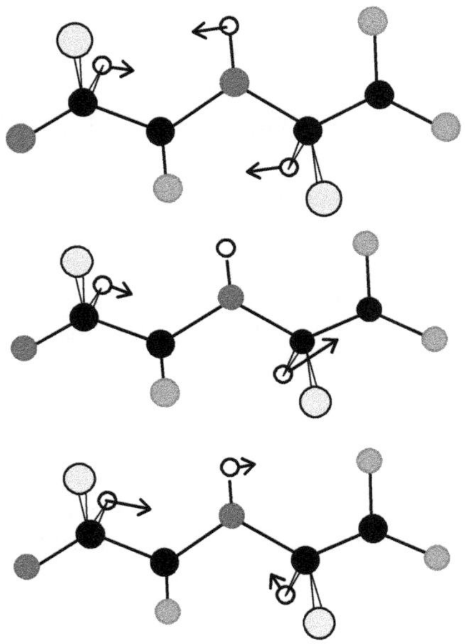

Figure 10.5 *Reassignment of the amide III vibrations of the peptide linkage. See text for detail. Red spheres: oxygen, blue spheres: nitrogen, gray spheres: methyl groups, and black spheres: carbon. The hydrogen atoms of the* $-CH_3$ *and* $-NH_3^+$ *groups are omitted for clarity.* (See insert for color representation of this figure.)

[11] have confirmed that the assignment presented here (the "extended amide III" assignment) explains the conformational sensitivity of this spectral region in terms of the geometry-dependent coupling of the C—H and N—H deformation coordinates. This is in contrast with IR and VCD studies where the amide I manifold, due to the interactions between individual C=O oscillators, presents enormous conformational sensitivity. This aspect is discussed in more detail in the following sections of this chapter.

The amide IV–VI vibrations were observed below 700 cm^{-1} in NMAA. As they are nonspecific toward protein secondary structure and their assignment is based mostly in nonpeptide model systems, their usefulness is quite restricted.

10.3 Conformational studies of peptides and polyamino acids

A plethora of vibrational spectroscopic data exists for various small linear and cyclic peptides. In the context of this book, these studies can impossibly be reviewed in totality. Thus, a few studies are highlighted to exemplify the breadth of structural information that can be obtained from small peptides. Dipeptides (i.e., two amino acids connected by a peptide linkage) will not give any structural information in the amide I region, as there is only one peptide linkage and the coupling mechanism that gives rise to conformationally sensitive information does not exist. However, some conformational information in dipeptides can be obtained from the amide III region, as discussed earlier (see Figure 10.6).

However, at the tripeptide level, interactions of the two amide linkages can be observed. A model system, *N*-acetyl-alanine-*N'*-methyl amide (AAMA) as well as L-Ala-L-Ala-L-Ala [$(\text{L-Ala})_3$], has two peptide linkages that can undergo transition dipole coupling of the two amide I vibrations, as discussed in Chapter 5. In a neutral aqueous (D_2O) solution of $(\text{L-Ala})_3$, two peaks are observed in the amide I region in both IR and Raman spectra

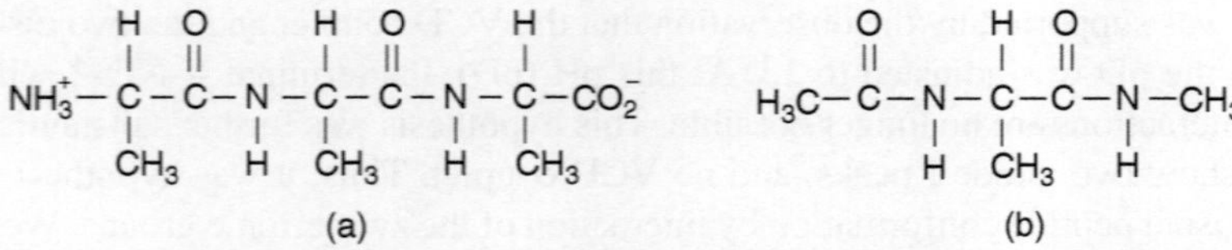

Figure 10.6 *(a) Structures of zwitterionic alanyl-alanyl-alanine ((L-Ala)$_3$) and (b)* N-acetyl-alanine-N′methyl-*amide (AAMA)*

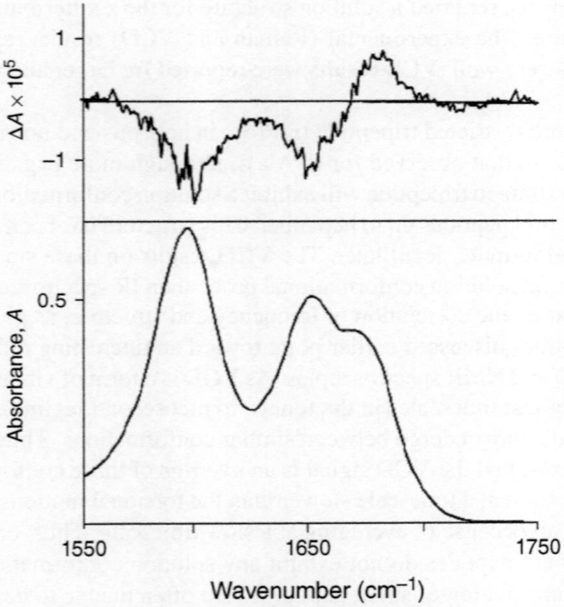

Figure 10.7 *Infrared absorption (bottom) and VCD spectra (top) of L-alanyl-L-alanyl-L-alanine in neutral aqueous solution (D$_2$O). The strong peak at 1595 cm^{-1} is the antisymmetric stretching mode of the ionized carboxylate group. Lee et al., 1989, [12]. Reproduced with permission from John Wiley & Sons*

(see Figure 10.7). The higher frequency peak at 1675 cm^{-1} is polarized in the Raman spectra, whereas the lower frequency peak at 1650 cm^{-1} is depolarized. Consequently, they were assigned as the symmetric and antisymmetric combination bands of the two amide I vibrations, respectively [12]. As expected from the exciton formalism discussed in Chapter 8, these two bands exhibit a conservative VCD couplet (Figure 10.7). These observations suggest that (L-Ala)$_3$ exists in neutral aqueous solution in a stable structure, which is unexpected for such a small peptide.

The intensity ratios of the symmetric and antisymmetric combination peaks, their exciton splitting, and the magnitude of the VCD intensities were used to search the conformational space of (L-Ala)$_3$ for any structure that reproduced these spectral parameters. Unfortunately, none of the generally accepted energy minima in the Ramachandran plot of peptides [13] reproduced the observed spectral parameters, and the only conformation that approximately reproduced the observation was a highly strained structure in which the zwitterionic groups were in relatively close proximity. Thus, we postulated that the observed structure may have been stabilized by Coulombic interactions of the ionic groups.

This hypothesis was supported by the observation that the VCD couplet and the two distinct amide I peaks disappeared when the pD was adjusted to 13. At this pH (pD), the terminal —ND_3^+ will be deprotonated, and zwitterionic interactions are no longer possible. This hypothesis was further substantiated by the fact that AAMA does not show two amide I peaks, and no VCD couplet. Thus, it was hypothesized that $(\text{L-Ala})_3$ is stabilized in an unusual peptide conformation by interaction of the zwitterionic groups. Woutersen and Hamm [14] reported for the same peptide at pD of 1 a much smaller exciton splitting, based on polarization-sensitive 2-D vibrational correlation spectroscopy; therefore, they arrived at a different solution conformation than the one derived in Ref. [12]. Nevertheless, both studies concluded that there was a solution structure to this peptide on the vibrational timescale. Similarly, a later VCD and Raman study [15], using more sophisticated algorithms for the data analysis, reported a solution structure for the zwitterionic species to be a left-handed, poly-proline II-like structure. The experimental (Raman and VCD) results reproduced the author's much earlier experimental results very well. VCD results were reported for larger alanyl peptides, $(\text{L-Ala})_n$, $n = 3$–6 as well [16].

The conformationally more restricted tripeptide $(\text{L-Pro})_3$ in aqueous and nonaqueous solutions also exhibited a VCD couplet, similar to that observed for $(\text{L-Ala})_3$, although more negatively biased [17]. This result indicates that an even more strained tripeptide will exhibit a solution conformation in aqueous and nonaqueous media that persist for the prolyl peptides up to heptamer. This structure has been identified as the poly-proline II structure, to be discussed in more detail later. The VCD results on these small peptides demonstrate that VCD is a much more powerful solution conformational probe than IR spectroscopy alone, which only reveals peptide conformation based on the correlation of frequency and structure, as given in Table 10.2.

The results on small peptides discussed earlier point toward an interesting difference between the solution structures deduced by VCD and NMR spectroscopies. As VCD is a form of vibrational spectroscopy, it probes peptide structures on a very fast timescale (in the femto- to picosecond regime), which is faster than the torsional motions that a peptide may undergo between similar conformations. Thus, VCD samples an ensemble of slightly different structures, and the VCD signal is an *average* of these conformations. NMR, on the other hand, samples the conformation at a timescale slower than the torsional motions, and at times may not detect a solution conformation at all because of averaging at a slow timescale. Thus, one may find statements in the literature indicating that small peptides do not exhibit any solution conformation at all, because techniques for monitoring solution conformation of small peptides were often unable to detect a preferred structure.

The aforementioned studies indicated that interactions between individual amide I vibrational transitions give rise to coupled amide I exciton states that explain the sensitivity of the amide I region toward secondary peptide structure. A generally accepted frequency–structure correlation for peptides in aqueous (H_2O and

Table 10.2 *Frequencies (cm^{-1}) for the amide I vibrations for different peptide secondary structures*

Structure	Amide I frequency	Amide I′ frequency[a]
α-Helix	1655	1652
β-Sheet	1635/1680	1630/1680
β-Turn	1675	1670
Disordered[b]	1655	1645

[a]The amide vibrations of a deuterated peptide linkage is referred to as the primed amide vibrations.
[b]The nature of the "disordered" peptide structure is explained in the following paragraphs.

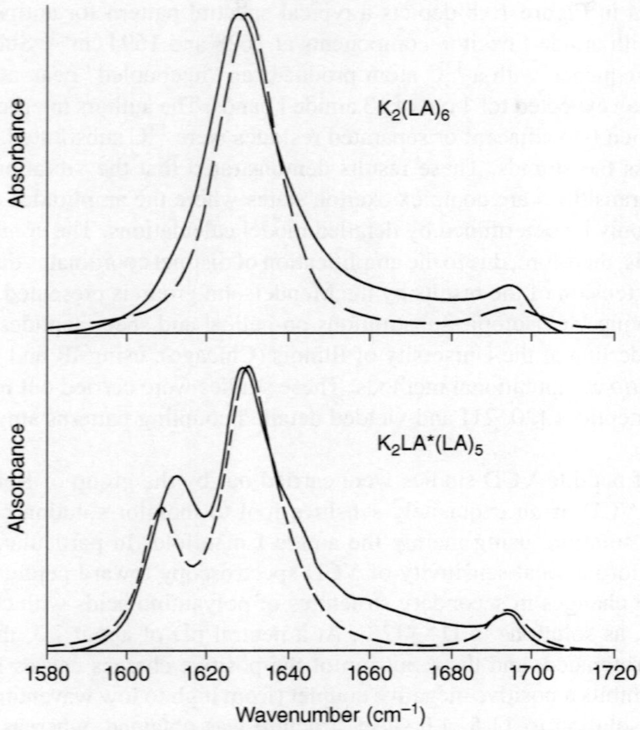

Figure 10.8 *Observed (solid lines) and simulated (dashed traces) infrared spectra of $K_2(LA)_6$ and K_2 $LA^*(LA)_5$. See text for detail. Brauner* et al.*, 2000, [18]. Reproduced with permission from American Chemical Society.*

D_2O) solution is presented in Table 10.2; however, this table is based primarily on empirical data, and until the 1990s, there were no good theoretical models to confirm these observations.

Thus, better models were needed to understand the conformational sensitivity of the amide I manifold. While the computations carried out for the $(Ala)_3$ studies discussed earlier [12] were rather primitive and included only dipolar coupling between individual amide I transitions, Mendelsohn and coworkers (Rutgers University, Newark) published an empirical computational approach to interpret IR band contours for peptides as well as proteins (see Section 10.4) [3, 18]. Their model included, in addition to transition dipole coupling, terms referred to as "through bond interactions," basically additional force constants describing the interactions of peptide vibrations separated by the C_α atom. In a pioneering study, they reported calculated and observed IR spectra of isotopically labeled peptides consisting of 14 amino acids, Lys-Lys-$(Leu\text{-}Ala)_6$ (also referred to by its one-letter code as $K_2(LA)_6$), which, in methanolic solution, form antiparallel β-sheet structures. These studies reproduced the splitting patterns observed in pleated sheet peptides and, furthermore, revealed the complicated coupling patterns between the amide I oscillators. This was accomplished by substituting one or more alanine or leucine residues by residues with a ^{13}C atom at the carbonyl carbon. As isotopic substitution at the carbonyl group shifts the amide I vibration by about 40 cm^{-1} to 1614 cm^{-1}, it effectively uncouples the substituted from unsubstituted amide vibrations. Before this work, similar studies involving α-helical peptides had demonstrated that ^{13}C substitution produced frequency and intensity patterns consistent with the heavier isotopic mass [19]. However, for the β-sheet structures assumed by $K_2(LA)_6$, even a single ^{13}C substitution produced unexpected spectral changes. This is shown in Figure 10.8.

The top set of spectra in Figure 10.8 depicts a typical spectral pattern for antiparallel β-sheet structures (see also Table 10.2) with amide I exciton components at 1628 and 1694 cm^{-1}. Substitution of the first Ala residue in the peptide sequence with a ^{13}C atom produces an "uncoupled" peak at 1611 cm^{-1}, but with an intensity much larger than expected for 1 out of 13 amide I bands. The authors interpreted this observation and similar observations when two adjacent or separated residues were ^{13}C substituted, in terms of complicated coupling patterns across the strands. These results demonstrated that the vibrational states created by the interaction of amide I transitions are complex exciton states where the amplitudes of each of the individual C═O oscillations can only be determined by detailed model calculations. The conformational sensitivity of the amide I "manifold" is, therefore, due to the amplification of distinct coordinates that is highly conformation dependent. A further extension of the results by the Mendelsohn group is presented in Section 10.4.

Similar studies involving ^{13}C isotopic substitutions on helical and sheet peptides were carried out by the research groups of Keiderling at the University of Illinois (Chicago), using IR and VCD spectral studies, in conjunction with *ab initio* computational methods. These studies were carried out for α-helical, β-sheet, and poly-proline II helical peptides [20, 21] and yielded detailed coupling patterns similar to the ones obtained by Mendelsohn's group.

The great majority of peptide VCD studies were carried out by the group of Keiderling who was instrumental in establishing VCD as an exquisitely sensitive tool to monitor solution conformation of peptides and peptide models in solution, using mainly the amide I manifold. In particular, his group was the first to demonstrate the conformational sensitivity of VCD spectroscopy toward peptide secondary structure by monitoring pH-induced changes in secondary structures of polyamino acids with charged side chains, such as poly-L-lysine (PLL), as solutions in D_2O [22]. At a neutral pD of about 7.3, the amide function of the lysine side groups is protonated, and the repulsion of the positive charges causes PLL to assume a "disordered" structure that exhibits a positive–negative couplet (from high to low wavenumber, see Figure 10.9(a)). Raising the pD of the solution to 11.5, a β-sheet structure was obtained, whereas at very high pD (in the presence of methanol), an α-helical conformation was obtained. In general, α-helical peptides exhibit a conservative, negative–positive couplet (from high to low wavenumber, see Figure 10.9(b)). Dukor and Keiderling [4] interpreted the VCD signal of the disordered state in terms of segments of left-handed helices, similar to the left-handed 3_1 helix, in a "disordered" arrangement. These VCD studies demonstrated that the "disordered" state possesses short range order at the vibrational timescale. This state exhibits opposite VCD signals than the right-handed α-helix, which suggested that the "disordered" state has a left-handed conformation. This result was not unexpected because electronic circular dichroism (CD) features of this disordered state can be described, to a first approximation, as the mirror image of the CD spectrum of the α-helix.

By the late 1980s, when these studies were carried out, state-of-the-art computers did not permit the computation of force fields or VCD intensities of sufficiently large peptide segments (such as an octapeptide) to reproduce the observed changes in IR and VCD spectra. Thus, very approximate VCD calculations using the exciton approach described in Chapter 8 were carried out for α-helical and several left-handed peptide structures. Results of a comparison between observed and computed VCD spectra for poly-L-tyrosine are shown in Figure 10.9 [2]. The computational model for the α-helix used typical conformational angles for this structure ($\varphi = -55°$, $\psi = -55°$). Two quite different left-handed structures, one with $\varphi = 60°$, $\psi = 30°$ (short dashes in Figure 10.9(a)) and the other with $\varphi = -110°$, $\psi = 120°$ (long dashes in Figure 10.9(b)) gave very similar computed VCD spectra, both in reasonable agreement with the observed spectra. The former of these structures corresponds to a left-handed 3_1 helix, whereas the latter falls in the general energy minimum of β-pleated sheets. This structure has a slight twist to account for the left-handed helicity.

Since these early studies, Keiderling and coworkers [21, 23] have used *ab initio* quantum mechanical calculations to refine the simple model presented earlier and reproduce the observed VCD features of a number of protein secondary structures. In particular, Keiderling and coworkers redefined the state that had been referred to as a "disordered structure" by researchers using different structural methods. This disordered state, also

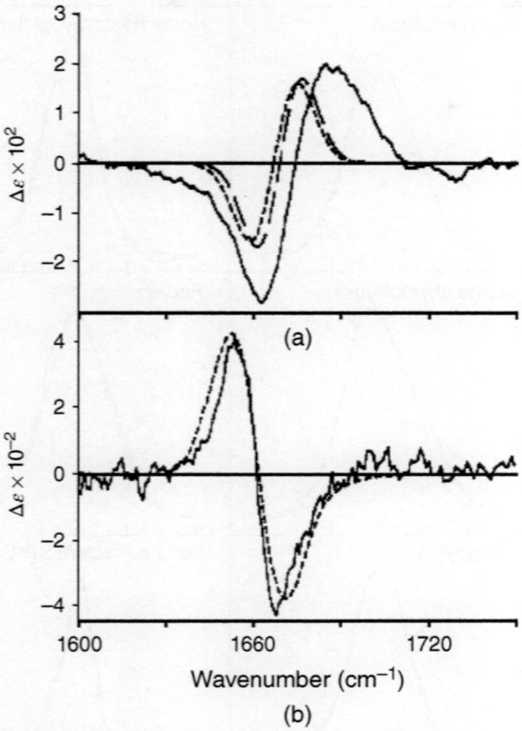

Figure 10.9 *(a) Observed (solid line) and computed (dashed lines) VCD features of "disordered" state of poly-L-tyrosine. (b) Observed (solid line) and computed (dashed line) VCD features of α-helical state of poly-L-tyrosine. See text for detail. Birke* et al., *1992, [2]. Reproduced with permission from American Chemical Society*

referred to as "random coil," exhibits signals in chiroptical techniques (electronic and VCD) that are approximately mirror images of the signals observed for the α-helical conformation [4]. The fact that there are CD and VCD signals in this "disordered" state dispels the notion that there is no structure, at least at the timescale of very fast transitions. In particular, VCD studies have shown unambiguously [24] that denatured peptide structures that have been referred to as "disordered structures" often exhibit left-handed poly-proline II-like helical segments that exhibit opposite VCD patterns than the right-handed α-helix.

VCD exhibits much more distinct signatures for the established secondary structures and appears to be sensitive to the occurrence of turns as well. Thus, VCD has become a method of choice for the analysis of the secondary structure of peptides and proteins (see Figure 10.10).

10.4 Protein spectroscopy: IR, VCD, Raman, resonance Raman, and ROA spectra of proteins

Infrared spectroscopy of proteins has been recently summarized in an excellent and comprehensive review by Barth [25]. This paper reviews the basic principles, including the transition dipole coupling model to explain the conformational sensitivity of IR spectroscopy; furthermore, it contains the most comprehensive

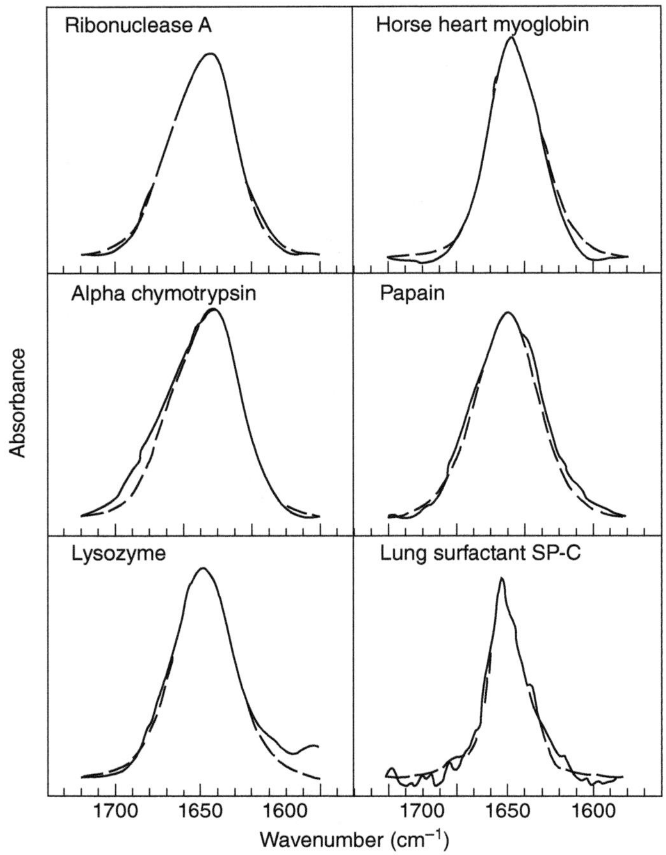

Figure 10.10 *Observed (solid lines) and simulated (dashed lines) amide I′ infrared band envelopes for several proteins as solutions in D_2O. Brauner* et al.*, 2005, [3]. Reproduced with permission from American Chemical Society*

compilation of amino acid side group vibrations in peptides and proteins, and a thorough discussion of the amide I and II manifolds.

Many of the principles discussed for peptides and polyamino acids in the previous section can be applied to proteins as well. Amide I spectral studies in IR spectroscopy are generally carried out as solutions in D_2O to avoid interference with the water deformation vibration. The resulting IR amide I band contours have been analyzed by Brauner *et al.* [3], in a computational approach discussed before (see Figure 10.10). Here, the coordinates of the peptide linkages of the proteins were taken from the protein structural databank, and a computational model was utilized that incorporated "through space" (dipolar coupling) as well as coupling through valence bonds and through hydrogen bonds. The observed amide I′ band contours were reproduced to a remarkable degree.

The first VCD spectra of proteins as solutions in D_2O were published by Keiderling's group [26]. A large number of proteins have been studied since then, with all the work being done at Keiderling's laboratory. These proteins have included myoglobin, hemoglobin, chymotrypsin, papain, lysozyme, and ribonuclease. One major conclusion can be drawn from these studies, namely that the secondary structures apparent from

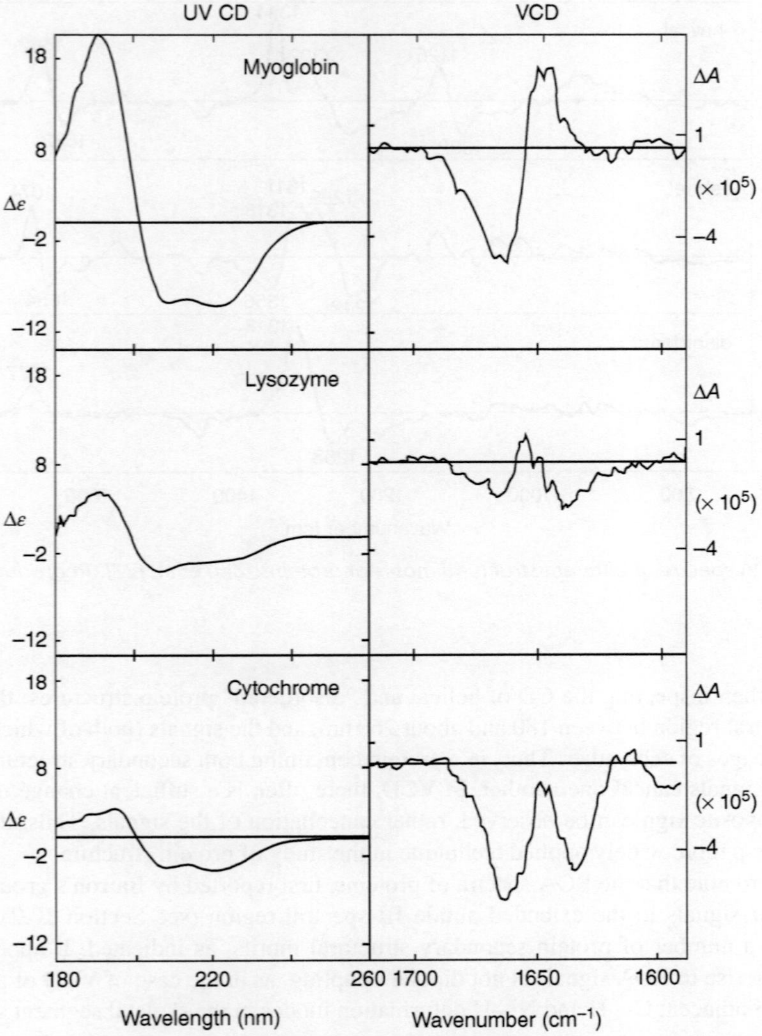

Figure 10.11 *Comparison of UV-CD (left column) and VCD (right column) results for selected proteins.* *Pancoska* et al., *1989, [26]. Reproduced with permission from American Chemical Society*

a cursory comparison of UV-CD and VCD spectra may be very different for some proteins. In particular, myoglobin, lysozyme, and cytochrome exhibit UV-CD spectra that one would classify by eye as mostly α-helical, but their VCD spectra contain spectral patterns that are distinctly nonhelical, with the exception of myoglobin (see Figure 10.11).

A quantitative interpretation of the amide I′ VCD region was achieved using factor analysis methodology that yielded reliable percentages of α-helical, β-sheet, poly-L-proline II, and other contributions. The factor analytical data appear to agree well with solid phase X-ray structures and to a somewhat lesser extent with UV-CD data. This may be, in part, due to the difficulties in analyzing UV-CD, because the observed bands are broader than the exciton splitting; consequently, considerable cancellation of CD intensities occurs. This

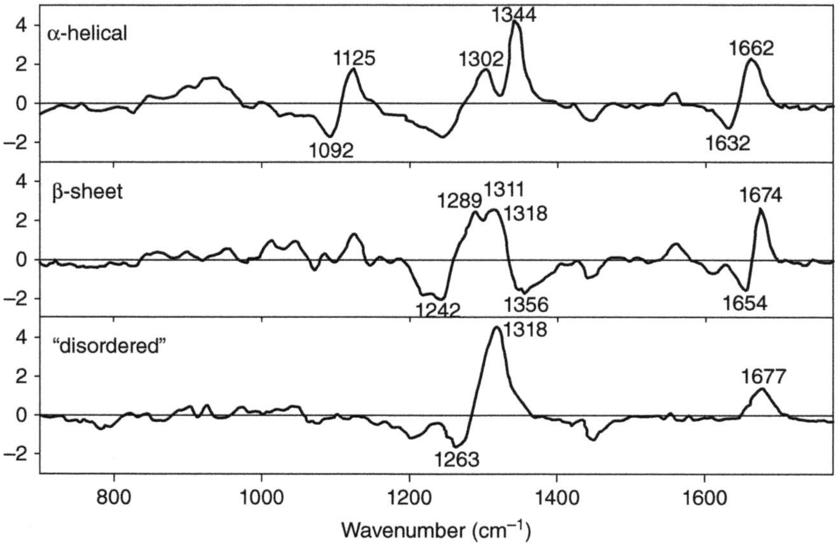

Figure 10.12 *ROA spectra of different structural motifs of proteins. Zhu* et al. *[27]. Reproduced with permission from Elsevier*

can be realized when inspecting the CD of helical and "disordered" protein structures: the spectral features occur over a spectral region between 180 and about 260 nm, and the signals (both of which are bisignate) are roughly mirror images of each other. Thus, in a protein containing both secondary structural motifs, the main effect is that the signals cancel one another. In VCD, there often is a sufficient change in band frequencies that signals of opposite sign can be observed, rather cancellation of the signals. Thus, it is anticipated that VCD will become a more widely applied technique in the study of protein structure.

It is interesting to note that the ROA spectra of proteins, first reported by Barron's group in Glasgow [27], showed the major signals in the extended amide III spectral region (see Section 10.2). This is shown in Figure 10.12 for a number of protein secondary structural motifs, as indicated. It appears that the major mechanism giving rise to ROA signals is not dipolar coupling, as in the case of VCD of proteins, but rather, the interactions of adjacent C—H and N—H deformation modes in the skeletal segment shown below:

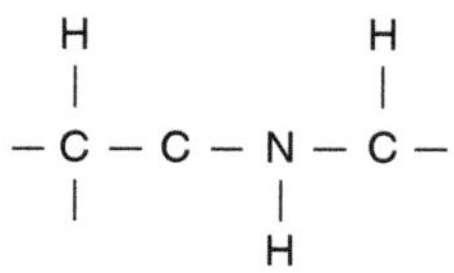

In very recent papers, ROA intensity calculations have been reported [10] for the major secondary structural motifs, modeled by oligopeptides with about 20 amino acid residues. The original observations by Barron's group (see Figure 10.12) were confirmed computationally and can be summarized as follows: both α-helical and β-sheet structures exhibit a positive/negative couplet (from high to low wavenumber) in the amide I region, at 1662/1632 cm^{-1} and 1674/1654 cm^{-1}, respectively. In addition, β-sheet structures exhibit negative intensities between 1200 and 1250 cm^{-1} in the extended amide III region that are less pronounced in other secondary structures. The fact that computational methods are available that permit correlation of observed and calculated ROA (and, for this matter, VCD) intensities will allow local peptide and protein secondary structure to be monitored and interpreted.

Protein secondary structure can also be probed with deep UV resonance Raman spectroscopy [28, 29]. Similar to the ROA results introduced earlier, the main conformational differences between various protein secondary structural motifs are observed as frequency and intensity variations in the amide III manifold. These studies have demonstrated the coupling patterns of the C—H and N—H coordinates as a function of the dihedral angle between them and have confirmed the empirical predictions by Lord [8] about the conformational sensitivity of the amide III vibration.

In Chapter 7, the advantages of carrying out difference spectroscopic measurements were discussed in the context of time-resolved Fourier transform infrared (FTIR) studies. In Figure 7.2, spectral differences observed for the hydrolysis reaction of Ras-GTP (guanosine triphosphate, GTP) to Ras-GDP (guanosine diphosphate, GDP) were shown; the authors of this work demonstrated that spectral intensity changes on the order of milliabsorbance units could be reliably detected. In the example discussed in Figure 7.2, protein conformational changes could be detected in addition to the expected changes in the phosphate region of GTP and GDP. Even smaller FTIR difference spectra were observed by Moss and Sage [30] in electrochemically induced changes in the oxidation state of cytochrome c; again, the changes were so small that they would have been unobservable by standard spectroscopic measurement, and only synchronous detection methods allowed their observation.

Similarly, Raman difference spectroscopy, using a split cell that was shuttled between two sampling positions, was employed by Callender and coworkers [31] to study substrate binding to enzymatic sites in order to elucidate protein conformational changes upon binding. Given the lower sensitivity of Raman spectroscopy, it is amazing what detail of spectral differences could be observed by solution phase Raman difference spectroscopy, and how much structural information on enzyme–substrate binding was available from these studies.

Carey [32] used a Raman microscopic approach (see Chapter 11) and difference methods to study protein structure in a method that was referred to as "Raman crystallography." The overall goal of this work was the investigation of structural changes in proteins upon binding with a substrate or other perturbations; however, in contrast to the Raman difference spectroscopic work described in the previous paragraph, Carey's work used protein crystals suspended in mother liquor in the focal point of a Raman microscope (see Chapter 11). By creating a crystal of a protein–ligand complex and suspending it in a drop of mother liquor without the ligand, the conformational changes in the protein could be followed as a time-dependent spectral change when the ligand diffused out of the protein crystal. In a reverse experiment, a crystal of pure protein was suspended in the mother liquor of the protein, with a ligand added to the liquor. Here, the diffusion of the ligand into the crystal, and subsequent binding to the protein could be observed. This method offered enormous advantages over other structural tools. As compared to solution phase Raman spectroscopy, Raman crystallography offers the advantage of much higher protein concentrations in the crystal than in solution. Furthermore, as a difference measurement is being carried out, the sensitivity toward small spectral changes is enormously enhanced. Finally, the concentration of the substrate is quite high in the protein–ligand crystal; thus, the spectral changes in the volume excited by the laser are much enhanced.

The experiment was carried out by suspending the crystal in a drop of mother liquor hanging from a cover slip at the focus of the Raman microscope, and collecting the backscattered light through the microscope objective. As the focus of a laser beam in a Raman microscope is diffraction limited (see Chapter 11), crystals measuring a few tens of micrometers in size can be studied.

10.5 Nucleic acids

10.5.1 Structure and function of nucleic acids

The nuclei of eukaryotic cells contain deoxyribonucleic acids (DNAs) that carries the genetic code of the cell. The genetic code is the chemically stored information that is transmitted by a cell, during cell division, to the next generation of cells.

DNA is a polymer of hundreds of thousands of monomers known as nucleotides for which a schematic structure is provided in Figure 10.13. Each nucleotide consists of a deoxy ribose (a five-membered sugar, shown in light brown in Figure 10.13) and an ionized phosphate group (yellow) that form phosphodiester linkages between the deoxy ribose units [33].

Each deoxy ribose is connected to one of the four "bases," which are aromatic, heterocyclic purine or pyrimidine derivatives. In DNA, bases denoted by the symbols A (adenine, green), C (cytosine, red), G (guanine, blue), and T (thymine, purple, see Figure 10.13) are found. The genetic information is stored *via* the sequence of the four different bases on the DNA strand. Two DNA strands are linked to each other through hydrogen bonding, as shown in Figure 10.13, to form a helical, double-stranded structure. It was found that in natural

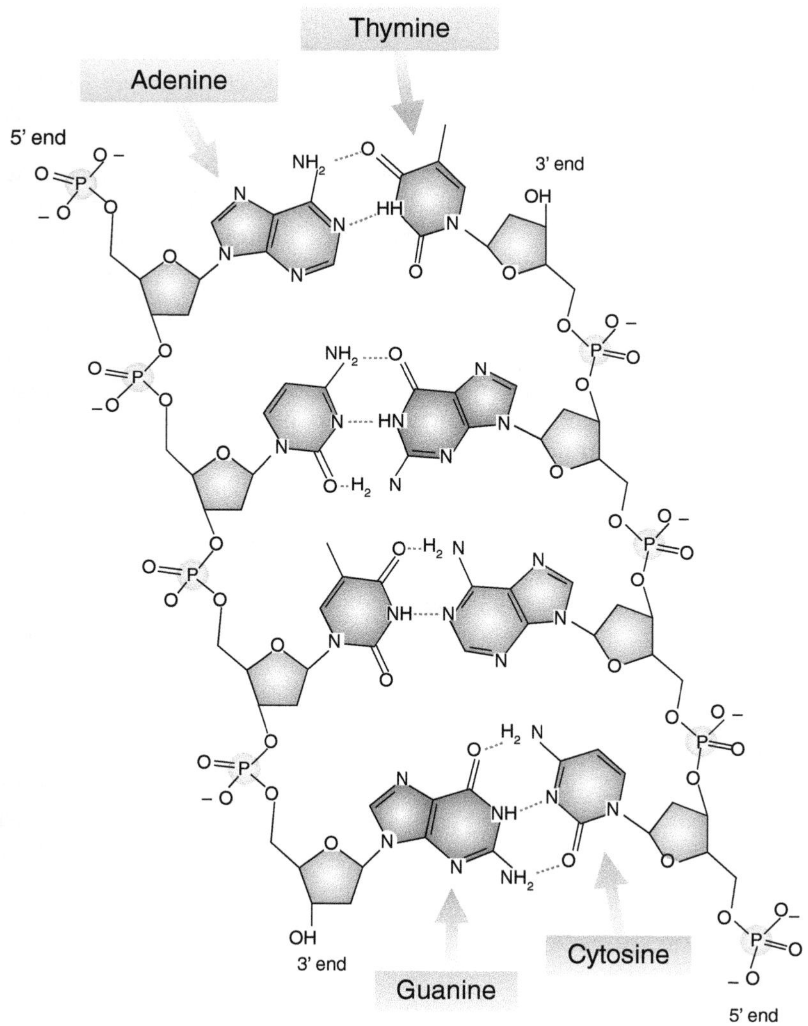

Figure 10.13 *Chemical structure of a DNA segment. Wikipedia, 2014 [33]. Reproduced with thanks to Madeleine Price Ball under the terms of the CC0 Creative Commons license.* (See insert for color representation of this figure.)

double-stranded DNA, the base C on one strand is always opposed by and hydrogen bonded to a G base on the other strand. Similarly, A and T form base pairs between opposing strands. Thus, the chemical information is stored in duplicate, because the complementary character of the bases on either strand contains basically the same information. The sequence of a DNA fragment is given from the 5′ to the 3′ end (*cf.* Figure 10.13).

The sequence of bases in one strand codes for the sequence of amino acids in a protein. Three base pairs code for one amino acid; thus, the sequence of three bases is referred to as a "codon." For example, the sequences CGC codes for the amino acid arginine (Arg) to be incorporated next in a protein. The process of decoding the information stored in DNA proceeds by partial unwinding of the DNA, construction of a complimentary ribonucleic acid (RNA) (see the following paragraphs) sequence (messenger ribonucleic acid or mRNA) in a process known as transcription, transport of the mRNA out of the nucleus of the cell into ribosomes in the cytoplasm, and construction of a new protein (translation) with the mRNA serving as a template. During translation, small RNA molecules known as transfer ribonucleic acid (tRNA) transport specific amino acids to the ribosome; each tRNA contains a complimentary three-base pair recognition region that binds to the tRNA. This insures that the proper amino acid as specified by the genetic code is in the correct position to create the peptide sequence.

All RNA types are macromolecules very similar to DNA, except that the sugar in RNA is a ribose, rather than a deoxy ribose (see Figure 10.14). This explains the names of ribonucleic acid (RNA) and deoxy ribonucleic acid (DNA). Furthermore, the base thymine in DNA is replaced by uracil in RNA (see Figure 10.14, structures (c) and (d)). In double-stranded RNA, one therefore finds GC and AU base pairs.

The most common form of DNA in the crystalline, fibrous, and solution phases is the so-called B-form, where two strands of DNA are wrapped into a right-handed helix. The direction of the two strands in the double helix is opposite. In the B-form, there are 10 base pairs per turn of the helix, and the bases are stacked perpendicularly to the helix axis, about 3.3 Å (0.33 nm) apart. The solution structure, that is, the shape a DNA or a model oligonucleotide assumes in solution, has been one major area of investigation in modern biophysical research. In particular, different secondary structures that are manifested as various different helical forms of DNA and its tertiary structures are being investigated in order to understand the mechanism of certain physiological processes involving DNA. One of the most fascinating aspects of these processes is the recognition of certain base sequences by enzymes with phenomenal specificity. At this point, the question arises whether or not a given sequence assumes a static or dynamic structure ever so slightly different from the rest of the molecule to permit recognition. These are the kinds of questions one wishes to answer by studying spectroscopic characteristics of DNA. *In vivo,* DNA is found wrapped around histone protein scaffolds. The DNA–histone complex itself is further coiled (DNA condensation) to form chromosomes. In analogy to peptides and proteins, this is referred to as tertiary structure.

Another common right-handed helix is the A-form, typically assumed by DNA under dehydrating conditions. The A-form is also found in double-stranded RNA. The different structural preferences of DNA and RNA are determined by the different conformation ribose and deoxy ribose assume (see the following sections). A particularly fascinating form of DNA is the left-handed double-stranded Z-form where the helicity

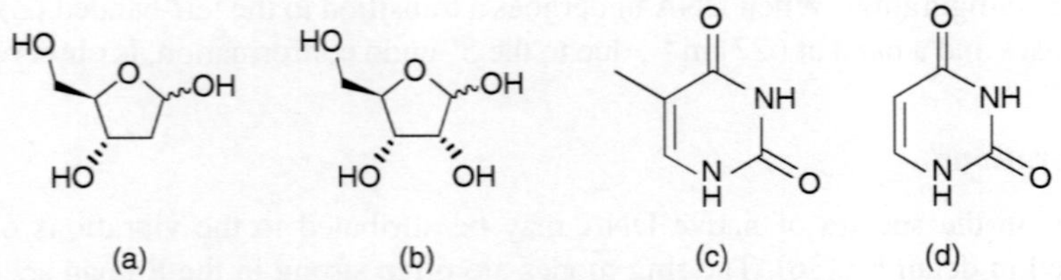

Figure 10.14 *Chemical differences between DNA and RNA: (a) deoxyribose, (b) ribose, (c) thymine, and (d) uracil*

is reversed to a left-handed form. This structure forms spontaneously in solution at high salt conditions or very low concentrations of certain polyamines, such as spermidine. It is believed that Z-DNA segments occur during DNA transcription to provide torsional strain relief.

Efforts to investigate the solution structures of nucleic acids using vibrational spectroscopy date back to the early 1970s. In analogy to the discussion of peptide spectroscopy, one can divide nucleic acids into logical segments for model studies. These segments are the ribose or deoxy ribose, the phosphodiester linkage, and the bases.

10.5.2 Phosphodiester vibrations

In DNA, the link between the various ribose units is a phosphodiester unit, shown below:

$$C - O - PO_2^- - O - C$$

There are two different phosphorus–oxygen bonds around the P atom, namely, the O—P single bonds of the actual phosphodiester linkage and the bonds between phosphorous and the other two oxygen atoms, which have multiple bond character (bond order of 1.5) and bear a delocalized charge of −1. The negative charge is balanced by counterions such as Na^+. The central grouping around the phosphorous atom is often (incorrectly) referred to in the literature a "phosphate group." The IR and Raman spectra of the $-PO_2^--$ group show distinct peaks at about 1235 and 1085 cm^{-1} that have been assigned to the antisymmetric and symmetric stretching coordinates, respectively. The phosphodiester vibrations can also be described in terms of symmetric and antisymmetric combinations of the P—O stretching coordinates: an intense Raman line at about 800 cm^{-1} has been assigned to be due to the symmetric – O—P—O – diester stretching mode, and a band at only slightly higher wavenumber (−830 cm^{-1}) to the antisymmetric stretching mode [34]. In addition, two C—O stretching vibrations are observed at 1017 and 1058 cm^{-1}. These vibrations, in particular the phosphodiester symmetric stretching vibration, are sensitive to DNA conformation as well (*vide infra*).

10.5.3 Ribose vibrations

The vibrations of the ribose or deoxy ribose moiety are very difficult to assign, because of the enormous mixing of C—C and C—O stretching coordinates within the five-membered ring and the mixing of these vibrations with those of the ring-external C—O stretching coordinates. In the Raman spectra, (deoxy)ribose modes are observed at 830, 895, 917, 975, 1144, 1448, and 1462 cm^{-1} [35]. Some of these modes are sensitive to the backbone conformation. This is because the pucker of the five-membered ring varies as the conformation of the nucleic acid changes. Particularly important are the conformations about the 2′ and 3′ carbon atoms. These can be in the endo and exo forms. In the former, the 2′-endo form, the 2′ carbon atom is endo (puckered out of the sugar plane and on the same side of the plane as the 5′ carbon). In the 2′-exo form, the 2′ carbon is puckered out of the plane and on the opposite side of the plane than the 5′ carbon atom. The lowest frequency band listed earlier for the ribose vibration is believed to be sensitive to the existence of the B-form with the ribose C2′ in endo configuration. When DNA undergoes a transition to the left-handed (Z) form, the intensity of this peak decreases and a band at 627 cm^{-1}, due to the 3′-endo conformation, is observed (see Table 10.3).

10.5.4 Base vibrations

Most Raman lines in the spectra of native DNA may be attributed to the vibrations of the bases, which have been analyzed in detail by [36]. The ring modes are often strong in the Raman spectra because of the π-character of many of the bonds, but the IR spectra also show strong vibrations, mainly because of the C═O stretching vibration at about 1650 cm^{-1}. Although one should expect the vibrations of the ring to be fairly

Table 10.3 Frequencies of conformation-specific marker bands in model polynucleotides[a]

Z-form	B-form	Assignment
627		G, 3′-endo
	835	BK, 2′-endo
1355	1362	G
1424	1421	G, BK
	1530	G

[a]From Ref. [35].
G, guanine and BK, backbone.

well-defined group frequencies and therefore not very sensitive to the conformation of the nucleic acid, one finds that certain ring modes, particularly those that also involve the glycosidic bond to the ribose, are very sensitive to the conformation. A number of the marker bands are listed in Table 10.3.

10.6 Conformational studies on DNA and DNA models using IR, Raman, and VCD spectroscopies

In order to simplify the problem that exists with naturally occurring DNA, with its enormous size and the variability in sequence, many model studies have been carried out on systems such as poly(dG-dC) · poly(dG-dC) or poly(dG) · poly(dC). Here, the standard nomenclature is employed that indicates that the nucleotides are composed of deoxyribose units (d) or ribose units (r). The former of the two species discussed earlier contains an alternating d(· · GCGC · ·) sequence in each strand, whereas the latter has d(· · CCCC · ·) in one strand and d(· · GGGG · ·) in the other. The dot between the strands implies a hydrogen-bonded, double-stranded structure.

In such model systems, the characteristic vibrations due to the A, B, and Z conformations could be established by recording the Raman spectra under carefully controlled conditions and monitoring the structure *via* other techniques, such as CD. The major conformational marker bands are listed in Table 10.3. Most noticeable is the intensity change of a strong vibration at 681 cm^{-1} in B-form DNA, which originates from a guanine ring vibration coupled with the vibration of the ribose when it is in the C2′-endo conformation. This band all but disappears when the polymer switches to the Z-conformation [35]. In addition, the stacking of the bases can be followed conveniently in Raman spectroscopy *via* hypochromism, an effect that describes an intensity decrease between denatured and B-form DNAs. Hypochromism is commonly observed for nucleic acids in the UV spectral region and is due to stacking interactions. Similarly, owing to the change in extinction upon stacking, Raman hypochromism is observed. A vibration at 1245 cm^{-1}, assigned to a cytosine ring mode, becomes much less intense when G- and C-containing polymers undergo a transition from a disordered state (e.g., at elevated temperature) to the base-stacked double helical state.

Infrared spectroscopy has been used for conformational studies to a lesser extent, because of the difficulties in collecting IR absorption data from aqueous solutions. However, studies on the dehydration of DNA and RNA films from the author's laboratory were reported in 2001 [37], and compared to solution phase IR spectra. A more recent study [38], using mainly attenuated total reflection (ATR) sampling methodology, has shown that IR spectroscopy detects conformational sensitive bands as well. This is shown in Figure 10.15. Owing to the inherently wider spectral peaks in IR, as compared to Raman spectra, it is advantageous to carry

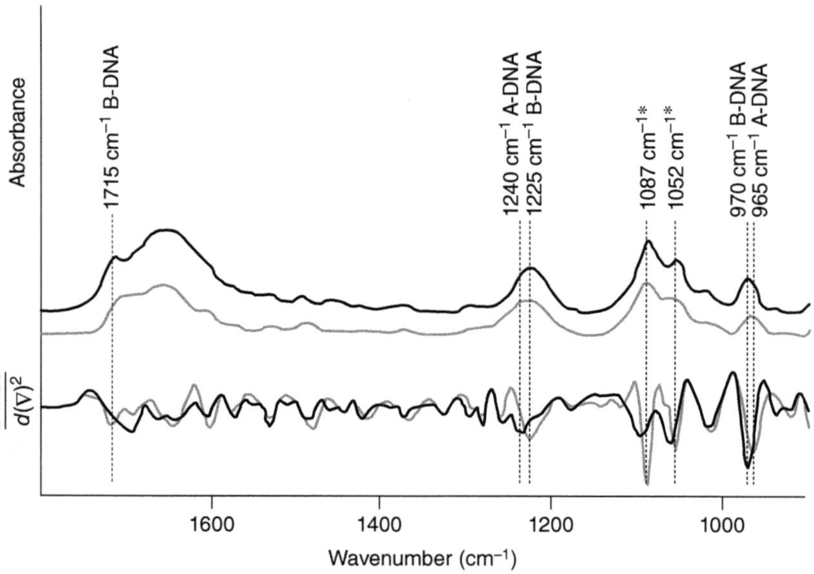

Figure 10.15 *Infrared absorption (top) and second derivative spectra (bottom) of fully hydrated (black) and dehydrated (gray) forms of double-stranded DNA. Asterisks denote bands that lose intensity upon B → A transition. Whelan* et al.*, 2011, [38]. Reproduced with permission from Oxford University Press*

out the analysis of these spectra on their second derivatives (see Chapter 12). Conformational sensitivity is exhibited by different bands in IR and Raman spectroscopies. This is not unexpected, because similar behavior was observed for peptides and proteins as well (see earlier). In IR spectroscopy, the C═O stretching vibrations of the nucleotide bases, which are arranged in a fixed geometry in double helical DNA or DNA model, interact strongly by a dipolar coupling mechanism similar to the one responsible for the coupling of the amide vibrations. This coupling produces distinct, conformation-dependent VCD signals between 1550 and 1750 cm^{-1}, to be discussed next.

As these modes are localized on the planar bases that are achiral, the VCD is nearly entirely due to the dissymmetric coupling (see Chapter 5) of transitions in these groups; this coupling is sensitive to the geometry between the groups. In analogy to the discussion of the VCD of various peptide structural motifs, the VCD of the common DNA structures will be introduced first.

Single-stranded RNAs, such as poly(rU) or poly(rC) (see Figure 10.16), exist in aqueous solution at low temperature as single-stranded, base-stacked, helical polymers [39, 40]. They exhibit a positively biased, negative/positive (from high to low wavenumber) VCD couplet in the carbonyl stretching vibration of the bases. Poly(rA), which does not contain any carbonyl groups, exhibits a similar but somewhat smaller VCD signal at lower frequency. These signals were attributed to coupling of C═O and C═C double bond stretching vibrations of the bases, which interact with each other by dipolar coupling. This aspect was demonstrated by the group of Wieser by heating (melting) model RNA systems and observing a disappearance of the VCD features [41].

The negative/positive couplet is a hallmark of right-handed helical nucleic acid structures, and was observed in single-stranded as well as in double-stranded RNA and DNA models. Although the single- and double-stranded RNA models exhibited nearly identical VCD patterns, the IR spectra differed significantly (*cf.* Figure 10.16). This indicates that sufficient order exists even in single-stranded polymers for dipolar

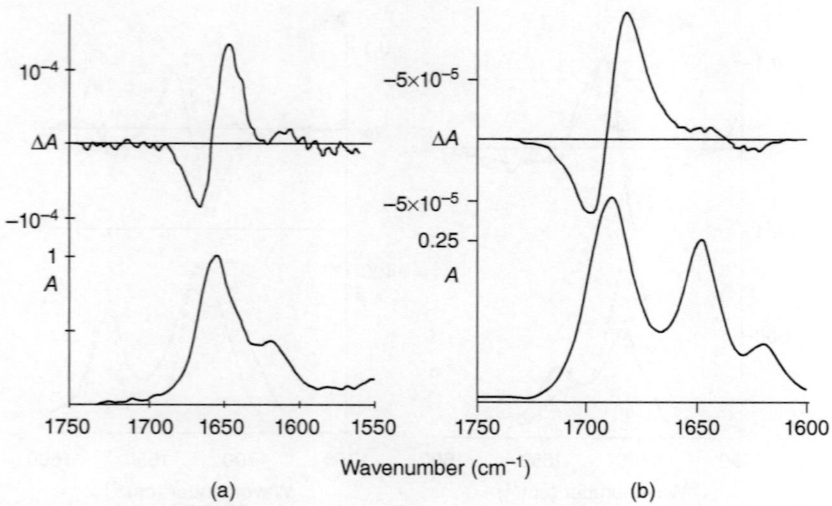

Figure 10.16 *VCD of (a) single-stranded poly(rC) Annamalai & Keiderling, 1987, [39]. Reproduced with permission from American Chemical Society and (b) double-stranded poly(rC) · poly(rG). Diem, 1993, [40]. Reproduced with permission from John Wiley & Sons*

coupling to occur and produce exciton-type VCD features. The signals shown in Figure 10.16(a) were, incidentally, the first ever observations of VCD in poly-nucleotides, and were reported from the laboratory of Keiderling [39]. Poly(rC)·poly(rG), which forms a solution conformation of the A family, exhibits a positive/negative couplet with an additional small low-frequency feature (Figure 10.16(b)) under a strong peak in the IR spectrum at 1650 cm^{-1}. This peak is much weaker in the single-stranded polymer. All VCD and IR absorption spectra of DNA and RNA model systems reported here were acquired as solutions in D_2O (to reduce contributions of the water deformation mode) and cacodylate buffer.

The VCD of a double-stranded poly(dG-dC)·poly(dG-dC) in the B-form is shown in Figure 10.17(a). B-form VCD may be described by a negative VCD signal at 1700 cm^{-1}, a positive peak at 1682 cm^{-1}, and a negative shoulder at 1660 cm^{-1}. The corresponding absorption peaks are at 1690, 1655, and 1625 cm^{-1}. The VCD of poly(dG)·poly(dC) in the B-form is similar to that of poly(dG-dC)·poly(dG-dC). Figure 10.17(a) also shows calculated VCD and IR features, based on canonical DNA structural parameters, using the formalism developed in Eqs. 1.5.21 and 1.8.18. The agreement between observed and computed VCD features is very good for this DNA analogue, indicating that the major contribution to the VCD is simple dipolar coupling.

The spectra of poly(dA-dT)·poly(dA-dT) and poly(dA)·poly(dT), as well as the corresponding RNA analogues, show more complex VCD patterns than the G–C polymers. In poly(dA)·poly(dT) and poly(rA)·poly(rU), see Figure 10.17(b), the highest frequency absorption peak exhibits very little VCD intensity, and a large negative/positive couplet (1675/1650 cm^{-1}) is followed by a positive/negative couplet (1640/1625 cm^{-1}), see Figure 10.17(b). In poly(dA-dT)·poly(dA-dT), there is an additional negative/positive VCD couplet at about 1700 cm^{-1} under the absorption peak, which shows no signal in poly(dA)·poly(dT) [42].

Significantly different spectra were observed for small DNA models, such as the self-complimentary tetramers 5′(CGCG)3′, 5′(GCGC)3′, 5′(CCGG)3′, and 5′(GGCC)3′. This is shown in Figure 10.18. The differences between the VCD spectra of the four tetramers can be explained qualitatively by noting that the most dominant interactions giving rise to VCD intensities are those between adjacent base pairs in a double strand. As base fraying will occur to some extent at the ends of the tetramers, the VCD spectra are basically determined by the center two base pairs. The orientation of the four carbonyl groups in the center two base

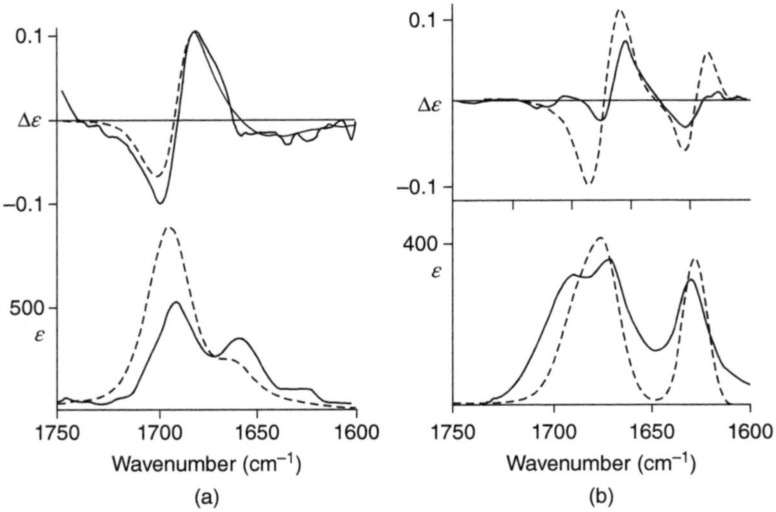

Figure 10.17 *VCD (top) and IR (bottom) spectra of (a) of poly(dG-dC) · poly(dG-dC) Diem, 1993, [40]. Reproduced with permission from John Wiley & Sons and (b) poly(rA) · poly(rU). In both graphs, solid traces represent observed and dashed traces computed data. Zhong* et al., *1990, [42]. Reproduced with permission from American Chemical Society*

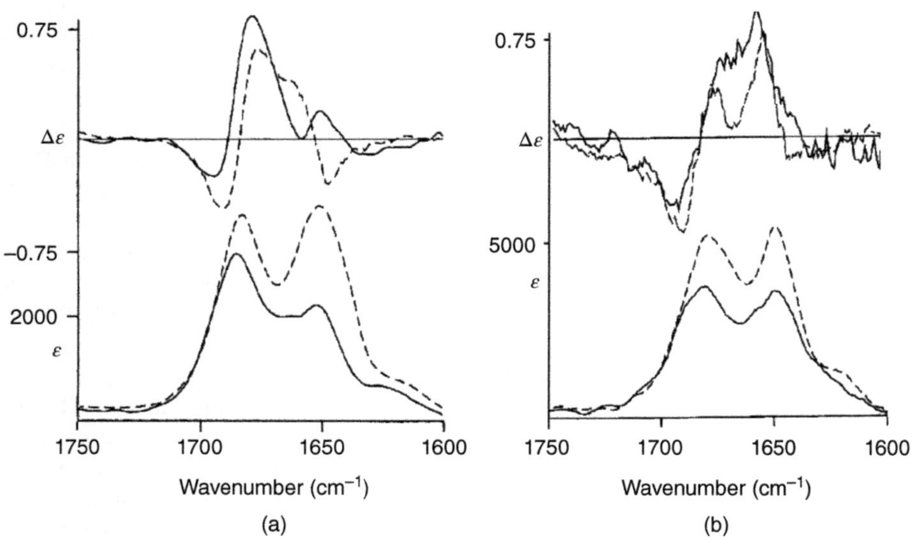

Figure 10.18 *Observed VCD (top) and absorption (bottom) spectra of (a) 5′d(CGCG)3′ (solid trace) and 5′d(GCGC) (dashed trace) and (b) 5′d(CCGG)3′ (solid trace) and 5′d(GGCC) (dashed trace). Diem, 1993, [40]. Reproduced with permission from John Wiley & Sons*

pairs is either such that all four carbonyl groups are nearly parallel or are twisted by nearly 90°. Owing to the handedness of the twist in the second case, a large optical activity is induced by the dipolar coupling of the transitions. Thus, it is not surprising that the observed VCD features of 5′d(CGCG)3′ and 5d(GCGC)3′ are different. Similar arguments can be made for the nonalternating tetramers.

Similar results were observed for even shorter chain length, d(CG) and d(GC). The results suggested that even these very small fragments remain hydrogen-bonded and form base-paired structures. In these two segments, the orientations of the four carbonyl groups differ drastically, and consequently entirely different VCD patterns are observed. In double-stranded 5′d(CG)3′, the four carbonyl groups are virtually all parallel, albeit in different planes. Thus, the induced chirality is low, but the coupling is large, as manifested by the large splitting in the IR absorption spectra. In double-stranded d(GC), on the other hand, the second set of carbonyl groups is twisted nearly 90° with respect to the first (lower) set of dipoles. This arrangement results in large induced chirality with a sign pattern similar to the one observed in poly(dG-dC)·poly(dG-dC) and $d(CG)_5$ [42]. The arrangement pattern of dipoles found in d(CG) and d(GC) is prototypical for B-type double helixes consisting of CG bases: there will be, between consecutive base pairs, either the "near parallel" alignment of dipoles or the "near 90° twist."

High ionic strength solutions (e.g., aqueous solutions containing high concentrations of salts) are known to induce a phase transition to a helical, left-handed form (Z-form) in ·· CGCG ·· sequences. VCD detects this phase transition by a reversal of the VCD couplets. The VCD spectrum of the Z-form of poly(dG-dC)·poly(dG-dC) in 3 M NaCl exhibits a positive peak at 1677 cm^{-1} followed by a broad negative peak with minima at 1658 and 1643 cm^{-1} [43]. The corresponding IR absorption peaks occur at approximately 1690 and 1650 cm^{-1}.

A similar reversal of the VCD couplet was reported by the Calgary group [44], see Figure 10.19, when heavy metal ions were added to DNA analogues. The observed Z-form spectra depend on the choice and concentration of the counterion used to induce the B → Z transition, and different left-handed structures were observed when the Mn^{2+}/phosphate concentration ratio exceeded 1.1. VCD in the symmetric phosphate stretching region (1087 cm^{-1}) and (presumably) the P—O—C stretching vibration at 973 cm^{-1} was also reported by these authors.

Triple-stranded RNA, incorporating Hoogstein base pairs, can be formed by poly(rA)·poly(rU)·poly(rU) at low temperature. Yang and Keiderling [45] reported for these polymers an overall VCD signal similar to that observed for double-stranded poly(dA)·poly(dT), namely, a large negative/positive–positive/negative couplet with an additional, positive low-frequency band. The spectrum of the triple helix was also observed when poly(rA)·poly(rU) was slowly heated, and unwinding of two double helices into a triple-stranded polymer and a single-stranded polymer occurred. This study demonstrated nicely the conformational sensitivity of VCD at the phase transition.

10.7 Lipids and phospholipids

A third class of important biological compounds is lipids, which are the molecular constituents of biological membranes. Membranes are assemblies of molecules, such as phospholipids that are held in their positions by noncovalent forces, such as polar head group interactions, van der Waals, and hydrophilic and hydrophobic forces. Phospholipid molecules typically consist of two nonpolar, saturated, or partially unsaturated fatty acids, typically with chain lengths between 14 and 24 carbon atoms that are linked to glycerol *via* ester linkages. The third position of the glycerol carries a (negatively charged) phosphoester linkage to a positively charged head group, as shown in Figure 10.20. There exist a wide variety of phospholipids that differ in the fatty acids and polar head groups attached to the glycerol moiety. The head group shown in Figure 10.20 is known as choline, and the entire structure is called phosphatidylcholine.

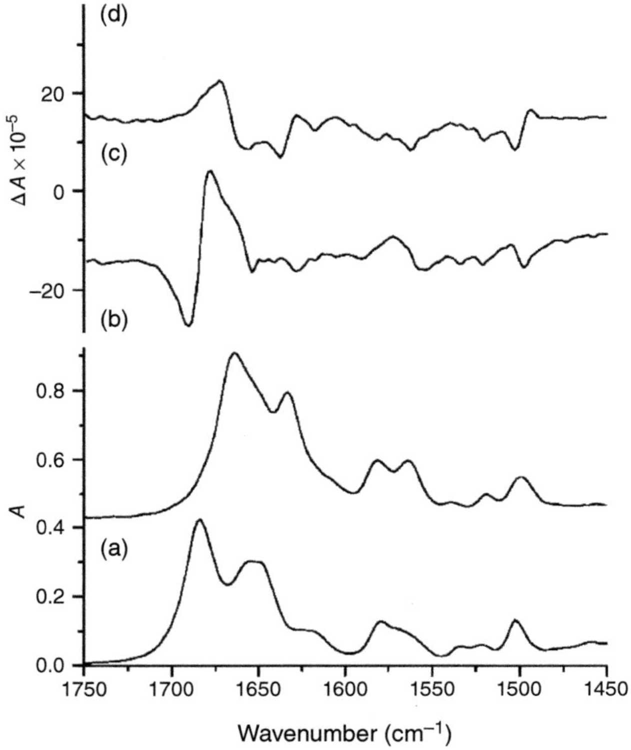

Figure 10.19 *IR absorption (a, b) and VCD (c, d) spectra of 5′d(GC)$_{20}$3′ in the absence (a, c) and presence of Mn^{2+} ions. Andrushchenko* et al. *[44]. Reproduced with permission from Taylor & Francis*

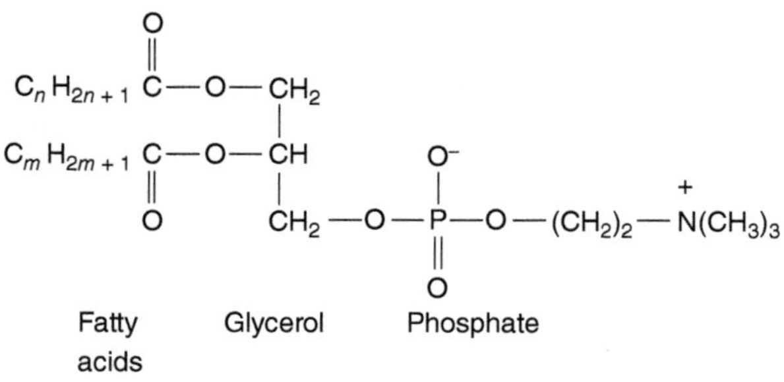

Figure 10.20 *Chemical structure of a typical phospholipid molecule*

The individual phospholipid molecules typically stack, as indicated in Figure 10.21, to form bilayer membranes. In these, the hydrophilic, charged head groups of each phospholipid are solvent (water) exposed, whereas the hydrophobic alkane chains exclude water and are held in place by van der Waals (dispersion) forces. In lipid bilayers shown in Figure 10.21, which are between 40 and 80 Å (4 and 8 nm) thick, two layers of lipids are arranged such that the zwitterionic and polar head groups point toward the aqueous phase, whereas the fatty acid tails form a hydrophobic region that points toward the fatty acids of the second layer.

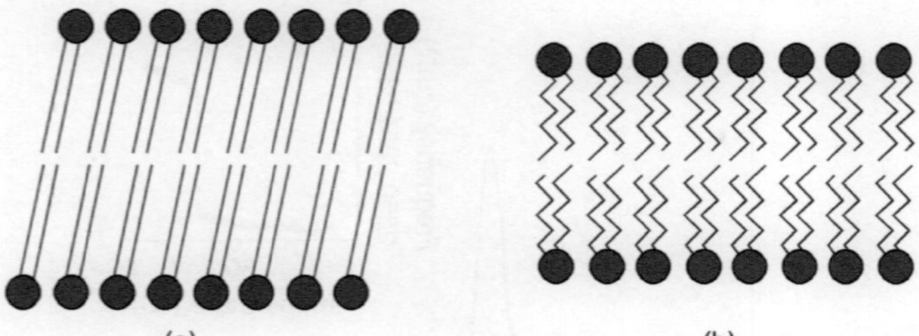

Figure 10.21 *Schematic of the structure of lipid bilayer membranes. (a) Gel phase, with all* trans *polyethylene chains. (b) Liquid-crystalline phase with partial* gauche *polyethylene chains. Diem, 1993, [40]. Reproduced with permission from John Wiley & Sons*

Lipid bilayers can form lamellar structures, in which concentric, spherical bilayers are arranged like onion skins, separated by aqueous regions. Alternatively, liposomes can be formed that consist of a single spherical bilayer structure that can be hundreds of nanometers in diameter.

The aliphatic chains of each fatty acid residue can be in an all-trans conformation. This structure is referred to as the gel phase. If the aliphatic chain contains *gauche* C—C groups, a liquid-crystalline phase of the phospholipid is obtained. In contrast to peptides, proteins, and nucleic acids, natural membranes are inhomogeneous mixtures of many compounds held together by noncovalent forces. Consequently, X-ray crystallography cannot be used as a structural tool. Vibrational spectroscopy is an extremely powerful and noninvasive technique to monitor certain structural properties of membranes, such as the conformational changes accompanying the gel–liquid crystal phase transition. Among vibrational techniques, Raman studies were used originally to study membranes [46, 47], but with modern FTIR and reflectance techniques, IR techniques have been applied as well.

The vibrational spectra of membrane constituents are dominated by the spectral features of the fatty acid chains that exhibit strong antisymmetric (2921 cm^{-1}) and symmetric (2850 cm^{-1}) methylene stretching vibrations in both Raman and IR spectra (see Figure 10.22). The frequencies and intensities of these vibrations depend on the phase (gel or liquid crystalline) of the hydrocarbon chain, and show sharp, "melting curve"-like temperature dependence, as shown in the inset of Figure 10.22 [48, 49].

The terminal methyl group exhibits the symmetric and antisymmetric stretching modes at 2936 and 2962 cm^{-1}, respectively. For phosphatidylcholine with different fatty acids, dispersed in aqueous solution to produce lamella, different gel–liquid crystal transition temperatures can be observed. These transition temperatures vary from 23.8 °C for C14 chains to 73.8 °C for C22 chains, with very sharp, cooperative phase transitions.

Another diagnostic vibration [50] of saturated side chains is the methyl scissoring (symmetric methylene deformation) mode at about 1466 cm^{-1} that is split into a doublet (1464/1471 cm^{-1}) at low temperature in ceramides (another class of lipids that occur in skin and brain tissue), but coalesces into one peak at 1467 cm^{-1} above the phase transition temperature at about 62 °C (see Figure 10.23). The skeletal optical mode at 1128 and 1064 cm^{-1} (the term *optical mode* implies vibrations perpendicular to the chain, whereas the term *acoustical mode* implies a vibration in the direction of the chain), also show the gel–liquid crystal transition. Similarly, the —CH_2— wagging vibration and its progression at 1360 cm^{-1} and the rocking vibration at 720 cm^{-1} are prominent spectral markers for side chain order.

Furthermore, IR studies on completely deuterated fatty acid side chains showed that the carbon–deuterium stretching vibrations can be analyzed similarly. In a multicomponent system, for example, which may contain dispersions of lipids and membrane proteins, the use of deuterated fatty acids allows an analysis of the lipid

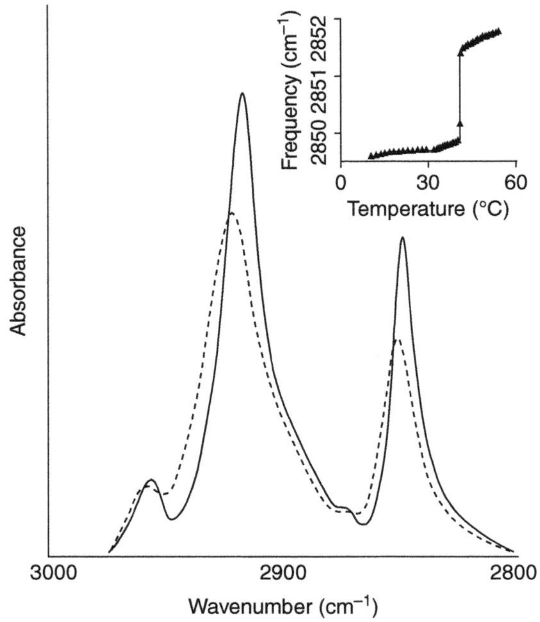

Figure 10.22 *Temperature dependence of the infrared spectra of 1,2-dipalmitoyl-sn-glycero-3-phospho-cholin (DPPC) in the C—H stretching region. The solid line represents the spectrum of DPPC in the gel phase at 10 °C and the dashed line the spectrum of DPPC in the liquid-crystalline phase at 50 °C. Inset: The relationship between frequency and temperature for the CH_2 symmetric stretching band of DPPC. Dluhy* et al.*, 1985, [48]. Reproduced with permission from Canadian Science Publishing*

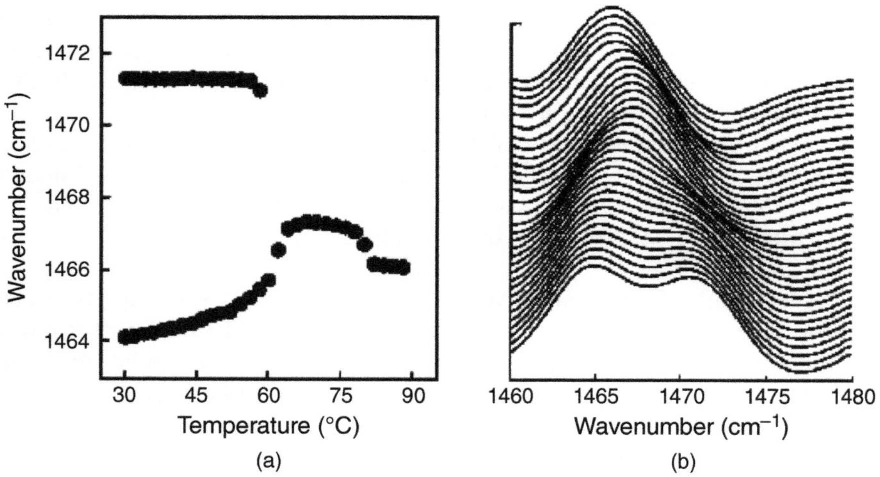

Figure 10.23 *(a) Temperature-dependent frequencies of the methyl scissoring vibration of a ceramide. (b) Plot of the inverted second derivative methyl scissoring vibration of a ceramide. Moore* et al.*, 1997, [50]. Reproduced with permission from American Chemical Society*

structure to be carried out even in the presence of peptides and proteins (e.g., see Carey [51], Chapter 8, and references therein]).

The ester linkage between the fatty acids and the glycerol moiety gives rise to strong C═O stretching vibrations in the spectra of lipids and phospholipids between 1700 and 1750 cm^{-1}. The high wavenumber peak is associated with the sn-1 carbonyl group, whereas the lower wavenumber peak is associated with the sn-2 position [52–54]. This splitting arises because of the conformational nonequivalence about the C1—C2 bonds of the sn-1 and sn-2 chains that adopt *trans* and *gauche* conformations, respectively, and through possible differences in the extent of hydration.

In phospholipids, the phosphodiester linkage, C–O–PO_2^-–O–C, of the head group (*cf.* Figure 10.20) gives rise to several strong vibrations in IR absorption, analogous to those in nucleic acids. The antisymmetric and symmetric stretching modes for the –PO_2^-– group are found near 1230 and 1085 cm^{-1}, respectively. Relatively variable and nondescript O–P stretching modes are found in the 900–800 cm^{-1} region. The vibrations of the polar head groups of phospholipids are not much affected by the temperature-dependent phase transitions in the fatty acid side chains or the ester linkages.

Recently, many studies have focused on the interactions between membranes and membrane-bound proteins. Located at the surface of membranes, or partially inserted into the membrane or even spanning it completely, are a number of proteins responsible for transport, signaling, or other catalytic functions. The locations of the protein components on the membrane surface, or in the membrane, are random, and the mechanisms of how a protein binds to the membrane or inserts itself through the membrane are the subjects of many recent studies. IR spectroscopy is an ideal method for examining the perturbations of the lipid bilayer that are introduced by the presence of integral polypeptides and proteins [49].

The C—H antisymmetric and symmetric stretching frequencies (*cf.* Figure 10.22) can be used to monitor the perturbations in membrane structure upon protein incorporation. It was found [55] that below T_m, certain proteins behaved in a similar manner to cholesterol; that is, they caused an increase in the disorder of the bilayer, preventing the acyl chains from attaining the all-trans conformation at low temperatures. Above the bilayer phase transition temperature, a reduction in the proportion of gauche conformers with respect to the pure lipid bilayer was observed.

10.8 Epilogue

The preceding discussion introduced the wide variety of spectral studies that have been carried out since the 1970s on biological model systems such as proteins, nucleic acids, and lipids. These studies paved the way for many of the studies referred to in the latter chapters of this book, Raman and IR (micro-)spectroscopies of human cells and tissue. Without a detailed understanding of the spectra and spectral variations of the spectra of cellular compounds, the biomedical applications of vibrational spectroscopy would have been impossible.

Thus, from the viewpoint of a researcher involved in medical diagnostics *via* vibrational spectroscopy, the author is deeply indebted to all the researchers mentioned earlier, and countless unmentioned colleagues, who have enabled the research in Chapters 13 and 14 to be carried out successfully.

References

1. Miyazawa, T., Shimanouchi, T. and Mizushima, S.-I. (1958) Normal vibrations of N-methylacetamide. *J. Chem. Phys.*, **29** (3), 611–616.
2. Birke, S.S., Agbaje, I. and Diem, M. (1992) Experimental and computational infrared CD studies of prototypical peptide conformations. *Biochemistry*, **31**, 450–455.
3. Brauner, J.W., Flach, C.R. and Mendelsohn, R. (2005) A quantitative reconstruction of the amide I contour in the IR spectra of globular proteins: from structure to spectrum. *J. Am. Chem. Soc.*, **127**, 100–109.

4. Dukor, R.K. and Keiderling, T.A. (1991) Reassessment of the random coil conformation: VCD study of proline oligopeptides and related polypeptides. *Biopolymers*, **31** (14), 1747–1761.
5. Diem, M., Oboodi, M.R. and Alva, C. (1984) Determination of peptide conformation via vibrational coupling: application to diastereoisomeric alanyl dipeptides. *Biopolymers*, **23**, 1917–1920.
6. Roberts, G.M., Lee, O., Calienni, J. *et al.* (1988) Infrared vibrational circular dichroism in the amide III spectral region of peptides. *J. Am. Chem. Soc.*, **110** (6), 1749–1752.
7. Oboodi, M.R., Alva, C. and Diem, M. (1984) Solution-phase Raman studies of alanyl dipeptides and various isotopomers: a reevaluation of the amide III vibrational assignment. *J. Phys. Chem.*, **88**, 501–505.
8. Lord, R.C. (1977) Strategies and tactics in the Raman spectroscopy of biomolecules. *Appl. Spectrosc.*, **31** (3), 187–194.
9. Diem, M., Lee, O. and Roberts, G.M. (1992) Vibrational studies, normal-coordinate analysis, and infrared VCD of alanylalanine in the amide III spectral region. *J. Phys. Chem.*, **96** (2), 548–554.
10. Weymuth, T. and Reiher, M. (2013) Characteristic Raman optical activity signatures of protein ß-sheets. *J. Phys. Chem. B*, **117**, 11943–11953.
11. McColl, I.H., Blanch, E.W., Gill, A.C. *et al.* (2003) A new perspective on beta-sheet structures using vibrational Raman optical activity: from poly(L-lysine) to the prion protein. *J. Am. Chem. Soc.*, **125** (33), 10019–10026.
12. Lee, O., Roberts, G.M. and Diem, M. (1989) IR vibrational CD in alanyl tripeptide: indication of a stable solution conformer. *Biopolymers*, **28**, 1759–1770.
13. Ramachandran, G.N., Ramakrishnan, C. and Sasisekharan, V. (1963) Stereochemistry of polypeptide chain configurations. *J. Mol. Biol.*, **7**, 95–99.
14. Woutersen, S. and Hamm, P. (2000) Structure determination of trialanine in water using polarization sensitive two-dimensional vibrational spectroscopy. *J. Phys. Chem. B*, **104** (47), 11316–11320.
15. Eker, F., Cao, X., Nafie, L. and Schweitzer-Stenner, R. (2002) Tripeptides adopt stable structures in water. A combined polarized visible Raman, FTIR, and VCD spectroscopy study. *J. Am. Chem. Soc.*, **124**, 14330–14341.
16. Kocak, A., R. Luque, and M. Diem, The solution structure of small peptides: an IR CD study of aqueous solutions of (L-Ala)n [n = 3,4,5,6] at different temperatures and ionic strengths *Biopolymers*, 1998. **46**: 455–463
17. Dukor, R.K., Keiderling, T.A. and Gut, V. (1991) Vibrational circular dichroism spectra of unblocked proline oligomers. *Int. J. Pept. Res.*, **38** (3), 198–203.
18. Brauner, J.W., Dugan, C. and Mendelsohn, R. (2000) 13C isotope labeling of hydrophobic peptides. Origin of the anomalous intensity distribution in the infrared amide I spectral region of beta-sheet structures. *J. Am. Chem. Soc.*, **122**, 677–683.
19. Decatur, S.M. (2000) IR spectroscopy of isotope-labeled helical peptides: probing the effect of N-acetylation on helix stability. *Biopolymers*, **54**, 180–185.
20. Huang, R., Kubelka, J., Barber-Armstrong, W. *et al.* (2004) Nature of vibrational coupling in helical peptides: an isotopic labeling study. *J. Am. Chem. Soc.*, **126**, 2346–2354.
21. Bour, P. and Keiderling, T.A. (2005) Ab initio modeling of amide I coupling in antiparallel beta-sheets and the effect of 13C isotopic labeling on infrared spectra. *J. Phys. Chem. B*, **109**, 5348–5357.
22. Yasui, S.C. and Keiderling, T.A. (1986) Vibrational circular dichroism of polypeptides. 8. Poly(lysine) conformations as a function of ph in aqueous solution. *J. Am. Chem. Soc.*, **108**, 5576–5581.
23. Kubelka, J., Kim, J., Bour, P. and Keiderling, A. (2006) Contribution of transition dipole coupling to amide coupling in IR spectra of peptide secondary structures. *Vib. Spectrosc.*, **42**, 63–73.
24. Chi, H., Lakhani, A., Roy, A. *et al.* (2010) Inter-residue coupling and equilibrium unfolding of PPII helical peptides. Vibrational spectra enhanced with 13C isotopic labeling. *J. Phys. Chem. B*, **114**, 12744–12753.
25. Barth, A. (2007) Infrared spectroscopy of proteins. *Biochim. Biophys. Acta*, **1767**, 1073–1101.
26. Pancoska, P., Yasui, S.C. and Keiderling, T.A. (1989) Enhanced sensitivity to conformation in various proteins. Vibrational circular dichroism results. *Biochemistry*, **28**, 5917–5923.
27. Zhu, F., Tranter, G.E., Isaacs, N.W. *et al.* (2006) Delineation of protein structure classes from multivariate analysis of protein Raman optical activity data. *J. Mol. Biol.*, **363**, 19–26.
28. Chen, X.G., Schweitzer-Stenner, R., Asher, S.A. *et al.* (1995) Vibrational assignments of trans-N-methylacetamide and some of its deuterated isotopomers from band decomposition of IR, visible, and resonance Raman spectra. *J. Phys. Chem.*, **99**, 3074–3083.

29. Asher, S.A., Ianoul, A., Mix, G. *et al.* (2001) Dihedral psi angle dependence of the amide III vibration: a uniquely sensitive UV resonance Raman secondary structural probe. *J. Am. Chem. Soc.*, **123**, 11775–11781.
30. Moss, D. and Sage, J.T. (1999) Oxidized-minus-reduced FTIR difference spectroscopy of cytochrome c crystals, in *Spectroscopy of Biological Molecules: New Directions*, pp. 73–74, Kluwer Academic Publishers, Dortrecht, The Netharlands.
31. Deng, H. and Callender, R. (2001) Vibrational studies of enzymatic catalysis, in *Infrared and Raman Spectroscopy of Biological Materials* (eds H.U. Gremlich and B. Yan), Marcel Dekker, Inc., New York, pp. 477–515.
32. Carey, P.R. (2006) Raman crystallography and other biochemical applications of Raman microscopy. *Annu. Rev. Phys. Chem.*, **57**, 527–554.
33. Wikipedia (2014). DNA. 04/02/2014, http://en.wikipedia.org/wiki/DNA (accessed 3 January 2015).
34. Prescott, B., Steinmetz, W. and Thomas, G.J. Jr., (1984) Characterization of DNA structures by laser Raman spectroscopy. *Biopolymers*, **23**, 235–256.
35. Thomas, G.A. and Peticolas, W.L. (1984) Sequence dependence of conformations of self-complementary duplex tetradeoxynucleotides containing cytosine and guanine. *Biochemistry*, **23**, 3202–3207.
36. Tsuboi, M., Takahashi, S., and Harada, I. (1973) *Infrared and Raman spectra of nuclei acid vibrations in the base residues: in Physico-Chemical properties of Nucleic Acids*, 91–145. J. Duchesne, Editor, Academic Press, London.
37. Pevsner, A. and Diem, M. (2001) Infrared spectroscopic studies of major cellular components. II. The effect of hydration on the spectra of nucleic acids. *Appl. Spectrosc.*, **55** (11), 1502–1505.
38. Whelan, D.R., Bambery, K.R., Heraud, P. *et al.* (2011) Monitoring the reversible B to A-like transition of DNA in eukaryotic cells using Fourier transform infrared spectroscopy. *Nucleic Acids Res.*, **39** (13), 5439–5448.
39. Annamalai, A. and Keiderling, T.A. (1987) Vibrational circular dichroism of poly(ribonucleic acids). A comparative study in aqueous solution. *J. Am. Chem. Soc.*, **109**, 3125–3132.
40. Diem, M. (1993) *Introduction to Modern Vibrational Spectroscopy*, Wiley-Interscience, New York.
41. Andrushchenko, V., Blagoi, Y., van de Sande, J.H. and Wieser, H. (2002) Poly(rA) • Poly(rU) with Ni2+ Ions at different temperatures: infrared absorption and vibrational circular dichroism spectroscopy. *J. Biomol. Struct. Dyn.*, **19** (5), 889–906.
42. Zhong, W.X., Gulotta, M., Goss, D.J. and Diem, M. (1990) Infrared VCD in DNA models. II. Experimental and theoretical results for B-family polymers. *Biochemistry*, **29**, 7485–7495.
43. Gulotta, M., Goss, D.J. and Diem, M. (1989) IR vibrational CD in model deoxyoligonucleotides: observation of the B-Z phase transition and extended coupled oscillator intensity calculations. *Biopolymers*, **28**, 2047–2058.
44. Andrushchenko, V.V., van de Sande, J.H., Wieser, H. *et al.* (1999) Complexes of (dG-dC)20 with Mn2+ ions: a study by vibrational circular dichroism and infrared absorption spectroscopy. *J. Biomol. Struct. Dyn.*, **17** (3), 545–560.
45. Yang, L. and Keiderling, T.A. (1993) Vibrational CD study of the thermal denaturation of poly(rA) · poly(rU). *Biopolymers*, **33** (2), 315–327.
46. Mendelsohn, R., Sunder, S. and Bernstein, H.J. (1975) Structural studies of biological membranes and related model systems by Raman spectroscopy: sphinomyelin and 1,2 dilauroyl phosphatidylethanol amine. *Biochim. Biophys. Acta*, **413**, 329–340.
47. Mendelsohn, R., Sunder, S. and Bernstein, H.J. (1976) The effect of sonification on the hydrocarbon chain conformation in model membrane systems: a Raman spectroscopic study. *Biochim. Biophys. Acta*, **419**, 563–569.
48. Dluhy, R.A., Moffatt, D., Cameron, D.G. *et al.* (1985) Characterization of cooperative conformational transitions by Fourier transform infrared spectroscopy: application to phospholipid binary mixtures. *Can. J. Chem.*, **63**, 1925–1932.
49. Lee, D.C. and Chapman, D. (1986) Infrared spectroscopic studies of biomembranes and model membranes. *Biosci. Rep.*, **6** (3), 235–256.
50. Moore, D.J., Rerek, M.W. and Mendelsohn, R. (1997) FTIR spectroscopy studies of the conformational order and phase behavior of ceramides. *J. Phys. Chem. B*, **101**, 8933–8940.
51. Carey, P.R. (1982) *Biochemical Applications of Raman and Resonance Raman Spectroscopies*, Academic Press, New York.
52. Bush, S.F., Levin, H. and Levin, I.W. (1980) Cholesterol-lipid interactions: an infrared and Raman spectroscopic study of the carbonyl stretching mode region of 1,2-dipalmitoyl phosphatidylcholine bilayers. *Chem. Phys. Lipids*, **27**, 101–111.

53. Mushayakarara, E. and Levin, I.W. (1982) Determination of acyl chain conformation at the lipid interface region: Raman spectroscopic study of the carbonyl stretching mode region of dipalmitoyl phosphatidylcholine and structurally related molecules. *J. Phys. Chem.*, **86**, 2324–2327.
54. Levin, I.W., Mushayakarara, E. and Bittman, R. (1982) Vibrational assignment of the sn-1 and sn-2 chain carbonyl stretching modes of membrane phospholipids. *J. Raman Spectrosc.*, **13**, 231–234.
55. Cortijo, M., Alonso, A., Gómez-Fernández, J.C. *et al.* (1982) Intrinsic protein-lipid interactions infrared spectroscopic studies of gramicidin A, bacteriorhodopsin and Ca2·+·–ATPase in biomembranes and reconstituted systems. *J. Mol. Biol.*, **157**, 597–618.

11

Vibrational Microspectroscopy (MSP)

11.1 General remarks

While in conventional light microscopy the image contrast is produced by variations in the refractive index and the overall optical density (OD) of the sample, variations in chemical composition, at near constant gross optical parameters, constitute the major image contrast in infrared (IR) and Raman microscopy. This is, of course, due to the enormous fingerprint sensitivity of vibrational spectroscopy toward molecular structure and composition. Thus, one can envision a sample of one material that differs in molecular conformation over some spatial dimension of the sample, for example, a protein film that was denatured at some area of the sample. Although the spot of denatured protein has still the same chemical composition as the surrounding protein, and still exhibits very similar gross optical properties such as refractive index or absorptivity in the visible range of the spectrum, its vibrational spectrum will be different due to the conformational change, and such a change can be detected – spatially resolved – by infrared microspectroscopy (IR-MSP, also referred to as infrared microscopy) or Raman microspectroscopy (RA-MSP, also known as Raman microscopy). Thus, the pioneers in vibrational microspectroscopy (MSP) who had envisioned this new contrast mechanism created microscopic tools that truly are a "chemist's microscope" since no other technique can visualize chemical changes over microscopic dimensions without the use of external labels as well as vibrational MSP.

Owing to this enormous chemical fingerprint sensitivity, the use of vibrational MSP has skyrocketed, and IR and Raman microspectrometers now constitute a sizable portion of the instrument market. In particular, in biomedical applications nearly all work is being carried out microscopically. This chapter introduces the instrumentation and methodology of MSP, and the biomedical applications are discussed in Chapters 13 and 14.

Both RA-MSP and IR-MSP create huge datasets, since hyperspectral datacubes are being collected during these experiments. A hyperspectral datacube (also known as spectral hypercube) is a three-dimensional dataset best envisioned by the schematic drawing shown in Figure 12.1: for each pixel, defined by x and y coordinates and pixel size, an entire vibrational spectrum is collected, often containing hundreds or thousands of spectral intensity points. Typical x and y dimensions of the sample often measure in the hundreds of coordinate points in both x and y directions. Thus, a typical dataset may contain between 10,000 and several millions of spectral vectors, each vector containing about 1000 intensity points. Analysis of such massive amounts of data requires data analysis and visualization algorithms that are discussed in Chapter 12.

Modern Vibrational Spectroscopy and Micro-Spectroscopy: Theory, Instrumentation and Biomedical Applications, First Edition.
Max Diem.

The present chapter starts with a short review of optical microscopy, followed by a discussion of RA-MSP, since it is a straightforward extension, in terms of the instrumentation, of visible and fluorescence microscopy. Next, a near-field Raman technique is introduced that allows spectroscopy to be carried out below the diffraction limit. This technique, known as tip-enhanced Raman spectroscopy (TERS) holds enormous potential for single-molecule spectroscopy. IR-MSP instrumentation is discussed including Fourier transform infrared (FTIR) MSP, synchrotron-based methods as well as the most recent developments in quantum cascade laser (QCL)-based MSP instrumentation. Finally, a near-field IR technique is introduced as well.

11.2 General aspects of microscopy

On the subject of microscopy, there are probably more books and reviews written than on vibrational spectroscopy. Thus, it is impossible and impractical to review the subject at a level of detail provided in these works, and the discussion here is at a very elementary level. The reader is referred to review articles in the literature for more details on basic microscopy [1]. The most basic form of a microscope is the "bright field" microscope in which visible light is focused onto the sample *via* a condenser, which is usually situated below the sample specimen. The light transmitted by the sample is collected through a microscope objective. Other forms of light microscopy include dark field and phase contrast microscopy. In the former, the parallel beam of light is blocked, *via* a "patch stop" before the condenser, such that only a narrow ring (annulus) of light is incident on the sample. This light is blocked by an aperture to enter the objective lens, such that only the light scattered by the sample enters the objective. In phase contrast microscopy, the phase difference in the scattered light (as compared to the background transmission) is used to suppress the background light by phase shifting it and subsequently reducing its intensity by a gray filter. The IR microscopes to be discussed later use simple bright-field illumination in transmission mode, and a somewhat more complicated arrangement in reflection measurement.

In a compound (bright field) microscope, an image of the illuminated specimen ("object") is collected by the objective lens (or microscope objective), which creates a real, magnified image of the specimen ("image 1") inside the microscope tube. The magnification of the objective in standard optical microscopes used in pathology are typically 4×, 10×, 20×, and 40×, but higher magnification objectives are available. Since the intermediate image is real, it can be observed by inserting a sheet of transparent material, such as tracing paper, at the focus of the objective.

This real image is subsequently magnified by a second lens (the eye-piece lens) that produces an inverted enlarged virtual image. The total magnification of this virtual image is the product of the magnification of the objective and the eye-piece lenses. While the optical diagram in Figure 11.1 represents a "finite tube length" design, most modern microscopes employ the concept of "infinity correction," in which a second lens (the "tube lens") is placed between the objective and the real image. This creates a region of parallel light propagation between the objective and the tube lens and allows the optical designer greater flexibility in the positioning of the real intermediate image.

In bright-field microscopy, one may assume that the microscopic specimen is a complex, two-dimensional grating with apertures and lines of various sizes. An image formed from such a specimen, therefore, will consist of superpositions of individual diffraction patterns of the apertures and lines. For a point source, the image formed will be the single point-spread function, which may be considered the diffraction pattern of a single aperture. This point-spread function, also known as the Airy-function, is shown in Figure 11.2. The details of a microscopic specimen will be resolved if the microscope objective collects the zeroth and first order of the light diffraction pattern shown in Figure 11.2 [2]. This, in turn, defines the lateral resolution d_{lat}. Since the Airy function depends on the wavelength λ, the lateral distance between two

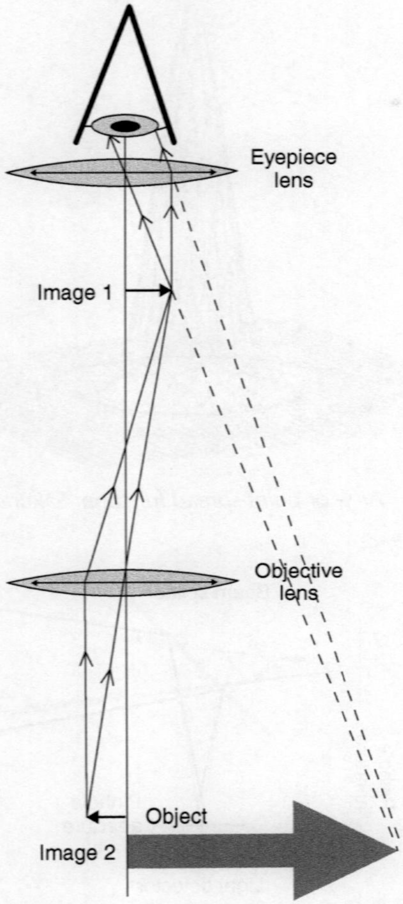

Figure 11.1 *Compound microscope. See text for details. From en.wikipedia.org/wiki/Optical_microscope*

adjacent objects that each create their own Airy disk and that can still be resolved is given by the Rayleigh criterion

$$d_{\text{lat}} = \frac{0.61\lambda}{\text{NA}} \tag{11.1}$$

In Eq. 11.1, NA denotes the numeric aperture of the objective, defined by

$$\text{NA} = n \sin\theta \tag{11.2}$$

Here, n denotes the refractive index of the optical material used to construct the objective and θ is the arctan of the lens radius $D/2$ divided by its focal length f:

$$\theta = \arctan\frac{D}{2f} \tag{11.3}$$

Index matching between the sample and the objective, *via* a drop of oil or water that touches the "immersion objective," increases the numerical aperture.

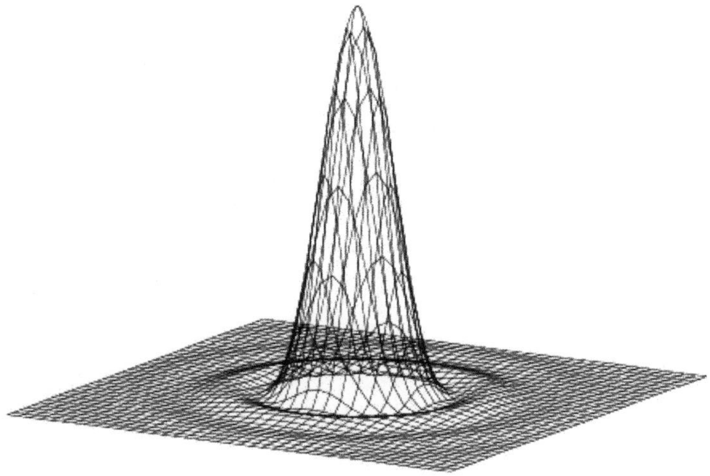

Figure 11.2 Airy- or point-spread function. Sakurambo 2009 [2]

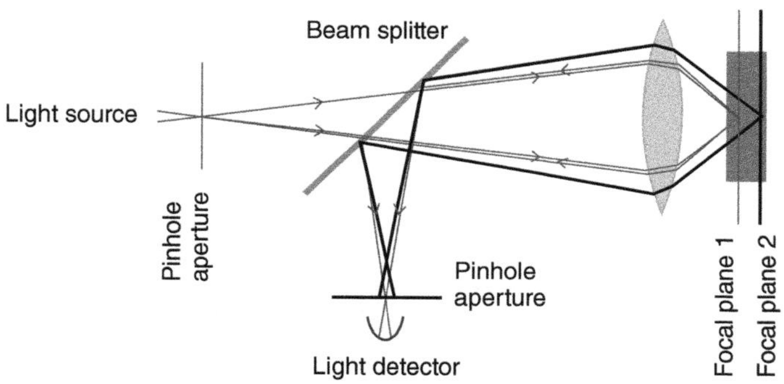

Figure 11.3 Confocal illumination. See text for details

Raman microscopy is often carried out as a confocal measurement, which implies that the light used to excite the sample passes through a pinhole. Similarly, the scattered light is collected and passed through a second pinhole. It is obvious from Figure 11.3 that light scattered from two different focal planes will not be able to pass the second pinhole, since the image from focal plane 2 is defocused at the pinhole. This confocal arrangement permits optical sectioning of the sample, since different focal planes can be brought into a position such that light scattered from one focal plane only reaches the detector.

The axial (depth) resolution d_{ax} is given by

$$d_{ax} = \frac{2n\lambda}{(\mathrm{NA})^2} \tag{11.4}$$

resulting in a theoretical depth resolution for a water ($n = 1.33$) immersion objective with NA = 1 between about 1300 and 1700 nm, for 488 and 633 nm excitation, respectively. The corresponding lateral resolution, defined by Eq. 11.1, would be between about 300 and 435 nm for the same excitation wavelengths and numeric apertures. The depth resolution can be established nicely by collecting the Raman scattering of a layered sample, such as a film laminate consisting of polyethylene terephthalate (PET), ethylene vinyl

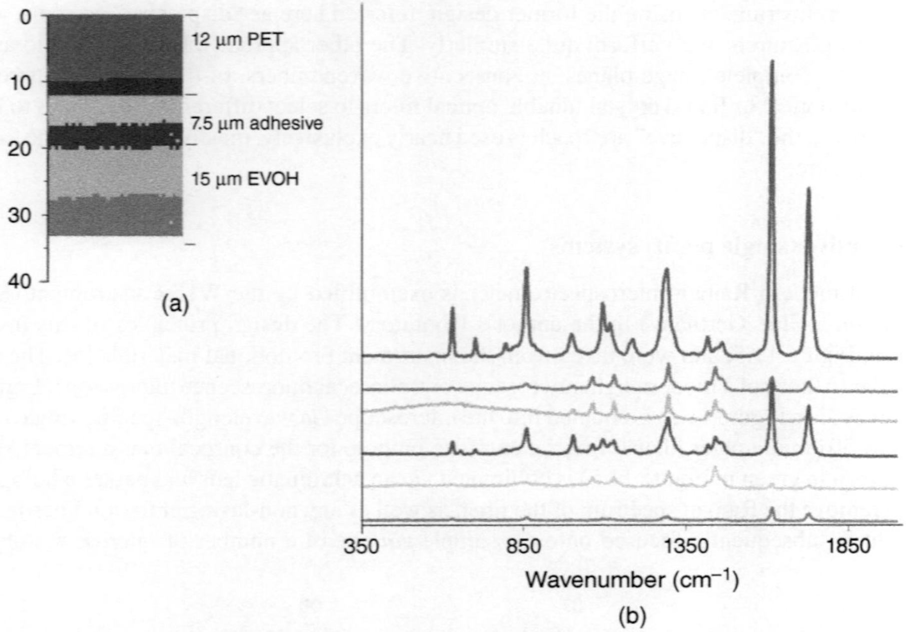

Figure 11.4 *(a) Cluster analysis results of a depth scan into a film laminate. (b) Mean cluster spectra for the distinct polymer and adhesive layers. (*See insert for color representation of this figure.*)*

alcohol (EVOH) over low-density polyethylene (Figure 11.4), with several adhesive layers in between. This film laminate is used widely in the food packing industry, and offers an excellent sample system to demonstrate the depth resolution of a confocal Raman system. Figure 11.4 presents results of hierarchical cluster analysis (HCA; see Chapter 12) from a depth scan dataset of the film laminate.

By using HCA for analysis, one does not predetermine any intensity criteria that define the boundaries between the layers; and it is obvious by inspection of the image that diffusion of the adhesive layer has occurred, which mixes spectral patterns to some extent.

The upper PET layer is supposed to be 12 µm thick, but beyond 10 µm depth, some signals of the adhesive appear superimposed on the PET spectra. Up to the depth of the EVOH layer, the Raman image agrees reasonably well with the known thickness of the layers [3].

Everall [4] has shown recently that the axial resolution is actually compromised in many-layer systems by the different refractive indices of the sample, and that a substantial error is encountered at depths exceeding 20–25 µm. Thus, for accurate depth probing, the interactions of both the incident laser light and the scattered photon with the interfaces between the multilayers have to be modeled. Cells and tissues studies at the Laboratory for Spectral Diagnosis (LSpD, see Chapter 14) rarely exceeded 10 µm in thickness; consequently, it is believed that the *X–Z* and *Y–Z* optical sections are reasonably accurate in terms of the depth sampled.

11.3 Raman microspectroscopy (RA-MSP)

In the following section, two designs for Raman microscopes are presented. One of these follows closely the design used in high-resolution visible microscopes or fluorescence microscopes, whereas the other utilizes an imaging approach. Instruments using the former design, referred here as "dispersive" systems, are available from several manufacturers and perform quite similarly. The other approach, here referred to as "imaging" approach, acquires complete image planes, at consecutive wavenumbers, of the Raman scattered radiation, and uses acousto-optical or liquid crystal tunable optical filters to select different color planes to be acquired. For biological work, the "dispersive" approach is used nearly exclusively, mostly because of the lower average laser power exposure.

11.3.1 Dispersive (single point) systems

The design of a modern Raman microspectrometer is exemplified by the WITec instrument (Model CRM 2000, WITec, Inc., Ulm, Germany) in the author's laboratory. The design principles of this instrument are summarized in Figure 11.5, and were taken from the instrument promotional materials [5]. These principles are quite similar to those of a modern light microscope or a confocal fluorescence microscope. Light from various laser sources (1 in Figure 11.5) is coupled into the microscope *via* wavelength-specific single mode optical fibers (2). This 50 µm diameter fiber forms the entrance pinhole for the confocal measurement. The incident laser light (shown in green in Figure 11.5) is collimated *via* an achromatic lens and passes a holographic band pass filter, to remove the Raman spectrum of the fiber, as well as any non-lasing emission lines from the laser. The laser light is subsequently focused onto the sample *via* one of a number of microscope objectives (3).

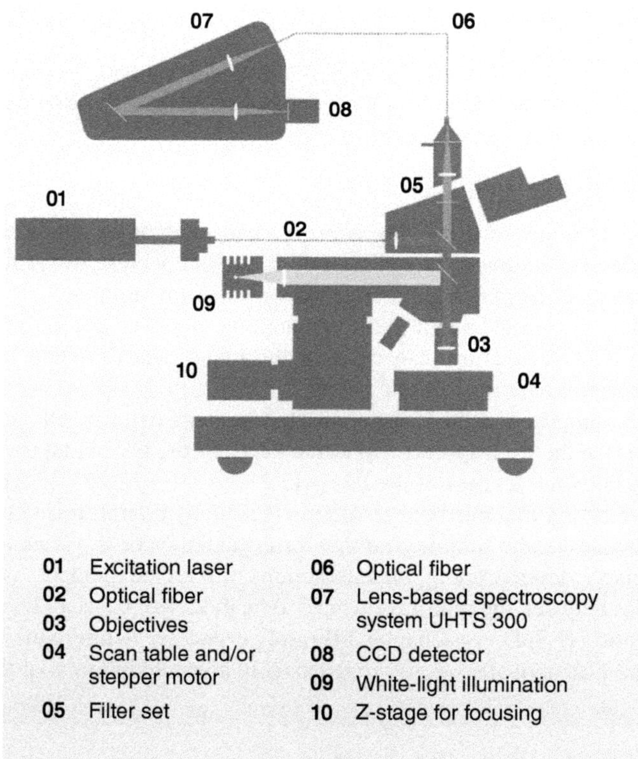

Figure 11.5 *Schematic of a dispersive Raman microscope (WITec, Inc., CRM 200). See text for details. WITec, 2014, [5]. Reproduced with permission from WITec.* (See insert for color representation of this figure.)

For most high-resolution work on tissues and cells, a Nikon Fluor (60 × /1.00 NA, WD = 2.0 mm) water immersion or a Nikon Plan (100 × /0.90 NA, WD = 0.26 mm) objective was used. Here, WD denotes the working distance of the objective.

The sample is located on a piezoelectrically driven microscope scan stage (4) with *X–Y* resolution of about 3 nm and a repeatability of ±5 nm, and *Z* resolution of about 0.3 and ±2 nm repeatability. Samples generally were spin deposited onto, or directly grown onto, optically flat CaF_2 windows, which are immersed into water, buffer, or cell culture medium if the water immersion objective is used. Visual images of the sample can be observed with white light excitation (9), shown in yellow in Figure 11.5, and projected on a camera. For best results, biological or biomedical samples are moved very slowly, but continuously, during Raman data acquisition through the laser focus; typically, the microscope stage was scanned at velocities of about 300 nm/300 ms. This prevents photodecomposition of the biological sample systems.

Raman backscattered radiation is collected through the microscope objective, and passes a holographic long pass filter (5) to block Rayleigh scattering and reflected laser light, before being focused into a multimode optical fiber (6) that constitutes the second pinhole for the confocal measurement. The light emerging from the multimode fiber is dispersed by a monochromator (7), incorporating a turret with three interchangeable gratings (1800 mm^{-1}, blazed at 500 nm; 600 mm^{-1}, blazed at 500 nm; and 600 mm^{-1}, blazed at 750 nm). The light is finally detected by a back-illuminated deep-depletion, 1024 × 128 pixel charge-coupled device (CCD) camera (8) operating at −82 °C. Spectral resolution depends on the excitation wavelength and grating groove density; furthermore, the spectral resolution varies significantly over the spectral range projected onto the CCD detector (see Chapter 4). For 488 nm excitation and the 600 mm^{-1} grating, the optical resolution varies from about 5.5 cm^{-1}/data point at about 200 cm^{-1} to about 3.3 cm^{-1}/data point at 4500 cm^{-1}. The 1800 mm^{-1} grating compresses the wavenumber coverage to about 1400 cm^{-1}. Owing to the nonlinear data point spacing of the resulting spectral vector, all data were "linearized," as described in Chapter 12, to constant wavenumber increments. This is advantageous when multivariate data analysis is being carried out for different laser excitation wavelengths.

11.3.2 Micro-Raman imaging systems

An alternative approach to collect Raman microspectral images has been pursued and commercialized by ChemImage, Inc. In their instrument, the exciting laser is defocused to illuminate an area as large as about 35 μm on edge (with a 100× objective); in other words, nearly an entire cell as shown in Figure 14.3 would be illuminated simultaneously. The scattered light from the entire illuminated area is collected and passed through a large aperture liquid-crystal tunable filter (LCTF) that selects one wavelength of the scattered photons and detects this light *via* a large, square, or rectangular CCD detector. Since the LCTF can be tuned rapidly to different wavelengths, a spectral hypercube can be collected image plane by image plane. This is in contrast to the dispersive systems discussed earlier where for each pixel point, the entire scattered spectrum is acquired simultaneously. The ChemImage system offers a significant multiplex advantage, since an image (or color) plane consisting of thousands of pixels can be collected simultaneously in a few seconds. Furthermore, since the LCTF can skip wavelength ranges of the scattered light that present no useful signal, a hypercube can be collected that is devoid of the non-informative regions.

However, large area illumination has its own problems. In the dispersive system, the laser power density at the focus reaches about 10^6 W/cm^2. This power density is responsible for a fairly strong Raman signal such that an entire spectrum can be recorded in about 300 ms. A simple calculation shows that in order to achieve the same excitation power density, a laser with about 4 W laser power would have to be focused into the sample. Most biological samples would decompose quickly when globally illuminated by such laser powers that can be tolerated in the diffraction size spots of dispersive systems.

11.4 CARS and FSRS microscopy

Two nonlinear Raman techniques were introduced in Chapter 6, coherent anti-Stokes Raman spectroscopy (CARS) and femtosecond stimulated Raman scattering (FSRS). Both these effects have been well understood but were of little practical importance until their observation through microscopes was reported [6].

In CARS, the sample is illuminated by two lasers, one of them with a fixed wavelength usually referred to as the pump laser ω_P or ω_1, and a second tunable or broad-band laser, referred to as the Stokes frequency ω_S or ω_2. Coherent mixing of the electric fields of these lasers results in a photon at the anti-Stokes frequency being emitted:

$$\hbar\omega_{\text{AS}} = \hbar(2\omega_P - \omega_S) = \hbar(\omega_P + \omega_M) \quad (6.20)$$

where $\hbar\omega_M$ is the energy of one of the molecule's vibrational modes. The emission of the anti-Stokes photon occurs only if the wavelength of the tunable Stokes laser fulfills the condition

$$\hbar\omega_S = \hbar(\omega_P - \omega_M). \quad (6.21)$$

Since the molecular system is left in the original state after the creation of the CARS photon, the wave vectors need to add up to zero. This leads to the phase-matching condition for CARS, which can be written as

$$\boldsymbol{k}_p - \boldsymbol{k}_s + \boldsymbol{k}_p - \boldsymbol{k}_{\text{AS}} = 0 \quad \text{or} \quad \boldsymbol{k}_{\text{AS}} = \boldsymbol{k}_p - \boldsymbol{k}_s + \boldsymbol{k}_p \quad (6.23)$$

The anti-Stokes photons leave the sample as a collimated, coherent laser beam, for which the spectral resolution is given by the line width of the exciting lasers.

CARS spectroscopy has taken a huge step forward when it was discovered in microscopic measurements [6] where in the large solid angle of the light cone in a microscope objective, the phase-matching angle is always fulfilled [7]. Thus, a typical CARS microscope can be constructed according to the schematic shown in Figure 11.6 [8]. Here, the laser pulses at ω_p and ω_s are combined at the lower dichroic mirror (DM), and are raster scanned, *via* a rotating prism scanner, over the sample, after being focused by a microscope objective in the inverted microscope setup shown. The light scattered if the energy and phase matching conditions for CARS generation are fulfilled is directed *via* a second DM toward the detector. An instrument as shown in Figure 11.6 can produce CARS images at video rates, and using the same excitation/detector geometry, can detect two photon fluorescence as well as second harmonic generation.

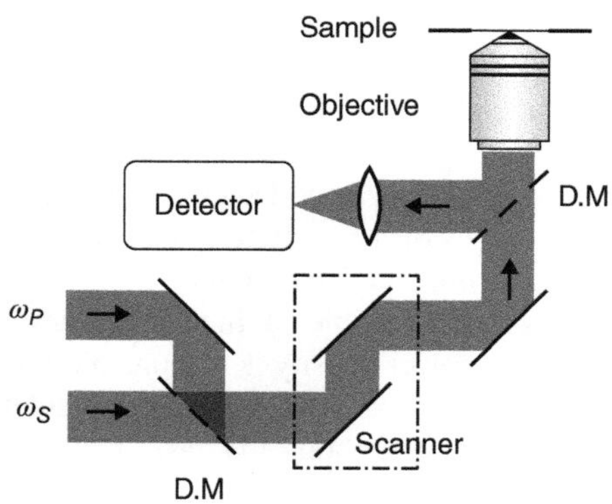

Figure 11.6 Schematic of a CARS microscope. See text for details. Cheng, 2007, [8]. Reproduced with permission from Society for Applied Spectroscopy

In an elegant optical arrangement reported by Kano [9], femtosecond laser pulses from a Ti:Sapphire laser were split to obtain pulses that were used as CARS pump pulses after filtering them to about 20 cm^{-1} band width. The other part of the split beam was directed to a photonic crystal fiber to create coherent super-continuum pulses for Stokes excitation. In this way, CARS spectra of 2000 cm^{-1} width could be collected simultaneously in a microscope set-up (*cf.* Figure 6.8).

Owing to the presence of line shape distortions by non-resonant background, and the nonlinear dependence of the CARS signal on sample concentrations, biological imaging studies are now often carried out *via* FSRS [10, 11]. Here, broadband femtosecond pulses are mixed coherently with a narrow pump laser frequency, and the same frequency mixing described before for stimulated Raman scattering takes place (see Chapter 6). The major difference in FSRS is that all frequencies contained in the broadband pulse simultaneously can mix with the pump pulse; thus, the entire Raman spectrum can be probed at once. However, the observation of FSRS signals is made more difficult since it requires the detection of the increase in intensity of the probe pulse in the presence of the pump pulse. Thus, high-frequency modulation techniques are applied to observe this intensity increase synchronously. The research group of Xie at Harvard University has exploited microscopically observed FSRS for a number of pioneering spectroscopic studies of biological and biomedical interest [12, 13]. In particular, neuron bundles in mouse brain and depth-resolved mouse skin structures could be identified in the FSRS images taken in the C—H stretching region [7].

11.5 Tip-enhanced Raman spectroscopy (TERS)

A completely different form of high-resolution spectroscopy is afforded by a method known as TERS that was developed at the ETH in Zürich [14]. TERS is a near-field optical technique closely related to scanning near-field optical microscopy (SNOM), but uses the principles of surface-enhanced Raman spectroscopy (SERS) as a read-out mechanism. SERS was introduced in Chapter 6, and it was pointed out in this discussion that SERS long suffered from poor reproducibility, since the Raman scattering enhancement critically depends of the surface morphology of the nanoparticle substrate responsible for the electric field enhancement (see Figure 11.7).

In TERS, this problem is avoided by using an extremely sharp metal tip that is brought within a few nanometers from the analyte surface. This tip may have a radius of curvature of a few tens of nanometers, and produces a sharp signal increase when brought within close contact of the surface. The motion of the tip is controlled by the same cantilever technology used in atomic force microscopy (AFM) or SNOM (see Figure 11.7). Since the same tip is scanned over the sample surface during an experiment, the signal enhancement is highly reproducible during a given experiment. However, tip-to-tip variations are still large.

One of the most amazing applications of TERS is the direct optical sequencing of a strand of RNA by TERS [15]. The authors argue that the size of the TERS tip simultaneously samples about six nucleobases. Distinct TERS signals could be observed when the tip was moved along the RNA strand in steps of about 100 nm.

This technique has created a new scientific field best described as nanospectroscopy. In RA-MSP, one may find between a few thousand to a few ten thousand molecules in the focal volume of the laser (about 10^{-16} L), whereas in TERS this number is reduced to between 10 and 100 molecules. At this spatial resolution, different problems of reproducibility of the measurement arise, since sample heterogeneity affects the results obtained from this technique.

11.6 Infrared microspectroscopy (IR-MSP)

As in the case of Raman microspectrometers where dispersive (point-by-point) and global (imaging) data collection schemes were discussed, IR-MSP can be carried out using two approaches. One of them is an imaging

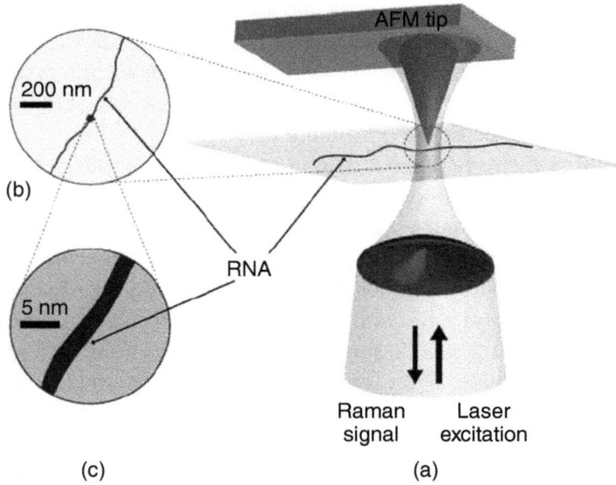

Figure 11.7 Principles of TERS spectroscopy. (a) Schematic of sampling geometry in TERS. The sharp tip is touching the sample from top, while laser radiation is focused on the sample from below. Raman scattering is observed through the same objective that focuses the beam. (b) Comparison of the sample spot size and the size of the sample, here a single-strand RNA molecule. (c) Further expansion of sampled spot size. Bailo and Deckert, 2008, [15]. Reproduced with permission from John Wiley & Sons

approach quite similar to that discussed in Section 11.3.2, where a large area detector directly measures the absorbance of thousands of pixels illuminated by a tunable laser, typically a quantum cascade laser (QCL). This approach does not require a wavelength sorting device such as an interferometer or a monochromator, since the tunable laser will provide discrete wavelength for excitation. The other approach uses broadband IR sources and interferometers to produce an interferogram of the light attenuated by the sample for each detector element. This method is called FTIR imaging.

11.6.1 Fourier-transform infrared imaging systems

As indicated, these instruments use broadband IR radiation, from a globar or synchrotron source, and send the IR radiation through an interferometer as discussed in Chapter 3 and Appendix 4. Subsequent to the interferometer, the light is focused into the sample, using Cassegrain objectives. The light transmitted (or reflected) by the sample is detected *via* a single point or an array detector. Owing to the slow data acquisition rates of single detector systems, nearly all modern IR imaging systems use either large area (about 128 × 128 pixels) military grade focal plane array (FPA) detectors or smaller (16 × 1 or 8 × 2) element detectors.

In the following discussion, the PerkinElmer Spectrum One – Spotlight 400 is discussed as an example of a modern IR imaging microspectrometer. This system consists of a Spectrum One FTIR spectrometer bench coupled to a Spectrum Spotlight 400 IR microscope. The combination of these two components is henceforth referred to as the PE400. This totally integrated imaging IR microspectrometer incorporates an 8 × 2 element HgCdTe (MCT) array detector and a single point, 100 μm × 100 μm MCT detector mounted on the same Dewar. Each of the detector elements in the array detector measures 25 μm on edge. Both detectors operate in photoconductive mode at liquid nitrogen temperature. The D^* of each element in the array detector exceeds 4.5×10^{10}cm $Hz^{1/2}W^{-1}$. The detectors were designed for use with 1300 K sources typically used in IR spectroscopy, and cover the spectral range down to 720 cm^{-1}. The single point MCT detector even allows data collection down to 650 cm^{-1}.

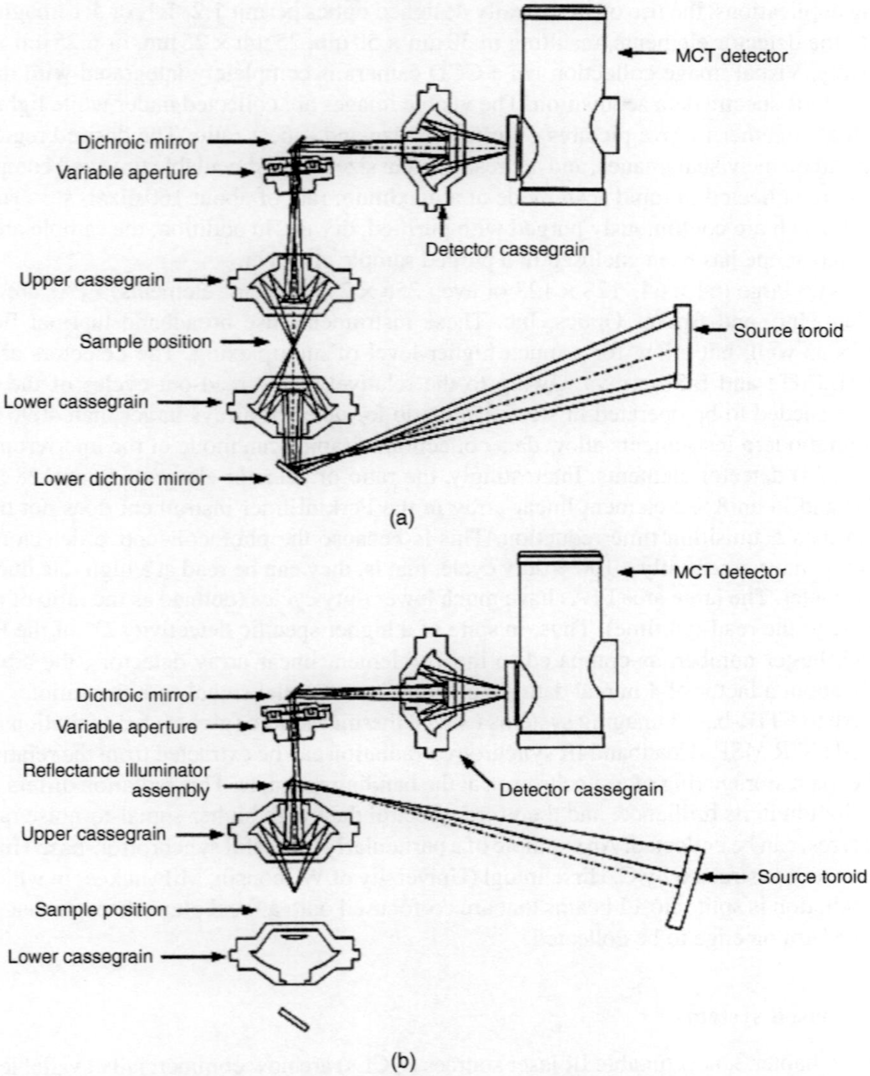

Figure 11.8 *Optical diagram of a typical FTIR-based infrared imaging system in (a) transmission and (b) reflectance mode. See text for details. From Ref. [16]*

The optical path is diagramed in Figure 11.8 for the PerkinElmer instrument [16]. For transmission measurements, light from the interferometer is focused by the condenser into the sample. The transmitted light is collected by the Cassegrain microscope objective. The symmetrically arranged objective and condenser provide an image magnification of 6× (at 1:1 imaging), and each has a numerical aperture of 0.58. Subsequently, light passes a variable aperture before being focused by a third Cassegrain onto the detector. In reflectance (panel b) configuration, light from the source is focused *via* a small mirror into one half of the objective Cassegrain, while the reflected light is collected by the other half, passed through the variable aperture before being focused on the detector.

For imaging applications, the use of specifically designed optics permit 1:2, 1:1, or 4:1 imaging of the sampled area onto the detector elements, resulting in 50 μm × 50 μm, 25 μm × 25 μm, or 6.25 μm × 6.5 μm pixel size, respectively. Visual image collection *via* a CCD camera is completely integrated with the microscope stage motion and IR spectra data acquisition. The visible images are collected under white light illumination, and are "quilted" together to give pictures of arbitrary size and aspect ratio. The desired regions for the IR maps are selected on the visual images, and are restricted in size only by available time and computer memory. Spectral maps are collected in rapid scan mode at a maximum rate of about 160 pixels s^{-1}. The microscope and the optical bench are continuously purged with purified, dry air. In addition, the sample area in the focal plane of the microscope has been enclosed in a purged sample chamber.

Instruments with large (64 × 64, 128 × 128 or even 256 × 256 detector elements) FPAs are manufactured by both Agilent, Inc. and Bruker Optics, Inc. These instruments use broadband thermal IR sources and interferometers as well, but allow for a much higher level of multiplexing. The detectors are available as photovoltaic HgCdTe and InSb arrays. Owing to the relatively slow read-out cycles of the detectors, the interferometers needed to be operated in step-scan mode for detector arrays larger than 4096 (64 × 64) elements, although modern instruments allow data collection in rapid scan mode of the interferometer for up to 16,384 (128 × 128) detector elements. Interestingly, the ratio of detector elements in an 128 × 128 element array (16,384) and in an 8 × 2 element linear array in the PerkinElmer instrument does not translate into a factor 1000 in data acquisition time reduction. This is because the photoconductive detectors in the 8 × 2 element detectors have essentially a 100% duty cycle, that is, they can be read at a high rate during the sweep of the interferometer. The large area FPAs have much lower duty cycles (defined as the ratio of the actual data integration time to the read-out time). Thus, in spite of a higher specific detectivity D^* of the FPA elements, and their much larger number, as compared to the 16 element linear array detectors, the advantage of the former is only about a factor of 4 in real data acquisition time at equal signal-to-noise ratio.

An alternative to FTIR-based imaging systems using a thermal source (globar) for excitation is synchrotron radiation-based FTIR MSP. Broadband IR synchrotron radiation can be extracted from the relativistic electron bunches in the main storage ring of a synchrotron at the bending magnets. This radiation differs from standard thermal IR radiation in its brilliance, and therefore, spectral data with higher signal-to-noise ratio, and using smaller pixel sizes, can be collected. An example of a particularly powerful synchrotron-based imaging system is the spectrometer constructed by C. Hirschmugl (University of Wisconsin, Milwaukee) in which the original synchrotron radiation is split into 12 beams that are co-focused onto a focal plane detector that permits pixels of as low as 0.54 μm on edge to be collected.

11.6.2 QCL-based systems

As indicated in Chapter 3, new tunable IR laser sources (QCLs) are now commercially available with a tuning range of about 900 cm^{-1} that covers the IR fingerprint region, 900–1800 cm^{-1}, which is the most diagnostic region for spectral diagnostic applications (see Chapter 13). The brilliance of such QCL-based sources is orders of magnitude higher than that of thermal sources; thus, IR imaging microscopes can be build that use room temperature detectors even larger than the FPAs described above. Also, the tunability of the QCLs avoids the use of an interferometer. The resulting IR imaging microscopes are more similar in character to the Raman imaging machines described in Section 11.3.2 in that an entire image plane is collected at one QCL wavelength, and that subsequent planes, collected at different wavelengths, are used to construct the familiar spectral patters at a given pixel. An instrument with a large detector and 900 cm^{-1} tuning range was introduced early in 2014 by Daylight Solutions, Inc. These instruments are likely to revolutionize the collection of IR microspectral hypercubes.

11.7 Sampling strategies for infrared microspectroscopy

While nearly all RA-MSP data are collected in backscattering geometry (see Chapter 4) in an inverted or non-inverted arrangement, several modalities exist for IR-MSP data acquisition. These mirror, in part, the different data collection methods in standard, macroscopic FTIR spectroscopy (see Chapter 3), and are discussed next. This discussion emphasizes the biomedical data acquisition of cells and tissue, which is the subject of Chapter 13.

11.7.1 Transmission measurement

For transmission measurements, the IR microscope is set up for transmission geometry, that is, the light from the interferometer is focused by the Cassegrain condenser, passes the sample, is collected *via* a Cassegrain objective, and focused on the detector, see Figure 11.9(a). Cells and tissue samples can be measured easily *via* IR-MSP since the thickness of these samples matches the requirements for obtaining good signal-to-noise spectra. Thus, cellular samples can be grown directly onto IR-transparent samples such as CaF_2, or cells can be spin-deposited (cyto-centrifuged) onto such substrates.

The cytoplasm of a dried or cultured cell may only be about 1 μm thick and give a maximal signal, in the amide I band, of 0.1 absorbance (OD) units, whereas the nucleus presents a much thicker sample layer and may exhibit intensities of about 0.5 OD units. If the cells are dried and fixed, they need not be covered during data acquisition.

Unfortunately, CaF_2 is at this time the only suitable substrate material for transmission measurements. Other window materials listed in Table 3.2 are either water soluble, which precludes them for both cell culture work and cyto-centrifugation, or toxic or expensive; or their refractive indexes are so high that the light losses at the window interface are prohibitive. Even CaF_2 windows should not be immersed in phosphate buffer solution for extended periods of time due to the precipitation of $Ca_3(PO_4)_2$ and $Ca(HPO_4)$ deposits. The main disadvantage of CaF_2 windows or slides, however, is their high cost. Studies such as the spectral cytology screening (see Chapter 13) would be prohibitively expensive if measurement were to be carried out on CaF_2 microscope slides, which retail for about $40 (or more) a piece, and if a large-scale screening was being carried out that required hundreds of samples. Thus, many IR-MSP studies have been carried out using coated reflective slides using a procedure known as "transflection" measurement, to be discussed next.

11.7.2 Transflection measurement

Alternatively, IR microspectral data can be collected in reflectance. The sample is deposited onto a completely reflective substrate, such as an aluminized or gold-coated microscope slide. Such sample substrates are not

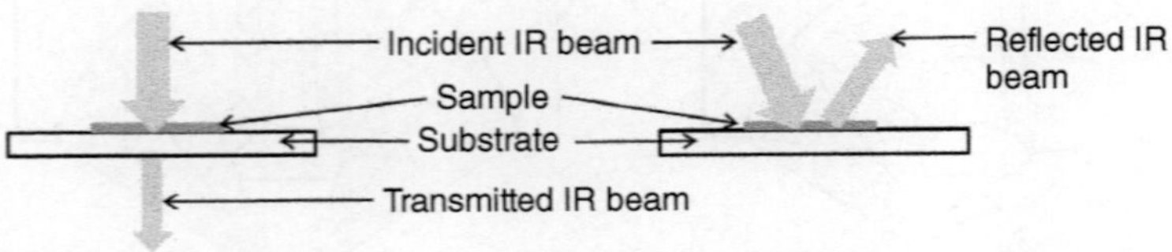

Figure 11.9 Schematic of transmission (left) and transflection measurements (right)

very suitable for subsequent histopathology or cytopathology. Thus, spectral histopathology (SHP) and spectral cytopathology (SCP) are often carried out using the so-called low emissivity ("low-e") slides, also known as MirrIR slides or Kevley slides.

These slides consist of glass with a thin silver coating and an inert overcoat. These slides are completely reflective in the IR spectral region and nearly completely transparent in the visible region; thus, samples on these substrates can be imaged in the IR spectral region, and subsequently stained and imaged with visible light to provide a correlation between classical and SHP. Unstained tissue sections on the low-e slide are used for IR data acquisition (molecular stains would interfere with data collection); subsequently, they are stained and cover-slipped and used for classical histopathological or cytopathological analysis.

All reflectance measurements, being carried out on "low-e" or metallic slides, do exhibit a standing wave artifact that was recently pointed out by P. Gardner's group ([17], see Chapter 12). As this artifact causes confounding spectral variations related to the sample thickness, low-e slide should be used with care, and data should be corrected for possible signal distortions due to this standing wave effect. Methods to do so have been discussed in the literature [18].

11.7.3 Attenuated total reflection (ATR)

As discussed in Chapter 3, ATR is based on the interaction of the evanescent wave with the sample; the evanescent wave is created when light impinges on the interface between an optically more dense material and a less dense material at an angle exceeding the critical angle (Eq. 3.38). Using special ATR Cassegrain objectives, shown schematically in Figure 11.10, it is possible to retract the ATR crystal such that the sample can be viewed in a standard reflection setup, where half of the objective is used for illumination, and the other for light collection. Subsequently, the ATR crystal is brought in contact with the sample and into the focal point of the Cassegrain [19]. There are several major advantages to ATR MSP. First, the high refractive index of the ATR crystal changes the numeric aperture of the objective according to Eq. 11.2; thus, ATR MSP affords significantly better spatial resolution than standard transmission or reflection MSP. Second, since the evanescent wave is confined to the direct vicinity of the crystal/sample interface, the Mie and resonance Mie scattering discussed in Section 12.3.2 are largely absent. Unfortunately, pressure of contact needs to be carefully controlled in ATR microscopy, and the physical contact of the ATR crystal with the sample requires

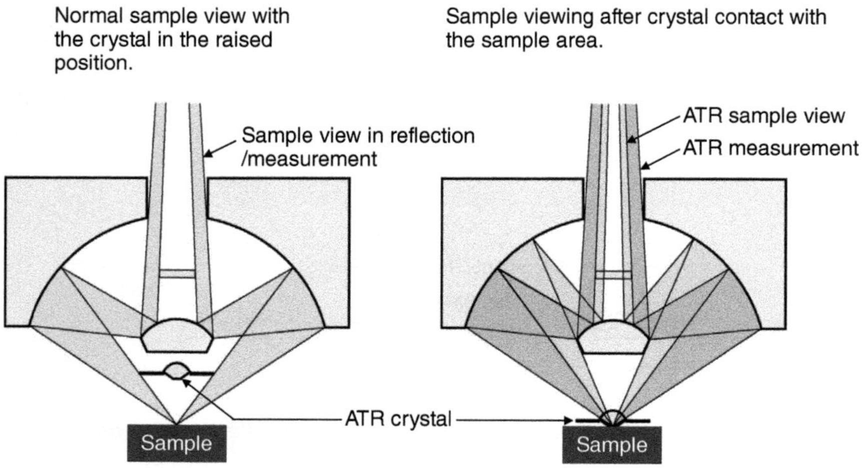

Figure 11.10 *Schematic of a Cassegrain ATR objective. See text for details. Jasco [19]*

careful cleaning of the contact area after each measurement to avoid contamination of the sample. An excellent review of ATR mircospectral imaging was recently published by Kazarian and Chan [20].

11.8 Infrared near-field microscopy

In Section 11.5, a near-field Raman technique, TERS, was introduced that bypasses the Rayleigh criterion (Eq. 11.1) and allows spectra to be collected with a spatial (lateral) resolution that is determined by the curvature of the tip that provides the Raman enhancement. Similarly, a tip-based IR near-field microscopic technique exists that was referred to as atomic force microscopy used in the infrared (AFMIR) by the researchers who first described the effect [21]. In AFMIR, an AFM tip is used in contact mode with the sample that is mounted on an IR transparent prism. IR pulses from a free-electron laser (FEL) with pulse lengths of 8 μs at about 25 Hz cause local heating of the sample if the pulses are being absorbed by the sample. The local heating causes a damped oscillation in the cantilever tip of the AFM probe. This oscillation, detected *via* a four quadrant detector, corresponds to the Fourier frequency of the vibrational transitions. By tuning the FEL through the mid-IR spectral range of interest, the IR spectrum of a single bacterial particle could be recorded at a lateral resolution of about 0.1 μm and a sample thickness of about 450 nm. As in the case of TERS spectroscopy, the lateral resolution in AFMIR no longer depends on optical parameters but rather on the size and mechanical properties of the AFM tip. Together, TERS and AFMIR represent two new spectral techniques in the field of nanospectroscopy.

References

1. Davidson, M.W. and Abramowitz, M. (2003) *Optical Microscopy*. http://micro.magnet.fsu.edu
2. Wikepedia. Point Spread Function, http://en.wikipedia.org/wiki/Point_spread_function (accessed 3 January 2015).
3. Xiao, C., Flach, C.R., Marcott, C. and Mendelsohn, R. (2004) Uncertainties in depth determination and comparison of multivariate with univariate analysis in confocal Raman studies of a laminated polymer and skin. *Appl. Spectrosc.*, **58** (4), 382–389.
4. Everall, N.J. (2004) Depth profiling with confocal Raman microscopy, part I. *Spectroscopy*, **19** (10), 22–27.
5. WITec Inc. (2014) *WITec Product Catalogue*, WITec Inc., Editor, Ulm.
6. Cheng, J.-X., Volkmer, A., Book, L.D. and Xie, X.S. (2001) An epi-detected anti-Stokes Raman scattering (E-CARS) microscope with high spectral resolution and high sensitivity. *J. Phys. Chem. B*, **105**, 1277–1291.
7. Evans, C.L., Potma, E.O., Puoris'haag, M. *et al.* (2005) Chemical imaging of tissue in vivo with video-rate coherent anti-Stokes Raman scattering microscopy. *Proc. Natl. Acad. Sci. U.S.A.*, **102** (46), 16807–16812.
8. Cheng, J.-X. (2007) Coherent anti-Stokes Raman scattering microscopy. *Appl. Spectrosc.*, **61** (9), 197A–208A.
9. Kano, H. (2008) Molecular vibrational imaging of a human cell by multiplex coherent anti-Stokes Raman scattering microspectroscopy using a supercontinuum light source. *J. Raman Spectrosc.*, **39**, 1649–1652.
10. Ploetz, E., Laimgruber, S., Berner, S. *et al.* (2007) Femtosecond stimulated Raman microscopy. *Appl. Phys. B*, **87**, 389–393.
11. Lee, S.-Y., Zhang, D., McCamant, D.W. *et al.* (2004) Theory of femtosecond stimulated Raman spectroscopy. *J. Chem. Phys.*, **121** (8), 3632–3642.
12. Freudiger, C.W., Min, W., Saar, B.G. *et al.* (2008) Label-free biomedical imaging with high sensitivity by stimulated Raman scattering microscopy. *Science*, **322**, 1857–1861.
13. Saar, B.G., Freudiger, C.W., Reichman, J. *et al.* (2010) Video-rate molecular imaging in vivo with stimulated Raman scattering. *Science*, **330**, 1368–1370.
14. Stoeckle, R.M., Suh, Y.D., Deckert, V. and Zenobi, R. (2000) Nanoscale chemical analysis by tip-enhanced Raman spectroscopy. *Chem. Phys. Lett.*, **318**, 131–136.
15. Bailo, E. and Deckert, V. (2008) Tip-enhanced Raman spectroscopy of single RNA strands: towards a novel direct-sequencing method. *Angew. Chem. Int. Ed.*, **47**, 1658–1661.

16. PerkinElmer Inc. (2004).Private Communication, The Perkin Elmer Corporation
17. Bassan, P., Lee, J., Sachdeva, A. *et al.* (2013) The inherent problem of trans-flection-mode infrared spectroscopic microscopy and the ramifications for biomedical single point and imaging applications. *Analyst*, **138**, 144–157.
18. Miljković, M., Bird, B., Lenau, K. *et al.* (2013) Spectral Cytopathology: new aspects of data collection, manipulation and confounding effects. *Analyst*, **138**, 3975–3982.
19. Jasco Benelux B.V. The Clear-View Series of FT-IR Microscope ATR Objectives, http://www.jasco.nl/clearview (accessed 3 January 2015).
20. Kazarian, S.G. and Chan, K.L.A. (2013) ATR-FTIR spectroscopic imaging: recent advances and applications to biological systems. *Analyst*, **138**, 1940–1951.
21. Dazzi, A., Prazeres, R., Glotin, F. and Ortega, J.M. (2005) Local infrared microspectroscopy with subwavelength spatial resolution with an atomic force microscope tip used as a photothermal sensor. *Opt. Lett.*, **30** (18), 2388–2390.

12

Data Preprocessing and Data Processing in Microspectral Analysis

12.1 General remarks

This chapter presents a review of the methodology for analyzing large spectral data sets typically available from vibrational microspectral measurements. These data sets typically are comprised of tens of thousands to millions of spectra. The analysis of such large data sets both requires and allows the use of multivariate methods, as it is impossible to analyze thousands of spectra by hand, and the size of the data sets allows the multivariate methods to work at their best.

However, these methods certainly are also applicable to macroscopically acquired data. In fact, early applications of multivariate methods of analysis were utilized by Keiderling and coworkers [1] to analyze vibrational circular dichroism (VCD) data from proteins and correlate them to secondary structure features. Similarly, Haaland *et al.* [2] were the first to apply principal component analysis (PCA) to infrared (IR) spectral data sets of mammalian cells. At present, virtually all microspectral data are analyzed by multivariate methods; thus, it is appropriate to introduce these methods in this chapter. The discussion to be presented here is aimed at introducing data analysis methods specific to vibrational spectra and is organized as follows. First, general methods of data preprocessing are introduced. These methods aim to improve the data quality and to assure that multivariate methods can be successfully applied to these data. Next, the discussion will focus on several spectroscopic effects that confound the integrity of spectral data, in particular, of IR data sets. Methods are introduced to ameliorate these confounding factors. Subsequently, unsupervised methods of data segmentation are discussed. These methods use the inherent spectral variance in a data set, without any input on the origin of the variance, to group or cluster the data set.

In the following section on supervised methods, the spectral variance is correlated with prior knowledge of the outcome (metadata), and methods and algorithms are derived that allow for supervised classification, for example, to allow the construction of computer-based self-learning diagnostic algorithms (machines). Finally, the correlation of spectral changes with continuously varying parameters, such as spectral variations with time, is introduced. This method is commonly employed in "two-dimensional" FTIR spectroscopy.

Until now, most data preprocessing and processing have been performed using software developed individually by each research team, typically using the MATLAB environment that incorporates many of the required functions. However, a few attempts have been made to produce standard software packages that

Modern Vibrational Spectroscopy and Micro-Spectroscopy: Theory, Instrumentation and Biomedical Applications, First Edition.
Max Diem.

incorporate the most frequently used and required functions and tasks encountered in vibrational microspectral data analysis. The first commercial package, the Cytospec® software [3], developed originally in the author's laboratory, provided a number of preprocessing and data analysis features. This program package was subsequently expanded to allow processing of much larger data set, and distributed briefly as freeware under the name of *Vi*brational *Che*mometrics (ViChe).

12.2 Data preprocessing

For the following discussion, the data are assumed to be contained in a three-dimensional matrix of dimensions x, y, and n (the spectral hypercube or hyperspectral data cube). Here, x and y denote the coordinate from which a pixel spectrum was acquired. Each pixel spectrum is stored as a vector of length n, where n is the number of measured spectral intensities at given wavelengths or wavenumbers [4]. If the intensity values in a spectral vector are equidistant, it is sufficient to store final and initial wavenumbers, and the wavenumber spacing between the intensity points, as well as n measured spectral intensities. If the data are not equally spaced, for example, in typical Raman spectral hypercubes, the intensities are stored with the corresponding wavenumber values. In the WITec instrument discussed in Chapter 11, for example, the wavenumber values are stored as one 1024-point vector, one element for each CCD element. It is assumed that this wavenumber vector holds for each spectrum in the data set. Thus, the spectral raw vectors contain 1024 intensity values as well, one intensity value being associated with a wavenumber value. A schematic representation of a hyperspectral is presented in Figure 12.1.

Each instrument manufacturer utilized its own data format; thus, it is necessary to incorporate in the data analysis software routines that can read the raw instrument data, and convert them into a convenient 3D matrix. It is assumed that each spectrum in the data set has equal spacing between data points, and common low and

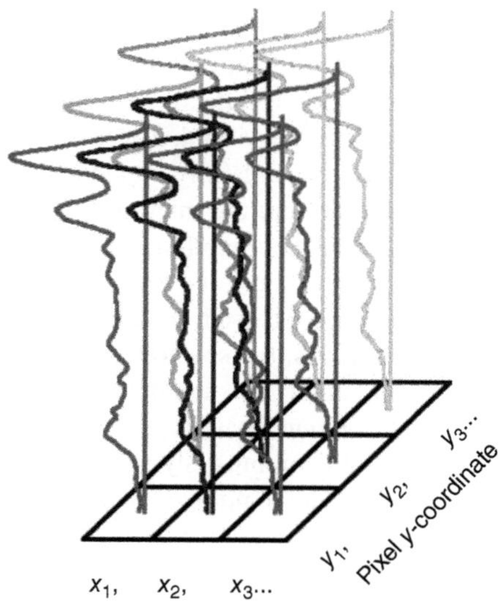

Figure 12.1 *Schematic representation of a hyperspectral data cube. Diem* et al., *2013, [4]. Reproduced with permission from John Wiley & Sons.* (See insert for color representation of this figure.)

high wavenumber (wavelength) limits. Typically, the intensity values are stored as 4 or 8 byte floating point numbers.

Next, a few commonly used spectral preprocessing routines are introduced. Some of them are specific to Raman microspectral data set, whereas others are specific to FTIR data sets. Thus, it may not be necessary to apply all the listed procedures to a given data set; also, the order in which to apply these functions is somewhat variable.

12.2.1 Cosmic ray filtering (Raman data sets only)

Raman spectra acquired on instruments equipped with CCD detectors (see Chapter 4) frequently are contaminated by sharp intensity spikes that arise when cosmic rays interact with individual CCD detector elements, thereby discharging the capacitor at a given detector element and causing large spurious intensity peaks. Their effects may be mitigated by carrying out and co-adding several short duration acquisitions rather than one longer one, and in the co-addition process check for intensities, at each diode location, that deviate greatly from those of the other acquisitions.

If such cosmic spikes are found in a stored hyperspectral data set, the best way to remove them is *via* a spatial filter. This can be accomplished by consecutively selecting each spectrum in the data set, and comparing each intensity point to the surrounding pixel intensities. Sharp spikes in the spectrum can thus be identified, because one may assume that the real variations in pixel intensities are much more subtly with the spatial coordinates than the cosmic spikes.

12.2.2 Linear wavenumber interpolation (Raman data sets only)

In contrast to most FTIR-based microspectrometer systems, where data point spacing is constant over the spectral range, dispersive Raman instruments using CCD detectors often present data at data point intervals that vary between the low- and high-frequency ends of the spectrum. This is, as discussed in Chapter 4, because the diode size is constant over the dimension of the CCD element, but the linear dispersion of a monochromator changes as a function of the diffracted angle. Depending on the focal length of the monochromator and the grating parameters, the data point spacing may vary between 3 and 5 cm^{-1} from one end of the spectrum to the other. Interpolation to linear wavenumber increments may be accomplished by inputting the observed wavenumber and intensity vectors, and a new linear wavenumber vector into an interpolation routine available, for example, in MATLAB. It is important to have spectra represented with linear wavenumber spacing when smoothing or integrating spectra, or comparing spectra taken with different excitation wavelength in Raman spectroscopy.

12.2.3 Conversion from transmittance to absorbance units (some infrared data sets)

This step may be required because some instruments store spectra in the mapping file as raw transmission spectral vectors. This conversion is carried out *via* Eq. 1.111.

12.2.4 Normalization

In IR spectroscopy, in general, and in IR microspectral data sets, in particular, the observed spectral intensities depend on the thickness of the sample, according to the Beer–Lambert law. Typically, samples are between 5 and 10 μm in thickness, but if the sample was prepared using a microtome or other cutting device, the sample thickness may not be constant over the dimension of the sample. Thus, there may be an overall bias in the observed intensities of the spectral data set. Several normalization procedures exist that mitigate the sample thickness errors. One of them is the so-called min–max normalization procedure in which the minimum spectral value is arbitrarily set to 0 (or any other convenient number) whereas the maximum intensity is set to 1 (or any other convenient number).

A more commonly applied normalization is the so-called vector normalization in which the summed squared intensity of a spectrum is set to unity. This is accomplished by dividing every spectral intensity point by a normalization factor that is obtained by summing all squared intensity values of the spectrum. Often, a spectrum is mean centered before normalization.

When processing confocal Raman data sets (see Chapter 14), it is often advantageous to use unnormalized data because the overall spectral intensity may contain valuable information on sample composition and morphology.

12.2.5 Noise reduction

Noise is an omnipresent problem in vibrational microspectroscopy, and methods to improve the signal-to-noise (S/N) ratio are most important features in data processing.

12.2.5.1 Noise reduction during data acquisition

Instrumental (white) noise can be reduced during data acquisition by selecting longer data integration times, or co-adding multiple instrument scans. In both cases, the S/N increases with the square root of the number of co-added scans or integration time: four co-added scans will improve the S/N ratio by a factor of 2. Similarly, selecting an integration time of 4 over 1 s on a CCD detector will produce, in principle, an S/N ratio that is better by a factor of 2. However, the increased integration time will only reduce the white noise; read-out noise in CCD detectors is independent of the data acquisition times.

12.2.5.2 Noise reduction in Fourier space

In Chapter 4, noise reduction in Fourier space was introduced, and many instrument software packages contain Fourier transform (FT)-based methods to truncate the high-frequency Fourier components that are responsible for spectral noise. Fourier smoothing can be carried out by multiplying the observed interferogram with a function that reduces the high-frequency contributions relative to the center burst. Kauppinen *et al.* [5] showed that the best function to accomplish noise reduction with minimal signal distortion is the sinc function:

$$\mathrm{sinc}(x) = \frac{\sin(x)}{x} \tag{12.1}$$

Of course, Fourier smoothing can be applied retroactively by carrying out a fast Fourier transform (FFT), multiplying both the real and imaginary parts by a sinc function, and performing a reverse transformation.

12.2.5.3 Savitzky–Golay smoothing

Among the non-FT-based noise reduction algorithms, the method by Savitzky and Golay [6] is likely the most frequently used noise reduction procedure. This is accomplished by applying a "sliding smoothing window" to the data vector. The smoothing coefficients C_i for the sliding window algorithm were first derived by Savitzky and Golay, and are shown, for the case of an 11-point smoothing window, in Figure 12.2 and are listed in Ref. [6]. These coefficients are applied, successively, to all data points in the original spectra, Y_j, and the results

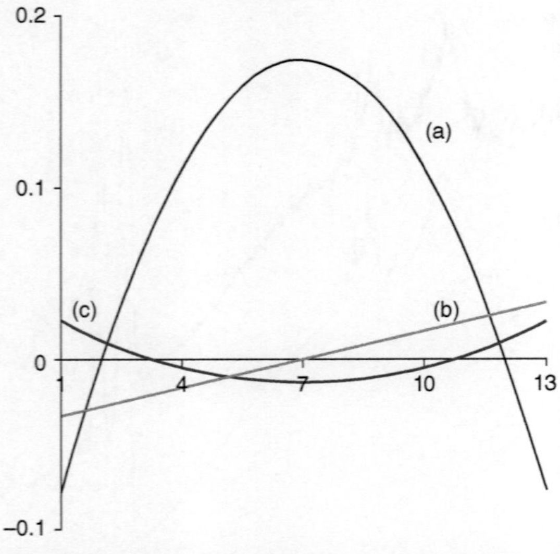

Figure 12.2 *Graphic representation of quadratic Savitzky–Golay smoothing/derivative functions. (a) 13-point smoothing, (b) 13-point first derivative, and (c) 13-point second derivative*

of the multiplication/additions shown below replace the current center data point in each window, as shown in Eq. 12.2:

$$Y_j^* = \frac{\Sigma_{i=-m}^{i=m} C_i Y_{j+i}}{N} \tag{12.2}$$

Here, Y_j^* represents the *j*th data point of the smoothed array, *N* is a normalization constant listed with the smoothing coefficients, and the summation is from $-m$ to $+m$, where *m* is the half width of the smoothing window, −1/2. The data vector must be represented in linear wavenumber increments; otherwise, the smoothing window will produce distorted spectra.

12.2.5.4 Noise-adjusted principal component analysis (NA-PCA)

An elegant method of noise reduction in hyperspectral data sets was originally developed by the remote sensing community [7] and applied to IR data sets by Bhargava and coworkers [8]. This method utilizes the noise, measured from areas from which spectra are collected, but that are not occupied by tissue or cellular material, to define the instrumental noise during data acquisition (or close in time to the actual time of data acquisition). In particular, this method creates a noise profile of the instrument as a function of wavenumber. This is important because most IR spectrometers exhibit a noise profile that is significantly worse at high wavenumber than at lower wavenumber, as shown in Figure 12.3. This is because solid-state detectors have a fairly sharp cut-on close to the band gap and reduced sensitivity at wavenumbers significantly larger than the band gap. Thus, a noise data set is created by selecting pixels spectra, in a hyperspectral data cube, that are devoid of signals.

Similarly, a signal data set is created by selecting pixel spectra that are due to areas occupied by sample. From these data sets, noise and signal correlation matrices are constructed (see Section 12.4.1). The noise correlation matrix is subsequently used to create noise-adjusted signal principal components (PCs) from which spectra with reduced noise are created. Details of this procedure are presented in the discussion of PCA later in this chapter.

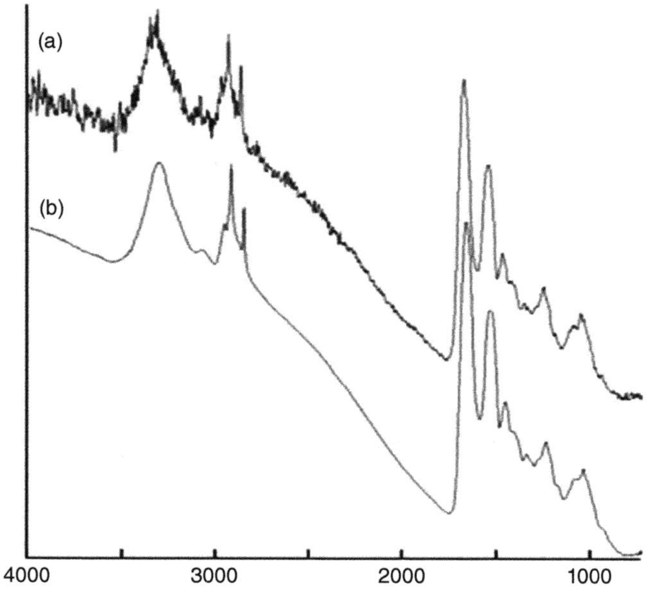

Figure 12.3 Effect of noise-adjusted principal component noise reduction. Spectrum (a) before and (b) after noise reduction

The effect of noise-adjusted principal component analysis (NA-PCA) is demonstrated in Figure 12.3. The top trace depicts a raw spectrum, with the superimposed instrumental noise [9]. As can be seen, the noise profile of the instrument is much worse in the high wavenumber region. The bottom trace shows the spectrum after NA-PCA. The improvement in S/N ratio, in general, is about an order of magnitude. Such an improvement would require a very broad Savitzky–Golay smoothing window function that would seriously degrade the signal quality as well. Furthermore, the difference in noise profile at the opposite ends of the spectra would overcompensate one end of the spectrum over the other end.

12.2.6 Conversion of spectra to second derivatives

Much of the data processing to be discussed in Section 12.4 works best if the data are converted into first or second derivatives. This statement has been criticized by many colleagues in the past, and during one round-table discussion at a conference in Japan, two well-known vibrational spectroscopists presented arguments pro- and con-derivatization. In particular, the discussion centered around the problem whether or not data can be improved, at all, by mathematical procedures applied after the data are collected. This discussion is similar to that involving zero-filling in FT spectroscopy (see Section 3.3) and seemingly boils down to the question whether one ever can get something for free.

Figure 12.4(c) shows an IR absorption spectrum of a tissue pixel, and panel (d) the same spectrum after second derivative computation. The absorption spectrum obviously exhibits a background slope that aggravates any intensity comparison, for example, the computation of the correlation coefficient (see Section 12.4.1) of two spectra. The second derivative spectrum is devoid of this background, and displays reduced band width that, in turn, helps distinguish spectra of similar appearance. This is shown in Figure 12.4(a) and (b). Here, a single Gaussian band is compared to a band envelop that consists of two Gaussian bands with a shift less than

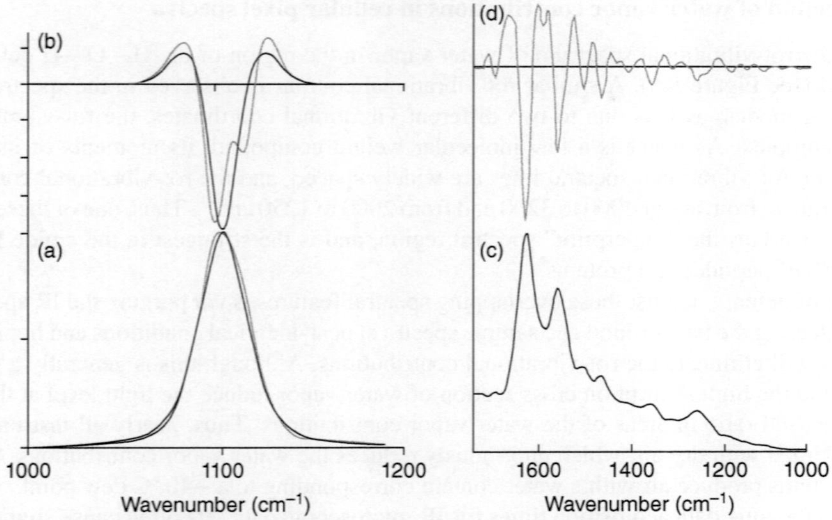

Figure 12.4 (a) Simulated spectral band, consisting of one (black) and two (gray) band envelopes. From Ref. [10]. (b) The enhanced spectral resolution for the second derivative spectra enables a better discrimination of the two cases. (c) Observed raw absorption spectrum of a cell or tissue pixel. (d) Second derivative of spectrum shown in (c), demonstrating narrower bands and elimination of sloping baseline

the half width of the bands. The correlation coefficient between these bands is close to unity [10]. When second derivatives of these band shapes are taken, the splitting becomes obvious, and the correlation coefficient is about 0.9, allowing for a better discrimination of the two traces. The increased resolution, discrimination of similar spectra, and removal of the background come at the expense of S/N of the data.

Derivatives can be computed readily using the Savitzky–Golay algorithm and sets of different smoothing derivative coefficients [6]. These are shown in Figure 12.2(b) and (c). Second derivatives can also be computed in Fourier space by multiplying the interferogram by a parabolic function centered about the zero path difference (ZPD). It should be emphasized that the second derivative spectra introduce an intensity distortion in the sense that the peaks are no longer directly proportional to the original intensity, but are depending on the curvature of the original peaks. That is, a sharp peak and a broad peak with the same peak intensity will have different second derivative intensities. However, the second derivative spectra will reduce, to some extent, the distortion brought on by the "standing electromagnetic wave effect" discussed in the following section.

12.3 Reduction of confounding spectral effects

In IR spectroscopy and microspectroscopy, there are several physical effects that confound the observed spectra. These have been referred to in the past as "artifacts"; but as they are real and well-understood physical phenomena, the "artifact" label should be avoided. Three effects are discussed here: the water vapor rot-vibrational spectral contributions, Mie and resonance Mie (R-Mie) scattering and their effects on observed spectra, and the standing wave effect on reflectance slide. All these effects may confound observed spectra to a significant degree.

12.3.1 Reduction of water vapor contributions in cellular pixel spectra

In Chapter 5, the rot-vibrational spectrum of water vapor in the region of the H—O—H deformation mode was introduced (see Figure 5.6). A similar rot-vibrational contour is observed in the spectral region of the O—H stretching modes; as it is due to two different vibrational coordinates, the rot-vibrational envelope is even more complex. As water is a low molecular weight compound, its moments of inertia are small; consequently, the rot-vibrational spectral lines are widely spaced, and the rot-vibrational contours occupy a broad spectral range, from about 4000 to 3200 and from 2000 to 1250 cm^{-1}. Thus, one of these rot-vibrational spectra largely overlaps the "fingerprint" spectral region, and is the strongest in the amide I and II regions (see Chapter 10) of peptides and proteins.

The first line of defense against these overlapping spectral features is *via* purging the IR spectrometer with dry air, and collecting the background and sample spectra at near-identical conditions and hoping that ratioing the two spectra will eliminate the rot-vibrational contributions. Although this is generally a good approach, the long path and the high absorption cross section of water vapor reduce the light level at the detector, and thus, reduce the S/N ratio in areas of the water vapor contributions. Thus, nearly all instruments are either evacuated or purged with dry air, which enormously reduces the water vapor contributions. Commercial air purification systems produce air with a water content corresponding to a −40 °C dew point.

Nevertheless, the long data acquisition times for IR microscopic data sets often cause slight changes in the water vapor content in the instruments. Consequently, spectra collected at the beginning and the end of a 1-h data acquisition often exhibit different water vapor contributions. When ratioed against the same background, positive and negative water vapor spectral contributions are observed superimposed on different spectra of the data set. Even when these water features are not detectable by the naked eye, one frequently finds that higher loading vectors of a PCA (see Section 12.4) are dominated by water vapor contributions. Simple subtraction of a scaled water vapor background is difficult because the water vapor spectra are temperature dependent (increasing temperature shifts the P and R branch intensity maxima away from the Q-branch, see Section 5.4); furthermore, the scaling of the water background is difficult because the required intensity scale factor is difficult to establish in the presence of noise.

An elegant multivariate method to reduce the remaining water vapor contributions was introduced by Bruun *et al.* [11]. This method is based on collecting water vapor spectra under a variety of different conditions, including temperature and concentration variations. These changes were reduced by PCA (see Section 12.5) into two PCs, the water vapor mean spectrum and a first variance spectrum that contains the most prominent changes to the mean spectra. Subsequently, both these spectral vectors are used in an approach referred to as extended multiplicative signal correction (EMSC) [12] to model the water vapor contributions in a spectral data set, and subsequently subtract it. This procedure reduces residual water vapor signals by over an order of magnitude.

12.3.2 Mie and resonance Mie scattering

One of the major confounding effects that plagued IR microspectroscopy has been termed R-Mie scattering by the group that first described it [13]. This paper followed a number of observations from the author's and other laboratories that some data sets exhibited unusual band profiles that resulted from the interference of absorptive and reflective line shapes; these line shapes were briefly introduced in Section 3.1.

As pointed out in the discussion in Section 3.1, the absorption process and the dispersion of the refractive index are coupled processes that are governed by the complex refractive index:

$$\eta = n - i\kappa \tag{3.1}$$

In standard FTIR spectroscopy, it is assumed that these contributions can be separated completely by the phase correction algorithm of the FTIR software (see Section 3.3), and that a pure absorption spectrum can be obtained.

In 2005, two papers from the author's laboratory [14, 15] demonstrated that Mie scattering can contribute an undulating background superimposed on the spectra of some human cells. In another paper, published the same year [16], the reduction of the dispersive band shapes in Fourier space was suggested. The former of these papers paved the way toward the formulation of the concept of R-Mie scattering by a group of spectroscopists in the United Kingdom and France and the first real correction algorithm of these spurious effects [17], whereas the second one laid the foundations for another algorithms to correct for the R-Mie effect, based on an iterative phase correction. In the following discussion, the Mie scattering mechanism is introduced briefly. Subsequently, the process leading to a mixing of the real and imaginary parts of the spectra, mediated by Mie scattering, is introduced. Finally, two methods that are used to reduce the R-Mie effect are presented.

12.3.2.1 Mie scattering

Mie scattering is a well-known physical phenomenon [18] that describes the interaction of electromagnetic radiation with dielectric or metallic spherical particles that have sizes about the same as the wavelength of light. This scattering is described by a scattering cross section Q_{sca} given by

$$Q_{sca} = \frac{1}{\rho^2} \sum_{m=1}^{\infty} (2m+1) \left(|a_m|^2 + |b_m|^2 \right) \quad (12.3)$$

Here, ρ is a size parameter, defined by

$$\rho = 4\,\pi r\, \frac{n_{12} - 1}{\lambda} \quad (12.4)$$

In Eq. 12.4, r is the radius of the scattering sphere, n_{12} the ratio of the refractive indices of the spheres' material and the surroundings, and λ the wavelength of light. The scattering coefficients a_m and b_m in Eq. 12.3 are complicated expressions in the Riccati–Bessel and Hankel functions that themselves contain half-integer Bessel functions that account for the undulating wavelength dependence of Mie scattering.

When truncating higher terms in the sum described in Eq. 12.3, an approximation of the Mie scattering expression can be obtained as follows:

$$Q_{sca} = 2 - \frac{4}{\rho}(\sin\rho) + \frac{4}{\rho^2}(1 - \cos\rho) \quad (12.5)$$

A plot of this equation, derived independently by Walstra [19] and van de Hulst [20], is shown in Figure 12.5, along with a plot of the Mie scattering cross section calculated according to the full theory (Eq. 12.3).

12.3.2.2 Resonance Mie (R-Mie) scattering

If the Mie scattering produced nothing but the undulating background shown in Figure 12.5, its effect on the observed spectra would be relatively minor and could be corrected relatively easily. However, the ratio of the refractive indices that determine the size parameter (defined in Eq. 12.4) undergoes anomalous dispersion in a region of absorption. Thus, the refractive index of the medium (assuming that the refractive index of the surroundings stays constant) enters the equation for the Mie scattering, producing strongly distorted spectra as shown in Figure 12.6. Here, trace (a) shows the pure Mie scattering curve, in a much narrower spectral region, for a particle with a radius of 5 μm and a refractive index of 1.5 (a mean value assumed for proteins).

Trace (c) shows a pure absorption spectrum and trace (b) a simulated spectrum obtained by explicitly including the dispersion of the refractive index, obtained by Kramers–Kronig transform of the absorption spectrum, in the equation for Mie scattering (Eq. 12.5 or 12.3).

It is obvious that the R-Mie-contaminated spectrum is useless for data analysis, because its distortion is so severe that no diagnostic algorithm would consider it to be equivalent to the pure absorption spectrum shown

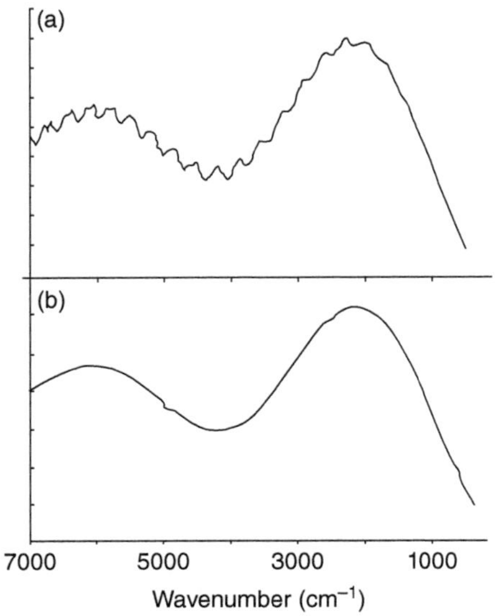

Figure 12.5 *(a) Mie scattering curves calculated for the full theory and (b) using approximate theory described in Eq. 12.5. Romeo* et al. *[15]. Reproduced with permission from Elsevier*

in Figure 12.6(c). Thus, the design of algorithms for the correction of these effects was an absolute necessity. In fact, during the Spec2008 conference in Sao Paulo, Brazil, participants exhibited a distinct pessimism about the future of medical FTIR microspectroscopy because of the prevalence of these spectral distortions.

12.3.3 Correction of dispersive band shapes

In two pioneering papers, researchers from several group of synchrotron FTIR users [17] reported how reflection processes and Mie scattering can mediate the mixing of dispersive and absorptive band shapes [13, 21]. Shortly thereafter, a correction algorithm for R-Mie-based effects was published by Bassan *et al.* [17] that utilized R-Mie spectra that were derived from the anomalous dispersion of the refractive index, using Eq. 12.5; a typical R-Mie spectrum obtained from this equation is shown in Figure 12.6(b). This spectrum explicitly includes the distortions of band profiles owing to the anomalous dispersion of the refractive index (R-Mie contributions) as well as pure Mie scattering. The Bassan correction algorithm used a protein (matrigel) reference spectrum to compute an approximate refractive index spectrum (*via* the Kramers–Kronig transform) and R-Mie contributions, using Eq. 12.5, for spherical particles of different sizes. The R-Mie spectra were compressed *via* PCA and used to correct all spectra in the data set using EMSC [22] to fit the best R-Mie components. Subsequently, each corrected spectrum was used, instead of the matrigel reference spectrum, as a reference spectrum and the procedure was used to correct all spectra in the data set with reference spectra that approached the real spectra more and more closely. The algorithm was applied to data sets of IR spectra of individual cells and – as shown by the authors – converged to produce real, unperturbed cell spectra. However, this algorithm took hours of computation time; furthermore, if it was terminated after too few iterations, the results tended to underestimate the spectral differences between classes of spectra.

Another approach was introduced by the author's group [23] based on the well-documented "phase correction" approach that is widespread in standard FTIR spectroscopy (see Section 3.3). The phase correction approach is based on the concept that the complex FT separates the real and imaginary parts of spectra or

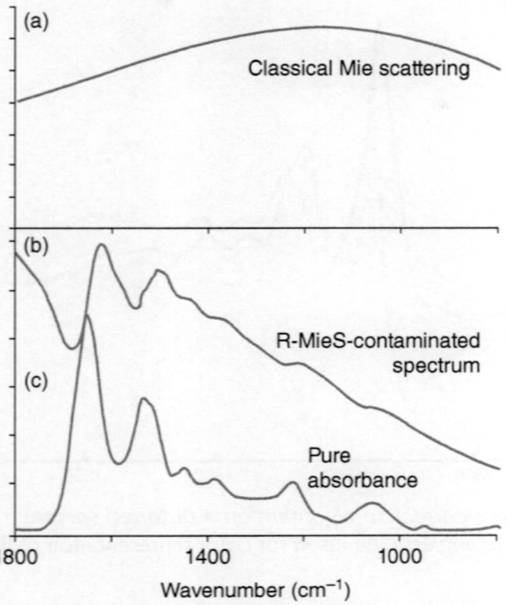

Figure 12.6 *(a) Mie scattering curve for a spherical particle with a radius refractive discussed in the text. (b) Resonance Mie scattering of the same particle, considering the dispersion of the refractive index in the mid-IR region, calculated via Kramers–Kronig transform of the pure absorption spectrum. (c) Pure absorption spectrum of a protein sphere*

interferograms by varying the phase angle between them. In classical FTIR spectroscopy, the collected interferogram is generally asymmetric about the ZPD peak; such a "chirped" interferogram gives, upon forward FFT, a spectrum that contains a mixture of reflective and absorptive band shapes, as shown in Figure 3.12 [24]. Nearly all commercial FTIR instruments use the Mertz phase correction method [25] for which the instrumental phase angle is determined experimentally, and is used to correct the spectra.

A phase correction-based approach was attempted [16] earlier but worked only intermittently owing to some computational and theoretical problems. Recently, a revised phase correction algorithm was implemented for fast, reliable, and elegant correction of reflective band contributions. In short, the distorted spectra, expanded to the desired frequency range and de-noised by NA-PCA (see Section 12.2.5.4), are transformed back into interferogram space by the finite Hilbert transform, which can be implemented by a truncated reverse FT [26]. The resulting real (Re) and imaginary (Im) interferograms are phase shifted by a trial phase θ by a similarity transform according to

$$\begin{bmatrix} \mathrm{Re}' \\ \mathrm{Im}' \end{bmatrix} = \begin{bmatrix} \cos\theta & \sin\theta \\ -\sin\theta & \cos\theta \end{bmatrix} \begin{bmatrix} \mathrm{Re} \\ \mathrm{Im} \end{bmatrix} \tag{12.6}$$

Phase-corrected spectra are computed by complex forward FFT of ($\mathrm{Re}' + i\mathrm{Im}'$). The "best" phase is assumed to be the one that produces the highest intensity, or the highest frequency, in the corrected spectra.

The justification for this choice is that the addition of a reflective component will nearly always reduce the peak intensity [24] and shift the peak to lower wavenumber.[1] This is shown in Figure 12.7. In principle, a wavenumber-dependent phase angle can also be computed, as in the Mertz algorithm, from a low-resolution

[1]This statement is true only for ranges of particle size and refractive indexes relevant for IR-MSP of biological samples.

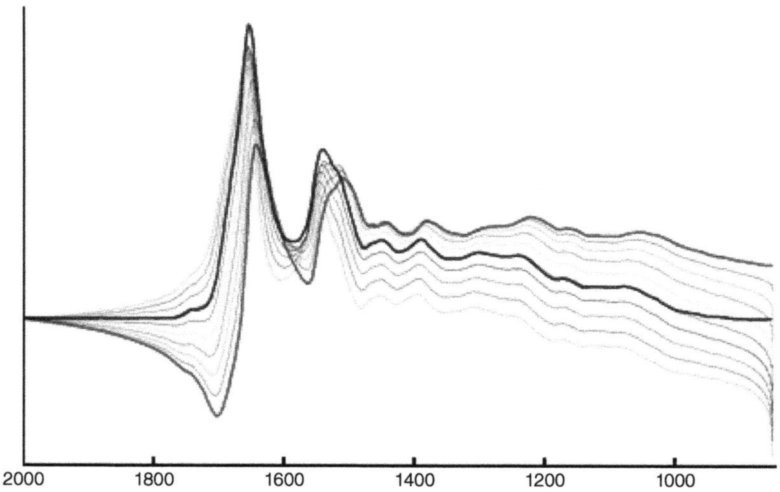

Figure 12.7 *Effect of the phase correction algorithm on a distorted spectrum (red) to the corrected spectrum (blue) using different trial phase angles.* (See insert for color representation of this figure.)

interferogram obtained by truncating the FFT to fewer data points. The phase correction algorithm is best carried out using second derivative data.

All data manipulation and analysis was carried out, using software developed in house using the MATLAB (Mathworks, Natick, MA) platform. Most of the data analysis routines are contained in a software package referred to as "*ViChe*" (see Section 12.1), which includes all of the preprocessing and multivariate imaging reconstruction algorithms, for example, PC and hierarchical cluster analysis (HCA) imaging.

12.3.4 Standing wave effect

Recently, several research groups [27–29] have pointed out a problem specifically associated with the use of low-e slides (see Chapter 11) for IR microscopy of biomedical samples. This problem arises from the fact that the reflection from the metallic (Ag) surface of the low-e slides creates a standing electromagnetic wave on the side of the low-e slides on which the sample is mounted. Thus, the sample is exposed not only to the incident wave but also to the standing wave created by reflection, which causes a distortion of the observed transflection spectrum. This distortion manifests itself by a large enhancement of absorption in the high wavenumber region of very thin samples [9, 27]. The origin of this effect can be understood from the following considerations. The first node of the standing wave above the reflective surface depends on the wavelength of radiation; in IR microspectroscopy, these wavelengths typically range from 2.5 μm or 4000 cm^{-1} to 12.5 μm or 800 cm^{-1}. Therefore, the first node of the electromagnetic wave coincides with the thickness of the sample at 2000 cm^{-1} (5 μm) when the sample thickness is 5 μm, see Figure 12.8(a). At 1000 cm^{-1} (10 μm), the magnitude of the standing wave exhibits a maximum at this distance (5 μm) from the surface Figure 12.8(b). Thus, the change in absorbance with slight variations in sample thickness will be very different at these two wavelengths, and the sample will, in general, not obey the Beer–Lambert law (Eq. 1.113).

To estimate the effect of the standing electromagnetic wave on the absorbance spectrum, one may simply integrate the square of the electromagnetic field between the surface and an arbitrary distance d from the

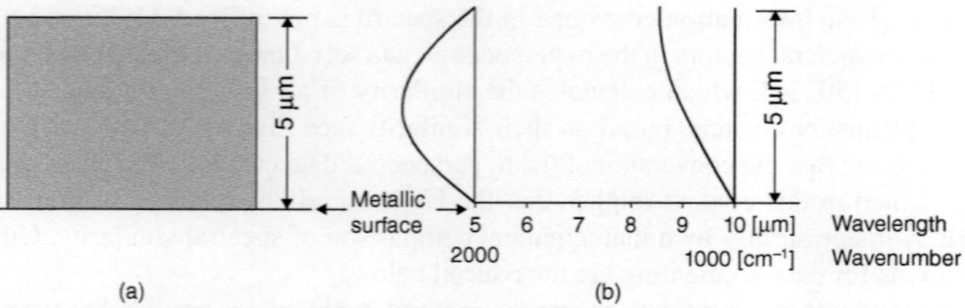

Figure 12.8 *(a) Model of the cross section of a 5-μm-thick slab of tissue on metallic surface. (b) Amplitude of standing electromagnetic wave on metallic surface as a function of wavelength for a 5-μm-thick sample.* Miljković et al. *[9]. Reproduced with permission from The Royal Society of Chemistry*

surface for all wavelengths of interest:

$$\int_0^d E_0^2 \sin^2\left(\frac{2\pi}{\lambda}\right) x\, dx \tag{12.7}$$

The results of such a simulation for a sample thicknesses of $d = 5$ μm and normal incidence reveal that there is about a 30% error in absorbed intensity owing to the standing wave effect. The resulting spectral distortion is quite severe, even over the relatively narrow wavenumber range used in biomedical spectroscopy, and will aggravate a comparison of two spectra collected at slightly different sample thicknesses.

However, for non-normal incidence in a 0.6 NA (see Chapter 11) microscope objective, the effect will be significantly smaller because the node will not be a sharp plane but a diffuse region due to the varying angles of incidence. The artifact can be further reduced by using mean-centered, vector-normalized second derivative spectra rather than the absorbance spectra. Second derivative spectra depend not only on the absolute intensities of peaks but also on the curvature (which is related to the half width of each band that is unaffected by the standing wave effect). Thus, the manifestation of the standing wave error can be reduced significantly by vector normalization and computation of spectral derivatives. In addition, by vector normalizing spectral regions individually in different spectral ranges (e.g., from 1750 to 1480 cm^{-1}, the "protein region," and from 1480 to 900 cm^{-1}, the "nucleic acid/lipid region"), one can further enhance spectral discrimination by reducing the dependence on absolute intensities [9].

12.4 Unsupervised multivariate methods of data segmentation

The goal of spectral diagnostic methodology is the recognition of tissue spectral features associated with and specific to disease and to utilize these features to aid in medical diagnoses. As pointed out earlier, the raw data contained in a spectral hypercube cannot, *a priori*, be used for this purpose, and methods need to be developed to accomplish this task. Most of the researchers in the field of optical diagnosis *via* vibrational spectroscopy have adopted a diagnostic scheme that will be introduced in Chapter 13 (see Figure 13.2). This scheme includes data preprocessing, unsupervised data segmentation, and finally supervised classifiers to render a diagnosis. Both the data segmentation and classification are carried out by multivariate methods, as opposed to univariate methods. Univariate methods utilize one piece of spectral information from each spectrum, for example, one intensity point or the width of one band. Univariate methods are counterproductive in the sense that most of the information contained in the spectra is not utilized. Multivariate methods, in contrast, utilize the entire spectral vectors in the hyperspectral data sets. One commonly used multivariate method is unsupervised HCA [30, 34], which calculates the similarity of all spectra in a data set and assigns color codes to spectral groups, or clusters, based on their similarity (see Figures 12.12(b) and 13.3). It should be emphasized at this point that the conversion of the hyperspectral data cube into the HCA image is completely "unsupervised," which in this context implies that the HCA algorithm needs no training, but arrives at the pseudo-color HCA image strictly by a mathematical comparison of spectral similarity. Other unsupervised, multivariate methods for data segmenting are introduced below.

In unsupervised methods, no input data except the spectral hypercube is provided to the algorithm, and the class membership is calculated from the variance (or other property) of the data set. These methods provide less quantitative information in terms of (bio)chemical changes between the spectral classes, but do not require any prior knowledge of sample composition, and compositional changes. Unsupervised methods may be classified into factoring methods that detect certain factors, or latent variables, within the data, and clustering methods that segment data by similarity. Both approaches are introduced in the following sections.

12.4.1 Factor methods

Several factor methods have been developed to decompose data sets into a bilinear model of variables. Factor methods are unsupervised and can reveal small, but recurring spectral differences in a data set. In PCA [36], for example, a set of new spectra, known as principal components (PCs) or loading vectors is calculated from the covariance matrix of the entire data set. Subsequently, the original data set is reconstructed from these PCs and abundance coefficient (scores). It is assumed that similar spectra in the original data set require the same PCs and scores to reconstruct the spectra. When plotting these scores against each other, a "scores plot" is obtained that can show whether spectra are related to each other, or not. Linear discriminant analysis (LDA), another covariance-based method, aims at providing optimum class separation. Vertex component analysis (VCA) [31] and N-FINDR [51] decompose a hyperspectral data set into linear combinations of endmember spectra that are considered the pure "component" spectra. These methods are introduced in the following section.

12.4.1.1 Principal component analysis (PCA)

PCA is a well-established multivariate data analysis method [36] ideally suited to detecting small, re-occurring spectral variations in large data sets containing uncorrelated variations. For PCA, the entire spectral data set, containing n spectra, is written as one matrix **S**. In this matrix, each column represents one spectrum $S(\nu_k)$ of m intensity data points. The correlation matrix is constructed from the spectral matrix **S** according to

$$\mathbf{C} = \mathbf{S}\,\mathbf{S}^{\mathbf{T}} \quad (12.8)$$

or

$$C_{kl} = \sum_{i=1}^{n} S^i(\nu_k) S^i(\nu_l) \quad (12.9)$$

C is an $m \times m$ matrix, in which the off-diagonal terms C_{kl} are the correlation coefficient between intensity values at wavelengths ν_k and ν_l, summed over all spectra. The diagonal elements of this matrix C_{kk} contain the squared sum of all intensity data points $S^i(\nu_k)$ in one spectrum i, summed over all spectra in the data set. If vector-normalized, mean-centered spectral data are used, the correlation matrix equals the covariance matrix of the data set.

Diagonalization of the intensity correlation matrix, according to

$$\mathbf{P}^{\mathbf{T}}\mathbf{C}\,\mathbf{P} = \mathbf{\Lambda} \tag{12.10}$$

yields the eigenvector matrix **P**, from which "principal components" **Z** are calculated according to

$$\mathbf{Z} = \mathbf{SP} \tag{12.11}$$

The eigenvalues Λ express the variance contained in each of the PCs. Thus, from the viewpoint of linear algebra, the PCs are the original spectra expressed in a rotated coordinate system, based on the maximum variance of the original spectra. Subsequently, each of the original spectra $S(\nu)$ is expressed in terms of the new PCs (also known as "loading vectors")

$$\mathbf{S}' = \alpha\mathbf{Z} = \mathbf{P}^{\mathbf{T}}\mathbf{Z} \tag{12.12}$$

or

$$S_i'(\nu) = \Sigma_p \alpha_{ij} Z_j(\nu) \quad j = 1 \cdots p \tag{12.13}$$

where the "scores," α, are given by

$$\alpha = \mathbf{P}^{\mathbf{T}} \tag{12.14}$$

In general, one finds that a large fraction of the total spectral variance is contained in the first few "loading vectors." Typically, about 15 loading vectors contain more than 99% of the variance. Thus, the spectral expansion given by summation in Eq. 12.13 can be truncated after the pth term, where p is the number of relevant loading vectors or PCs. Thus, the score matrix α**,** which determines how much each PC contributes to each spectrum, will have the dimension $(p{\cdot}n)$. This method results in a much-reduced size of the data set, because all spectra are expressed in terms of a few (typically <15) basis functions and a "score vector" of about p entries.

Similar spectra exhibit similar scores α, which may be used to discriminate, or group, spectra. This is accomplished by plotting the values α_i and α_j (i.e., the contribution of PC_i and PC_j to each spectrum) against each other, where each data point represents one spectrum. The resulting two- or three-dimensional "scores plot," will be used in Chapters 13 and 14 to indicate that significant variations exist in a data set. A typical "scores plot" for a data set of spectra containing three classes is shown in Figure 12.9.

PCA has also been used in the author's laboratory [30] as an unsupervised imaging tool. To this end, the scores of all spectra for one PC are scaled to fall between 0 and 255 such that they can be described by an 8-bit byte, where 0 (black) denotes zero and 255 (white) denotes the highest abundance of a PC. These values can be converted into a monochrome intensity map of the imaged object. Up to three monochrome maps for different PCs can be combined into a red, green, or blue (RGB) image that yields a pseudo-color map with mixed colors indicating the intensity values of each of three PCs. PCA imaging, as described here, is a soft clustering method in that a given pixel can have contributions from more than one PC and can show up as a mixed color in the spectral maps. An example of a PCA image is shown in Figure 12.10(a).

As indicated earlier, PCA also can be used as a noise-reducing method, as follows. When inspecting the loading vectors (Eq. 12.12), it becomes obvious that the first 15 or so loading vectors contained most of the signal and data variance, whereas higher loading vectors are just uncorrelated noise. Thus, a tempting approach at noise reduction based on this principle is to reconstruct spectra by co-adding the first 20 or 30 PCs.

However, higher PCs do contain a small amount of relevant spectral variance, and ignoring them in the spectral reconstruction can induce some error, in particular, when further analyzing the resulting spectra by multivariate methods. A better approach was introduced by Reddy and Bhargava [8], based on some prior work

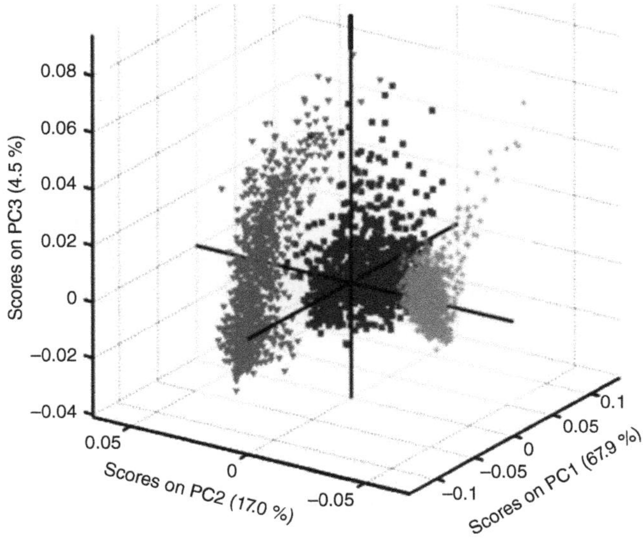

Figure 12.9 *Example of a 3D "scores plot" of spectral data, demonstrating the separability of a data set in a space of vastly reduced dimensionality (3 dimensions* vs. *the original 450 dimensions). (*See insert for color representation of this figure.*)*

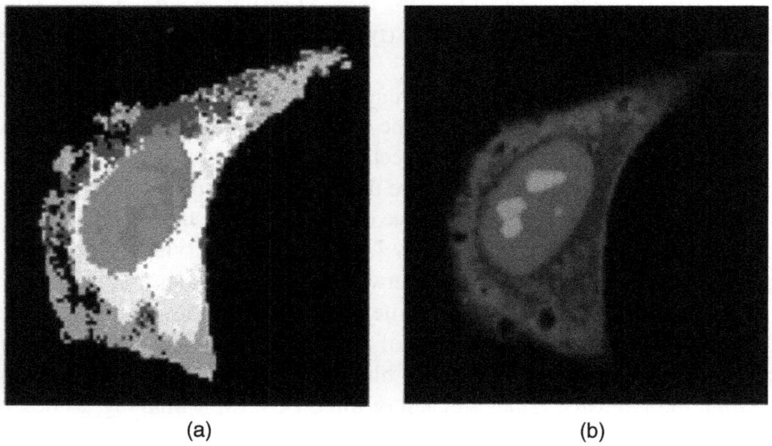

Figure 12.10 *(a) PCA-based and (b) VCA-based images of a HeLa cell, constructed from a Raman hyperspectral data set. See text for details. Miljković* et al. *[30]. Reproduced with permission from The Royal Society of Chemistry. (*See insert for color representation of this figure.*)*

from the remote sensing community [7]. This approach is generally referred to as "noise-adjusted principal component analysis," and works as follows.

One assumes that the data set contains pixels that do not represent real sample data, but are taken from positions that are devoid of cells or tissues. These pixels represent the true instrument noise at the time of data collection and accurately reflect the wavelength (or wavenumber) dependence of the noise. A simple routine can be written that determines the presence of sample at a given pixel location, and segments the overall data set into a data matrix $\mathbf{S_D}$ and a noise matrix $\mathbf{S_n}$. Alternatively, an area of no sample may be imaged before data acquisition to obtain a noise data set. In analogy to Eq. 12.8, noise and signal covariance matrices are calculated according to

$$\mathbf{C_n} = \mathbf{S_n}\,\mathbf{S_n^T} \text{ and } \mathbf{C_D} = \mathbf{S_D}\,\mathbf{S_D^T} \tag{12.15}$$

Both the noise and signal covariance matrices are based on mean-centered spectral vectors. Next, the noise covariance matrix is diagonalized:

$$\mathbf{E^T}\,\mathbf{C_n}\,\mathbf{E} = \,\boldsymbol{\Delta}_\mathbf{n} \tag{12.16}$$

and a noise-renormalization (or "noise-whitening") matrix **F** is calculated according to

$$\mathbf{F} = \mathbf{E}\,(\boldsymbol{\Delta}_\mathbf{n})^{-1/2} \tag{12.17}$$

Then

$$\mathbf{F^T C_n\,F} = \,(\boldsymbol{\Delta}_\mathbf{n})^{-1/2}\,\mathbf{E^T}\boldsymbol{\Delta}_\mathbf{n}\,\mathbf{E}\,(\boldsymbol{\Delta}_\mathbf{n})^{-1/2} \,=\, \mathbf{I} \tag{12.18}$$

This step is known in signal processing circles as a de-correlation transform that transforms a set of noise vectors into white noise contributions because the noise eigenvector elements F_i are all uncorrelated and have unit variance.

A "noise-adjusted" data covariance matrix is calculated according to

$$\mathbf{C_{adj}} \,=\, \mathbf{F^T C_D\,F} \tag{12.19}$$

and subsequently diagonalized

$$\mathbf{G^T C_{adj}\,G} = \,\Lambda_{\mathrm{adj}} \tag{12.20}$$

The eigenvector matrices **F** and **G** are combined to yield the matrix **H**

$$\mathbf{H} = \mathbf{FG} \tag{12.21}$$

that allows the computation of the "noise-adjusted principal components" according to

$$\mathbf{Z'_{adj}} \,=\, \mathbf{H^T}\,\mathbf{S_D} \tag{12.22}$$

in complete analogy to Eq. 12.11. Here, the eigenvectors (loading vectors) are no longer arranged in order of decreasing variance, but in order of increasing noise contribution. Noise-reduced spectra are then computed in analogy to Eq. 12.12 or 12.13 by summing the first 30–50 noise-adjusted eigenvectors.

12.4.1.2 Vertex component analysis (VCA)

In VCA, each spectrum in a data set is expressed as a linear combination of "pure" component spectra that are obtained by spectral unmixing [31]. The "pure component spectra" are obtained as the outliers of the spectra represented in m-dimensional color space (where m, as defined before, is the number of intensity points in a spectral vector). If a spectrum is recorded at two frequencies, or channels, v_1 and v_2 only, each spectrum can be decomposed into intensities along the two channels v_1 and v_2. If two spectra are identical, they will have two identical vectors; two spectra having the same "composition" but different total intensities will have superimposed vectors of different lengths. For spectra recorded at three channels or frequencies, a quiver of vectors is obtained whose endpoints lie on a triangle. The vectors (spectra) passing through the corners (vertexes) of this triangle are the "endmembers," which may be considered the "pure component spectra." For spectra collected at four color points, the endpoints of the vectors lie in a tetrahedron. In general, the progression from point to line to triangle to tetrahedron, and so on, is known in geometry as a simplex. A simplex describes a p-dimensional hull, or polytop, that possesses $p + 1$ corners.

The principle behind VCA is the assumption that the most extreme vectors in the original representation are "pure component spectra" or endmembers. VCA proceeds from the original dimensionality m and reduces the dimensionality step-by-step by a process known as "orthogonal subspace projection" [31]. In each of these steps, the problem is reduced from a p-simplex to a $(p - 1)$-simplex. The process is terminated when the number of endmembers reaches a predetermined level. This "spectral unmixing" described so far does not necessarily find the true spectral basis sets but will decompose the spectra into a few basis spectra, or endmembers that represent the degree to which spectra can be unmixed [32].

Subsequently, all spectra S in a data set are expressed as linear combinations of the endmember spectra E:

$$S = EA + W \tag{12.23}$$

where E represents the matrix that is constructed from the endmembers, A is a matrix that determines the fractional contribution α of each endmember to each spectrum, and W accounts for arbitrary fluctuations such as noise. The fractional abundances α are normalized to 1:

$$\sum_{i-1}^{P} \alpha_{ij} = 1 \tag{12.24}$$

for a given spectrum.

VCA results can be visualized, as discussed earlier for PCA imaging, by scaling the abundance values for a given endmember to range between 0 and 255, and converting them into a monochrome intensity map of the imaged object. An overlay of these RGB color maps yields a pseudo-color map with mixed colors indicating the contributions of each of three endmembers. The same data set shown in Figure 12.10(a), analyzed by VCA, is shown in Figure 12.10(b). In this figure, the nucleoli (orange), the cell nucleus (red brown), mitochondria-rich areas of the cytoplasm (dark brown), and the cytoplasm (blue hues) can be clearly detected. More details can be found in the original literature [30].

12.4.1.3 Linear discriminant analysis (LDA)

LDA is a method that classifies or separates a data set by projecting the data in a direction that maximizes class separability. This best can be understood from an example taken from a web-based tutorial [33], shown in Figure 12.11. This graph presents a data set consisting of three classes, indicated by the red crossed, blue circles, and green triangles. Although these three classes are readily seen to form different clusters in x space, neither the x_1 nor the x_2 axis are suitable dimensions to separate the data set. Thus, a linear transformation is sought that will provide the best separation of the data in a space of reduced dimensionality. The line marked "b" in Figure 12.11 represents the direction of the best possible separation in one dimension, whereas the direction of the line marked "a" offers the worst separation. LDA, in contrast to PCA, does not provide features (such as the loading vectors) that indicate what spectral differences were used to provide the separation but strictly provides a coordinate transformation to achieve the best separation.

Both PCA and LDA provide classification of a data set by reducing the dimensionality of the problem and express the data in a new coordinate system: in PCA, this coordinate system is based on the variance of the data, whereas in LDA, it is based on the best class separability.

In LDA, a discriminant function is sought that defines the direction of best separation. In the two-dimensional case shown in Figure 12.11, this function is defined by a linear transformation of the data into a new coordinate system $y(y_1, y_2)$ that provides the best separation of the data. This transform is given by

$$\boldsymbol{y} = \boldsymbol{w}^T \boldsymbol{x} \tag{12.25}$$

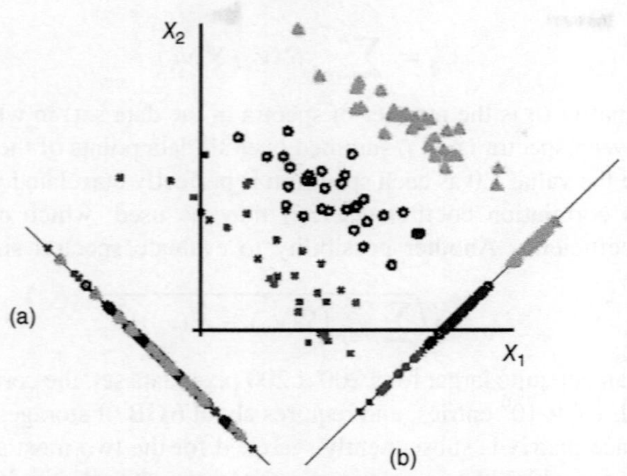

Figure 12.11 LDA plot of a data set containing three classes. See text for details. Farag and Elhabian, 2008, [33]. Reproduced with permission from A. A. Farag. (See insert for color representation of this figure.)

LDA proceeds by computing the "within-class scatter matrix" S_W as the sum of the covariance matrices (Eq. 12.8) for each class. Here, each class' scatter matrix S_i is given by

$$S_i = (x - \overline{x})(x - \overline{x})^T \tag{12.26}$$

where $\overline{x}$ is the mean of all values x. This is directly analogous to the covariance matrix defined earlier for PCA and defines the variability of data in the original coordinate system $x(x_1, x_2)$.

In addition, the "between-class scatter matrix" S_B, defined as

$$S_B = (\mu_1 - \mu_2)(\mu_1 - \mu_2)^T \tag{12.27}$$

is computed. The best discriminant direction J is given by

$$J = \frac{w^T S_B\, w}{w^T S_W\, w} \tag{12.28}$$

LDA often gives complementary information when compared to PCA and is an easily implemented, fast, and unsupervised method for data segmentation.

12.4.2 Data segmentation by clustering methods

12.4.2.1 Agglomerative hierarchical cluster analysis (AHCA or simply HCA)

Agglomerative hierarchical cluster analysis (AHCA) is a method for data segmentation based on finding the smallest "distances" between items such as spectra, where the term "distances" may imply Euclidean or Mahalanobis distances, or correlation coefficients. AHCA is a classification rather than an imaging method, but it can be used to construct pseudo-color maps from hyperspectral data sets based on cluster assignments [34, 35]. In subsequent chapters in this book, this method is referred to by its most commonly used abbreviation, HCA.

HCA starts by computation of the **spectral** correlation matrix, C' according to

$$\mathbf{C'} = \mathbf{S^T S'} \tag{12.29}$$

or

$$C'_{ij} = \sum_{N=1}^{n} S^i(\nu_N)\, S^j(\nu_N) \tag{12.30}$$

This matrix is an $(n{\cdot}n)$ matrix (n is the number of spectra in the data set) in which the off-diagonal terms C'_{ij} are the correlation between spectra i and j, summed over all data points of the spectral vector $S(\nu_N)$. The diagonal elements all have the value 1.0 as each spectrum is perfectly correlated with itself.

Alternatively, Pearson's correlation coefficients [36] may be used, which are mean-centered, vector-normalized correlation coefficients. Another possibility to evaluate spectral similarity numerically is by Euclidean distances:

$$D_{ij} = \sqrt{\sum_{N=1}^{n} \{S^i(\nu_N) - S^j(\nu_N)\}^2} \tag{12.31}$$

The correlation matrices can get quite large: for a 200 × 200 pixel data set, the correlation matrix is of dimension 40, 000 × 40, 000 with 1.6×10^9 entries, and requires about 6 GB of storage.

The correlation or distance matrix is subsequently searched for the two most similar spectra coefficients, that is, two spectra i and j for which the correlation coefficient is closest to unity. Subsequently, these two spectra are merged into a new object, and the correlation coefficient of this new object and all other spectra is recalculated. The process of merging is repeated, but the items to be merged may be spectra, or spectra, and merged objects. In the merging process, a membership list is kept that accounts for all individual spectra that are eventually merged into a cluster. Ward's algorithm [37] is used most commonly for the merging of spectra because it gives highly homogeneous clusters. Once all spectra are merged into a few clusters, color codes are assigned to each cluster, and the coordinates from which a spectrum was collected is indicated in this color. In this way, pseudo-color images are obtained that are based strictly on spectral similarities. Mean cluster spectra may be calculated that represent the chemical composition of all spectra in a cluster. In addition, a "dendrogram" (a tree-like structure) is produced that indicates at what level of similarity two spectra or items were merged.

A typical example of a pseudo-color image obtained by HCA of a Raman hyperspectral data set (the same as the one shown in Figure 12.10) is shown in Figure 12.12(b). Here, the cluster colors are assigned arbitrarily, yet the image represents a true distribution of similar spectra, and the unsupervised discrimination of subcellular organization is excellent. A more detailed interpretation of the images shown in Figures 12.10 and 12.12 is given later in this section.

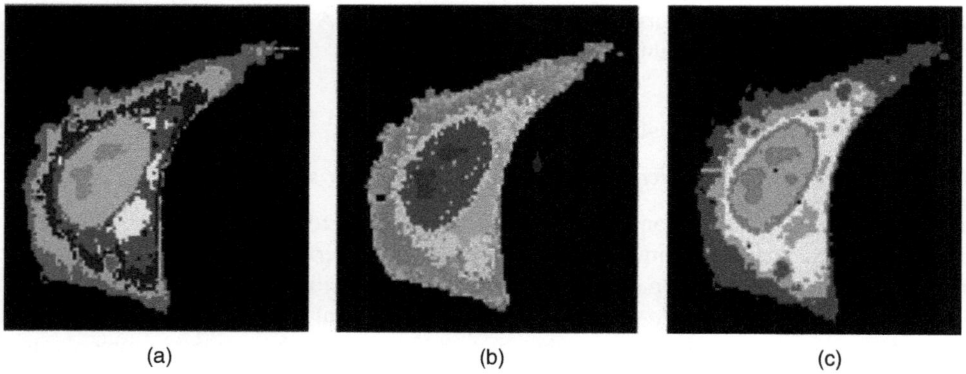

Figure 12.12 *(a) KMCA, (b) HCA, and (c) DCCA-based images of the same Raman data set shown before in Figure 12.10. See text for the different subcellular regions depicted. Miljković* et al. *[30]. Reproduced with permission from The Royal Society of Chemistry.* (See insert for color representation of this figure.)

12.4.2.2 K-Means cluster analysis (KMCA)

K-means cluster analysis KMCA is a another method of segmenting n spectra of a data set into k clusters such that each spectrum belongs to the cluster with the nearest (closest) mean [38]. It is a method that requires the operator to select (guess) an initial number k of clusters in the data set.

KMCA minimizes the sum of distances between spectral vectors S_i^J and cluster centroids m_k

$$\sum_J (S_i^J - m_k) \tag{12.32}$$

where J spectral vectors originally are randomly assigned to belong to a given cluster k.

Next, the KMCA algorithm determines whether or not each spectrum is assigned to its nearest (closest) cluster centroid. If this is not the case, the spectrum will switch its cluster membership and is reassigned into the cluster nearest to it. Subsequently, the centroids of the "old" and the "new" cluster are recalculated as the mean position of all spectra in the clusters. Here, "new" and "old" refer to the two clusters to which the spectrum was added or from which it was removed, respectively. This process is repeated until no spectrum is reassigned during a pass that involves all spectra in the data set, or until the homogeneity of the clusters does not improve beyond a preset limit.

KMCA is an unsupervised, crisp clustering method that requires relatively short computational times and is, therefore, applicable to large data sets. KMCA provides good discrimination of spectral classes [36, 39]. As with all crisp clustering methods, the spectra in the resulting k clusters may be averaged to produce mean cluster spectra of vastly improved spectral quality. In fact, the spectral quality of the resulting mean cluster spectra is sufficiently high to permit interpretation of the biochemical changes. Once all spectra are merged into a few clusters, color codes are assigned to each cluster and the coordinates from which a spectrum was collected is indicated in this color. The color assignment is based strictly on similarity and requires no fitting to any reference spectra. An example of an image created by KMCA is shown in Figure 12.12(a).

12.4.2.3 Divisive correlation cluster analysis (DCCA)

The divisive correlation clustering algorithm (DCCA) was first introduced as a means for detecting patterns in gene-expression data, to produce clusters with high correlation and biological significance [40]. DCCA differs from agglomerative methods in that with divisive methods, the spectra in a given data set are initially treated as a single cluster and are split into subsequent clusters as the algorithm proceeds. Agglomerative methods, on the other hand, begin with the assumption that each individual spectrum in the data set is a cluster, and the algorithm groups these single clusters (spectra) together based on a distance measure of similarity, until all the spectra belong to one cluster. For spectral data sets of biological samples, it is rare that more than 10–15 clusters are required to adequately represent the spectral variance, and it is, of course, much faster to divide 10,000 spectra into 15 clusters than to combine 10,000 spectra, one at a time, into 15 clusters. The DCCA is a completely unsupervised technique as well in that no *a priori* information, for example, the number of clusters or information about the spectral data, is required for the analysis.

The DCCA algorithm utilizes the correlation coefficient matrix for measuring similarity/dissimilarity between the two spectra S^L and S^M, and groups spectra based on a measure of attraction or repulsion, defined as

$$\text{if} \begin{cases} C_{LM} > \theta \text{ there is an "attraction" between } S^L \text{ and } S^M \\ C_{LM} < \theta \text{ there is a "repulsion" between } S^L \text{ and } S^M \end{cases} \tag{12.33}$$

where θ is the "correlation threshold," CT.

Vibrational spectroscopic data of biological components within cells are highly correlated; therefore, one typically uses CTs between 0.9 and 0.98 in order to produce clusters that adequately differentiate the spectra of the different biological components in the cell. For each iteration, the algorithm selects the two spectra S^L and S^M in a cluster with the most different correlation coefficients between them. The selected cluster is split into two new clusters in such a way that the spectra S^L and S^M are placed in different clusters. The remaining spectra are then divided between the two new clusters based on their correlation with S^L and S^M, that is, the spectra that have a higher correlation coefficient with S^L than with S^M are placed in the cluster containing S^L. The placement of each spectrum into a given cluster is then checked by comparing the correlation coefficient of the spectrum with the average correlation coefficients of the spectra in both clusters.

If the placement is inappropriate, the spectrum will be moved to the cluster to which it has the highest correlation. The process of partitioning spectra into clusters, placement checking, and correction continues until all the spectra have been grouped into clusters and no repulsion exists within a cluster (according to the defined CT). Visualization of the resulting cluster structures is achieved as in the case of HCA. A more detailed description of the methods can be found in the literature [30, 40].

A comparison of three clustering methods on the same data set shown in Figure 12.10 is presented in Figure 12.12. These images will be discussed in more detail in Chapter 14. At this point, just a short description of the images is presented. As an example, Figure 12.12(b) is elaborated upon.

The data set shown was collected from a fixed HeLa cell attached to a CaF_2 window and immersed in a drop of buffer solution. Ten thousand individual Raman spectra were collected from pixels about 300 nm in diameter, at a grid spacing of 500 nm. The areas shown in black represent the Raman spectra of pure buffer solution. In Figure 12.12(b), the green, elliptical area corresponds to the nucleus, and the dark blue areas inside the nucleus are the nucleoli. The salmon and yellowish areas correspond to mitochondria-rich areas of the cytoplasm, whereas the light blue and olive areas are other regions of the cytoplasm. Given the similarity of spectral features of nucleus and nucleolus, it is amazing that all five unsupervised image analysis methods shown in Figures 12.10 and 12.12 differentiate these two spectral regions. Notice, however, that the actual selection of the color in which each region is represented is arbitrary and cannot be compared from one to the other methods. For reasons that are not completely understood, but most likely having to do with the noise characteristics of the data set, VCA (Figure 12.10(b)) and DCCA (Figure 12.12(c)) yielded the most detailed pictures of subcellular structures in this data set, even hinting on delineating nuclear membrane features.

For IR data sets of tissue, HCA generally gives the most detailed images, particularly if the mean S/N ratio of the data set is very high (>1000:1) and second derivative spectra can be utilized. However, the computation times required for HCA makes this method impractical for data sets larger than about 100,000 pixels. For very large data sets, KMCA gives very good results as well. In these methods, operator input is required only to select the clustering endpoint, but not for supplying any training data. In HCA, between five and eight clusters generally produce images that represent tissue structure and disease states very well. The same number of clusters also gives good representation of subcellular features in the Raman data of the cell shown earlier. In addition, the three clustering methods and VCA yield mean cluster or endmember spectra, respectively, which allow correlation of the observed clusters with real changes in the sample's biochemical composition.

12.5 Supervised multivariate methods

12.5.1 Discussion of sensitivity, specificity, and accuracy

While the previously discussed multivariate methods analyze a data set without the use of reference spectra, supervised multivariate methods are trained to recognize the similarity of a spectral pattern to other spectral patterns that were used to train the algorithm. Consequently, supervised algorithms can be used to analyze (classify) tissue pixels or individual cells and render a diagnostic output. These methods, by necessity, are trained on and tested against data for which the outcome is known; thus, they can be used to obtain basic

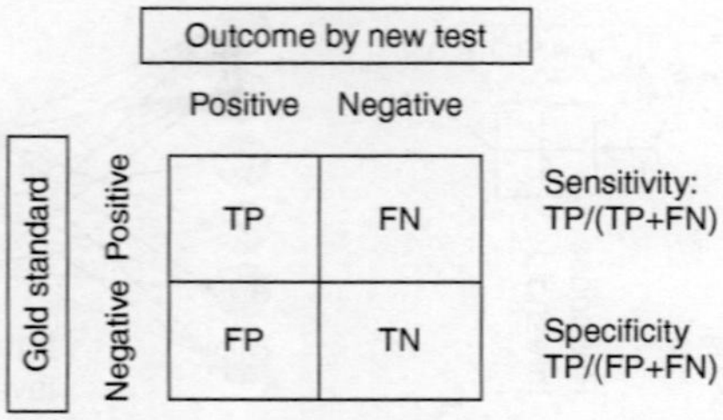

Figure 12.13 *Confusion matrix and visualization of TP, TN, FP, and FN predictions*

statistical results such as sensitivity, specificity, and accuracy. Before discussing these algorithms *per se*, the terms sensitivity, specificity, and accuracy, as used in the context of biomedical statistics, need to be introduced because in common analytical language, these terms have different meanings.

Sensitivity and specificity are defined in terms of the true positive (TP), true negative (TN), false positive (FP), and false negative (FN) outcome of a diagnostic, supervised algorithm, as follows:

$$\text{Sensitivity} = \frac{\text{TP}}{\text{TP} + \text{FN}} \tag{12.34}$$

$$\text{Specificity} = \frac{\text{TN}}{\text{FP} + \text{TN}} \tag{12.35}$$

The TP, TN, FP, and FN values are often collected in a format generally known as the "confusion matrix," shown in Figure 12.13.

(Sometimes, the axes of the confusion matrix are reversed, and the "gold standard" results appear on the horizontal axis.) Nevertheless, the important points of this plot are that the sensitivity is defined as the agreement, or positive finding (TP) of the new method with the gold standard, divided by total positive findings (TP + FN) from the gold standard, whereas the specificity is defined as the agreement, or negative results (TN) between the gold standard and the new method, divided by the total negative results (FP + TN) from the gold standard. The positive and negative predicted values are similarly defined as

$$\text{Positive Predicted Value} = \frac{\text{TP}}{\text{TP} + \text{FP}} \tag{12.36}$$

$$\text{Negative Predicted Value} = \frac{\text{TN}}{\text{FN} + \text{TN}} \tag{12.37}$$

The accuracy of a predictive algorithm is defined as the mean of sensitivity and specificity.

12.5.2 Soft independent modeling of class analogy (SIMCA)

Soft independent modeling of class analogy (SIMCA) is a supervised method of data classification based on PCA that is carried out separately for the different classes of spectra to be analyzed [36]. For the training of the algorithm, the first two or three PCs of each class are retained. Subsequently, the data in each set are projected onto the plane or hyperplane formed by the PCs, and the distance is calculated for each spectrum from the plane or hyperplane. A model is formed in which the distances from each measurement, or spectrum, to the plane, or hyperplane, determine whether or not it belongs to one class or the other.

When applying the model to an unknown data set, each new spectrum is projected onto each of the planes or hyperplanes established during training of the model, and the distances are calculated. An observation is assigned to the model class if its distance from the model is below the statistical limit for the class. In SIMCA, an unknown spectrum may belong to more than one class, depending on the distances to the hyperplanes.

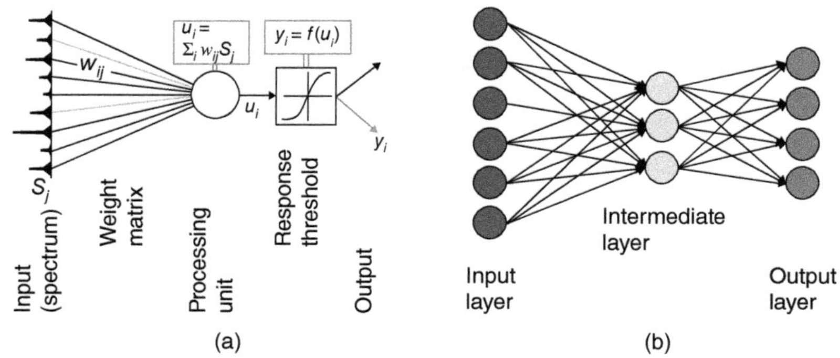

Figure 12.14 *(a) Schematic of an ANN. (b) Schematic of a multilayer ANN. See text for details*

12.5.3 Artificial neural networks (ANNs)

Artificial neural networks (ANNs) have done very well in the analysis of IR spectral data [41]. ANNs are supervised multivariate methods for data analysis [36] and are modeled after the neural patterns found in mammalian brains. In an ANN, input neurons – in this case, spectral vectors consisting of N intensity points S_j – are linked to the outputs y_i by a weight matrix w_j, see Figure 12.14(a):

$$u_i = f\Sigma_j w_{ij} S_j \tag{12.38}$$

This "linkage" of information occurs through programmable logical elements that resemble neuron connections in the brain. This linking of cells is a self-learning process that is largely invisible to the operator; therefore, ANNs have the ability to learn tasks such as spectral classification from training sets. The cells are the simplest decision-making units, or classifiers, in an ANN. Complex, multilayer ANNs for the analysis of spectral data sets with a number of possible outputs (e.g., tissue types) are a network of fully connected nodes, see Figure 12.14(b).

For each node, the response u_i at the summation point of all input neurons is defined by:

$$u_i = \Sigma_j w_{ij} S_j \tag{12.39}$$

or

$$y_i = f(u_i) \tag{12.40}$$

f in Eqs. 12.38 and 12.40 represents a sigmoidal response function, which determines the response for a single neuron. This function is necessary because the input vectors S_j, in the data set, may exhibit quite different intensity values; therefore, u_i may vary for each input vector S_j. The response function is generally of the form

$$f(u) = \tanh(u) \text{ or } f(u) = 1/(1 + e^{-u}) \tag{12.41}$$

For each of the possible classifications, y_i, there will be a weight vector **W** with j entries. These vectors **W** are initialized to contain random numbers. During the training phase of the ANN, the weights are adjusted *via* a feedforward algorithm until they provide the proper connection between spectral features and the desired diagnosis.

For an ANN algorithm to recognize a particular spectral feature and associate it with a diagnosis, the ANN is trained with a large number of spectra for each diagnostic category, including spectral sets from different samples, but the same diagnostic category. The number of spectra included in the training set is determined

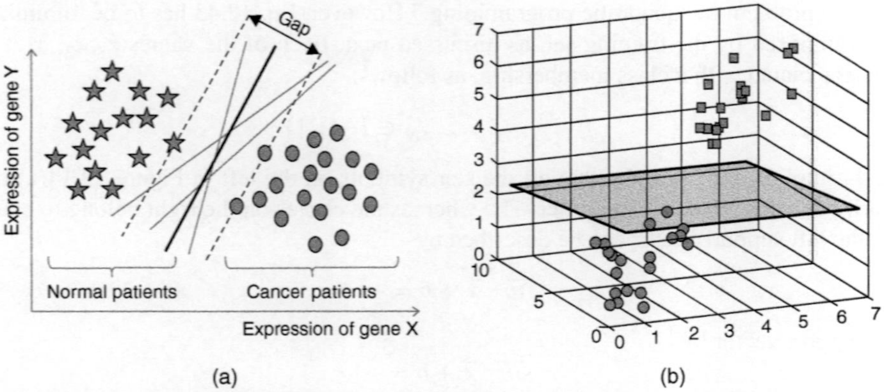

Figure 12.15 (a) Representation of observed values, the separator (black line) and the support vectors (dashed lines) in 2D space. (b) Representation of a hyperplane separating observed values in 3D space. From Ref. [43]

by the smallest number of spectra in a class. Covariance-based feature selection is used to define the spectral areas of highest discriminant content.

ANNs have faced unduly harsh criticism by some researchers, based on some poorly designed previous applications with spectacular failures in some military applications. In the biomedical field, they performed poorly in the analysis of visual images of individual cervical cells to detect abnormal cells in Pap smears [42]. Their lack of acceptance is also based on the fact that the two training cycles of ANNs do not give exactly the same results, because the initial weight matrix is based on random numbers.

12.5.4 Support vector machines (SVMs)

Support vector machines SVMs are trained (supervised) algorithms whose operation can be visualized most readily in a two-dimensional case, see Figure 12.15(a) or by Figure 12.15(b) for a three-dimensional case. This figure shows the results of measurements represented by a scatter plot [43]. These measurements may be a spectral feature that differs for two classes of spectra, or – as shown – the expression of a particular gene that differs between normal and cancerous patients. The heavy black line in Figure 12.15(a) is the best separator of the two classes and is uniquely defined by the two "support vectors," shown as the dashed lines in Figure 12.15(a).

The task at hand for computing the best separator is then reduced, in the two-dimensional case, to find a line that is the furthest removed from any of the data points. This is achieved by maximizing the distance D between the two support vectors (or parallel hyperplanes in three-dimensional space), where the distance, or gap, is defined by

$$D = \frac{|b_1 - b_2|}{\|\vec{w}\|} \tag{12.42}$$

Here, $\|\vec{w}\|$ is the length of the vector separating the support vectors or hyperplanes and b_1 and b_2 are scalar quantities, along the direction of $\vec{w}$, denoting the distance of the support vectors from the separator. Instead of maximizing D in Eq. 12.26, one can minimize $\|\vec{w}\|$. This is accomplished by minimizing the expression

$$\frac{1}{2}\,\|\vec{w}\|^2, \tag{12.43}$$

which can be carried out by "greedy algorithms" that recursively finds the minimum, in this case, of the quadratic expression given in Eq. 12.43. As this equation contains a quadratic expression, solving this equation is referred to as a problem in "quadratic programming." However, Eq. 12.43 has to be minimized subject to the constraints imposed by the training set, as discussed next. Each of the values $x_1, x_2, x_3, \ldots, x_N$ in the training set is associated with a class membership, as follows:

$$y_1, y_2, y_3, \ldots, y_N \in \{-1, 1\} \tag{12.44}$$

Equation 12.44 implies, for example, that all the star symbols on the left in Figure 12.14(a) belong to one class (arbitrarily designated by the identifier −1), whereas the circles on the right belong to class designated as +1. Then, the left support vector can be described by

$$\vec{w} \cdot \vec{x} + b = -1\,, \tag{12.45}$$

and the right support vector by

$$\vec{w} \cdot \vec{x} + b = +1, \tag{12.46}$$

and the separator by

$$\vec{w} \cdot \vec{x} + b = 0 \tag{12.47}$$

Thus, the function given in Eq. 12.27 must be minimized subject to the constraint

$$y_i(\vec{w} \cdot \vec{x} + b) - 1 \geq 0 \tag{12.48}$$

In cases with more than two dimensions, the best separation will be achieved by placing multidimensional planes (hyperplanes) between the classes, as shown in Figure 12.14(b).

One of the most intriguing features of SVMs is the ability to achieve separability of data by switching the "kernel function." These functions are used to project data into multidimensional space in which separation of the observables is possible, even if such a separation is impossible in 2D or 3D space. The best way to visualize the effect of the kernel function is shown in Figure 12.16. In the applications of SVMs to lung cancer classification discussed in Section 13.4, only linear kernel functions were used.

12.5.5 Random forests (RFs)

A multivariate classifier based on a large number of decision tress is known as the random forest (RF) method [44]. Decision trees are sequential steps similar to the one shown in Figures 12.17 and 13.5. In the analysis of

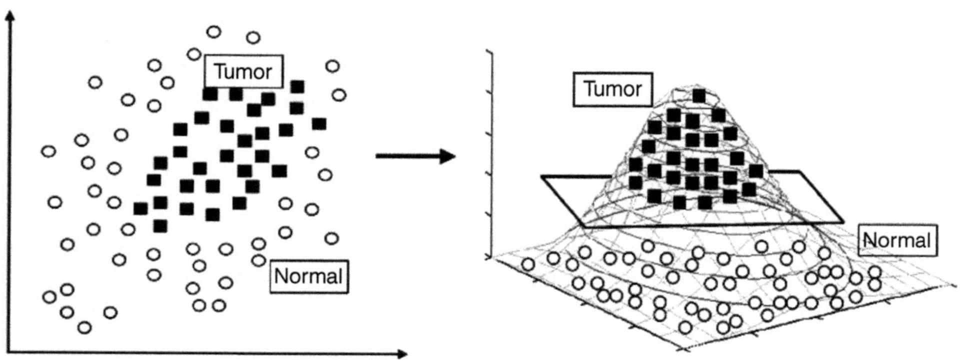

Figure 12.16 *Data that are not linearly separable becomes separable once transformed into the feature space of the kernel function. From Ref. [43]*

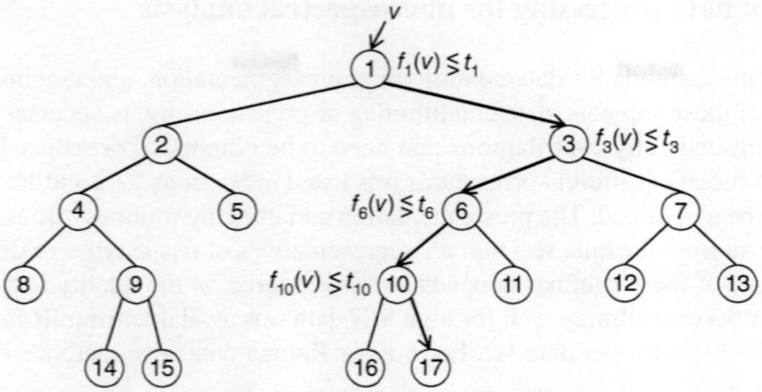

Figure 12.17 *Decision tree in a random forest. From Ref. [45]*

a spectral data set, these decisions could be based on the response of the data at an individual wavenumber, for example, and a decision tree node could be a binary or higher decision. In Figure 12.17 [45], a single tree is shown into which a piece of data, for example, a spectrum, v, is input. At node 1, a decision is made whether or not one aspect of v, for example, an intensity value at a given wavenumber, exceeds or is smaller than a threshold, t_1. This leads to two possibilities, node 2 and node 3. The process is repeated until the desired endpoint (true or false) is reached at points 16 and 17. These endpoints are referred to as leaves. In the training phase, different input subsets are split into trees, each subset using different features of the spectra. In the end, an ensemble decision is reached, based on a consensus of the decisions of all the different trees (the forest).

RFs offer the advantage that the tree structure readily reveals what features were selected by the algorithm to arrive at a classification, a feature that is not available from SVMs and ANNs. Moreover, the RF algorithm is a process that can easily be parallelized because each tree, in principle, can run in a separate processor. Thus, RFs are trained very quickly.

12.5.6 Cross-validation

In supervised data analysis, that is, when the correct outcome of the analysis is known for at least part of the data set, it is desirable to know how the trained diagnostic algorithm would fare against unknown data. This is particularly so in a new field, where the data sets are always restricted in size owing to the availability of patient data. In order to avoid overfitting the data, cross-validation is frequently employed. For spectral data sets, the number of patients normally is small, but the number of spectra for each patient may be large. The cross-validation described below can be carried out on a *per-patient* or a *per-spectrum* basis.

Two common forms of cross-validation are introduced briefly. One involves splitting the data (normally by patient) into two independent, matched sets, referred to as *training set* and *validation* (*or test*) *sets*. A diagnostic algorithm, trained *via* the *training set*, is subsequently applied to the *test set*, and the accuracy of prediction is determined. Subsequently, the original data set is split into new training and validation sets, and the process is repeated. The overall accuracy is the average of all runs using different training and validation sets.

Another frequently used validation method is the "leave-one-out-cross-validation" (LOOCV). As the name implies, LOOCV uses all but one patient in the data set for training and applies the trained classifier to the patient left out of the training set. This process is repeated until all patients were left out once in the training process.

12.6 Summary of data processing for microspectral analysis

In the previous sections, methods for data conditioning, presegmentation, and diagnostic classification were introduced. The first of these subjects, data conditioning or preprocessing, is necessary because spectral data may be noisy and confounded by contributions that need to be eliminated or reduced to allow for a reliable analysis of the data. A menu of different procedures was listed in Sections 12.2 and 12.3 from which a preprocessing sequence can be assembled. The presegmentation and imaging routines discussed in Section 12.4 aim to extract salient spectra from the data sets that are representatives of tissue types or disease state by unsupervised methods. The use of these routines depends so some degree on the quality and quantity of input data. For example, HCA works exceedingly well for high S/N data sets available from IR imaging, but is restricted in size to about 100,000 spectra per data set. For noisier Raman data sets, methods of color decomposition seem to work better.

At this stage in Raman and IR microspectroscopic applications to medical problems, the final step, namely the supervised classification, is of extreme importance, for it is only this step that can be used to demonstrate sensitivity and specificity of the spectral methods and compare them to classical approaches. The main supervised techniques, ANNs, SVMs, and RFs, have performed with similar success in different laboratories, and the latter two have emerged as the most commonly applied techniques. However, their success critically depends on the quality of input data sets: both the homogeneity of the spectral data sets of the disease classes and the accurate annotation of their pathology are of prime importance. Both these aspects are discussed in more detail in Chapter 13.

12.7 Two-dimensional correlation methods in infrared spectroscopy (2D-IR)

The final section of the chapter on data analysis deals with a slightly different problem, namely two-dimensional (2D) correlation methods in FTIR spectroscopy. At first sight, this subject may seem unrelated to the data manipulation methods introduced in earlier sections of this chapter, as most applications of 2D-IR have dealt with the analysis of periodic changes to the IR spectra [46, 47], for example, by applying a periodic stress/strain to the sample, and not with microspectral data sets. Thus, they could have been discussed in the experimental methods of IR spectroscopy (Chapter 3).

However, any change in the spectra, periodic or not, can be analyzed by correlation methods [48, 49], and the following discussion will focus on nonperiodic change because these are more relevant to biomedical problems than the modulation experiments reported in Ref. [47]. Nonperiodic, for example, slow changes in spectral properties could also be analyzed by other methods, such as PCA.

In analogy to 2D-NMR spectroscopy, 2D-IR spectroscopy is based on the cross-correlation function between variations in the time-dependent spectral intensities (absorbances) $A(\nu_1, t)$ and $A(\nu_2, t)$. Although the perturbation of the spectra is presented here as a time-dependent change, the perturbation can be any other variable (pressure, temperature, etc.) that changes the spectra. The resulting synchronous and asynchronous contributions, $\Phi(\nu_1, \nu_2)$ and $\Psi(\nu_1, \nu_2)$, respectively, are plotted as two-dimensional intensities against two wavenumber axes. The synchronous and asynchronous contributions are related to the FTs of the spectral intensities:

$$\Phi(\nu_1, \nu_2) + i\Psi(\nu_1, \nu_2) = \frac{1}{\pi t} \int_0^\infty Y_1(\omega)\, Y_2^*(\omega)\, d\omega \tag{12.49}$$

In Eq. 12.49, the frequency-domain components $Y_1(\omega)$ and $Y_2^*(\omega)$ are obtained by FT of the time-dependent spectra according to

$$Y_1(\omega) = \int_{-\infty}^{\infty} A(\nu_1, t)\, e^{-i\omega t}\, dt = \mathrm{Re}[Y_1(\omega)] + i\, \mathrm{Im}[Y_1(\omega)] \tag{12.50}$$

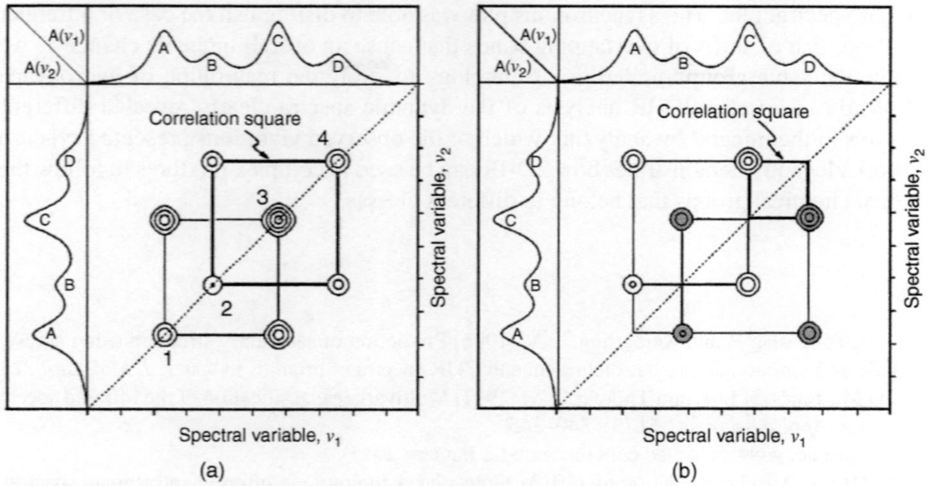

Figure 12.18 *(a) Schematic synchronous 2D-IR spectrum, with 1D reference spectra shown along axes. (b) Corresponding asynchronous spectrum. See text for details. Noda, 1993, [48]. Reproduced with permission from Society for Applied Spectroscopy*

and

$$Y_2^*(\omega) = \int_{-\infty}^{\infty} A(\nu_2, t)\, e^{i\omega t}\, dt = \mathrm{Re}[Y_2(\omega)] + i\,\mathrm{Im}[Y_2(\omega)] \tag{12.51}$$

The synchronous correlation intensity $\Phi(\nu_1, \nu_2)$ expresses the coherence or correlation between the time-dependent variations displayed at two different wavenumbers. This is shown in Figure 12.18(a). A 2D plot contains peaks on the diagonal (the so-called autopeaks) that are due to simultaneous or synchronous changes of intensities measured at ν_1 and ν_2. In Figure 12.18(a), the four autopeaks 1, 2, 3, and 4 will have different intensities than the one-dimensional peaks A, B, C, and D, with the intensities of the autopeaks determined by the extent of dynamic coupling. The synchronous autopeaks are always positive.

Peaks at off-diagonal positions of the synchronous map are referred to cross peaks. These peaks can be positive or negative. Positive cross peaks are due to simultaneous changes, in the same direction, of spectral intensities at the two wavenumber positions, which indicates that the origin of the two signals is coupled. Negative cross peaks indicate a coupled mechanism as well, but the changes at the two wavenumber positions occur in opposite directions. The "Correlation squares" drawn in Figure 12.18(a) indicate that peaks A and C, as well as B and D, are synchronously correlated.

The asynchronous map in Figure 12.18(b) consists of cross peaks only and represents noncorrelated or nonsynchronous (sequential) processes. The asynchronous map is antisymmetric with respect to the diagonal that implies that a positive cross peak to the left of the diagonal will have a negative counterpart on the right of the diagonal, and *vice versa.* Negative peak intensities in Figure 12.18(b) are indicated by gray shading. The correlation square in this panel indicates that peak A is asynchronously correlated with peaks B and D, as are peaks B and C, and peaks C and D.

Mendelsohn and coworkers [50] have modeled the synchronous and asynchronous 2D-IR surfaces for a number of spectral changes that may result from an external perturbation. These simulations include small band shifts, increase in band width, and combinations of shift and band width. They reported, for example, that for the band shift, two autopeaks and two cross peaks were found at the initial and final frequencies in the synchronous spectral plot. The asynchronous plot was able to distinguish the case of a frequency shift of a single peak from that of shifts of overlapping bands that cause an overall intensity change as well.

Subsequently, the same group applied this technology to study the maturation of hydroxyapatite over a time period of up to 3 weeks. 2D-IR analysis of the dynamic spectra clearly revealed different classes of phosphate groups in the mineral by analyzing which of the observed vibrations presented synchronous cross peaks. This work elegantly demonstrates how 2D-IR can be used in complex mixtures to follow the vibrations of near-identical chemical groups that belong to different classes.

References

1. Baumruk, P., Pancoska, P. and Keiderling, T.A. (1996) Prediction of secondary structure using statistical analyses of electronic and vibrational circular dichroism and FTIR spectra of proteins in water. *J. Mol. Biol.*, **259**, 774–791.
2. Haaland, D.M., Jones, H.D.T. and Thomas, E.V. (1997) Multivariate classification of the infrared spectra of cell and tissue samples. *Appl. Spectrosc.*, **51** (3), 340–345.
3. Lasch, P. Cytospec, www.cytospec.com (accessed 2 January 2015).
4. Diem, M., Mazur, A.I., Lenau, K., *et al.* (2013) Molecular pathology via infrared and Raman spectral imaging, in *Ex-Vivo and In-Vivo Optical Pathology* (eds M. Schmitt and J. Popp), Wiley-VCH Verlag GmbH. pp 45–103.
5. Kauppinen, J.K., Moffatt, D.J., Mantsch, H.H. and Cameron, D.G. (1982) Smoothing of spectral data in the Fourier domain. *Appl. Opt.*, **21** (10), 1866–1872.
6. Savitzky, A. and Golay, M.J.E. (1964) Smoothing and differentiation of data by simplified least-squares procedures. *Anal. Chem.*, **36** (8), 1627–1639.
7. Roger, R.E. (1994) A faster way to compute noise-adjusted principal components transform matrix. *IEEE Trans. Geosci. Remote Sens.*, **32** (6), 1194–1196.
8. Reddy, R.K. and Bhargava, R. (2010) Accurate histopathology from low signal-to-noise ratio spectroscopic imaging data. *Analyst*, **135**, 2818–2825.
9. Miljković, M., Bird, B., Lenau, K. *et al.* (2013) Spectral cytopathology: new aspects of data collection, manipulation and confounding effects. *Analyst*, **138**, 3975–3982.
10. Romeo, M.J., Boydston-White, S., Matthäus, C., *et al.* (2008) Infrared and Raman microspectroscopic studies of individual human cells, in *Vibrational Spectroscopy for Medical Diagnosis* (eds M. Diem, P.R. Griffiths and J.M. Chalmers), John Wiley & Sons, Ltd, Chichester, pp. 27–70.
11. Bruun, S.W., Kohler, A., Adt, I. *et al.* (2006) Correcting attenuated total reflection-Fourier transform infrared spectra for water vapor and carbon dioxide. *Appl. Spectrosc.*, **60** (9), 1029–1039.
12. Kohler, A., Kirschner, C., Oust, A. and Martens, H. (2005) Extended multiplicative signal correction as a tool for separation and characterization of physical and chemical information in Fourier transform infrared microscopy images of cryo-sections of beef loin. *Appl. Spectrosc.*, **59** (6), 707–716.
13. Bassan, P., Byrne, H.J., Bonnier, F. *et al.* (2009) Resonant Mie scattering in infrared spectroscopy of biological materials – understanding the 'dispersion artefact'. *Analyst*, **134**, 1586–1593.
14. Mohlenhoff, B., Romeo, M.J., Diem, M. and Wood, B.R. (2005) Mie-type scattering and non-Beer–Lambert absorption behavior of human cells in infrared microspectroscopy. *Biophys. J.*, **88** (5), 3635–3640.
15. Romeo, M.J., Mohlenhoff, B. and Diem, M. (2006) Infrared microspectroscopy of human cells: causes for the spectral variance of oral mucosa (buccal) cells. *Vib. Spectrosc.*, **42**, 9–14.
16. Romeo, M.J. and Diem, M. (2005) Correction of dispersive line shape artifact observed in diffuse reflection infrared spectroscopy and absorption/reflection (transflection) infrared micro-spectroscopy. *Vib. Spectrosc.*, **38** (1-2), 129–132.
17. Bassan, P., Kohler, A., Martens, H. *et al.* (2010) Resonant Mie Scattering (RMieS) correction of infrared spectra from highly scattering biological samples. *Analyst*, **135**, 268–277.
18. Mie, G. (1908) Beitrage zur Optik Trüber Medien, Speziell Kolloidaler Metallösungen. *Ann. Phys. (Leipzig)*, **25**, 377–452.
19. Walstra, P. (1964) Approximation formulae for the light scattering coefficient of dielectric spheres. *Brit. J. Appl. Phys*, **15**, 1545–1552.

20. van de Hulst, H.C. (1981) *Light Scattering by Small Particles*, Dover Publication, Mineola, NY.
21. Bassan, P., Byrne, H.J., Lee, J. *et al.* (2009) Reflection contributions to the dispersion artefact in FTIR spectra of single biological cells. *Analyst*, **134**, 1171–1175.
22. Kohler, A., Sulé-Suso, J., Sockalingum, G.D. *et al.* (2008) Estimating and correcting Mie scattering in synchrotron-based microscopic Fourier transform infrared spectra by extended multiplicative signal correction. *Appl. Spectrosc.*, **62** (3), 259–266.
23. Diem, M., Bird, B., and Miljković, M. (2011) Phase Correction to Compensate for Reflective Distortions of Optical Spectra, U.P. Office.
24. Griffiths, P.R. and De Haseth, J.A. (1986) Fourier transform infrared spectrometry, in *Chemical Analysis*, vol. **83** (eds P.J. Elving and J.D. Winnefordner), John Wiley & Sons, Inc., New York.
25. Mertz, L. (1967) Auxiliary computation for Fourier spectrometry. *Infrared Phys.*, **7** (1), 17–23.
26. Bertie, J.E. and Zhang, S. (1992) Infrared intensities of liquids. IX. The Kramers–Kronig transform, and its approximation by the finite Hilbert transform via fast Fourier transforms. *Can. J. Chem.*, **70**, 520–531.
27. Bassan, P., Lee, J., Sachdeva, A. *et al.* (2013) The inherent problem of transflection-mode infrared spectroscopic microscopy and the ramifications for biomedical single point and imaging applications. *Analyst*, **138**, 144–157.
28. Filik, J., Frogley, M.D., Pijanka, J.K. *et al.* (2012) Electric field standing wave artefacts in FTIR micro-spectroscopy of biological materials. *Analyst*, **137**, 853–861.
29. Wehbe, K., Filik, J., Frogley, M.D, *et al.* (2012) The effect of optical substrates on micro-FTIR analysis of single mammalian cells. *Anal. Bioanal. Chem*, **405**, 1311–1324 (2013)
30. Miljković, M., Chernenko, T., Romeo, M.J. *et al.* (2010) Label-free imaging of human cells: algorithms for image reconstruction of Raman hyperspectral datasets. *Analyst*, **135**, 2002–2013.
31. Nascimento, J.M.P. and Bioucas Dias, J.M. (2005) Vertex component analysis: a fast algorithm to unmix hyperspectral data. *IEEE Trans. Geosci. Remote Sens.*, **43** (4), 898–910.
32. Matthäus, C., Chernenko, T., Quintero, L. *et al.* (2008) Raman microscopic imaging of cells and applications monitoring the uptake of drug delivery systems. *Proc. SPIE Biophotonics: Photonic Solutions Better Health Care*, **6991**, 101–108.
33. Farag, A.A. and Elhabian, S.Y. (2008) *A Tutorial on Data Reduction: Linear Discriminant Analysis (LDA)*, University of Louisville.
34. Wood, B.R. Chiriboga, L. Yee, H., *et al.* (2004) FTIR mapping of the cervical transformation zone, squamous and glandular epithelium. *Gynecologic Oncol.*, **93** (4), 59–68.
35. Romeo, M.J., Boydston-White, S., Matthäus, C., *et al.* (2008) Vibrational microspectroscopy of cells and tissues, in *Biomedical Vibrational Spectroscopy* (eds P. Lasch and J. Kneipp), Wiley-Interscience, Hoboken, NJ, pp. 121–147.
36. Adams, M.J. (2004) Chemometrics in analytical spectroscopy, in *RSC Analytical Spectroscopy Monographs*, 2nd edn (ed N.W. Barnett), Royal Society of Chemistry, Cambridge.
37. Ward, J.H. (1963) Hierarchical grouping to optimize an objective function. *J. Amer. Stat. Assoc.*, **58** (301), 236–244.
38. MacQueen, J.B. (1967) Some methods for classification and analysis of multivariate observations, in *Proceedings of the 5th Berkeley Symposium on Mathematical Statistics and Probability*, University of California Press. PP 281–297.
39. Lasch, P., Haensch, W., Naumann, D. and Diem, M. (2004) Imaging of colorectal adenocarcinoma using FT-IR microspectroscopy and cluster analysis. *Biochim Biophys Acta*, **1688** (2), 176–186.
40. Bhattacharya, A. and De, R.K. (2008) Divisive correlation clustering algorithm (DCCA) for grouping of genes: detecting varying patterns in expression profiles. *Bioinformatics*, **24**, 1359–1366.
41. Schiffer, G., Udelhoven, T., Labischininski, H., *et al.* (2002) Application of FTIR spectroscopy in antibacterial drug research, in *Workshop on FTIR Spectroscopy in Microbiological and Medical Diagnostic*, Robert Koch Institute, Berlin.
42. The Secretary, M.S.A.C. (2003) Computer-assisted image analysis for cervical screening, T.C.o.A. Medical Services Advisory Committee, Editor 2003.
43. Statnikov, A., Hardin, D., Guyon, I., *et al.* (2009) *A Gentle Introduction to Support Vector Machinesin Biomedicine*.World Scientific Publishing, Hackensack, NY
44. Breiman, L. (0000) *Decision Trees and Random Forests*. http://www.stat.berkeley.edu/~breiman/RandomForests.
45. Miljković, M., 2012.
46. Noda, I. (1989) Two-dimensional infrared spectroscopy. *J. Am. Chem. Soc.*, **111**, 8116–8118.

47. Noda, I. (1990) Two-dimensional infrared (2D IR) spectroscopy: theory and applications. *Appl. Spectrosc.*, **44** (4), 550–561.
48. Noda, I. (1993) Generalized two-dimensional correlation method applicable to infrared, Raman, and other types of spectroscopy. *Appl. Spectrosc.*, **47** (9), 1329–1336.
49. Gericke, A., Gadaleta, S.J., Brauner, J.W. and Mendelsohn, R. (1996) Characterization of biological samples by two-dimensional infrared spectroscopy: simulation of frequency, bandwidth, and intensity changes. *Biospectroscopy*, **2**, 341–351.
50. Gadaleta, S.J., Gericke, A., Boskey, A.L. and Mendelsohn, R. (1996) Two-dimensional infrared correlation spectroscopy of synthetic and biological apatites. *Biospectroscopy*, **2**, 353–364.
51. Winter, M.E., *N-FINDR: an algorithm for fast autonomous spectral end-member determination in hyperspectral data*. Proc. SPIE, 1999. **3753**: p.266–275.

13

Infrared Microspectroscopy of Cells and Tissue in Medical Diagnostics

13.1 Introduction

The concept of using vibrational spectroscopic method as adjunct medical diagnostic tools dates back to the 1940s and 1950s when infrared (IR) spectroscopy itself was in its infancy [1, 2]; yet even then, forward-looking spectroscopists thought of the possibility of using the biochemical information contained in the spectral results, rather than the morphological information commonly used in classical cytopathology and histopathology, for medical diagnoses. However, it took until the first decade of the 21st century for the promise of spectral cytopathology (SCP, spectral diagnosis of cells) and spectral histopathology (SHP, spectral diagnosis of tissue) to materialize. Notwithstanding several reports from the 1990s that proclaimed spectral diagnostic successes, it took over a dozen of years of intense efforts to understand even the basic effects that confound IR spectroscopy of cells and tissues [3, 4], to develop the computational methods to detect the often minute changes in the spectra of cells and tissues with disease [5], and develop medically acceptable methods for the comparison between spectral and classical diagnostic results. Advances in the spectroscopic efforts were aided enormously by concomitant improvement in measurement technology in the early 2000s, and an explosive growth of computational power available to spectroscopists. Interestingly, in the eyes of the author, the increased computational power, along with the development of some fundamental theoretical underpinnings, were the most important developments to propel SCP and SHP toward the commercial realm.

The consequence of this last statement is the fact that spectral changes that are thought to be the consequences of disease or any other cell biological event and are visible to the naked eye are most likely not due to the anticipated effects, but due to the aforementioned confounding variations of spectral features based mostly on morphological changes within the tissue or the cells studied. Thus, one universally applicable and highly important result of the spectroscopic studies of cells and tissues is the realization that IR (micro)spectra are highly dependent on sample morphology: if the sample is not a homogeneous film, but consists of discrete particles, and if the particle size is approximately the same as the wavelength of the IR light, scattering effects will confound the observed IR spectra by mixing absorptive and dispersive line shapes, which was first documented by researchers in the field of biomedical applications of IR spectroscopy [4, 6–8], and was discussed in Chapter 12.

Modern Vibrational Spectroscopy and Micro-Spectroscopy: Theory, Instrumentation and Biomedical Applications, First Edition.
Max Diem.

In this chapter, mostly results from the author's laboratory at Northeastern University (the Laboratory for Spectral Diagnosis, LSpD) are discussed, both for SHP and SCP. However, many references to work by other laboratories have been included to permit the reader to gauge progress in this field worldwide.

13.2 Spectral histopathology (SHP)

13.2.1 Review of classical histopathology

Histopathology, the visual microscopic inspection of stained tissue sections from biopsies, has been the first line of diagnosis in the treatment of many diseases, in particular cancer. Modern histopathology dates back to the late 19th century, when R. Virchow first used microscopic techniques to gain insight into the changes in cellular morphology that accompany disease. Gross morphological changes that occur with the transition from healthy to diseased tissue include abnormal tissue architecture, the appearance of cells at locations where they are normally not found (metaplasia), the presence of inflammatory cells, abnormal cell morphology including changed nucleus-to-cytoplasm (N/C) ratio, abnormal nuclear membrane morphology, abnormal chromatin distribution, the number of nucleoli in the nucleus, the presence of unusually large number of mitotic figures indicating rapid cell division, and a plethora of other, often ill-defined features such as foaminess of the cytoplasm [3]. To arrive at these diagnostic descriptors, tissue sections, generally about 5 μm thick, are obtained by cutting, *via* a microtome, formalin-fixed, paraffin-embedded (FFPE) tissue blocks, or unfixed, flash-frozen tissue sections. However, in either case, the tissue sections present themselves as white or grayish deposits on microscope slides and show very little contrast under the microscope. Thus, stains with different affinity for cellular components have been used as early as the late 19th century to create a visual contrast in histopathology. The most common stain in histopathology is the hematoxylin/eosin (H&E) combination stain [4], which visualizes the basic proteins of the histones (mainly arginine) dark bluish or purple, whereas the eosin imparts a pinkish color to the cytoplasm of cells. One may think of the H&E staining procedure as a very basic "optical method of diagnosis," as the two stains bind differentially to different cellular structure, making these structures absorb light of different wavelengths, thereby permitting their distinction. Thus, classical histopathology, which uses the human eye as a detector and databases stored in the pathologists memory as a diagnostic discriminator, has been the earliest form of an optical method to detect disease.

In order to reduce the interobserver variability of classical histopathology, digital imaging acquisition (at one or more different visible wavelengths) [5] and subsequent computer analysis of the resulting image planes of histopathological specimens have been introduced and commercialized; these methods are generally referred to as "digital pathology" or "whole slide imaging" [6]. However, the analysis of such data sets is still carried out in the "morphology domain," rather than in the "spectral domain," as the few wavelengths at which the images acquired do not permit a true spectral discrimination, in particular, in the range of visible wavelengths. Furthermore, the images are based on external stains with low inherent discrimination between normal and abnormal structures in tissue.

Although histopathology has been extraordinarily successful as a first line diagnostic tool, modern medicine, especially oncology, requires more than the subjective descriptors available from classical histopathology in order to decide on a suitable treatment regimen for disease. This requirement gave rise to the development of new methodologies that explore the abundance and localization of specific target molecules, such as cancer markers. The most common of these cancer marker-specific techniques is immunohistochemistry (IHC), which was first described in 1942 [7]. This approach is generally described as a method of "molecular pathology," as it reduces the reliance on morphological features, and instead uses changes in chemical composition – the presence or overexpression of certain marker molecules – as a primary means of distinguishing normal from diseased tissues. In IHC, an optical chromophore such as a fluorescent label is attached to a mono- or poly-clonal antibody, to an antigen or other protein marker. Upon binding the antibody–chromophore complex to the target molecules in a cell or tissue, these binding sites can be detected by fluorescence microscopy. IHC has very high sensitivity and specificity, but it is restricted by the fact that only markers for which an antibody is applied can produce a detectable signal; that is, unknown cancer markers cannot be detected by this approach.

13.2.2 Spectral methods in histopathology

Thus, different approaches have been developed, mostly in the biophysical and spectroscopic communities. These may be based on detecting either inherent spectral signatures of cellular components or external stains and labels (as discussed earlier), but they are not restricted to the visible wavelengths of light. These methods include, for example, white light endoscopy or fluorescence endoscopy. Miss rates up to 25% have been reported for white light endoscopy that led to treatment of polyps without malignant potential and, consequently, without benefit but additional costs and risks to the patient [8]. Therefore, fluorescence-based techniques, such as the detection of inherent tissue markers, have gained acceptance mostly in endoscopic diagnoses and that use mainly ultraviolet excitation. Owing to the few tissue components that exhibit significant autofluorescence, tissues are often stained with specific fluorophores such as 5-aminolevulinic acid (ALA) [9] or indocyanine green (ICG) [10] to increase sensitivity and specificity, particularly for *in vivo* observations.

During the last 15 years, methods have been developed that can be viewed as molecular pathology methods that depend on imaging the entire genome, proteome, and metabolome of cells and tissues, rather than probing for the presence of selected markers [11, 12]. These methods rely on the detection of the inherent spectral signatures of the biochemical components in a pixel of tissue. This is possible using any of the modalities of vibrational spectroscopy that were introduced in Part I. These forms of vibrational spectroscopy may be described as "fingerprint" techniques, as every molecule exhibits a distinct and specific fingerprint spectrum that allows identification of the molecular compounds and quantitation of their abundance in a sample [13, 14].

13.2.3 Infrared absorption spectroscopy of cells and tissue: introductory comments

After the discussion of IR spectroscopy in Part I and the introduction of IR microspectral techniques in Chapter 12, it should come as no surprise that a volume element (voxel) of tissue exhibits an IR absorption spectrum. The x and y dimensions of such a voxel often are determined by the diffraction limit (see Eq. 11.1), whereas the thickness is determined by the thickness of a dried cell (typically a few micrometers) or the thickness to which a tissue section is microtomed. For SHP, the sample typically is cut to a thickness of 5–10 μm. SHP is being carried out on carefully dried samples, whereas SCP has been reported for both dry cells and living cells in aqueous medium [9].

The IR absorption spectrum of a voxel of tissue or a cell is expected to be a superposition of the IR absorption spectra of the individual cellular components, which include proteins (about 60% by dry mass), nucleic acids (25%), lipids, sugars, glycoproteins, and a few other compounds at much lower abundance. Typical IR absorption traces of some of these are shown in Figure 13.1(a) [10, 11].

The spectra shown in Figure 13.1(a) can be divided into two regions: the high wavenumber region (2500–4000 cm^{-1} or 4–2.5 μm) corresponds to absorptions due to light atom stretching motions: at 3400 cm^{-1}, the O—H stretching vibration is observed, whereas the N—H stretching vibration of the amide backbone (the "amide A" vibration) is found between 3270 and 3310 cm^{-1}, depending on the hydrogen bonding properties of this proton. C—H stretching vibrations of aromatic, olefinic, and aliphatic C—H bonds occur between about 2800 and 3050 cm^{-1}. The 700–1800 cm^{-1} region, frequently referred to as the "fingerprint" region, contains in the 1500–1700 cm^{-1} range the vibrations of the protein amide linkage, the "amide I" vibration at about 1650 and the "amide II" vibration at about 1550 cm^{-1} (see top three traces of

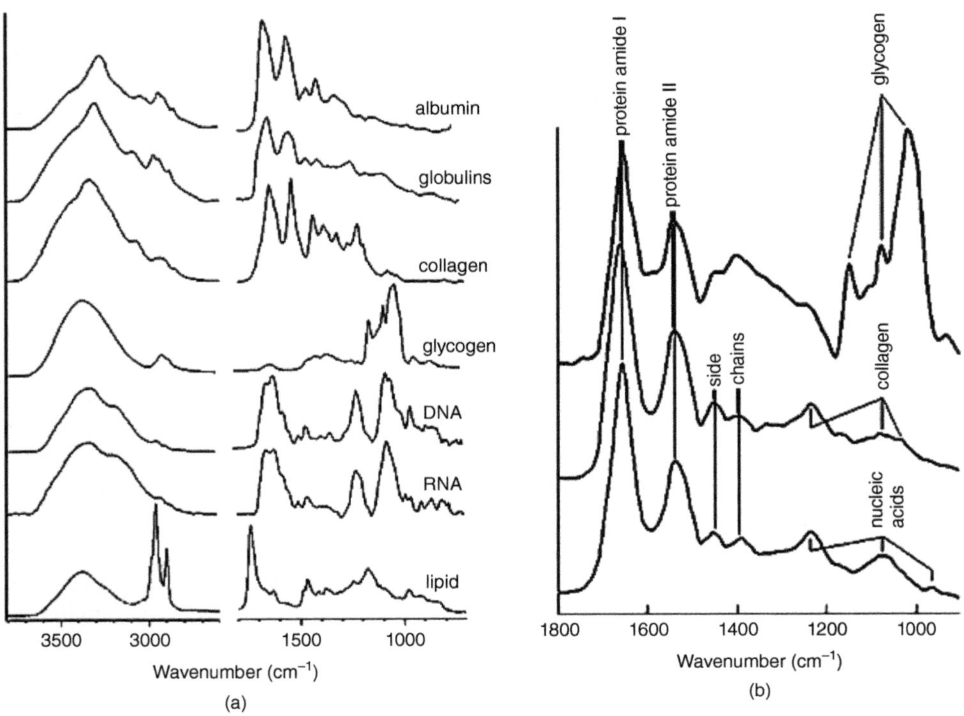

Figure 13.1 *(a) Infrared absorption spectra of major cellular compounds, collected from thin, dehydrated films. (From Ref. [10].) (b) Typical expanded infrared absorption spectra of different tissue pixels: (top) glycogen-rich cervical squamous tissue, (middle) connective tissue (stroma), and (bottom) typical cancer tissue. See text for details. Bird* et al.*, 2012 [15]. Reproduced with permission from Nature*

Figure 13.1(a)). For a listing and discussion of many protein vibrational frequencies, the reader is referred to Chapter 10.

Inspection of Figure 13.1(a) reveals that proteins with different secondary structure exhibit quite different IR absorption spectra, as discussed in Chapter 10. Other biomolecular components of cells and tissues are found in the lower four traces of Figure 13.1(a), including polymeric sugars (glycogen) that exhibit strong C—O stretching and C—O—H deformation vibrations between 1000 and 1200 cm^{-1}, DNA and RNA that can be identified by the vibrations of the phosphodiester vibrations (1095 and 1235 cm^{-1}) and aromatic base vibrations between 1600 and 1700 cm^{-1}, and phospholipids with distinct C—H stretching peaks between 2850 and 3000 cm^{-1} in addition to their own phosphodiester vibrations (see Chapter 10).

For a complex sample, such as human cells or tissues, the observed IR spectra are expected to be a superposition of all the spectra of the individual biochemical components. This is shown in Figure 13.1(b) for three different tissue types. The top trace shows a very weak band about 1740 cm^{-1} that is due to the ester linking of phospholipids [5, 12]. All three spectra are dominated by the protein amide I and II bands, discussed in detail in Chapter 10. The top trace also exhibits very strong sugar (glycogen) peaks at 1151, 1078, and 1028 cm^{-1}. Carbohydrate bands are also observed for glycoproteins, particularly in mucus. In the absence of glycogen (middle trace of Figure 13.1(b)), the protein amide III vibration is observed at 1235 cm^{-1}. In the stroma, the contributions of nucleic acids is very weak; however, characteristic signals for collagen (see Figure 13.1(a)) can be found in the amide III and 1000–1100 cm^{-1} region.

The antisymmetric and symmetric phosphodiester stretching vibrations of DNA, RNA, and phospholipids are observed at about 1235 and 1090 cm^{-1}. The intensity of these bands varies enormously between different tissue types, and with hydration and disease. In mature, stratified squamous cells, the spectral features of DNA/RNA are unobservable. However, in actively growing cells, nuclear DNA/RNA and cytoplasmic RNA were readily observed. These components were identified in cancer cells *via* difference spectroscopy before and after digestion with RNase and DNase [5, 13]. These studies showed unambiguously that IR microspectroscopy (MSP) is a sensitive tool to monitor cell activity.

The daunting task of utilizing spectroscopy for medical diagnosis is to decode the complex spectral signatures observed in cells and tissues. Modern methods of mathematical data analysis, aided by computer algorithms, have succeeded in this decoding process, and the spectral methods are now poised to enter the mainstream diagnostic arena. Preliminary results from a number of laboratories worldwide [16] have indicated that these novel spectral methods can distinguish between normal tissue types, disease types and stages, and even identify the primary tumors from spectral patterns observed in metastatic cells. In spite of the heavy use of computer-based algorithms, it is appropriate to discuss and assign a typical IR spectrum of a cell or tissue pixel. The 1500–1700 cm^{-1} region of the spectrum of a cell or tissue pixel is dominated by the protein amide I and amide II bands; both these bands split into sub-bands in the second derivative spectra. The amide I sub-bands are known to be due to exciton-like coupled states of mainly C=O stretching coordinates [14], see Chapter 10. Certain proteins, in particular the proteins of connective tissue (collagen), have sufficiently different IR spectra to allow detection of their spectral signatures with the naked eye. As discussed earlier, phosphodiester vibrations are observed at 1090 and 1235 cm^{-1}; the latter region also contains the amide III vibration. The carboxylate symmetric stretching vibration is sometimes observed just below 1400 cm^{-1}, and carbohydrate vibrations occur in the 1020–1160 cm^{-1} region.

13.3 Methodology for SHP

13.3.1 General approach

The course to be taken for successful SHP studies was first outlined in a series of pioneering papers by the group at the Robert Koch Institute, Berlin [16–19], and involved the following key steps: acquisition of high S/N spectral data, preprocessing including computation of derivatives and normalization to minimize instrumental and background artifacts, data presegmentation by unsupervised methods such as hierarchical cluster analysis (HCA), very careful annotation of diseased areas by a pathologist, and sufficiently large training data sets to construct a robust diagnostic algorithm. The diagnostic algorithm used were artificial neural networks (ANNs) trained on thousands of spectra [19] for which the disease state or tissue type was known by correlating HCA and the corresponding H&E images of the sample. This work laid the ground rules in SHP, and demonstrated that the patient-to-patient variations of the observed spectra were smaller than those due to disease classification or tissue type [16].

Unfortunately, the development of statistically significant data sets useful for clinical applications is time consuming and quite expensive. Therefore, many studies have terminated at the stage of pseudo-color HCA images, rather than diagnostic images. Nevertheless, these studies have revealed that the HCA-based pseudo-color images, at the level of proof-of-principle, are very valuable, and have opened the eyes of pathologists to the potential of spectral methods. Over the past decade, tissue sections from bladder, bone, brain, breast, cartilage, cervix, colon, esophagus, kidney, liver, skin, spleen, teeth, thyroid, and a few other organs have been studied by SHP. For a summary of some of these studies, the reader is referred to recent reviews [20–22]. Only a few studies have been published, or are underway, in which well-controlled and -documented patient data sets have been constructed, and the accuracy of SHP was directly compared to the gold standard [15].

As can be seen from the previous paragraph, classical histopathology is used as the gold standard to correlate SHP with primary medical diagnostics. Classical histopathology has good sensitivity and specificity in detecting cancers, but it is somewhat more ambiguous when it comes to grading or subclassification of disease. It also is an inherently subjective approach to diagnostics and lacks reproducibility and cannot easily be carried out *via* a quantitative and reproducible measurement. Furthermore, the detection of specific subtypes, for example, the overexpression of cancer genes, requires immunohistochemical stains and subsequent pathological analysis. SHP has the promise to enhance many of these aspects and combine morphological aspects and biochemical compositional information into a novel approach. In the development of this technique, efforts were made to arrive at a protocol for SHP that minimally interferes with the workflow in a pathological laboratory, because it is absolutely essential to incorporate standard classical histopathology results to correlate SHP results and to train multivariate algorithms for the automatic analysis and classification of tissue.

13.3.2 Sequence of steps in classical histopathology

The relevant steps involved in SHP procedures are discussed next. Classical histopathology is being carried out on (archived) FFPE tissue blocks, or flash-frozen tissue directly from the operation theater. For analysis, sections between 5 and 10 μm thick are cut off the blocks using a microtome or cryo-microtome and mounted on glass or quartz microscope slides. Subsequently, the slides are de-paraffinized, stained, and coverslipped. The de-paraffination step is not required for frozen sections. Histopathology departments utilize programmable machines to perform the de-paraffination and staining automatically. After being coverslipped, the slide is ready for the pathologist or for digital imaging.

13.3.3 Sequence of steps for spectral histopathology

For SHP, a sequence of steps different from those of classical histopathology needs to be employed. This sequence is summarized in Figure 13.2 and is discussed next.

13.3.3.1 Sample preparation for SHP

Tissue samples for IR data acquisition are prepared as follows. Tissue sections, from FFPE tissue blocks, or frozen tissue, are cut *via* a microtome and mounted on suitable substrates, generally an IR transparent material such as CaF_2 or an IR reflecting substrate such as the "low-e" slides discussed in Chapter 11. For IR-SHP, the thickness of the sample should not exceed about 5 μm if the sample is examined in transflection mode and not exceed 10 μm for transmission measurements.

After mounting tissue sections on appropriate substrates, FFPE slides may be de-paraffinized, washed, and air dried; alternatively, IR images can be collected and analyzed from tissues still in paraffin [23, 24]. The presence of paraffin reduces the gradients in refractive index and reduces Mie and resonance Mie scattering. Fresh frozen tissue need not be treated any further, and just needs to be thawed and air dried before data acquisition.

Formalin-fixation, paraffination, and subsequent de-paraffination procedures commonly used in standard histopathology laboratories will certainly change protein structure; on the other hand, these changes are sufficiently small that immunohistochemical agents still recognize specific protein structures and binding sites. In the past, several reports on fixation [25, 26] and comparisons of fixed and unfixed (flash-frozen) have appeared in the literature. The equivalence of the two approaches was demonstrated when two virtually identical IR imaging studies on a rat model of glioblastoma multiform were published [27, 28]; one using frozen tissue section and the other using FFPE sections. Although there were, of course, spectral differences between the two tissue preparations, both studies arrived at images that were quite comparable, and reached similar

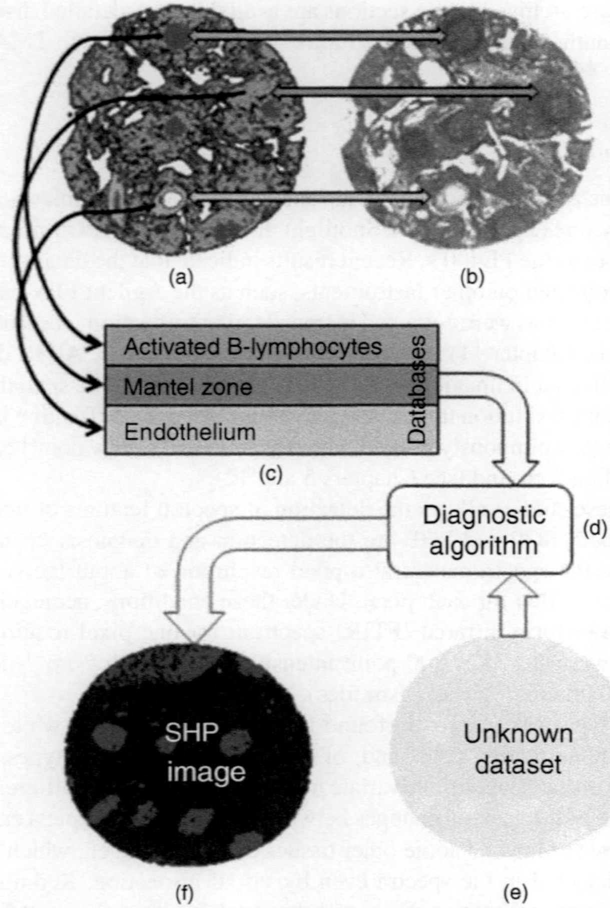

***Figure 13.2** Schematic of SHP process. (a) Pseudo-color image of tissue section obtained from hyperspectral data cube by HCA. (b) Photomicrograph of same tissue section, H&E-stained after IR data acquisition. (c) Databases extracted from HCA image during annotation process. (d) Diagnostic algorithm trained with database "C." (e) Hyperspectral data set from unknown sample. (f) Pseudo-color SHP image produced by diagnostic algorithm to distinguish activated B-lymphocytes from other tissue types. Diem* et al.*, 2013 [10]. Reproduced with permission from Wiley-VCH.* (See insert for color representation of this figure.)

conclusions. After IR data acquisition, slides may be stained and coverslipped using standard histopathological procedures. It is advantageous to digitally image the stained biopsy slides to allow comparison with the IR-based images; this method insures that SHP and classical histopathology are being carried out exactly on the same sample.

Many of the recent SHP studies have used commercial tissue microarrays (TMAs, see Figure 13.10) for samples [15, 29]. A TMA consists of between 50 and 120 individual tissue cores, each about 1–3 mm in diameter that have been punched out of paraffin-embedded tissue blocks and may be seen as prototypical examples of a given cancer type. These cores themselves are embedded in paraffin and sectioned to standard thickness. Thus, one can purchase a TMA that contains samples from dozens of patients or disease stages. The use of TMAs was pioneered by the group around Levin at the NIH [29] and has been adopted by several groups [20], including the LSpD. As these archived tissue sections are available with detailed disease diagnoses and often with disease outcome, the author believes that the future of SHP will be tied to TMA methodology for some time to come.

13.3.3.2 Data acquisition

Most of the IR spectroscopic results introduced in this chapter were acquired microscopically *via* one of three imaging IR microspectrometers (Spectrum One/Spotlight 400, PerkinElmer Corporation, Shelton, CT) at the LSpD, henceforth referred to as the PE400's. Recent results indicate that the data from the PE400 instruments are compatible with data collected on other instruments, such as the Agilent FPA-based machines.

At the LSpD, IR spectra of tissues were collected in transflection (reflection–absorption) mode from samples mounted on low-e slides (see Chapter 11) at a spectral resolution of 4 cm^{-1}. All IR data presented here (both for SCP and SHP) were collected in imaging mode at 6.25 μm pixel size. The spatial resolution of the PE400 was established using military resolution targets and was about twice the diffraction limit at 1600 cm^{-1}, about 12 μm. The instruments were continuously purged with dry air (−40 °C dew point) to reduce the atmospheric water vapor rot-vibrational background (see Chapters 5 and 12).

The pixel size used in these studies allows the detection of spectral features of items as small as a cellular nucleus. As the goals of both SCP and SHP are the detection and diagnosis of individual cancer cells, it is advantageous to operate the spectrometers at a pixel resolution of about the size of a cellular nucleus. Two interferograms were co-added for each pixel. Under these conditions, acquisition of a complete (about 700–4000 cm^{-1}) Fourier transform infrared (FTIR) spectrum for one pixel requires between 5 and 10 ms using the PE400. Spectra were stored as 800 point intensity vectors with 2 cm^{-1} data spacing from 800 to 4000 cm^{-1} in native PE400 imaging format (.fsm files).

In SHP, different tissue types are frequently found in one section, such as white and gray brain matters, stroma, epithelial layers, inflammatory cells, and, of course, diseased tissue types. In general, IR imaging techniques, combined with unsupervised multivariate methods, can detect the different tissue types, and allow a biochemical interpretation of the spectral changes between tissue types. A typical example is the detection of different protein classes: the stroma and some other tissues contain collagen, which has a very characteristic IR spectrum and can be detected in the spectra even by visual inspection. Keratin, a structural protein, is found in keratinizing squamous cell carcinomas, and the resulting "keratin pearls" were first described by Schultz *et al.* [30]. Similarly, parakeratosis (the deposition of keratin) in squamous epithelium was described by Wood *et al.* [31]. However, for the training of data sets for detailed clinical studies, it is essential to assign the pixel spectra, obtained by HCA, to morphologically distinct regions of the corresponding H&E images, a process that is referred to as "annotation" (see Figure 13.2). This annotation step, performed by a pathologist, requires high-quality visual images that need to be collected from the same specimen from which the IR data were collected.

13.3.3.3 Data preprocessing and presorting

The hyperspectral data cubes collected in a manner discussed earlier are subsequently preprocessed as described in detail in Chapter 12. The aim of the preprocessing methods is the elimination of pixels from the data sets that do not contain spectral information, to reduce unwanted signals (such as those of water vapor), improvement of the S/N ratio of the data set, reduction of resonance Mie scattering contributions, and converting the data set into second derivatives that are more sensitive to small spectral variations than the original absorption spectra.

Subsequently, the spectral data are converted automatically into pseudo-color images. For data sets less than about 40,000 (= 200 × 200) pixels, agglomerative HCA can be carried out in a few minutes computation

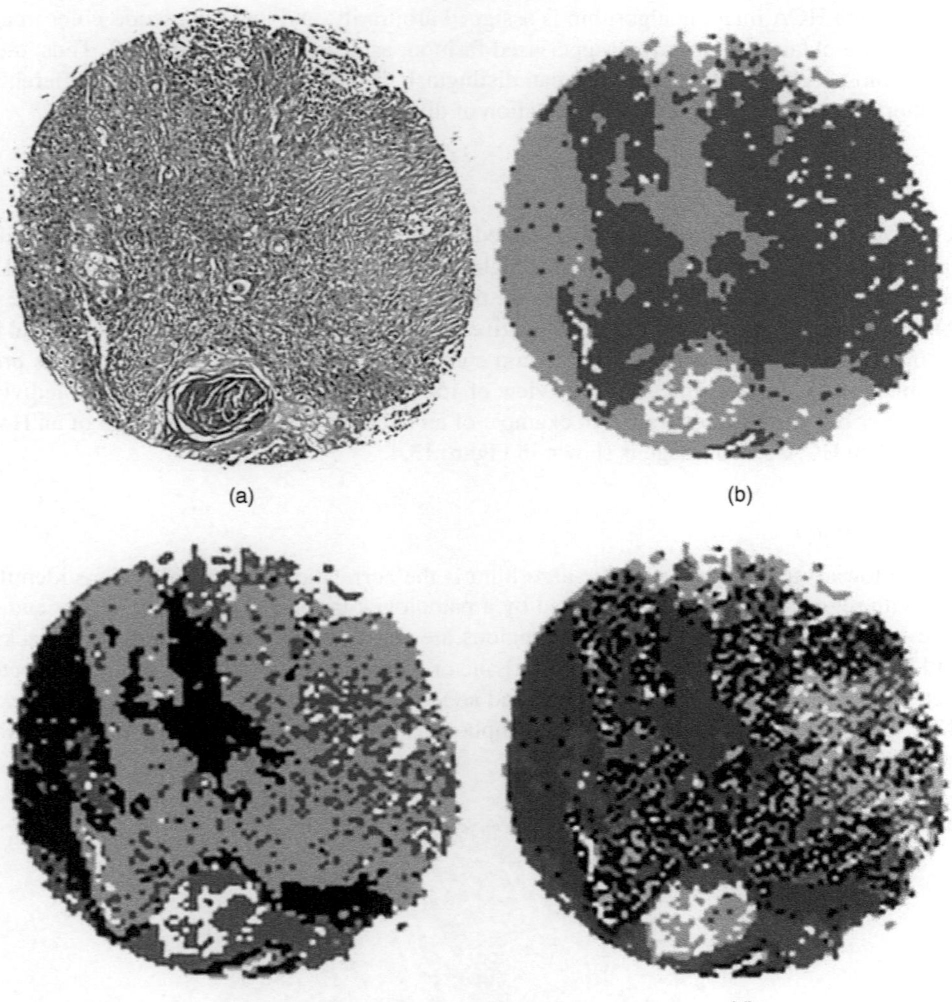

Figure 13.3 *(a) Photomicrograph of an H&E-stained TMA spot. (b) 2-cluster, (c) 4-cluster, and (d) 6-cluster HCA pseudo-color images of same tissue spot. See text for details.* (See insert for color representation of this figure.)

time on a high-end workstation. The problem with HCA is the large memory requirement for the distance or correlation matrix: for a 40,000 pixel data set, the memory requirement exceeds 16 GB; thus, for larger data sets, k-means clustering (see Chapter 12) is advantageous owing to the shorter execution time and reduced memory requirement. These clustering methods produce pseudo-color images as shown in Figure 13.3(b)–(d). Depending on the cluster level selected, different discrimination between tissue types can be achieved: in Figure 13.3(b), regions with disease (green) are discriminated from noncancerous regions at the two-cluster level. At the four-cluster level (c), keratin and low-grade SqCC (squamous carcinoma) are shown in gray, high-grade SqCC in blue, connective tissue in green, and inflammatory cells in red. At the six-cluster level (d), the keratin pearl is shown in green, low- and high-grade SqCC in red and purple, respectively, connective tissue in gray and dark blue, and inflammatory cells in light blue. It should be noted that the color for each cluster used by the HCA imaging algorithm is assigned arbitrarily, and that the pseudo-color images shown in Figure 13.3 are obtained in a truly unsupervised fashion, as discussed in Chapter 12. Thus, these images represent the most basic proof that IR MSP can distinguish different tissue types by the differences in their spectral patterns that themselves are a manifestation of different biochemical composition.

13.3.3.4 Visual imaging

After IR data acquisition, the tissue section is stained (typically using H&E stain) and imaged, using a visual microscope at 20× or 40× magnification and a high-resolution digital camera. This high optical resolution is needed to assure that the pathologist can inspect tissue architecture and cellular features at the customary resolution. Even at 20× magnification and the relatively small TMA spots discussed here, a visual image will be many times larger than the screen display on a computer. Thus, the individual image tiles produced by the visual microscope, with a typical field of view of 150 μm× 180 μm on edge, are stored individually and stitched together by real-time software. An example of an overlay of a photomicrograph of an H&E-stained tissue spot and an HCA-based image is shown in Figure 13.4.

13.3.3.5 Annotation

The next step toward training a diagnostic algorithm is the correlation of spectral regions identified by the HCA map with regions that can be recognized by a pathologist in the H&E images. To this end, these two images are superimposed and registered, and regions are selected (see green, yellow, and black free-form shapes in Figure 13.4) that indicate prototypical tissue or disease features. Spectra from the selected regions are extracted and tagged with a diagnostic code and added to a database, as indicated by the black arrows in Figure 13.2. These classes include activated B-lymphocytes (red areas in Figure 13.2), and endothelial cells

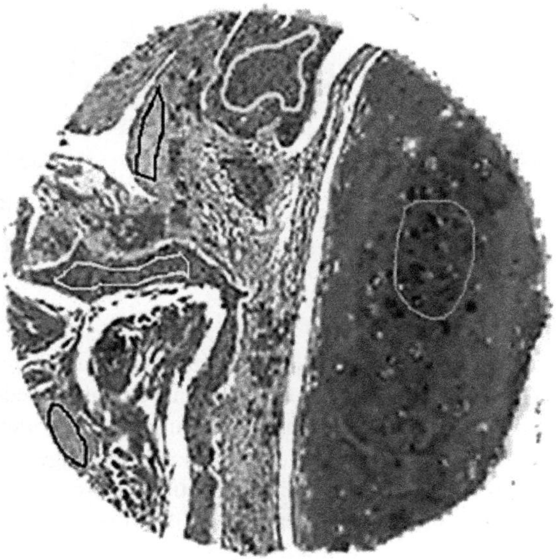

Figure 13.4 *Superposition of an H&E image and a HCA cluster image for annotation. The free-hand figures drawn by the pathologist identify regions of homogeneous tissue/disease state within one spectral cluster.* (See insert for color representation of this figure.)

(light blue areas). This annotation step is necessary only during the training and testing phase of a diagnostic algorithm, and it is omitted for the diagnostic analysis of an unknown sample.

It is essential that this annotation is carried out at an image resolution equivalent to what a pathologist is accustomed, generally at a screen magnification that is equivalent to a 20× or 40× objective. Furthermore, the pathologist needs to identify details and features that are typical for a tissue type or disease stage. Often, up to 20 individual annotation regions are required to capture the characteristic tissue types of a tissue spot that may measure about 2 mm in diameter.

The importance of this annotation step cannot be overemphasized, because the quality of the training and test data sets critically depends on the quality and quantity of annotated spectra. The annotation method presented here also insures the traceability of the spectra selected into training and test sets, because the origin of each pixel spectrum can be traced back to an image element of the H&E picture.

13.3.3.6 Diagnostic data analysis

As indicated earlier, the overall goal of the optical diagnostic methods is to provide medical information on a tissue section that is based on chemical composition and compositional variations, rather than on morphological criteria. In order to correlate the compositional changes with disease, self-learning mathematical algorithms are utilized that correlate spectral (compositional) changes with any other available diagnostic information, such as immunohistochemical or – particularly in the early stages of development – classical staining and morphological changes, and disease outcome. Thus, the self-learning algorithms (SLAs) for analyzing spectral data may be based on morphological correlations, although the goal of these efforts is to move away from the morphological descriptors and utilize strictly compositional aspects of the tissue samples.

After establishing databases of annotated spectra from many patients with the same disease diagnosis, algorithms are trained to recognize the distinct spectral features associated with a particular disease. Although it is possible to train SLAs that recognize several disease states and tissue types, it is advantageous to design an algorithm tree or a set of hierarchical algorithms that analyze the data in consecutive binary steps (decision trees), as indicated in Figure 13.5 for the case of a four-class system. Depending on the tissue type and disease state, it may be advantageous to distinguish the most different class of spectra first. In breast cancer (BC), for example, the most different tissue type may be adipose tissue, whereas in certain blood cancers, it may be advantageous to test for the spectral class of erythrocytes first. Subsequently, spectral classes with the next highest variance may be differentiated, for example, molecular subtypes of BC [32]. Thus, the diagnostic algorithm may be broken down into a set of binary decision steps. The order of the binary decision steps may be determined by classical methods of multivariate analysis *via* spectral covariance or by performing hierarchical clustering of the spectra in the annotated database.

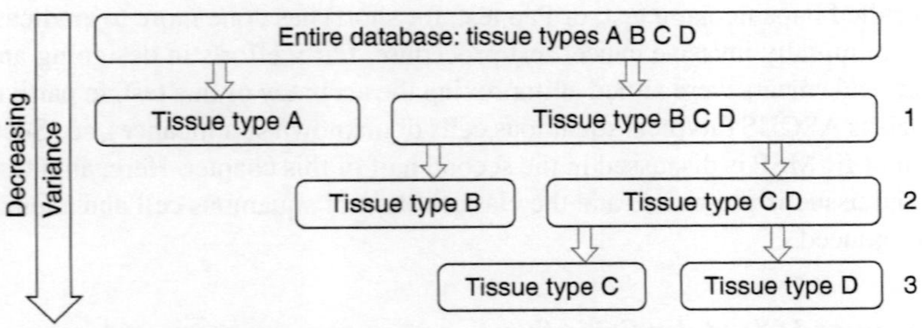

Figure 13.5 *Schematic of a hierarchical decision tree for diagnosis of a four-class problem. Each of the end-points (pure tissue type A, pure tissue type B, etc.) could be the starting point for further algorithms for a finer discrimination (tissue subtypes)*

The training of each of the binary decisions consists of the following steps, explained here for the case of a binary decision between two spectral classes, A and BCD (see Figure 13.5). First, a "feature selection" may be carried out that utilizes hundreds or thousands of spectra in each of the databases of classes A and BCD to determine which spectral intensities, or "features," can reliably differentiate between the two classes. Thus, feature selection can be viewed as a process that removes irrelevant features from the data. A typical fingerprint region spectral vector may consist of about 500 intensity points; typically about 1/10 of those are useful in discriminating the spectral classes. Several mathematical methods exist, for example, as MATLAB library functions (The Mathworks, Natick, MA) to carry out feature selection.

Next, a decision has to be made to select the SLA methodology. The major types of SLAs, as discussed in Chapter 12, are ANNs, support vector machines (SVMs), random forests (RFs), and others. Their performance, when properly trained and validated, is quite similar, and ANNs and SVMs have been used in the author's laboratory.

Typically, an SLA is trained on a training set comprising about one half of the total data set and is subsequently tested on the other half. Care must be taken that the training and test sets are balanced with respect to the number of patients and spectra. For example, the training set for the first-level algorithm in Figure 13.5 would consist of 6000 spectra from 12 patients for tissue class A and 6000 spectra of classes BCD, equally divided among classes B, C, and D (2000 spectra each, from four patients in each class). The SLA in the training phase "learns" to associate the selected spectral features with the diagnostic outcome. When applying such a trained algorithm to the test data set, the sensitivity and specificity (see Section 12.5.1) typically are better than 80%, often even better than 90%, if sufficiently large data sets are utilized, and if the annotation step (see earlier sections) is carried out sufficiently carefully by a pathologist to give pure class spectral databases.

13.4 Applications of SHP for the classification of primary tumors

In this section, the application of spectral methods to primary tumors is reviewed. Of the numerous publications on detecting cancers by SHP, four studies are highlighted because they deal with common cancers – breast, prostate, lung, and cervix – and because they involved studies with significant patient numbers. These studies also aimed at cancer grading and classification by SHP.

13.4.1 Cervical tissue and cervical cancer

Cervical tissue was one of the earliest targets for SHP and SCP, in part, because of the well-documented shortcomings of classical cytopathology of the cervix. That said it is important to note that cervical cancer screening (the so-called Papanicolaou test, or Pap test, for short) has done more to eradicate a particular cancer than any other minimally invasive cancer test procedure. Early efforts in designing an instrument-based method to read cervical smears were aimed at improving the accuracy of this test, in particular, for low-grade disease stages (such as ASCUS (atypical squamous cells of unknown significance), see Section 13.7.1). Cytological application of IR MSP is discussed in the second half of this chapter. Here, an introduction to SHP of multilayer epithelial tissue is presented, and the classification of squamous cell and adenocarcinoma (ADC) of the cervix is introduced.

13.4.1.1 Squamous and Glandular Epithelium

Stratified squamous epithelial tissue is a frequently found epithelium in the human body (nasopharyngeal and oral cavities, esophagus, urethra, vagina, cervix, and others). It is a multilayered structure (see Figure 13.6)

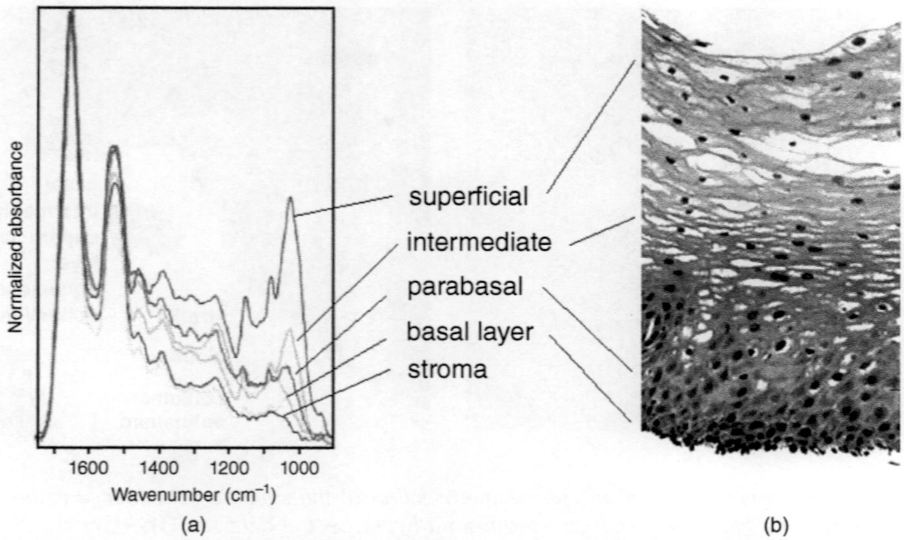

Figure 13.6 *(a) IR spectra of the different layers of normal squamous epithelial tissue. (b) Photomicrograph of H&E-stained epithelial tissue. (See insert for color representation of this figure.)*

consisting of a basal layer of actively dividing basal cells anchored to the basement membrane, beneath which one finds connective tissue (stroma). The daughter cells created by division of the basal cells form the parabasal layer and mature and migrate to the surface layer. In this process, their morphology and chemical composition change drastically. While the basal cells are roughly cuboidal in shape, about 15 μm on edge, with a large nucleus and very little cytoplasm, the mature stratified (flat) cells may measure up to 60 μm on edge and exhibit very small, pyknotic nuclei. They also accumulate glycogen for energy storage (except for the oral mucosal cells), but they are nearly devoid of cytoplasmic RNA signatures [12, 33]. The glycogen signal contributes a triad of peaks in the 1000–1200 cm^{-1} region, see Figure 13.6(a).

In contrast to squamous epithelium with its multilayer structure, glandular or columnar epithelium consists of a single layer of epithelial cell. These glandular cells are responsible for secretion and absorption of biochemical components into or from the spaces surrounding them. They are attached to the basement membrane from which they receive nourishment by diffusion. Many of the major organs are composed of glandular epithelium (breast, lung, stomach, colon, etc.) with specialized function.

In several organs (esophagus and cervix), some parts of the organs are covered with glandular epithelium, whereas others are covered with squamous epithelium. The ectocervix, for example, is covered by squamous epithelium, whereas the endocervix is covered by glandular epithelium. Similarly, the upper part of the esophagus consists of squamous epithelium, whereas the lower part (closer to the stomach) consists of glandular tissue. The region where the two tissue types abut is known as the transformation zone or squamo-columnar junction (SCJ). This area is particularly vulnerable to developing abnormalities, possibly by the attack of certain viruses that are oncogenic.

13.4.1.2 Normal cervical squamous and columnar epithelium and the SCJ

Several papers have reported the normal spectral changes within the layers of squamous tissue and the distinction of the underlying stroma from the squamous tissue [31, 34, 35]. The spectral differences here are quite

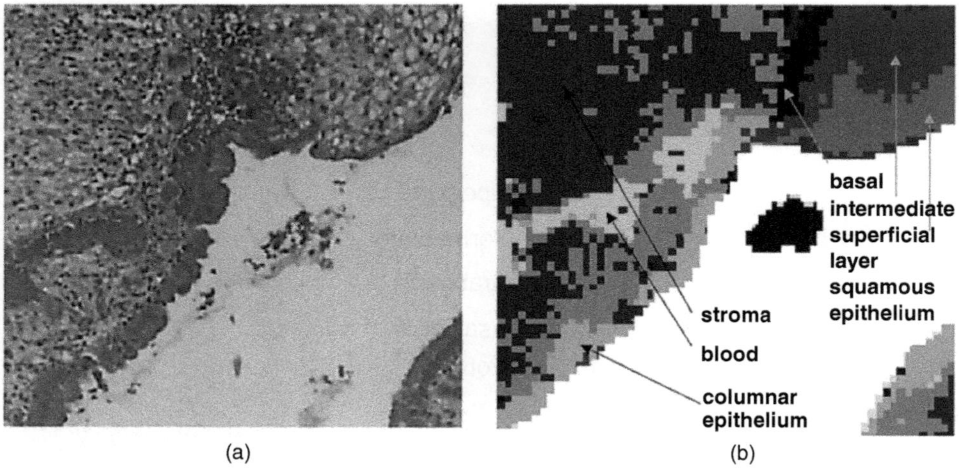

Figure 13.7 *(a) Photomicrograph of an H&E-stained section of the squamo-columnar junction in a cervical biopsy. (b) Pseudo-color HCA image constructed from a hyperspectral data set. Wood,* et al.*, 2004 [36]. (*See insert for color representation of this figure.*)*

pronounced (see Figure 13.6) and provided the first indication of the sensitivity of SHP. Figure 13.7 shows an IR-based pseudo-color image of the SCJ. Here, the glandular tissue (orange) is clearly differentiated from the squamous tissue whose layers (red, purple, and dark blue) represent the superficial, intermediate, and basal layers.

13.4.1.3 Squamous cell- and adenocarcinoma of the cervix

Spectral detection of cervical dysplasia and SqCC was reported by Steller *et al.* [35] and Wood *et al.* [36], but spectral characterization of cervical ADC was not reported until much later [37], partially because of the rare occurrence of this disease and because of the difficulties encountered with inflammatory cells.

Both squamous cell carcinomas and ADCs (cancers of glandular epithelium) can be readily distinguished from normal squamous or glandular tissues, respectively, by SHP. However, in cervical tissue, both cancers are often confounded by the presence of inflammatory cells. Steller *et al.* [35] reported spectral changes due to these cells in the stroma underlying a squamous cell carcinoma, but the infiltration of inflammatory cells was relatively mild. The sample of cervical ADC depicted in Figure 13.8, however, shows heavy infiltration of inflammatory cells that can be easily detected visually in (a). These inflammatory cells, shown in light blue and red hues in the HCA images shown in (b), are easily separated by HCA from the surrounding stroma; however, the distinction between ADC (green) and inflammatory cells is quite difficult. Like all images based on HCA, no reference data set is utilized in this image reconstruction process; rather, the image is based entirely on spectral similarities. Spectral classes obtained from HCA images, and pathological diagnoses of the cluster-based regions, have been used to train diagnostic algorithms for the automatic diagnosis of tissue sections.

Of particular interest was the question of whether adeno- and squamous cell carcinomas of the same organ could be differentiated by SHP. To this end, a TMA (see Section 13.3.3) that contained both cervical squamous cell- and ADC tissue cores was analyzed. In this study, a data set of 28 cores (14 ADCs and 14 squamous cell carcinomas) was created, as described earlier. As each tissue spot produced on the order of 100,000 spectra, the resulting data set was statistically significant, in terms of both patient and spectral numbers. Annotation

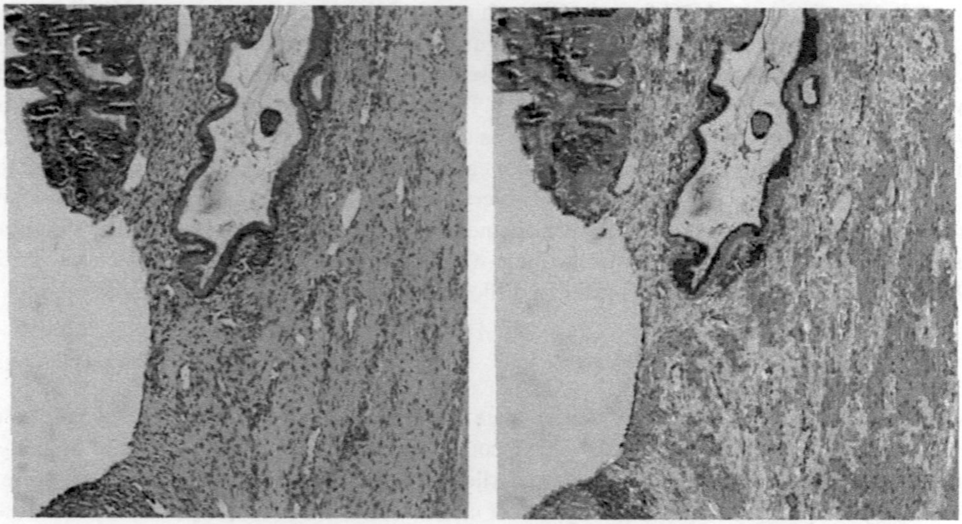

Figure 13.8 (a) Photomicrograph of an H&E-stained section of a cervical adenocarcinoma biopsy. (b) Overlay of HCA image on the H&E image. See text for details. Diem, et al., 2012 [37]. (See insert for color representation of this figure.)

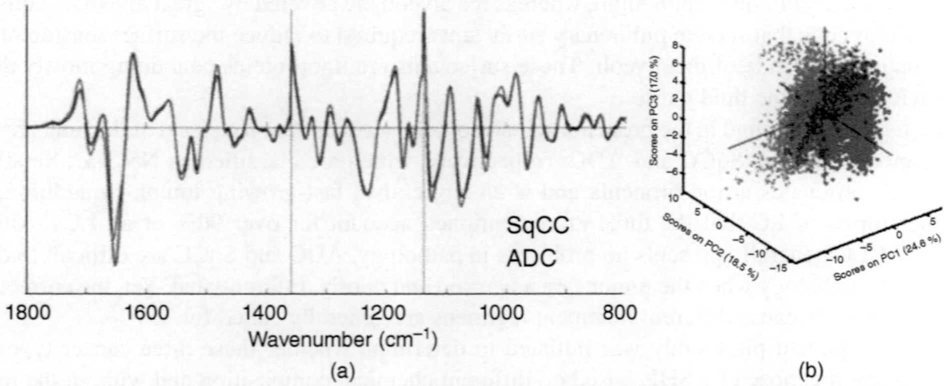

Figure 13.9 *(a) Comparison of mean second derivative spectra of cervical adenocarcinoma (green) and cervical squamous cell carcinoma (red). Note the normalization of the spectra in three separate regions (see text). (b) PCA scores plot of the separation of ADC and SqCC.* (See insert for color representation of this figure.)

was performed as described in Section 13.3. The resulting data set was analyzed using both unsupervised and supervised methods.

Figure 13.9 demonstrates that principal component analysis (PCA) of a data set containing thousands of ADC and SqCC spectra reveals a clear separation of the two cancerous classes, although the mean spectra are, at first sight, quite similar. When individual spectra show good separation as indicated in Figure 13.9(b), supervised algorithms generally can be trained quite readily to perform spectral diagnosis. Here, an ANN was trained that discriminated ADC from SqCC with a sensitivity of 98% and a specificity of 98%. Thus, it appears that cervical ADC and SqCC can be distinguished with very good accuracy by SHP. Note that the second derivative spectra shown in Figure 13.9 are separately vector normalized in three different regions (1800 to 1380, 1380 to 1220, and 1280 to 800 cm^{-1}). This procedure was shown to enhance the sensitivity of HCA by weighting the protein spectral region less, as compared to the low-frequency regions.

The importance of cancer classification into major cancer types may not be all that important in the case of cervical cancer, as the incidence of cervical ADC is very low. However, in other organs, ADC and SqCC occur with similar rates of incidence, and their proper diagnosis often is difficult, yet enormously important for the proper treatment. This is particularly so for the lung, where both ADC and SqCC – collectively referred to as nonsmall cell lung cancers (NSCLC) – present a difficult diagnostic problem, to be discussed next.

13.4.2 Lung cancer

In this section, results are presented that demonstrate that SHP can reliably differentiate cancers occurring in the same organ. Most SHP studies to date have concentrated on differentiating normal tissue types from cancerous tissue and inflammatory response, see earlier sections. The only exceptions to this were the early efforts to identify the primary tumors causing cancer metastases in the brain [38], in which spectral methods clearly could distinguish different cancer types and primary sites.

Here, a summary of recent work in lung cancer (LC) is presented. The aim of this study was to classify different cancer types, as well as their grade, by SHP. This information is of extreme importance in the care of a patient; thus, this is the first foray of SHP into more prognostic areas, which will help in therapeutic decisions and patient care.

LC originates in the epithelial tissue of the lung. The bronchia are covered by simple squamous epithelium and pseudo-stratified columnar epithelium, whereas the alveoli are covered by "great alveolar" cells, a special kind of glandular cells that secrete pulmonary surfactants required to reduce the surface tension of the aqueous layer coating the inside of the alveoli. These surfactants are lipoproteins containing mostly dipalmitoyl phosphatidylcholine as the lipid phase.

Given the tissue types found in the bronchia and alveoli, the most common cancers of the lung are squamous cell and adenocarcinomas, SqCC and ADC, respectively, which are classified as NSCLC. Small cell lung cancer (SCLC) originates at the bronchia and is an aggressive, fast-growing tumor. In addition, there are several other forms of LC, but the three aforementioned account for over 90% of all LCs. Although the diagnosis of SCLC generally presents no problems in pathology, ADC and SqCC are difficult to distinguish in classical histopathology when the tumors are advanced and poorly differentiated. Yet, the correct diagnosis is highly important, because different treatment regimens are generally called for.

In 2011, an 80-patient pilot study was initiated to determine whether these three cancer types could be differentiated and diagnosed by SHP, based on different chemical composition and without the use of morphological criteria or advanced methods such as IHC or gene array technology [39]. This study was recently expanded to include a much larger patient cohort and results from nonmalignant tumors. As this latter study is still ongoing, the following discussion is mostly based on the original 80-patient study.

In this pilot study, a TMA was used, which contained tissues from 10 subjects diagnosed as normal (NOR), 20 tissue spots from patients with ADCs, 10 bronchiolo-alveolar carcinomas (BACs, a precursor of ADC, now generally referred to as lepidic adenocarcinoma), 12 squamous cell carcinomas, 19 small cell carcinomas, and a few other types that were underrepresented and were not included in the study. A portion of the tissue cores, about 1.5 mm in diameter each, is shown in Figure 13.10 after staining. In the original study, a 1 mm × 1 mm area of the tissue core was selected for imaging, and yielded about 25,000 spectra each. The spectral data sets were preprocessed as explained in Chapter 12, and presorted *via* HCA. Subsequently, digital images of the H&E-stained tissue were taken and annotated by a lung pathologist to correlate visual tissue features (disease)

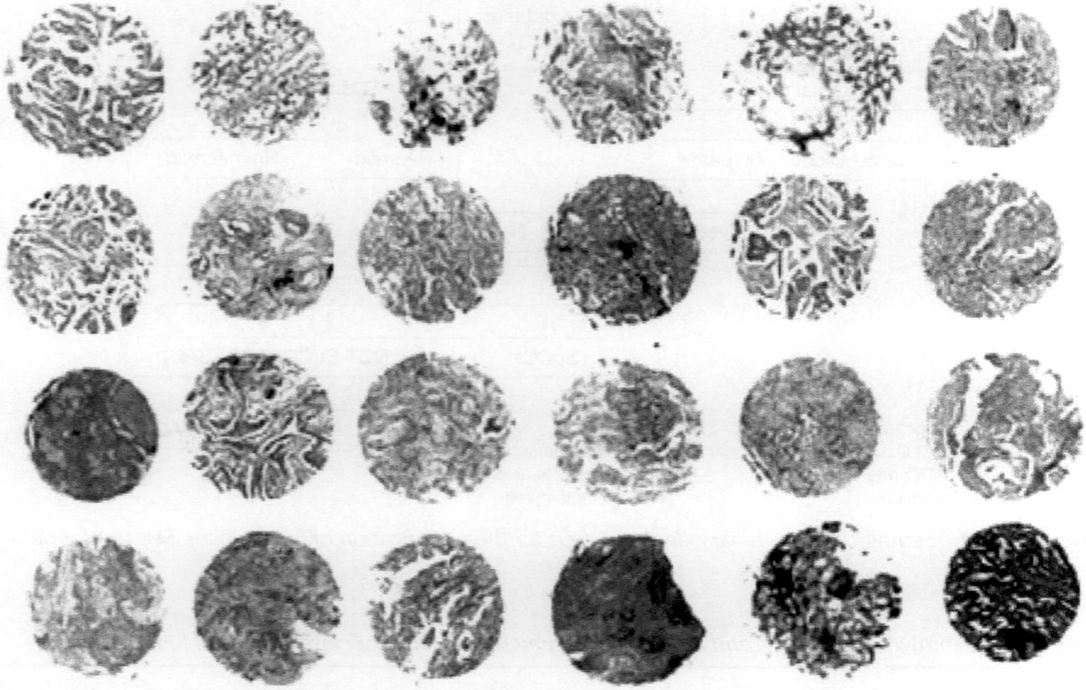

Figure 13.10 *View of a section of a stained tissue microarray (TMA). Bird,* et al.*, 2012 [15]. Reproduced with permission from Nature*

and cluster memberships. This resulted in a data set of about 106,000 spectra for which a detailed pathological diagnosis was available. Details of this study can be found in Ref. [15].

This data set was subsequently split, by tissue spot (patient), into a training set and a test set. The training set was used to train four consecutive (hierarchical) ANNs that were applied to the tissue spots of the test set to render a diagnosis. The ANNs were trained to classify, in a binary decision tree, the following classes: NORMAL *vs.* NOT NORMAL, SCLC *vs.* NOT SCLC, SqCC *vs.* NOT SqCC, and ADC *vs.* BAC, as shown in the flow diagram depicted in Figure 13.11. The order of the individual decision steps was determined from the dendrogram obtained by hierarchical clustering of the mean class spectra; this dendrogram indicated the "relatedness" of spectra and indicated which classification was based on the largest variance. In the subsequent study involving the much larger data set, the steps marked as "not yet attempted" in Figure 13.11 were carried out very successfully as well.

The top-level ANN (ANN level 1, see Table 13.1) was trained to distinguish NORMAL from NOT NORMAL spectra (equivalent to CANCER *vs.* NOT CANCER). To this end, the training set for NORMAL tissue included various noncancerous tissue types, for example, normal fibro-connective tissue spectra from cancerous tissue spots, as well as several other normal tissue features (endothelium and connective tissue) from normal tissue spots. The NOT NORMAL spectra were randomly selected from the patient-separated tissue spots representing the four cancer types. A total of 1848 NOT NORMAL spectra and 1840 NORMAL spectra were used in the training set, where the number of spectra used was determined by the smallest number of patient spectra in one of the cancer data sets. The top-level ANN distinguished the NORMAL from the NOT NORMAL spectra with a pixel-level sensitivity of 99.3% and a specificity of 94.4%, when applied to the

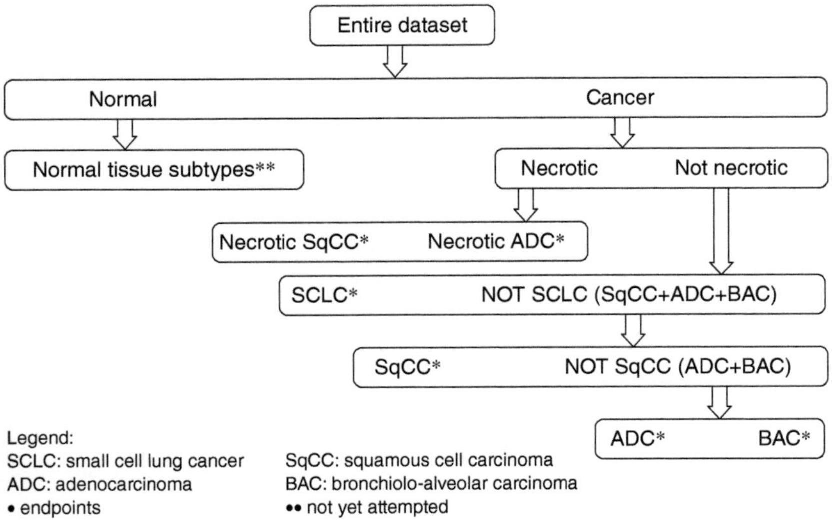

Figure 13.11 *Sequence of the binary decision steps for the classification of lung cancer. See text for details.*

Table 13.1 *Sensitivity, specificity, and accuracy of hierarchical ANNs for classification of lung cancers*

ANN level	Classification	Sensitivity (%)	Specificity (%)	Accuracy (%)
Level 1 ANN	NOR *vs.* NOT NORMAL	99.3	94.4	96.8
Level 2 ANN	SCLC *vs.* SCLC	91.2	98.0	94.6
Level 3 ANN	SqCC *vs.* NOT SqCC	90.4	95.0	92.7
Level 4 ANN	ADC *vs.* BAC	88.8	47.2	68.0

entire test set (see Table 13.1). This ANN used feature selection of the 65 most significant (second derivative) intensity points. The neural network topography (the number of hidden layers, nodes in the hidden layer, and the number of input features) affected the network performance only minimally; the values of sensitivity and specificity changed by less than 1%. Thus, the reported results are for ANN structures with one hidden layer that contained five nodes.

As there were insufficient spectra in the data set to train an ANN for necrosis, the necrotic spectra were removed from the data sets, although in the subsequent study, necrosis could be classified with excellent accuracy. The effect of necrosis on the observed vibrational spectra was first reported by Jamin *et al.* [40]; in these spectra, a strong shoulder of the amide I peak at about 1630 cm^{-1} was reported. This shoulder appears in the second derivative spectra as a distinct peak. The large spectral changes observed for necrosis indicate major changes in the protein composition of necrotic cells, as the 1630 cm^{-1} peak is associated with unfolded and precipitated proteins [15, 40].

The remaining data set was subject to a second-level ANN (level 2 ANN, see Table 13.1) to distinguish between SCLC and NOT SCLC, which includes SqCC, ADC, and BAC. To this end, 5280 spectra from the SCLC training set were selected randomly. The NOT SCLC training set consisted of 1760 spectra each, selected randomly from the three cancer classes, SqCC, ADC, and BAC. This second-level ANN distinguished

SCLC from NOT SCLC spectra from the test set (again at an individual pixel-level) with a sensitivity of 91.2% and a specificity of 98.0%. The neural network structure was the same as discussed earlier.

A third-level ANN (level 3 ANN, see Table 13.1) in the binary decision tree was trained for the distinction of SqCC from NOT SqCC (ADC and BAC). This distinction is clinically highly significant, because treatment options are quite different between adeno- and squamous cell carcinomas. Training of this algorithm was accomplished using 6640 SqCC spectra from the training set and 6640 NOT SqCC (ADC and BAC) spectra. For this ANN, a sensitivity of 90.4% and a specificity of 95.0% were obtained. Finally, a fourth-level ANN was trained to differentiate ADC from BAC. As BAC originates in the alveolar lining of lung tissue, it is considered a precursor of ADC. The pathological distinction between the two is based mainly on whether or not the neoplasm has penetrated through the basement membrane (ADC); if not, the disease is classified as BAC. From the viewpoint of the biochemical composition of the cancer cells, the two diseases differ mostly in the stage of disease; thus, SHP was not able to distinguish these two states reliably (with an acceptable sensitivity of 88.8%, but a low specificity of 47.2%). Larger data sets and even more careful annotation may increase the accuracy of this diagnosis as well.

To test how well the diagnostic algorithms discussed in the previous section performed on data sets that were not annotated at the pixel-level, entire tissue spots from the test set were subjected to ANN analysis. This was accomplished as follows. The raw data set from entire spots was preprocessed as discussed earlier. However, rather than presegmenting the data set by HCA, as was done for the annotation and pixel-based diagnostic tests, all pixel spectra of each spot were analyzed directly by the trained ANNs, and the pixel-level output of the ANN was converted into a graphical binary format. In this format, each pixel spectrum analyzed by the various level ANNs can have a binary output, "YES" coded in red or "NO" coded in green. Results were obtained for all spots in the test set; the results for several spots from TMA are presented as follows. These include spot H9 (NOR), E8 (SCLC), D2 (SqCC), A1 (ADC), and E7 (BAC).

Depending on the ANN, the red and green areas can have different diagnostic meanings, as shown in Figure 13.12. When the "Level 1 ANN" algorithm (CANCER *vs.* NOT CANCER) was applied to tissue spot H9, diagnosed as "normal" by pathology (a), the green areas imply "NO" or NOT CANCER, which is the correct diagnosis for this tissue spot. When the same Level 1 ANN was applied to spot E8 from the SCLC test database (b), the answer was "YES," or positive for cancer. When the same spot was analyzed by the level 2 ANN, SCLC *vs.* NOT SCLC, the result again was YES (positive for SCLC) as indicated by the red display in (c). The third row depicts applications of the level 1 ANN, level 2 ANN, and level 3 ANN (left to right) to tissue spot D2. Application of the level 1 ANN reported the fibro-connective tissue areas as "NO" (NOT CANCER) in green, and the cancerous regions in red (d). The level 2 ANN subsequently analyzed the CANCER areas, but determined that they were NOT SCLC, hence they were displayed in green (e). Level 3 ANN analyzed the cancerous regions, and found them to be positive (red) for SqCC. The fourth row in Figure 13.12 depicts results from the analysis of tissue spot A1, which was diagnosed as ADC by classical methods. The level 1 ANN properly defined most areas as cancerous (red) with just a few normal regions. Application of the level 2 ANN and the level 3 ANN both revealed negative results (green) when analyzing the cancerous regions, because the cancer was NOT SCLC and NOT SqCC. However, level 4 ANN recognized the cancer as ADC. Note that the areas shown in green in (g) appear blank in the output of the level 4 ANN, because the algorithm properly detects them as "NOT ADC." Finally, in row 5 of Figure 13.12, results are shown for spot E7 diagnosed as BAC. The majority of the spot was identified as cancerous except a region of blood vessel with red blood cells within its lumen in the upper left quadrant that were clearly diagnosed as normal by "level 1 ANN." Level 2 ANN, level 3 ANN, and level 4 ANN found "NOT SCLC," "NOT SqCC," and "NOT ADC," respectively, which were the correct diagnoses for a tissue spot diagnosed with BAC. In summary, Figure 13.12 demonstrates that a decision tree of hierarchical, binary ANNs can be used to analyze for the presence of various cancers. The consecutive applications of these algorithms requires less than 1 min once the training of the algorithms is accomplished.

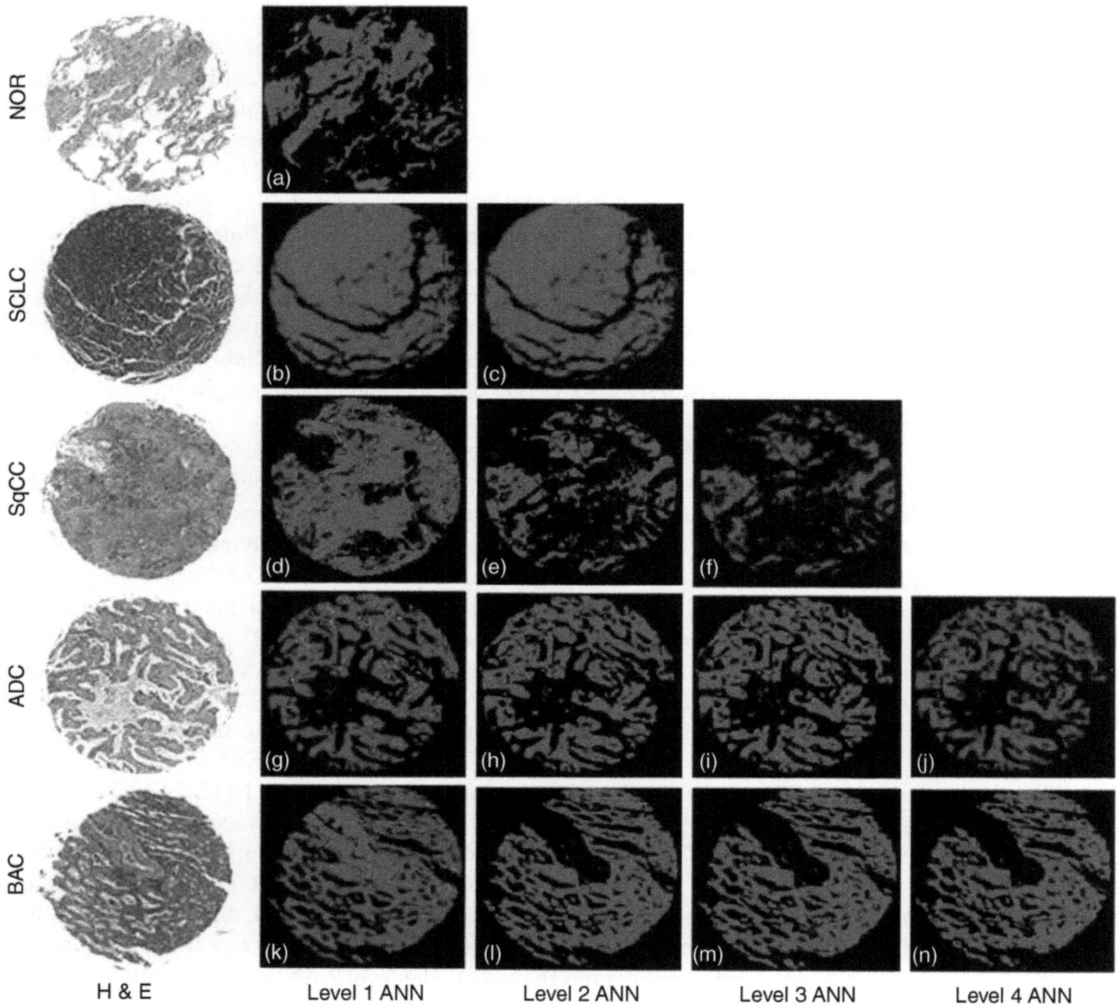

Figure 13.12 *Results of analysis of five different tissue spots by trained hierarchical ANNs. Left column: H&E images, second through fifth columns: successive ANN results, as indicated. See text for details. Bird,* et al., *2012 [15]. Reproduced with permission from Nature (*See insert for color representation of this figure.*)*

A heat-map was also generated [15] that demonstrated that the ANNs actually used real spectral features for the discrimination, and not some random features of the data set. This has been further confirmed in the aforementioned studies using a much larger patient data set. This study is also carried out in a multicenter arrangement to establish instrument and laboratory independence.

In the aforementioned extension of this study, the subclassification of ADC into the major subclasses (such as lepidic, acinar, papillary, micropapillary, and solid) was pursued as well. Results from this study that used more refined data collection and analysis protocols and SVMs for classifiers will to be published at a later date. This study indicated that the discrimination of cancer types and subtypes can be carried out very well by SHP. At this point, it appears that the availability of samples and their accurate annotation define the limits of SHP, rather than the inherent spectral ability to differentiate sample.

13.4.3 Prostate cancer

One of the first cancers studied in detail by SHP was prostate cancer, a frequent ADC that afflicts about 200,000 men/year in the United States. The pathological distinction between the usual slow-growing forms of prostate cancer from aggressive forms is difficult, and is based on the assignment of a Gleason score by a pathologist. Primary screening is by the prostate-specific antigen (PSA) that detects a blood-borne protein cancer marker. Thus, it is no surprise that the prostate was singled out by researchers at the National Institutes of Health (USA) as the subject of one of the largest studies to date of a given cancer by SHP [29]. This study also introduced the use of TMAs (see Sections 13.4.1 and 13.4.2) in SHP. In a subsequent paper by the same group [41], many of the statistical measures commonly used in biomedical studies, such as the receiver–operator curve (ROC), area under the receiver–operator curve (AUROC), and others, were used, for the first time, in the analysis of IR spectral tissue data.

These studied were based on 171,000 pixel spectra from 16 patients and 10 histological classes. Data were acquired using the same spectrometers and methodology described in the previous two sections; however, the SLA was not trained on spectra extracted from annotated regions as suggested by imaging methods such as HCA but on metrics such as intensities or intensity ratios of certain bands that were assigned by eye. Thus, this study did not use the same level of unsupervised data segmentation that was used in the author's laboratory.

Nevertheless, this study produced an impressive classification of the normal histological classes, with an overall accuracy of better than 96%. The reported accuracy of detecting prostate ADC, however, varied widely, probably because not all spectral features were used to construct the metrics for the diagnostic algorithms.

More recently, the group of Gardner (Manchester, UK) has imaged and analyzed large sections of prostate tissue *via* SHP, and also individual, cultured prostate cells (see Section 13.8.6). In their most recent study [42] on large areas prostate tissue sections, they used paraffinized tissue samples, a method that was first reported by the Reims group of Manfait [24]. The presence of paraffin matches the refractive index between tissue and tissue voids and, thereby, nearly eliminates the R-Mie effect, although standard Mie scattering still is observed. This paper also reports an enormous increase in data acquisition speed, reading out four complete, 128×128 pixel tiles a minute, each covering a field of view of about 0.7 mm on edge. A 1-cm^2 tissue section, under these conditions, yields about 3.2 million spectra. Such a data set, shown in Figure 13.13, cannot be analyzed by methods such as HCA any longer (the correlation matrix alone would occupy about 45 TB). Thus, the authors used a few metrics to abstract the spectra into a few parameters and trained an RF algorithm

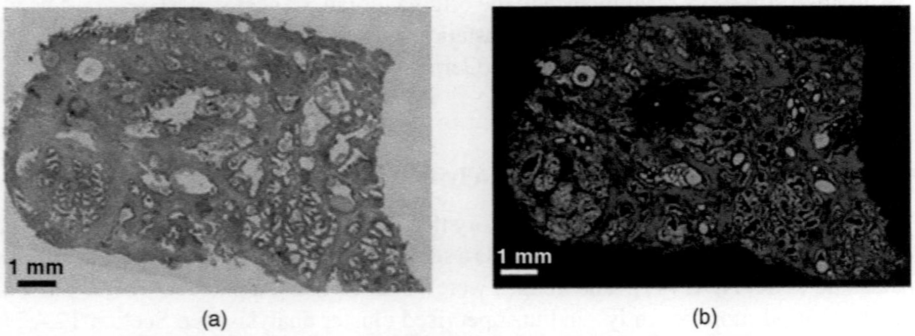

Figure 13.13 *(a) Photomicrograph of an H&E-stained large prostate biopsy section and (b) SHP-based pseudo-color image of same section, embedded in paraffin. The color scheme is as follows: epithelium: green, smooth muscle: purple, blood: red, extracellular matrix: pale yellow, fibrous stroma: pink, concretion: orange, and lymphocytes: blue. Bassan,* et al., *2014 [42]. Reproduced with permission of SPIE.* (See insert for color representation of this figure.)

(see Chapter 12) to analyze the data. The results of these studies were reported in the form of a "digitally stained" or "SHP-stained" image as shown in Figure 13.13. These images demonstrate that SHP, using only a few parameters extracted from the spectra, can create SHP-based pseudo-color images that distinguish the major, normal tissue classes. However, it remains to be seen whether or not this approach can classify and subclassify different cancer types, as shown earlier for lung tissue.

13.4.4 Breast cancer

In spite of the high incidence of BC, this disease has not yet been studied *via* SHP at the same level of detail than other cancers discussed earlier. This is, in part, because BC samples are difficult to study *via* SHP owing to the high fat content of breast tissue. Early efforts at the LSpD were inconclusive for large, flash-frozen tissue sections, but connective and fat tissue could readily be distinguished from the glandular structures in the lobules. More recently, Bhargava's group reported results from a multipatient, TMA-based study of BC tissue [20, 43]. One of these studies concentrated on rapid data acquisition and noise reduction *via* noise-adjusted principal component analysis (NA-PCA) (see Section 12.4.1) [43] and provided classification results for the distinction between stroma and epithelial tissue. The main argument raised in this paper was the need of methods to speed up data acquisition is well taken, and needs to be addressed before SHP can be used routinely in pathological laboratories.

More recently, the group of Goormaghtigh (Brussels) has analyzed the extracellular matrix surrounding ductal carcinomas [44]. Their findings mirror recent results from the author's laboratory and other groups [45] that normal tissue that is proximal or distal to cancerous tissue show slightly different spectral signatures. This was first observed by the group of Miller [46], who attributed these changes to the presence of certain metallo-proteinases that are known to occur around tumors and weaken the tissue structure of the tumor surroundings to facilitate angiogenesis around the tumor.

13.5 Application of SHP toward the detection and classification of metastatic tumors

Metastatic tumors can form when cancerous cells from a primary tumor site migrate, *via* the blood stream or the lymphatic system, to other parts of the body where they form new foci of the original primary tumor. The metastases still contain the genetic features of the primary organ from which they originate; thus, the metastatic tissue often still bears resemblance to the original organ. Cancer metastases have been studied by a number of research groups *via* SHP. Here, two metastatic cancers studied by the author's group are introduced, followed by a review of the work by the Dresden (Germany) group on identifying the nature of the primary tumor in brain cancer metastases.

13.5.1 Detection of colon cancer metastases in lymph nodes

One of the earliest results from the author's laboratory that demonstrated the detection of metastatic cancer in lymph nodes is shown in Figure 13.14, which depicts a section from an excised lymph node with a well-defined region of metastatic colon ADC [47]. The images were obtained using a pixel size of 25 μm × 25 μm (larger than what has been used more recently) and unsupervised cluster analysis (see Section 12.4.2) to construct the pseudo-color images. Thus, the images shown in Figure 13.14 are not based on any diagnostic input, but strictly use the similarity of the IR spectra in the data set to distinguish tissue types.

Figure 13.14(a) and (b) depicts an H&E-stained photomicrograph of an entire lymph node section, measuring about 8 mm × 8 mm in size, and the corresponding SHP pseudo-color image obtained by HCA, respectively. In Panel (b), the tumor (dark red) is differentiated well from the medullary sinus (light blue), the

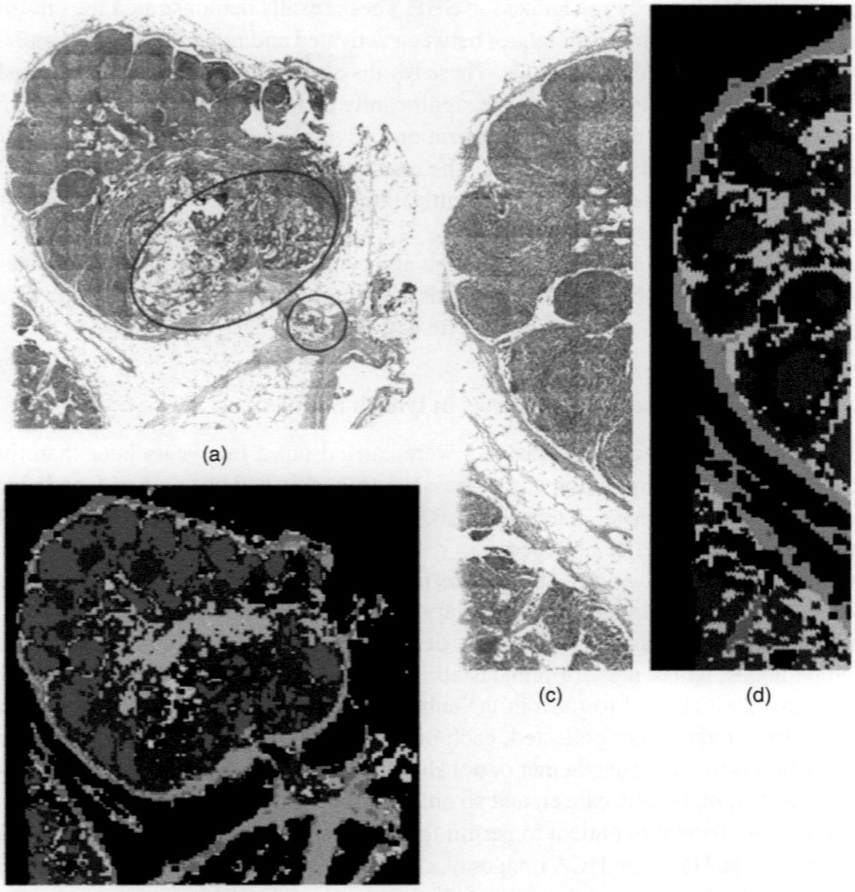

Figure 13.14 *(a) Photomicrograph of an 8 mm × 8 mm lymph node tissue section with a colon cancer metastasis (in blue ellipse and circle). (b) HCA-based pseudo-color image of section shown in (a) depicting different tissue types, including the adenocarcinoma (dark red). (c) Photomicrograph of partial section of same lymph node, showing reactive lymph node follicles. (d) HCA-based pseudo-color image of the section shown in (s); demonstrating the distinction between activated (dark brown) and nonactivated (dark blue) B-lymphocytes (dark blue) by SHP. Romeo, et al., 2005 [47].* (See insert for color representation of this figure.)

germinal centers (red), the T-lymphocytes (dark blue), and the capsule (yellow and blue-gray). Here, there is no pathological ambiguity in the diagnoses of cancer and normal tissue, and the IR-SHP results very well agree with the visual image. Panels (c) and (d) present enlarged sections of the same lymph node, containing only normal tissue. In Panel (d), the germinal centers (dark brown), surrounded by the mantle zone (dark blue), are clearly visible, indicating, for the first time, the spectral differentiation of B-lymphocyte activation: the germinal centers in a lymph node are activated by an antigen, here most likely originating from the cancer in the adjacent tissue. This B-lymphocyte activation is accompanied by high proliferation rates, which change the chemical composition observable in SHP. In Panel (d), the capsule of the lymph node (orange region) is also recognizable; this region consists of fibro-connective tissue that has a significantly different chemical composition that is easily recognized in SHP. These results demonstrated the enormous sensitivity of IR-SHP, because the biochemical differences between activated and nonactivated B-lymphocytes are quite subtle, yet are recognized by spectral methods. These results also demonstrated unambiguously that a region of ADC has biochemical characteristics that are significantly different from those of B- and T-lymphocytes and normal structures in lymph nodes. Characterization of lymphocyte activation also was the subject of a later combined Raman and IR-SHP studies [48]. The results presented here were analyzed before the R-Mie effect (see Section 12.3) was understood; consequently, the cluster analyses shown in Panels (b) and (d) were carried out omitting the amide I spectral range.

Nevertheless, this study demonstrated not only the ability of SHP to detect and delineate cancer in normal tissues but also to distinguish different normal tissue types (capsule, sinuses, and germinal centers) and describe subtle biological effects such as lymphocyte activation.

13.5.2 Detection of breast cancer metastases in lymph nodes

The efforts to detect and localize BC metastases were carried out a few years later than the colon cancer efforts; consequently, the data acquisition and processing procedure had been refined, and algorithms existed to correct the data for resonance Mie scattering. Also, the number of samples studied was much larger so that statically significant results could be obtained.

In this section, results of IR imaging studies are presented that used both supervised and unsupervised methods to detect BC micrometastases in the axillary sentinel lymph node [23, 49–52]. This is a significant medical problem because treatment of BC patients depends on the presence or absence of cancerous cells in the sentinel lymph nodes, where they form metastatic tumors. Metastases less than 2 mm in size are referred to as micrometastases, which tend to form in the subcapsular sinuses of the lymph nodes. In this study, 30 1 mm × 1 mm spectral images were collected, each one consisting of 25,600 spectra. The aim of this study was to use large data sets to establish whether or not HCA could reliably segment the raw data sets into spectra of capsule, lymphocytes, metastatic cancer, and so on, and whether the spectra of the metastatic cancers were sufficiently similar from patient to patient to permit their detection by a trained, diagnostic algorithm.

Figure 13.15 depicts the H&E and HCA images of a lymph node section that harbors a BC micrometastasis, about 0.2 mm in size. At the pixel resolution of 6.25 μm on edge, the detection of the small metastatic area poses no problem for SHP; however, the spectra from different micrometastases proved to be quite different, in particular, in the amide I spectral region, but in the lower wavenumber region as well. The author's laboratory demonstrated that the spectral differences were not due to different chemical composition but rather due to the R-Mie effect (Chapter 12) by comparing spectra collected from five different micrometastases from tissue sections still embedded in paraffin [23]. In this case, nearly identical spectra were observed for the metastatic regions. This is due to the better matching of the refractive index of areas occupied by tissue and areas devoid of sample; index matching enormously reduces the incidence of scattering effects that mix absorptive and reflective line shapes as discussed before. This emphasized the need to correct for R-Mie scattering that was performed, at this point, by an earlier algorithm from the author's laboratory [51] that was similar to the Bassan method [4] discussed earlier. After correction for these effects, the spectra of micrometastases were found to be sufficiently similar that diagnoses with trained algorithms were possible (see Figure 13.16).

Although the detection and diagnosis of large metastases, such as the one shown in Figure 13.14, in lymph nodes does not present any difficulty for a pathologist, the detection of micrometastases or small clusters of cancerous cells in lymph nodes *via* standard histopathology is a formidable task owing to the small size and the lack of distinguishing features of the abnormality within the lymph node tissue. Yet, the detection of these micrometastases is of prime importance to stage the spread of disease: if a lymph node is found to be free of metastatic cells, it is thought that the spread of cancer has been contained. On the other hand, a false-negative

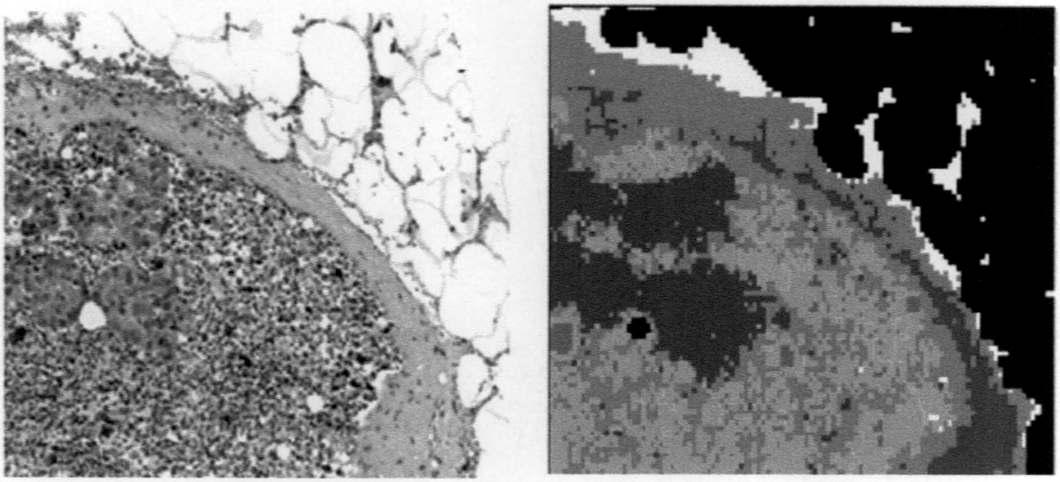

(a) (b)

Figure 13.15 *(a) Photomicrograph of a lymph node tissue section, measuring about 0.8 mm on edge. (b) HCA-based pseudo-color image of section shown in (a). The capsule (green, blue, and purple), the micrometastasis (dark blue), and the lymphocytes (gray, green, and salmon) regions are clearly distinguished. Bird,* et al., *2009 [50]. Reproduced with permission from Wiley-VCH. (*See insert for color representation of this figure.*)*

diagnosis, that is, a missed micrometastasis in a lymph node, presents too optimistic a diagnosis, and a more aggressive treatment should have been indicated.

In standard histopathology, excised lymph node tissue, typically from the sentinel lymph node (the lymph node closest to the tumor), is sectioned, stained, and analyzed by visual inspection under a microscope. If micrometastases are suspected, immunohistochemical agents/counter stains such as cytokeratin-specific stains [53] may be used to enhance the sensitivity of classical histopathology. However, this procedure is time consuming and costly and still relies on the human eye for detection. The problem is further complicated that many sections need to be diagnosed because the volume of a micrometastasis is small compared to the volume of a lymph node, and could easily be missed if just one section of a lymph node was analyzed.

In the study reported here, 30 1 mm × 1 mm areas of lymph node tissue adjacent to the capsule [54–56] were studied. Each lymph node in this study was excised during cancer surgery and was known to harbor micrometastases. However, not all sections actually contained micrometastases, because the sectioning process easily may miss small cancerous regions. SHP detected micrometastases in 5 of the total of 30 tissue sections, and all metastatic regions were confirmed by visual pathology. In addition, a micrometastasis was found in a section that was originally missed by classical histopathology, but later confirmed by this method. These results pointed to the feasibility of using spectral methods for routine pathological tasks, and to the fact that the sensitivity of an instrument and computer-based method can be higher than that of classical methodology. Results for three of the five tissue sections with micrometastases are shown in Figure 13.17.

13.5.3 Detection and classification of brain metastases

In two pioneering papers, Krafft *et al.* [57, 58] reported that SHP can distinguish and classify metastatic cancer in the brain. The primary tumors for these metastatic cancers were renal cell carcinoma (RCC), BC, LC, and colorectal carcinoma (CC), and the number of cases were seven, three, three, and four, respectively.

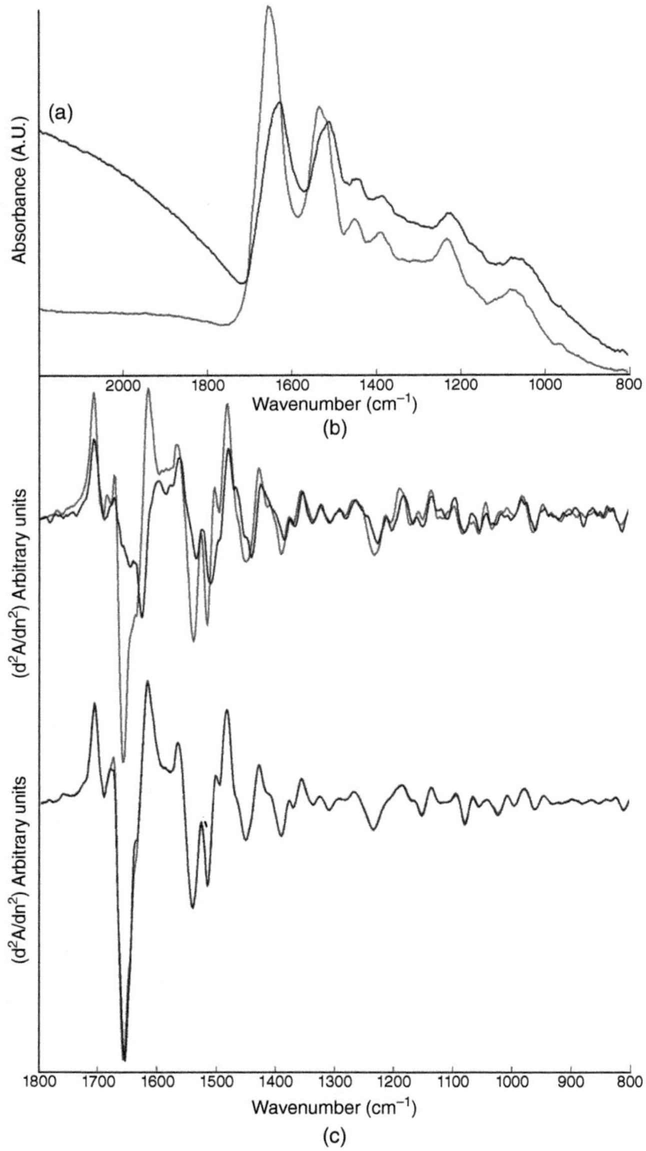

Figure 13.16 *(a) Absorption spectra of two different breast cancer micrometastases. (b) Corresponding second derivative spectra. (c) Spectra after R-Mie scattering correction. Bird* et al.*, 2008 [49]. (*See insert for color representation of this figure.*)*

Analyses were performed *via* k-means cluster analysis followed by soft independent modeling of class analogy (SIMCA) (see Chapter 12) analysis and linear discriminant analysis (LDA).

This analysis was carried out on flash-frozen and subsequently dried tissue sections. This study reported the classification, by primary tumor organ, with good accuracy, although the number of incorrect classification was high in 2 of the 17 patients. Yet, the importance of this work cannot be underemphasized, as it

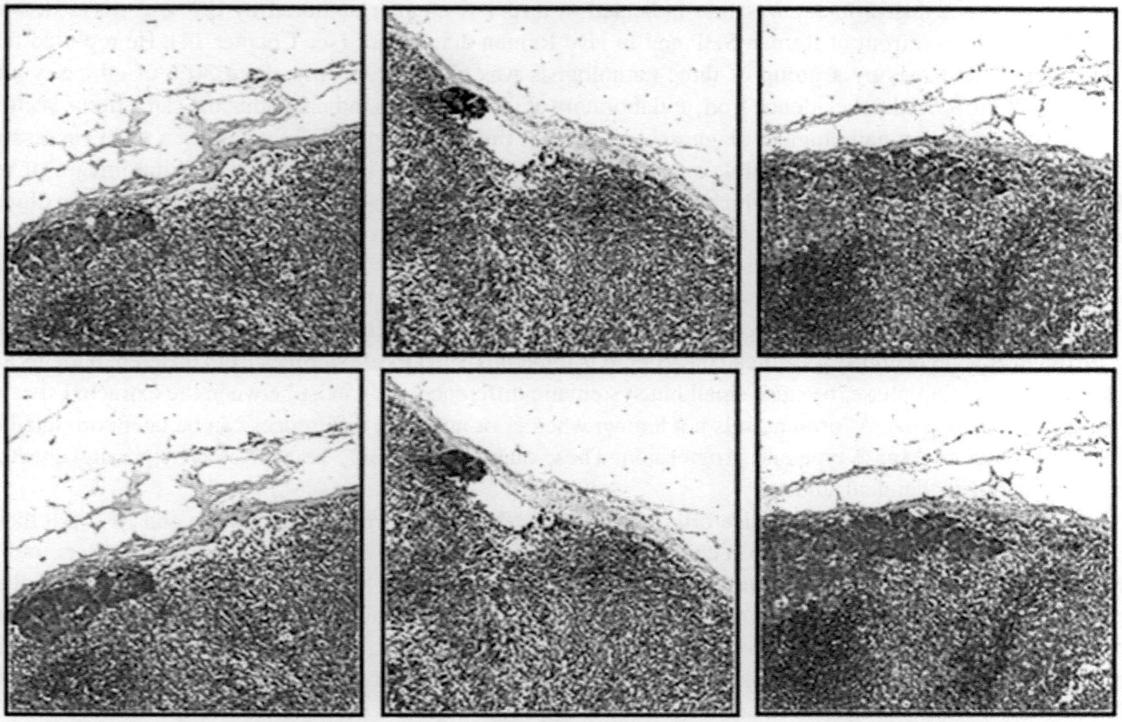

Figure 13.17 *Top row: photomicrographs of H&E-stained, 1 mm × 1 mm section of axillary lymph nodes harboring micrometastases. Bottom row: overlay of HCA-derived regions indicating tissue abnormalities. Bird* et al.*, 2009 [50]. Reproduced with permission from Wiley-VCH.* (See insert for color representation of this figure.)

demonstrated that SHP distinguishes not only healthy tissue types (gray and white brain matter), but ADCs from different primary tumors that have metastasized to the brain. This implies that there are subtle biochemical compositional changes that occur reproducibly between the same cancer types, namely ADC. When a metastatic tumor in the brain is found without any knowledge of the primary tumor, it is very difficult to determine the primary tumor only with classical histopathology. Thus, the fingerprint specificity of IR spectral methods offers a diagnostic advantage that few other techniques can match. This suggests that SHP may detect the presence or absence of distinct, tumor-related proteins (cancer markers) that are presently only detectable by immunohistochemistry.

13.6 Future prospects of SHP

The inherent sensitivity of vibrational spectral fingerprints toward changes in biochemical composition of tissue pixels makes SHP an ideal candidate for medical diagnostic imaging. With recent advances in data preprocessing, the increasing number of research groups involved in the field, and the generally good agreement between spectral and pathological results, it appears that SHP is poised to enter the mainstream diagnostic arena. The major obstacle to a broad application of SHP is, in the eyes of the authors, the difficulty in obtaining sufficient and reliable annotations to train diagnostic algorithms.

The severity of this problem was first indicated in a review chapter authored by Stone *et al.* [59], who has been on the forefront of Raman SHP and *in vivo* Raman diagnostics (see Chapter 14). He reported that a consensus diagnosis by a group of three pathologists was obtained in only about 30% of all cases presented to them. Anecdotal evidence from collaborators of the author has indicated that the same tissue section can produce different pathological diagnoses at different times. The approach that appears most successful in obtaining accurate and reproducible annotation involves the use of high-resolution digital images of the H&E-stained tissue sections onto which HCA images can be superimposed. This step requires that the digital H&E and the HCA images are exactly registered, and can be zoomed together. This approach permits the pathologist to annotate the spectral images based on single-cell features; that is, the pathologist can select the most typical features in a tissue section and correlate it to spectral features at the level of one or a few cells. Spectra from pathologist – annotated tissue areas are subsequently extracted from the data sets and used to train diagnostic algorithms. As indicated earlier, it is imperative that tissue samples from different patients are used in this training phase, because small but systematic differences may exist between the extracted spectra from different patients. At present, it is not known whether or not these differences can be later correlated to special aspects of disease type and progression. These questions can only be answered by parallel spectral and immunohistochemical studies.

The diagnostic algorithms of choice originally were ANNs, and more recently, SVMs and RFs. All these machine-learning algorithms (MLAs) can be used as binary (two-class) classifiers or multiclassifiers to differentiate more than two input classes. Recent studies comparing these algorithms have established that they perform at a similar level. The criticism for using these methods of analysis has often come from the corner of older, more conventional spectroscopists who never had to deal with a flood of spectral data measured in the millions, and buzzwords such as "overtraining" and the well-publicized failures of early applications of MLAs are repeatedly cited. These arguments can be countered as follows: First, there exist well-established rules in bioinformatics on the size of training and validation sets required to produce reliable algorithms; failure to adhere to these rules certainly may produce algorithms that are overtrained. Yet, any discriminatory algorithm, including the operation of the human brain, suffers from this shortcoming. In the latter case, the subjectivity of classical histopathology is certainly a manifestation of insufficient training.

Secondly, the well-published failures of MLAs of any kind can be traced back to insufficient training, such as omitting entire classes of possible inputs or conditions. Also, many tasks to which discriminate algorithms were applied, that is, the morphological discrimination of abnormal from normal exfoliated cells or for facial recognition, require the translation of certain features, such as nuclear-to-cytoplasm ratio of a cell, or the height *vs.* width ratio of a face, to be collected in metrics that are subsequently analyzed by the MLA. It is conceivable that these metrics lacked specificity for the task at hand and that the discrimination failed, not because of shortcomings of the discriminatory method, but a shortcoming of the input data. In SCP and SHP, the form of spectral results – one-dimensional vectors of intensity data at given wavenumber point – is ideal for an MLA and does not involve the constructing of metrics.

13.7 Infrared spectral cytopathology (SCP)

The beginning of this chapter discussed in some detail the applications of IR spectroscopic methods to study and diagnose human tissue. The same technology can also be applied to the study of individual cells, and this research area is henceforth referred to as spectral cytopathology (SCP). SCP bears the enormous advantage that cells are much more accessible than tissue section and offers an excellent prospect for cancer screening; however, cytological diagnoses, in general, are much more difficult and fraud with incorrect or ambiguous answers than pathology, as discussed in detail in the following paragraphs.

SCP also needs to be acknowledged as the subject that started the entire field of IR spectral diagnosis: IR spectroscopy was first applied to real biomedical problems in the 1980s at the Robert Koch Institute in Berlin by the group of Naumann [60–63] with the goal of identifying bacterial cells. This was the first successful foray into spectroscopy of living matter, in general, and cultured cells, in particular, and laid the ground work for all subsequent applications of IR and Raman spectroscopies to prokaryotic and eukaryotic organisms, tissues included. The early work by Naumann's group was driven mainly by aspects of developing a more rapid identification tool for bacterial cells. At that point, classical microbiological methods did exist for bacterial identification, but these methods often required many days of cell culturing before sufficient cell mass was available for typing. The newly developed spectral methods developed by Naumann and coworkers not only reduced the analysis time enormously but also defined many aspects of standardized data acquisition, data preprocessing, and data analysis that proved to be successful for the analysis of human cells and tissue as well [16, 32]. The original macroscopic data acquisition still required a few days of bacterial cultures to be carried out, but subsequent transition to IR microscopic methods reduced the time between harvesting a bacterial sample and positive identification to about a work day.

Starting in the last decade, efforts to reduce the required sample amount even further were started by several groups using Raman microscopy (see Chapter 14). Owing to the finer focus of the exciting light, it is possible to analyze and characterize a single bacterium, thus eliminating the bacterial culture altogether. Specialized instrumentation for the identification of multidrug-resistant strains of bacteria has been developed commercially [64], and methods to analyze bacterial cells in aqueous media [65], and by trapping the cells in an electric field, have been described. These enormously promising technical advances will be discussed in Chapter 14.

The following discussion on SCP starts with work performed at the LSpD on exfoliated cells. Work on exfoliated cells is nearly exclusive performed at the LSpD, with the only other papers in this field from the early studies of the McNaughton group at Monash University in Australia [66], and the early work by Cohenford and Rigas [67]. At the LSpD, there are now data sets comprising hundred thousands of individual cell spectra; this work was enabled by a method developed there for rapid acquisition of spectral data from individual exfoliated cells (the PapMap^{P} method).

However, SCP on cultured cells has been carried out at a number of laboratories worldwide. The author's laboratory, then still at the City University of New York, was the first to systematically investigate the properties of cultured cells, in particular spectral changes when a cell proceeds through the cell cycle [68, 69]. Similar studies now have been performed by a number of research groups, and papers investigating spectra differences between different cell types have been published. These results are reviewed in Section 13.9.

13.7.1 Classical cytopathology

The diagnostic approach described in the beginning of this chapter, histopathology, relies on sections of tissue obtained from biopsies. Unless the biopsy was taken from the skin, the collection of a biopsy requires a surgical procedure to remove the tissue material from the body. In the case of internal organs, these surgical procedures can be quite invasive, disruptive, and/or dangerous to the patient. Thus, a diagnostic branch of medicine seeks to reduce the risks and inconveniences of biopsy-based diagnoses by analyzing individual cells that can be obtained in a much less-disruptive way. This approach uses exfoliated or otherwise collected human cells, rather than tissue section, for diagnosis, and this approach is referred to as cytology or cytopathology.

Epithelial cells can readily be harvested from many internal organs by minimally invasive methods such as exfoliation with a cytobrush (see the following paragraph) for cervical, oral and nasopharyngeal, esophageal, and other applications. Cells can also be collected from body fluids (urine and sputum) or from lavages, or *via* thin needle aspiration (lung, thyroid, and pancreas). The resulting cell samples are generally cleaned by

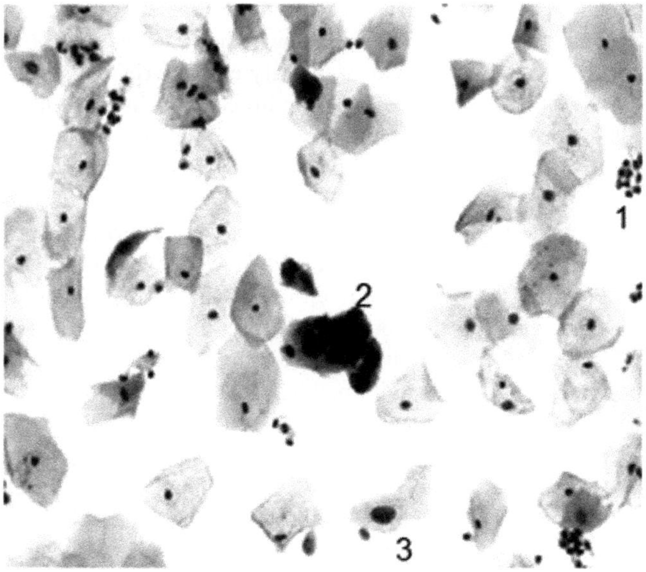

Figure 13.18 *Stained cytopathological sample of cervical epithelial cells. (1) Bacterial contamination, (2) clump of inflammatory cells, and (3) abnormal cell with enlarged nucleus. Schubert* et al.*, 2010, [76]. (*See insert for color representation of this figure.*)*

washing, pelleting (by centrifugation), and are subsequently deposited onto microscope slides, stained and finally, visually inspected under a microscope.

A typical cell sample collected by exfoliating with a cytobrush from the cervix or oral cavity may contain millions of cells. However, such an exfoliate is a mixture of many cell types, among them bacteria, the desired epithelial cells, cells of the immune system (leukocytes and lymphocytes), red blood cells (erythrocytes), and cell debris. A relatively clean cell sample, produced by a liquid-based deposition method, and subsequent staining *via* Pap stain [70] is shown in Figure 13.18. This figure shows about 50 stained cervical squamous cells that stain either pale blue or pale pink; in addition, this sample contains bacteria (1), polymorphonuclear leukocytes (PMNs, 2), cellular debris such as naked nuclei, and one "abnormal" cell (3), indicated by an enlarged nucleus. The difficulty in classical (visual) cytopathology is the detection of as few as a few percent of abnormal cells in a sample that may contain 1000–10,000 cells. Furthermore, an enlarged nucleus *per se* is also observed for cells from the lower layers of the cervical epithelium; thus further criteria such as the morphology of the nuclear membrane need to be invoked for a reliable discrimination of normal from abnormal cells (see Table 13.2). The level of abnormality also needs to be graded; for cervical cytology, the grades (in order of increasing severity) include reactive, low-grade dysplasia (low-grade squamous intra-epithelial lesion, LSIL), high-grade dysplasia (high-grade squamous intra-epithelial lesion, HSIL), carcinoma-in-site (CIS), and invasive cancer.

Classical cytopathology has relatively poor sensitivity and specificity (as defined in Chapter 12) even in the applications where the technique is most developed, namely in the cervical cancer screening arena. Interestingly, one should not take the low sensitivity and specificity of the Pap test as a failure of the method; quite contrary, no single screening test has reduced the incidence of and morbidity from a given cancer as much as the Pap test. Yet, for a single reading of a classical "smear," the overall accuracy (average of sensitivity and specificity) was typically less than 70%. Improvements were achieved by producing better samples: rather

Table 13.2 *Sensitivity and specificity of cytological criteria for predicting diagnoses of small cell lung carcinoma*

Cytomorphologic criteria	Sensitivity (%)	Specificity (%)
Small cell carcinoma		
Scant delicate cytoplasm	93.8	100.0
Cells 1.5 times the size of lymphocytes	100.0	88.5
Basophilic cytoplasm	73.5	88.1
Crush artifact	90.9	81.5
Single file arrangements	100.0	64.8
Inconspicuous nucleoli	56.7	71.7
High nuclear/cytoplasmic ratios	41.7	100.0
Squamous carcinoma		
Spindled tumor cells	0	51.6
Keratinized cytoplasm	0	50.8
Hyperchromatic smooth nuclei	5.9	50.8
Orangeophilic cytoplasm	6.7	52.5
Hyperkeratosis	0	56.5
Ink dot or pyknotic chromatin	53.6	68.8
Pearl formation	0	58.9
Malignant single cells	36.4	52.4
Intercellular bridges	0	59.5
Adenocarcinoma		
Macronucleoli	0	42.3
Foamy cytoplasm	0	41.2
Multiple nucleoli	0	40.0
Irregular nuclear membranes	29	30.0
Tumor giant cells	7.1	53.2
Coarsely granular chromatin	20.7	48.9
Gland formation	0	57.7
Nuclear overlap in groups	35.6	54.8
Papillary structures	0	59.5
Indistinct cell borders	42.1	68.4
Three-dimensional groups	39.3	60.0
Intracytoplasmic mucin	0	60.0

than smearing the exfoliated cells on a microscope slide, followed by staining, liquid-based methods were devised [72] that produced sparse monolayers of cells, which, after staining, provided a much clearer picture of individual cells.

The main difficulty with cervical and other cytopathological tests is the fact that the observer is faced with thousands of cells with normal morphology, with a few percent of abnormal cells mixed within. An abnormal cell may have a different nucleus-to-cytoplasm ratio (see Figure 13.18) than a normal cell, or manifest different nuclear membrane morphology. The visual detection of a few cells with such subtle differences, in an ensemble of many cells with normal morphology, is a daunting task that is made even more difficult by the fact that other cell types with different nuclear sizes may occur naturally within the sample. Furthermore, as the diagnosis is based on human inspection, there is low interobserver reproducibility in the low-grade

(ASCUS, see Section 13.8.3) and the high-grade (HSIL) cases of cervical abnormalities [73]. In the case of other organs, the accuracy of cytology is even lower, as demonstrated in Table 13.2. Here, the sensitivity and specificity of some of the criteria listed is below 50%, although it is expected that the trained eye of a cytologist can achieve higher accuracy by mentally combining several of the criteria.

13.7.2 Spectral cytopathology

One of the early goals of researchers involved in SCP was the developments of methods to aid in the diagnosis of cervical cell smears [53–55, 67] used for screening for cervical cancer (the so-called Pap test [38, 56]). The reason for selecting this subject was the high rate of false-positive and false-negative cytological results discussed earlier, and it was hoped that objective, spectral methods could improve upon the classical methods. In hindsight, SCP was probably not the best way to establish a novel technology, as will become clear in the remainder of this chapter. However, with diligent work by many research groups, it is now clear that SCP has a bright and hopeful future.

13.7.3 Methods for SCP

Here, the methodology utilized at the author's laboratory, the LSpD, is described. Other laboratories have used different approaches, particularly the laboratories using synchrotron-based methodologies. At the author's laboratory, an imaging approach was developed (the so-called PapMap^{P} method), in which the entire sample area was imaged, even areas between cells and not occupied by cells. These areas were subsequently used to construct a noise database to reduce the noise contributions in the pixel spectra from cells. Cellular spectra were subsequently reconstructed from individual pixel spectra of areas occupied by a cell. Typically, between 15 and 50 pixel spectra were used to construct one cellular spectrum. This approach permitted the collection of about one million of cellular spectra between 2010 and 2014.

13.7.3.1 Sample preparation for exfoliated cells

Clinical samples from patients with oral disease were harvested at Tufts Medical Center (TMC, Boston, MA) by swiping the suspected areas with a cytobrush that was immediately inserted into a vial filled with SurePath® solution. Oral cells from normal subjects were exfoliated by LSpD laboratory personnel as part of an oral cancer screening program at Northeastern University (under a local Institutional Review Board (IRB)). All cervical cells (normal and abnormal) were collected at TMC as well, and treated the same way. Esophageal cells were collected at by a collaborating group at the Gloucestershire Hospitals, NHS Foundation Trust in the United Kingdom. All exfoliated cells were treated exactly the same way: cells were exfoliated *via* cytobrushes that were immersed into SurePath fixative immediately after exfoliation. This fixation medium consists of 24% aqueous solution of ethanol and 1% each of methanol and isopropanol. The effects of fixation are discussed in Section 13.8.2. After repeated wash and centrifugation cycles, cells were spin-deposited *via* cyto-centrifugation onto low-e slides (see Chapter 11). Sparse, uniform samples of cells, which adhere to the substrate very strongly, could be produced this way.

Spin-depositing cells from a suspension onto a "low-e" slide, using cyto-centrifugation (CytoSpin, Thermo, Waltham, MA), work very well and produce a 5-mm diameter spot of cells. Depending on the concentration of cells in the original suspension and the deposition time, sparse monolayers of (unstained) cells can be produced in a few minutes. For relatively large squamous epithelial cells, one typically aims for about 50 cells/mm^2. Cells generally adhere well to the low-e slides. Cultured cells can be grown directly onto IR-compatible substrates. Cells may be stained after IR image acquisition for classical cytopathological analysis.

13.7.3.2 Data acquisition

Spectral data acquisition was carried out by imaging a 4.0 mm × 4.0 mm square region within the ca. 5 mm diameter deposition spot. This was accomplished by raster scanning the substrate through the focus of the IR beam of the PerkinElmer Spotlight 400 FTIR imaging systems (see Chapter 11). Data were acquired from the entire square sample area, including areas not occupied by cells. The instrument was purged with a continuous stream of dry air (−40°C dew point) to reduce atmospheric water vapor spectral contributions. The 16-element array detector of the PE400 was used; in this mode, images of specified areas were acquired where the pixel size from which each IR spectrum (the "pixel spectrum") was collected measured 6.25 μm × 6.25 μm in size. Data were acquired in reflectance mode of the spectrometer; in conjunction with the "low-e" sample substrates, this mode results in the collection of transflection (transmission–reflection) data as discussed in Section 11.7. The following instrument parameters were used: 4 cm^{-1} spectral resolution, Norton–Beer apodization, and one level of zero-filling. Two interferograms were co-added for each pixel and Fourier transformed (FT) to yield spectral vectors, each with a range 4000 to 700 cm^{-1} at a 2-cm^{-1} data interval.

Background spectra for all 16 detector elements were collected using 128 co-added interferograms to reduce the noise contributions of the background as much as possible. After FT, the sample and background spectra were automatically ratioed and stored as transmittance spectra. Each raw data set consisted of 409,600 spectra and occupied about 2.54 GB. The acquisition of a 4 mm × 4 mm sampling area required about 3 h, whereas the preprocessing of each data set required about 3 min, and typically yields between 200 and 1000 high-quality IR absorption spectra of individual cells.

13.7.3.3 Data preprocessing

The data preprocessing mirrors the procedures described for SHP earlier in this chapter and in Chapter 12. Raw data sets were imported into software written in-house in 64-bit MATLAB (The Mathworks, Natick, MA) for automatic preprocessing. The sequence of major data processing steps for each data set is summarized as follows and include (for details, the reader is referred to Ref. [39]):

- Construction of separate noise and signal data sets from areas not occupied by cell and areas occupied by cells, respectively, and subsequent NA-PCA (see Chapter 12)
- Reduction of water vapor contributions in cellular pixel spectra by a multivariate approach based on multiplicative signal correction (see Chapter 12)
- Construction of a binary mask to delineate areas occupied by cells from those of the blank substrate (part of the PapMap^P algorithm)
- Conversion to second derivatives [74] and expansion of spectra to 1800 to 778 cm^{-1} (fingerprint region)
- Correction of scattering effects [75] in cellular pixel spectra indicated by the binary mask by iterative phase correction
- Computation of cellular spectra from individual pixel spectra for each cell identified in the binary mask (PapMap^P) [76]
- Elimination of spectra due to noncellular materials in the sample
- Export spectra of individual cells, cell coordinates, and name tags to patient-based data structure.

In summary, the PapMap^P involves collecting data from the entire substrate, whether occupied by cells or not, constructing cellular spectra by co-adding pixel spectra from areas occupied by individual cells, and applying NA-PCA (with the instrument noise profile established from spectra collected from the substrate unoccupied by cells, see Chapter 12) to improve the signal-to-noise ratio of cellular spectra. The major advantages of the PapMap^P approach is enhanced and fully automatic data collection, and the elimination of bias in selecting cells for analysis. In a "point-and-shoot" approach used earlier, where unstained cells were selected visually for subsequent data acquisition, the operator had to decide whether a spot on the slide was actually a cell, cellular debris, or an artifact. In the PapMap^{P} method, the operator can decide, after staining of the slide, what constitutes a cell, and can adjust the limits (the number of pixels to be co-added) of the size of cells to be analyzed. This was particularly important for the analysis of esophageal cells, described in Section 13.8.5.

13.7.3.4 Data analysis

Both unsupervised and supervised methods of analysis were applied to the data sets created in the previous step. In subsequent sections, results are presented as two- or three-dimensional PCA plots (see Chapter 12). Alternatively, ANNs were trained for the supervised analysis of data sets; in these cases, the results are usually presented in the form of a confusion matrix or directly in terms of sensitivity, specificity, and overall accuracy.

13.7.3.5 Confounding factors

The IR spectra of cells are even more susceptible to confounding factors, such as distorted band shapes due to the R-Mie effect, than the spectra of tissue. This is due, in part, to the fact that in individual cells, the scattering and edge effects are amplified because cells are isolated, and the gradient of the refractive index at the cells boundary is not mitigated as in the case of tissue, where cells are in direct contact with neighboring cells. Bhargava and coworkers [77, 78] recently have shown the effect of the boundaries of arbitrarily shaped objects in an IR microscope; these effects are in addition to the previously discussed R-Mie distortions. Furthermore, the standing wave effects discussed in Chapter 12 may falsify observed intensities. To account for these effects, the data analysis was adjusted to account for these effects. The PapMap^{P} algorithm, for example, eliminates spectra from the edge of a cell and removes residual band distortions *via* phase correction. Finally, the PapMap^{P} algorithm allows the automatic selection of the size of cells to be analyzed; for squamous cells, one may assume that cells of similar size have similar thickness. This step reduces the chance of comparing stratified squamous and columnar cells in the same data set. Spectral dependence on the standing wave effect was further reduced by utilizing second derivative spectra only for analysis.

13.8 SCP results

This section is organized as follows. First, a short review of the early attempts in SCP is presented. This work focused on cervical cytology for reasons that are elaborated in Section 13.7. Subsequently, very recent results from the author's laboratory are presented. These results include a review of the fixation effects observed for exfoliated cells. This section should once and forever answer the questions about fixation that have been raised frequently. Next, recent results for cervical, oral, and esophageal cytology are presented.

13.8.1 Early results of SCP

The earliest attempts at utilizing IR spectroscopy for the diagnostic screening of exfoliated cervical cells date back to the early 1990s. At this point of time, IR MSP had not progressed to a level that permitted acquisition of spectra of individual cells in reasonable times; thus, cell pellets of unknown composition in terms of the cell types contained in the pellet were used as samples, and the measurements were carried out macroscopically. Results from these early efforts are exemplified by the PCA scores plot (M. Diem, unpublished results) shown in Figure 13.19. In this plot, each symbol represents one spectrum collected macroscopically from a cell pellet; the confirmatory diagnosis was by classical pathology. In retrospect, it is amazing that these early cell pellet results gave encouraging results, and it took quite a while to understand why these crude measurements

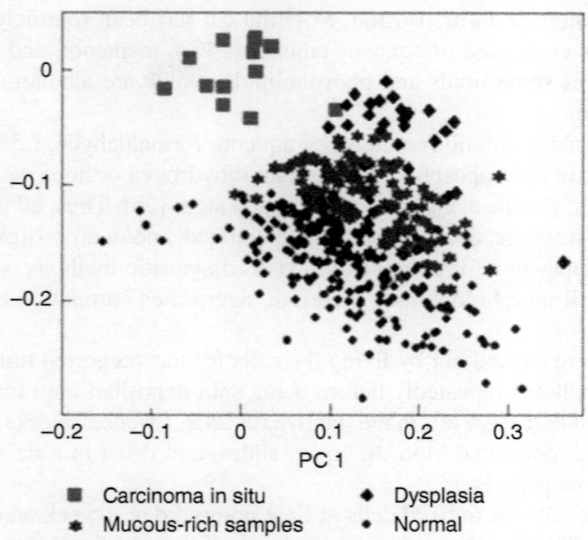

Figure 13.19 *PCA scores plot for macroscopically acquired cervical cell samples.* (See insert for color representation of this figure.)

showed any kind of discrimination, because – as pointed out in Section 13.7.1 – the number of abnormal cells in a dysplastic sample is very small.

The answer to this surprising finding was reported in 1998 by Cohenford and Rigas [67], when the first microscopically acquired data set of over 2000 individual cervical cells from over 20 subjects was collected. This pioneering study showed that in samples from patients with dysplasia and cancer, all cells, including superficial squamous cells with normal morphology, exhibited spectral patterns different from those of the same cell types from normal patients. Similar results had been reported by the group of A. Madevan-Jansen [118] (see Chapter 14) and attributed to effects known in pathology as "malignancy-induced changes" or "cancer field effect." These rather vaguely defined effects indicate that cells in the vicinity of a cancerous lesion all show a genetic abnormality, long before they show morphological abnormality. This observation explains the results shown in Figure 13.19 and permeates the results and conclusion of the following sections that deal with SCP of the oral, cervical, and esophageal mucosae.

13.8.2 Fixation studies

Before discussing the results obtained for oral, cervical, and esophageal screenings, the issue of spectral changes due to fixation is introduced. This issue has been the source of many questions and criticism of SCP and SHP results. Early studies [25, 80] have reported large spectral changes upon fixation, which could not be reproduced by other groups and were most likely due to changes in cell morphology, as some of these changes had the signature of R-Mie scattering. However, there doubtlessly exist spectral changes caused by treatment with some of the harsher fixation protocols; here, only the two most common tissue treatment methods are discussed.

A study was carried out, using exfoliated oral mucosa cells from laboratory volunteers, in which cells were left unfixed and dehydrated, fixed in the SurePath fixative (see the following paragraphs), or phosphate-buffered formalin. The fixatives used for this study were selected because of their common application in conventional cytopathology and SCP efforts [81–83]. The SurePath methodology is used by the medical collaborators at TMC, Boston, MA; thus, it has been routinely used in the LSpD as well. The SurePath solution is composed of aqueous ethanol (24%), methanol, and isopropanol (both 1%). The SurePath process removes some lipids and phospholipids, which are alcohol-soluble and renders proteins insoluble *via* dehydration.

Phosphate-buffered formalin solution contains 4% aqueous formaldehyde, 1.5% methanol, and less than 1% of both disodium hydrogen orthophosphate and sodium dihydrogen orthophosphate monohydrate. Formalin cross-links primary amino groups using a nearby nitrogen atom [25]. Thus, all procedures used here slightly alter the native protein's structure, to a precipitated, dehydrated, and/or cross-linked state. Although complete dehydration seems advantageous for composition-based diagnostic methods, such as SCP, it may produce significant changes in cell morphology and, therefore, aggravates correlation between SCP and traditional cytopathology.

The fixation studies were carried out by fixing the cells for the suggested time periods. Subsequently, the cells were washed and pelleted repeatedly before being spin-deposited onto low-e slides. Other aliquots of cells from the same volunteers were left in the fixative for 24 h, 1 week, 2 weeks, or 4 weeks. Another aliquot was immediately washed, deposited onto the low-e slides, and dried in a desiccator. These samples were rescanned at the same time periods.

Figure 13.20 depicts results for unfixed cells at time points 1 day, 1 week and 4 weeks to demonstrate the stability of unfixed cells. The raw absorbance spectra, see offset in (a) for clarity, are nearly indistinguishable and exhibit a similarity (correlation coefficient) of 0.99. Each trace is the average of about 50 individual cells. Even their second derivative spectra, shown superimposed in (b) and offset in Figure 13.20(c), are indistinguishable to the naked eye. However, they show slight separation in PCA, with the 1-day and 1-week spectra clustering together [71], but separate from the 4-week data. These plots show a slight water vapor component along PC1 (principal component), attributed to slow outgassing of the cells, and a slight change in the protein region in PC2, along which the early (<1-week-old) and the late (4-week-old) samples split. This indicates that unfixed, dehydrated cells are stable for at least a week. No morphological changes could be detected for these unstained cells over the 4-week time frame.

Figure 13.21 compares the results for the different fixation methods, as a function of time. Visual inspection suggests that, once fixed, the cells are stable over the 1-month time period. PCA again detects very small changes, for example, for the SurePath-fixed cells, along PC1, which is, as before, attributed to water vapor outgassing [71]. Between the fixation methods, there is one obvious difference in the spectra, as shown by the arrow in Figure 13.21. This is an interesting observation, because one would expect that the cross-linking agent, formalin, rather than the dehydrating alcohols in SurePath, would produce an additional band. PCA readily separates the different fixation methods. The conclusions to be drawn from this result are simply that one cannot mix fixation methods within one study. Finally, this study also followed spectral changes if the cells were left immersed in fixative for extended times. As expected, after leaving cells in fixative for 1 or 4 weeks, more pronounced spectral changes were observed. Thus, one concludes that cells should not be left in the fixative for extended time periods; in particular, correlation coefficients for the formalin-fixed cells dropped to 0.98 after 4-week exposure.

Another take-home message from these fixation studies is shown in Figure 13.22, which depicts PCA results of the three fixation protocols in comparison with the spectral changes due to oral precancer. This plot indicates that the changes induced by fixation methods, shown in red, green, and blue symbols are small in comparison to the changes induced by precancerous oral disease. Incidentally, the oral disease samples were fixed in SurePath solution.

Similar fixation studies were carried out for cultured cells as well and revealed nearly identical results. Thus, one may conclude that fixation effects are small, but do exist, and can be avoided by performing fixation under strict protocol, and adhering to one fixative.

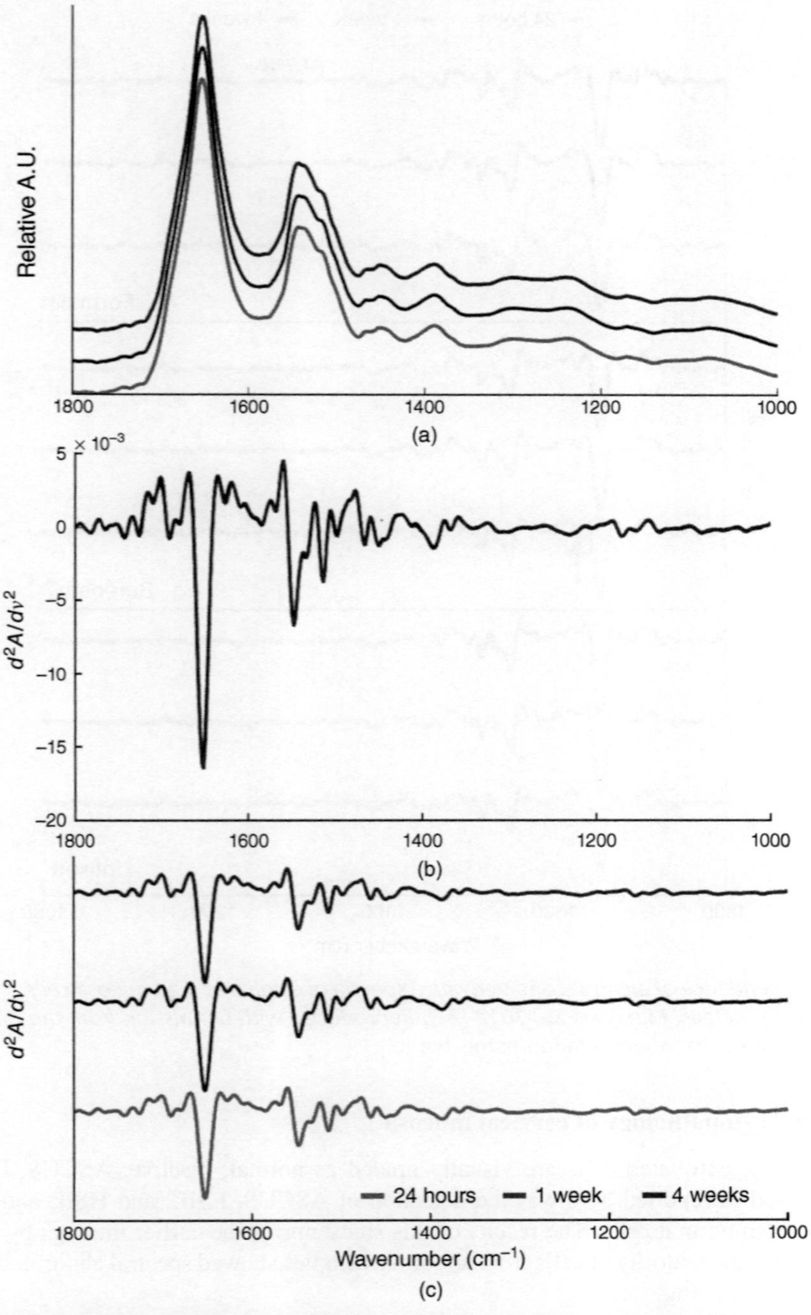

Figure 13.20 *Time evolution of unfixed oral mucosa cells (a) mean absorption spectra at the time points indicated, (b) overlay, and (c) stack plot of second derivative spectra, indicating no gross changes of unfixed cells for 4 weeks after exfoliation. Mazur,* et al.*, 2012 [71]. Reproduced with permission of the American Chemical Society. (*See insert for color representation of this figure.*)*

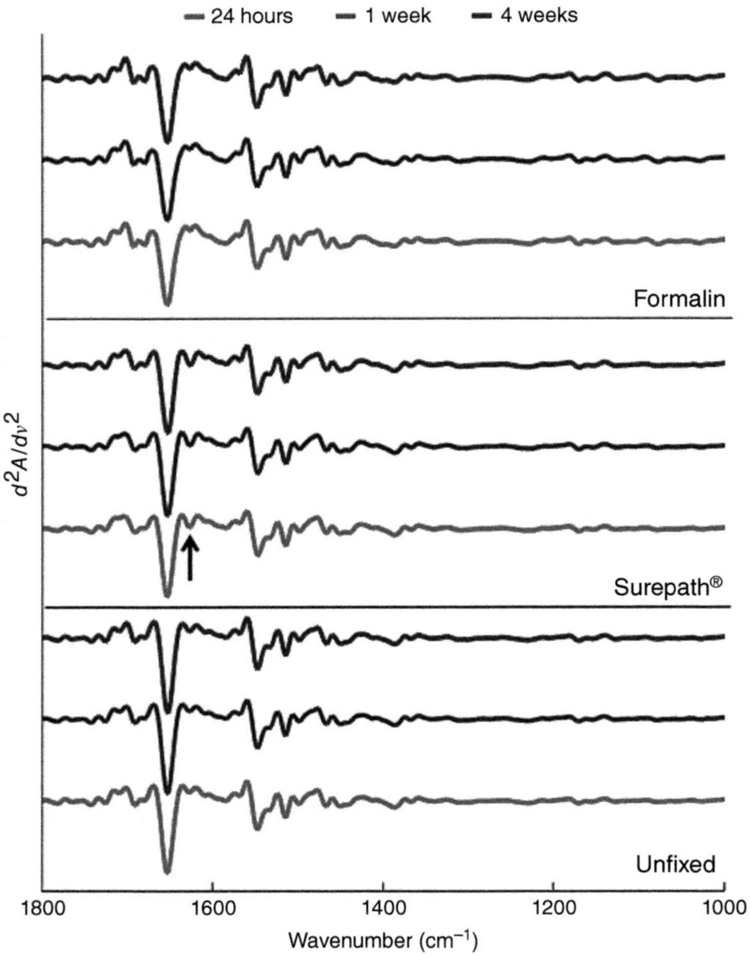

Figure 13.21 *Dependence of mean second derivative spectra of exfoliated oral mucosa cells on time and fixation protocol. See text for details. Mazur,* et al.*, 2012 [71]. Reproduced with permission from the American Chemical Society. (*See insert for color representation of this figure.*)*

13.8.3 Spectral cytopathology of cervical mucosa

In cervical cytology, exfoliated cells are visually graded as normal, reactive, ASCUS, LSIL, and HSIL. The aim of the studies reported here was the detection of ASCUS, LSIL, and HSIL and classification of these cell types from normal cells. The results of this study mirror the earlier findings by Cohenford *et al.* [55, 67], namely that the majority of cells from abnormal samples showed spectral abnormality, although they exhibited normal morphology.

Cervical cells are subject to monthly hormonal changes that cause maturation of the cells, accompanied by glycogen accumulation. The influence of these hormonal changes was investigated in earlier studies at the LSpD and other laboratories [82, 84, 85]. In order to eliminate spectral variance due to these effects, the studies reported here only included women using oral contraceptives that prevent maturation of the cervical mucosa. As the glycogen absorption bands mask a large part of the low-frequency (1000–1200 cm^{-1})

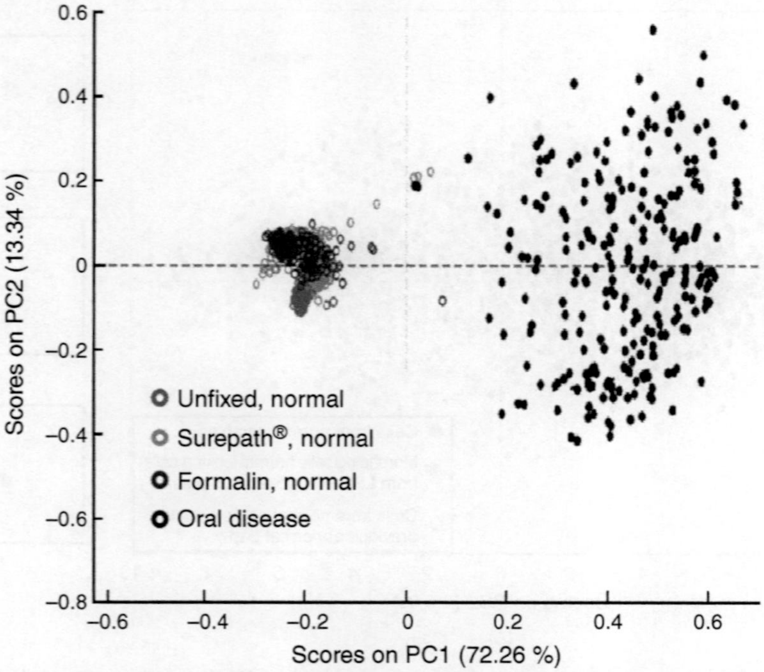

Figure 13.22 *PCA scores plot of spectral differences due to fixation (colored circles) and cellular abnormalities (black circles). Mazur,* et al.*, 2012 [71]. Reproduced with permission from the American Chemical Society. (*See insert for color representation of this figure.*)*

spectrum, valuable information in the nucleic acid phosphate stretching region (about 1090 and 1235 cm^{-1}) is rendered inaccessible. The changes in maturation patterns of cervical cells in response to menopause could be demonstrated nicely using SCP [86].

Figure 13.23 shows the results of PCA analysis of cells exfoliated from normal patients, and patients diagnosed with LSIL/HSIL [85]. These results mirror those obtained for oral cytology (see below) and confirmed that most of the cells from patients with dysplasia exhibit spectral abnormalities, although the cell morphology is normal (see cell images in Figure 13.23). Even more surprising is the fact that the cells from a patient with a prior diagnosis of HSIL, and subsequent treatment, still exhibit abnormal spectral patterns and cluster with the abnormal spectra. This lead to a further study, investigating the role of viral infection as the cause of the spectral changes (see below).

Figure 13.24 demonstrates that morphologically abnormal (dysplastic) cells, shown by the green squares) exhibit the same spectral features as morphologically normal cells from abnormal samples (red circles). Here, the spectra of eight individual cells with diagnosed abnormal morphology, from one patient, were analyzed via PCA, along with cells with normal morphology from the same patient. Since the cells with abnormal morphology cluster with the rest of the cells, one may assume that there are no spectral differences between these cell types, whereas the principal component plot (Figure 13.23) revealed spectral differences between morphological normal cells from patients with or without disease

The implications of this observation are quite far reaching in that detection of abnormality is not restricted to the few cells in a cervical exfoliate that exhibit abnormal morphology. Rather, SCP detects abnormal spectral signatures that are exhibited by most of cells, even if they still have normal morphology. The fact that the abnormality persist after treatment led to the exploration the possibility that the spectral changes (and,

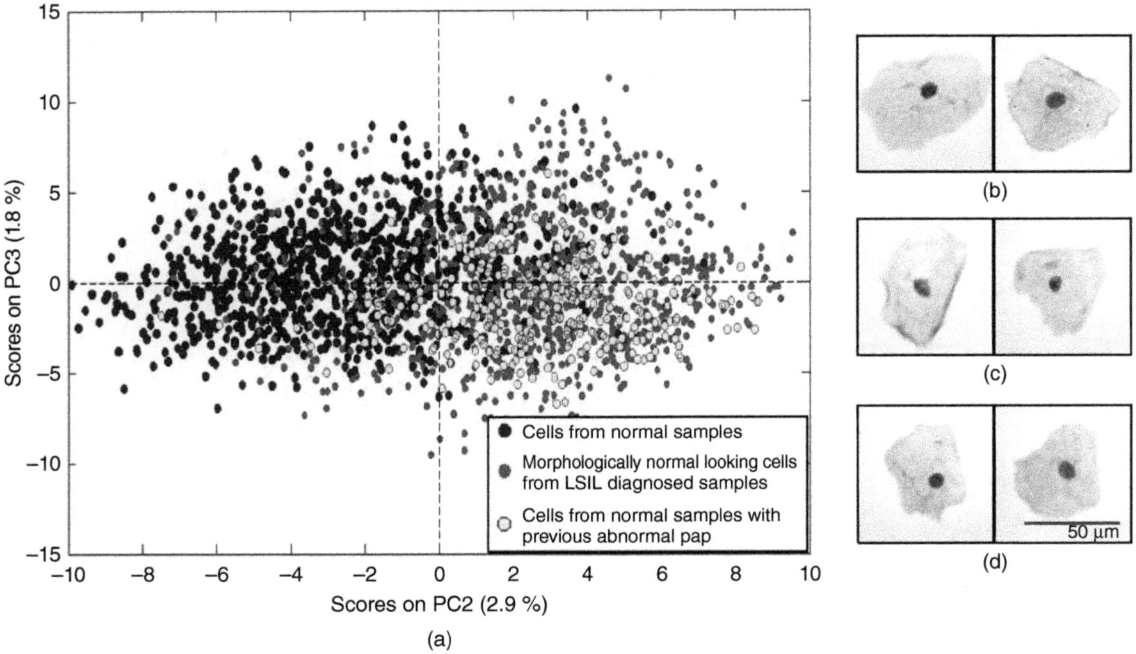

Figure 13.23 *(a) PCA scores plot of spectra of cervical squamous cells from five normal subjects (blue), from five patients diagnosed with LSIL (red) and two patients with history of HSIL (yellow). All cells exhibited normal morphology, as shown in panels (b), (c) and (d) for normal, LSIL and history of HSIL, respectively. (From Ref. [86]). (See insert for color representation of this figure.)*

therewith the "malignancy associated changes" [87] or "field cancerization" [88] mentioned earlier) are actually due to viral infection. In part, this thought was provoked by results from a patient with an acute *herpes simplex* infection of the oral cavity. In this case, most of the cells showed spectral abnormality, not only those that were so grotesquely deformed by the viral infection that they could be visually diagnosed by a cytologist [83]. As statistically over 95% of all cervical dysplasia cases occur along with (and likely are caused by) infection with the human papillomavirus (HPV), the possibility exists that SCP detects the infection by HPV in cervical cytology [89]. Similarly, oral dysplasia could be caused by the Epstein–Barr or HPV as well.

These observations may help explain the positive results reported for cell pellet studies (Figure 13.19): although the composition of the cell pellet in terms of contributions from superficial, intermediate, and parabasal cells, as well as PMNs and bacteria, was not known for these samples, the abundance of virally infected cells may have been responsible for the distinction of disease states. Efforts to answer the possibility of the involvement of viral infection toward the observed spectral changes are pursued next.

In order to explore the sensitivity of SCP toward viral infections, a study was undertaken, in which 48 samples were tested for high-risk human papillomavirus (hrHPV) infection *via* the Digene Hybrid Capture test (Qiagen, Valencia, CA). The spectral results were analyzed by SIMCA [85]. The result of a 10-sample training subset looked extremely promising, with good spectral separation of HPV-positive and HPV-negative samples. When applied to the remaining set of samples, a sensitivity of 88% was achieved, yet the specificity was only 43%. This implies that SCP was quite good at detecting the hrHPV strain when it was present (as determined by the Digene test), but not accurate when no hrHPV infection was present. This finding was attributed to the fact that low-risk human papillomavirus (lrHPV) infection is epidemic in the population of

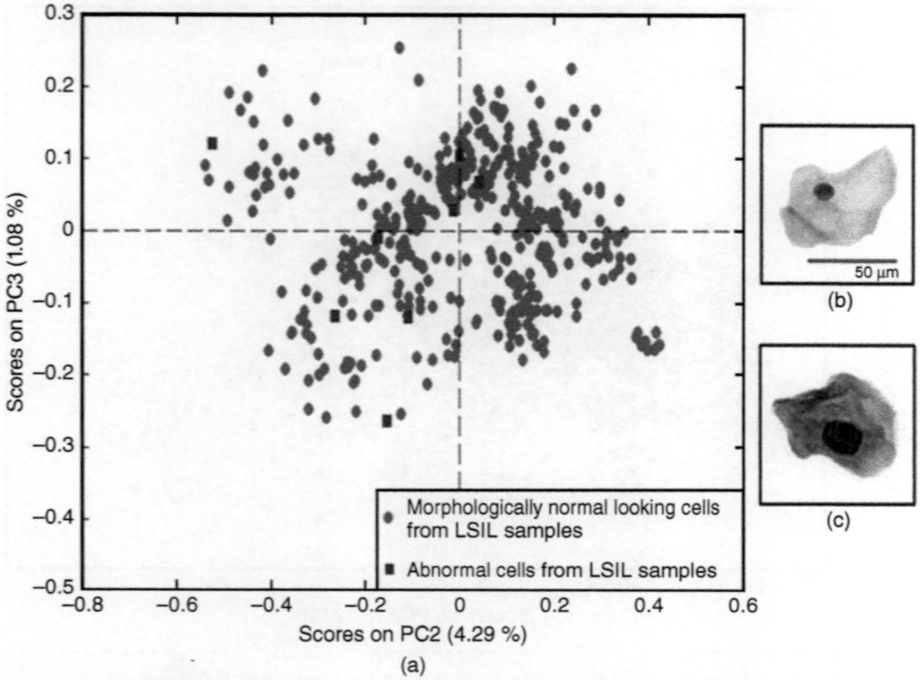

Figure 13.24 *(a) PCA scores plot of cells with normal morphology (red dots) and cells with abnormal morphology (green squares). (b) cell with normal morphology (c) cell with abnormal morphology (From Ref. [86]). (See insert for color representation of this figure.)*

women between 20 and 25 years of age, with infection rates of about 30%, or about the same as the infection rate with hrHPV [90]. Thus, it is quite likely that the samples that tested negative for hrHPV by the Digene test had lrHPV infection, which the spectral methods were not (yet) able to differentiate [85].

Inspection of the PCA loading vector indicated that the spectral changes, along which PCA distinguishes the HPV-infected cells from the normal cells, occurred in the protein spectral region by a distinct shift of the amide I band and the appearance of small shoulders. This leads to the conclusion that it is not a change in the viral DNA that is detected, but rather, different proteins expressed by the virus. Given the size of the viral genome (about 5000 base pairs) and the number of copies of the viral genome in a cervical cancer cell (maximally about 600 in the CaSki cell line, see Chapter 14), one arrives at a number of about 3 million base pairs added to the human genome in the case of HPV infection. The human genome consists of 3 billion base pairs; thus, it is impossible to detect this change with present methodology. A preliminary study to differentiate cervical cell lines (C33A, SiHa, HeLa, and CaSki) with different numbers of hrHPV viral copies was carried out [85], and indeed SCP could differentiate these cell lines quite easily. This study paralleled results from the Dublin group where Raman SCP was used to investigate these cell lines [91]. However, it is likely that both these studies detected different abundances of the viral coat protein in the cells, rather than changes in the DNA composition.

13.8.4 Spectral cytopathology of the oral mucosa

Most cancers that develop in the oral cavity are squamous cell carcinomas. The diagnosis of oral cavity and oropharynx cancers is difficult because these cancers often are asymptomatic until they have reached an

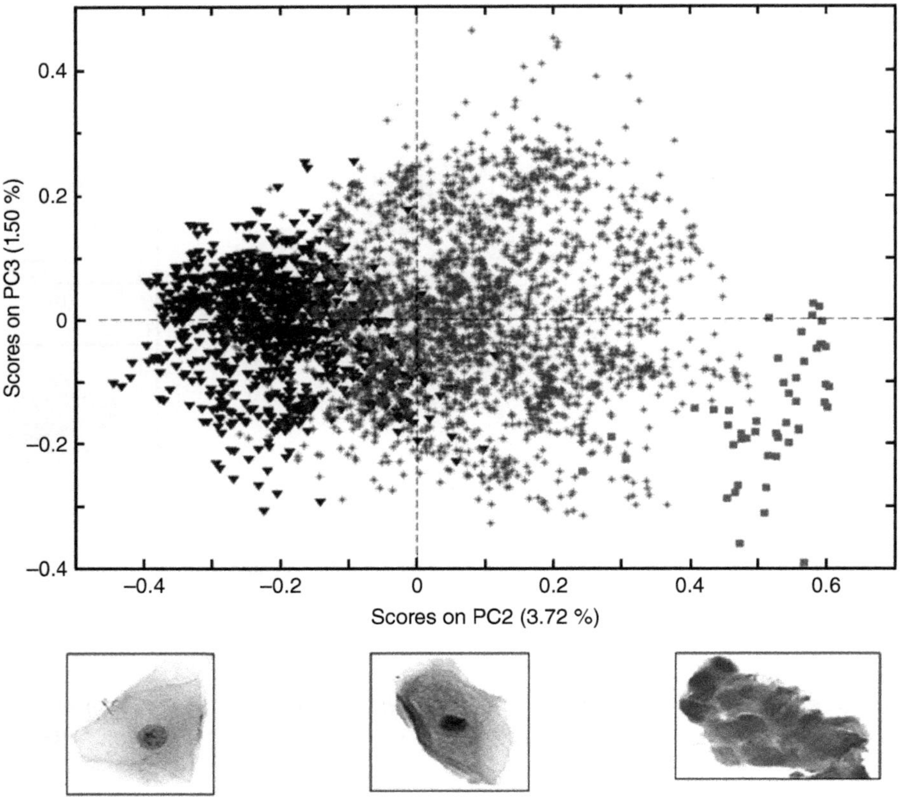

Figure 13.25 *PCA scores plot of oral SCP. The cell images represent a normal cell (left), a morphologically normal cell from an abnormal sample (middle), and a clump of cancerous cells. All cells were harvested from the tongue. Papamarkakis,* et al.*, 2010 [83]. Reproduced with permission from Nature.* (See insert for color representation of this figure.)

advanced stage. The 5-year relative survival rate for oral cavity cancer is 61%, whereas the 10-year survival is 50%. Once the cancers have reached advanced stages, they often spread to other areas within the body. Thus, a reliable screening method would be highly advantageous.

The first systematic study of SCP of the oral mucosa was published by the author's group and collaborators at TMC (Boston, MA) in 2010 [83]. As oral cancer is a disease that is quite prevalent in far eastern population, and because TMC services this population, a follow-up study was initiated, in which 93 (presumably) normal subjects from the campus at Northeastern University and 30 patients from TMC with diagnosed oral disease were included. The Northeastern University IRB approved this study so that volunteer samples could be collected from the members of the Northeastern community. The cells harvested at both institutions were treated exactly the same, as described in Section 13.7.3. The original and the follow-up studies confirmed the earlier findings that the majority of cells from patients with diagnosed oral squamous cell carcinoma exhibited abnormal spectra, even cells with normal morphology.

The original study [83] revealed that slightly different spectral patterns were observed for three different regions of origin in the oral cavity (tongue, mouth floor, and cheeks). Furthermore, cells from the tongue from normal subjects and dysplasia and cancer patients showed distinct separations as shown in Figure 13.25. In the scores plot of Figure 13.25, the normal cell spectra shown as blue symbols form a tight cluster. The cells

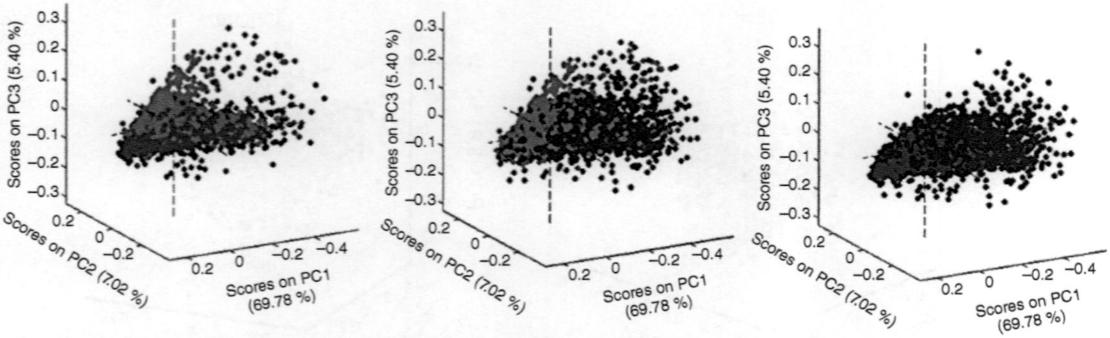

Figure 13.26 *PCA scores plot of cells from different anatomical regions of the oral cavity (blue: tongue, red: mouth floor, and green: cheeks). Diem* et al.*, 2014 [92]. Reproduced with permission from Wiley-VCH. (*See insert for color representation of this figure.*)*

harvested directly from a cancerous lesion of the tongue, shown in red, form a cluster that is well separated from the normal cells. Most interesting is the diffuse cluster represented by the green symbols. These spectra were from cells, which exhibited normal morphology, but they were collected from cancer patients from areas quite remote from the cancerous lesions, and from patients diagnosed with precancerous disease. These results further suggest a progression in spectral signatures from normal *via* dysplastic to cancer.

In the follow-up study (see earlier), samples were collected from 93 patients from three areas of the oral cavity, including the tongue, cheek, and mouth floor. Volunteers were asked to fill out an informational questionnaire that was developed to retrieve information such as gender, race, tobacco and medication usage, and oral disease history.

The spectral changes associated with different sample collection regions reported previously were confirmed in the follow-up study as well, albeit with less separation between the PCA clusters (see Figure 13.26). Although all attempts were made to compare matched data sets, for example, from nonsmokers only, the possibilities of confounding factors (the use of over-the-counter medication and oral contraceptives, history of viral infection, etc.), the patient heterogeneity in the end was large, and, therefore, the classification by anatomical region was not as good as before [83]. However, a sufficient number of samples did exist to allow a comparison between normal and clinical samples (see the following paragraphs) to be made for matched anatomical regions [92].

Figure 13.27 compares normal and clinical cells from two different anatomical regions. "Clinical," in this context, implies that the patients were diagnosed with oral cancer or dysplasia independently of this study, and that the samples collected from them contained mostly squamous cells with normal morphology in addition to a few abnormal cells. As discussed earlier, the majority of the abnormal cells exhibited different signatures than the normal cells. In this study, the number of "clinical" cell spectra was the limiting factor; thus, in the PCA results (Figure 13.25) and the ANN-based studies, an equal number of randomly selected normal cells was used.

The PCA plots shown in Figure 13.27 suggest a reasonable separation of clinical from normal cells. These plots are based on about 3900 spectra each for normal and clinical samples, from 17 and 22 patients, respectively (in the case of cancer of the tongue), and on about 630 spectra each for normal and clinical samples, from 4 and 4 patients, respectively (in the case of the cheeks). For the case of the tongue cells, an ANN was trained and repeatedly tested on 400 different, randomly selected spectra. This algorithm produced a sensitivity of 96% and a specificity of 94% for an overall accuracy of 95%.

In spite of the increased size of the data set of normal spectra, the desired results for detection of tobacco and over-the-counter medication use was inconclusive. The number of volunteers reporting previous infection

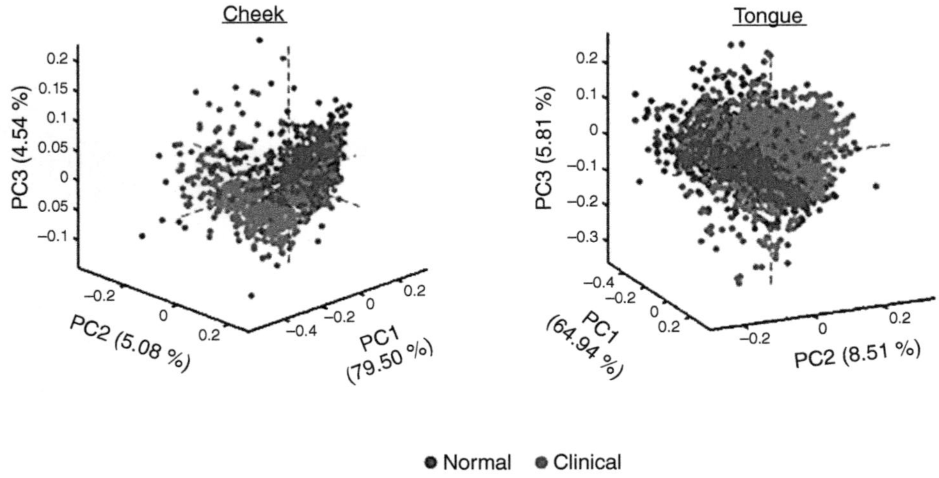

Figure 13.27 *PCA scores plot of clinical and normal samples from the cheeks and the tongue (see text for details). Diem* et al.*, 2014 [92]. Reproduced with permission from Wiley-VCH. (*See insert for color representation of this figure.*)*

with the Barr–Epstein virus (which causes mononucleosis) and with *herpes simplex* virus was too small to correlate the spectral results with any of these diseases. However, one unexpected result was found: the oral cells could distinguish female volunteers who used oral contraceptives from those who did not. The results for oral contraceptives show that the hormones effect the maturation of the squamous cells in the oral cavity that are known to express estrogen receptors. A more detailed review of the SCP study on the oral mucosa can be found in Ref. [92].

13.8.5 Spectral cytopathology of esophageal cells

A final SCP study from the LSpD is introduced next, namely the efforts to develop a method to aid in the diagnosis of esophageal disease. Cancer of the esophagus is one of the deadliest forms of cancer, and its diagnosis is very difficult. It generally develops in the distal part of the esophagus, and prolonged acid reflux disease is thought to be a major contributing factor. The exposure of the squamous epithelial tissue of the esophagus to stomach acid causes a metaplastic response where the squamous cells are replaced by columnar epithelium (Barrett's esophagus) that, in turn, can turn dysplastic and eventually develop into ADC. At the stage of Barrett's esophagus and dysplasia, treatment is possible; thus, it is imperative to properly diagnose the stages of disease. Presently, diagnosis is carried out by collection of many punch biopsies in an endoscopic procedure; typically, eight biopsies are collected at each depth increment within the esophagus. Thus, one procedure can yield between 30 and 50 individual biopsies, which present a substantial pathology workload. This and the uncertainty whether or not the punch biopsies were taken at a location actually affected by disease, spawned interest in alternative diagnostic methods. White light endoscopy and fluorescence endoscopy have been utilized, but neither method has had high accuracy in distinguishing Barrett's esophagus from dysplasia. Of the vibrational spectroscopic methods, *in vivo* endoscopic Raman imaging of the esophagus was first reported by the group of Puppels [93], at the Erasmus University, Rotterdam, and later by Stone and coworkers [59, 94, 95] at Gloucestershire Hospital in the United Kingdom. These studies proved to be enormously successful and will be reviewed in Chapter 14.

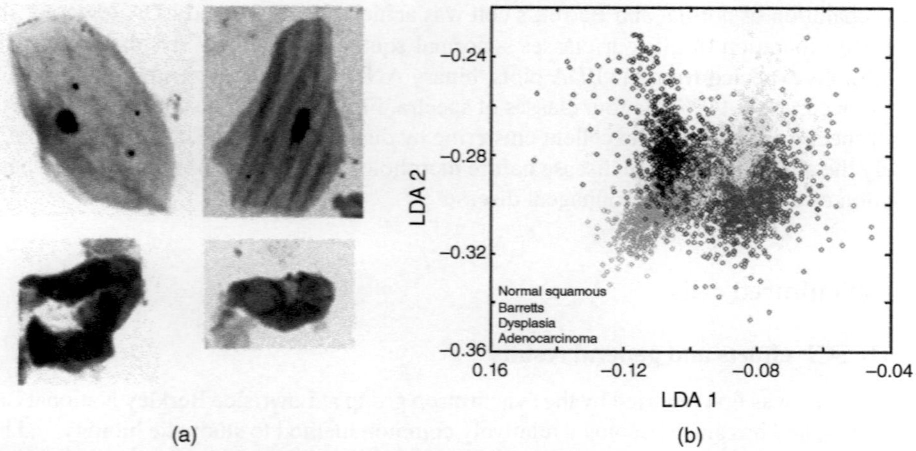

Figure 13.28 *(a) Top row: squamous cells from the esophagus and bottom row: glandular cells from the esophagus (Pap stain). (b) LDA plot of cell spectra from the esophagus.* (See insert for color representation of this figure.)

At the LSpD, a cytopathological approach toward detecting and classifying esophageal disease was initiated with the same medical group in the United Kingdom that had pioneered the *in vivo* Raman approach. Esophageal cells can be exfoliated during an endoscopic examination, and cells can be collected using a cytobrush. Another approach involves the patient swallowing a capsule with a string attached. In the stomach, the capsule dissolves and releases a small sponge that is retrieved through the esophagus. This latter method did not provide sufficient cells for SCP, whereas the cytobrush methods did. Raman endoscopy offers a point-by-point instantaneous diagnosis of the disease stage, whereas a cytology approach represents an integrating sampling technique because large portions of the esophageal surface can be sampled in one brushing.

Similar to the problems encountered with the cells from the urethra and the bladder, the columnar and squamous cells from the esophagus are quite variable in size. The squamous cells typically occupy an area as large as 4000 μm^2 (about 60 μm on edge), whereas typical columnar cells may occupy an area as small as 400 μm^2 (about 20 μm on edge, see Figure 13.28). It would have been nearly impossible to visually detect the small, unstained cells if data were to be collected using any approach other than the PapMap^P introduced earlier that allows many of the small cells to be captured and analyzed. To this end, the area defining a cell (e.g., the number of pixel spectra to be co-added to create a cell spectrum) was widened to between 10 and 100 pixels, corresponding to areas of 400–4000 μm^2. After staining, the smaller cells could be associated with individual columnar cells, as shown in Figure 13.28. Subsequent classical cytology identified some of them to be, indeed, dysplastic.

At present, there are over 4800 esophageal cell spectra in the data set, and many of them were individually diagnosed by classical cytopathology. Thus, this data set represents, at present, one of the largest annotated data sets of exfoliated cells. Of these, about 2340 are from 10 normal patients, about 1720 call spectra are from 8 patients diagnosed with Barrett's disease, and about 750 cell spectra are from 2 patients with Barrett's and dysplasia. Obviously, this data set is still much smaller than the data sets reported earlier for cervical and oral cells, although the number of individual cells with cytological diagnoses is much larger than in the previous studies.

The results follow patterns similar to the ones described earlier: First, most spectra of cells from patients with disease exhibit abnormal spectra. Second, a separation of cell type by diagnosed patient disease is possible.

A very clear separation of normal and Barrett's cell was achieved by PCA and LDA (data not shown). LDA also shows good separation of all four classes – normal squamous, Barret's, dysplasia, and ADC [92] see Figure 13.28(b). As expected from the LDA plots, binary ANN diagnostic algorithms were able to diagnose with very good accuracy (~90%) the four classes of spectra. Finally, a separate analysis (*via* PCA) of only the annotated cell spectra resulted in an excellent clustering by disease class. Here, the promise of SCP discussed earlier, namely the ability to diagnose disease before morphological changes occur, will add a new dimension to the distinction of early stages of esophageal disease.

13.9 SCP of cultured cells

13.9.1 Early SCP efforts and general results

SCP of cultured cells was first reported by the synchrotron group at Lawrence Berkley National Laboratory by Holman *et al.* [96], and has since become a relatively common method to study the biology and biochemistry of cells. Unfortunately, the changes in cell morphology are particularly pronounced in cultured cells: during interphase – the time between cell division – adherent cells exhibit an extended, thin cytoplasm with distinct pseudopods. During cell division, and particularly during mitosis, cells partially detach from the substrate and assume a near-spherical morphology. At this point, they can be shaken off the substrate, a method that has been used in cell-cycle synchronization. Trypsination of cells to remove them from the substrate has a similar "rounding" effect. During mitosis or immediately after trypsination, cells are particularly prone to show strong R-Mie scattering effects and, consequently, strongly distorted spectra.

SCP and high-resolution vibrational spectroscopic imaging of cells were pursued at the author's laboratory, starting with the first spatially resolved synchrotron-FTIR imaging studies by Lasch *et al.* [13], followed by digestion studies that demonstrated the distribution of nuclear and cytoplasmic components [97] and that RNA could be detected in the cytoplasm of actively growing cells [5]. This work was followed by cell-cycle studies [68, 69, 98] and the label-free imaging of processes during mitosis by Raman imaging (see Chapter 14). During these studies, it was realized that cultured cells, in all phases of the cell cycle and in the interphase, exhibit quite different spectral signatures than the terminally differentiated exfoliated cells from squamous epithelium. These spectral changes are manifested by stronger phosphate vibrational modes (see Section 10.5) at about 1090 and 1233 cm^{-1}, as well as distinct changes in the protein amide I region; these latter changes are particularly evident in the second derivative spectra (see Figure 13.29). It is interesting to note that the amide II vibration does not show any changes in the second derivative spectra.

Furthermore, it is interesting to note that the spectra of oral mucosa cells, Figure 13.29(a) are from normal, albeit terminally differentiated and inactive cells with small, pyknotic nuclei. The spectra of the HeLa cells are from actively growing cancer cells. It is not clear whether or not the changes in the amide I and the DNA/RNA phosphate bands are due to cancerous disease or merely because one cell type shown here is inactive, whereas the other is actively growing and dividing. Just about all cell spectra from cultured cells reported in the literature (see the following section) exhibit the spectral patterns shown in Figure 13.29(c) and (d), yet nearly all of these were either cancer or at least transfected cell lines; thus, the origin of the spectral differences between traces (a) and (b), on the one hand, and traces (c) and (d), on the other, cannot yet be unambiguously assigned to cancer. However, most cancer tissue spectra do not exhibit the low-frequency amide I shoulder shown in trace (d) as pronouncedly, whereas individual cancer cells did.

13.9.2 SCP of cultured cells to study the effects of the cell cycle and of drugs on cells

In the past decade, several SCP studies have been reported that investigated the effect of anticancer agents on cell cultures, the discrimination between closely related transfected cell lines, the confluence rate, and cell-cycle studies. Gaigneaux *et al.* [99] reported very slight spectral differences, detected *via* PCA, of a cancer

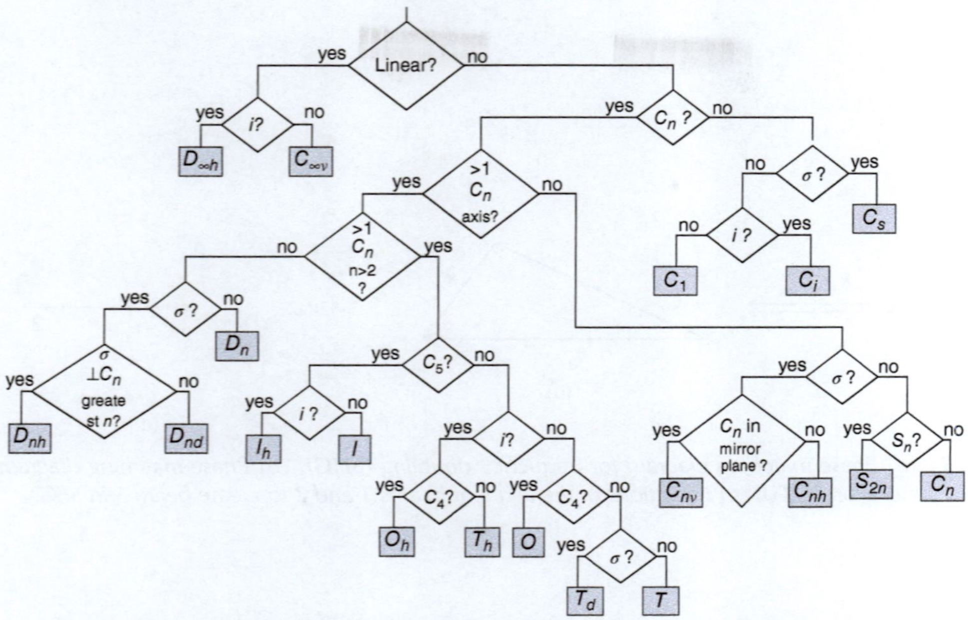

Figure 2.6 Scheme for classifying molecules into symmetry groups. From Ref. 2.

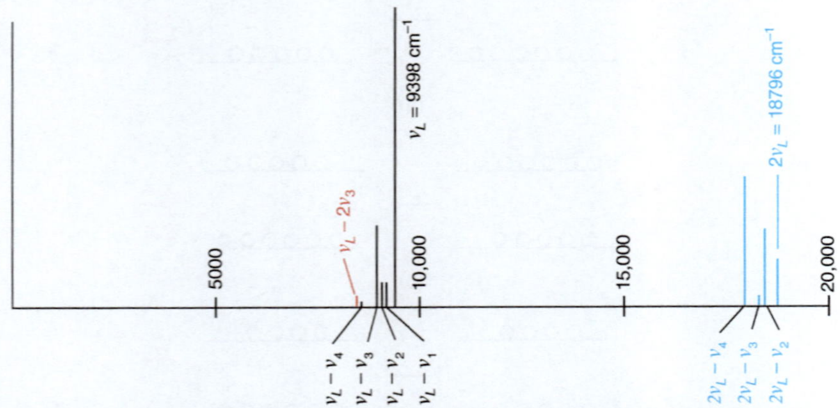

Figure 6.4 Schematic Raman and hyper-Raman spectra of liquid CCl_4, excited at 1.064 μm. See text for details.

Modern Vibrational Spectroscopy and Micro-Spectroscopy: Theory, Instrumentation and Biomedical Applications, First Edition.
Max Diem.

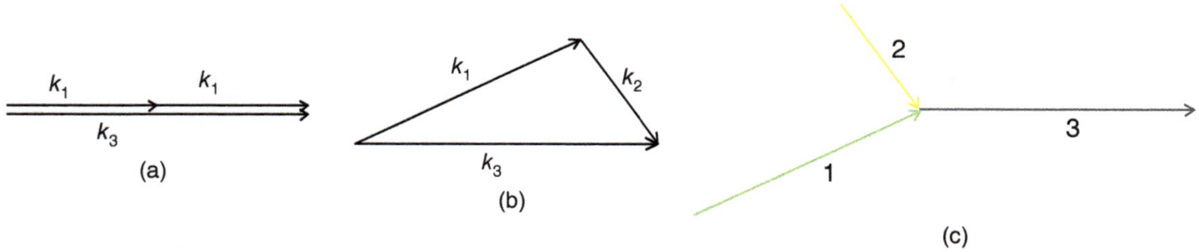

Figure 6.5 *(a) Phase-matching diagram for frequency doubling (SHG). (b) Phase-matching diagram for sum-frequency generation (SFG). (c) Directions of the incident beams 1 and 2 to create beam 3 in SFG.*

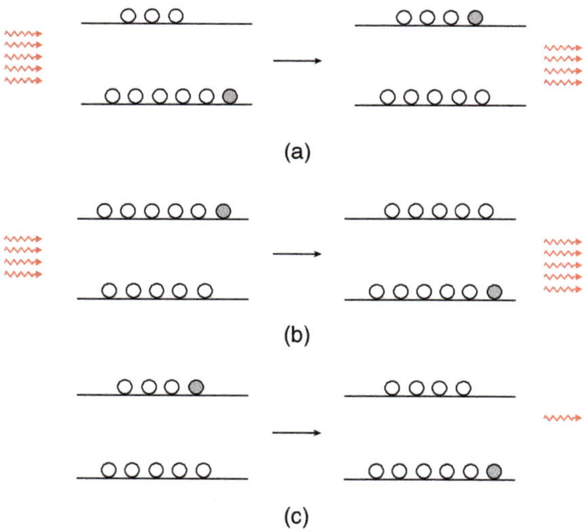

Figure 6.10 *(a) (Stimulated) absorption, (b) stimulated emission, and (c) spontaneous emission processes.*

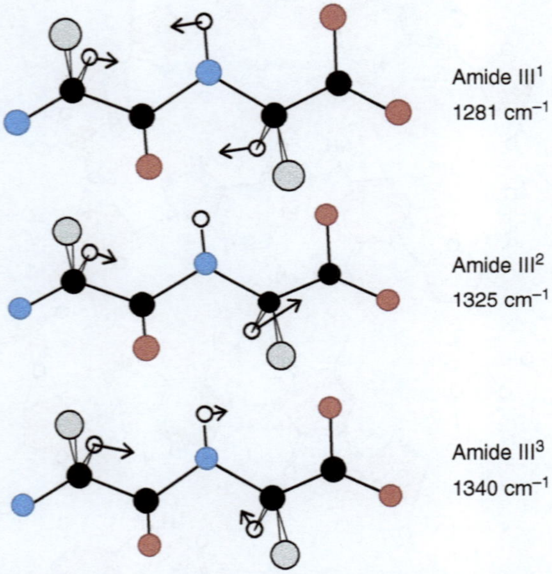

Figure 10.5 *Reassignment of the amide III vibrations of the peptide linkage. See text for detail. Red spheres: oxygen, blue spheres: nitrogen, gray spheres: methyl groups, and black spheres: carbon. The hydrogen atoms of the* $-CH_3$ *and* $-NH_3^+$ *groups are omitted for clarity.*

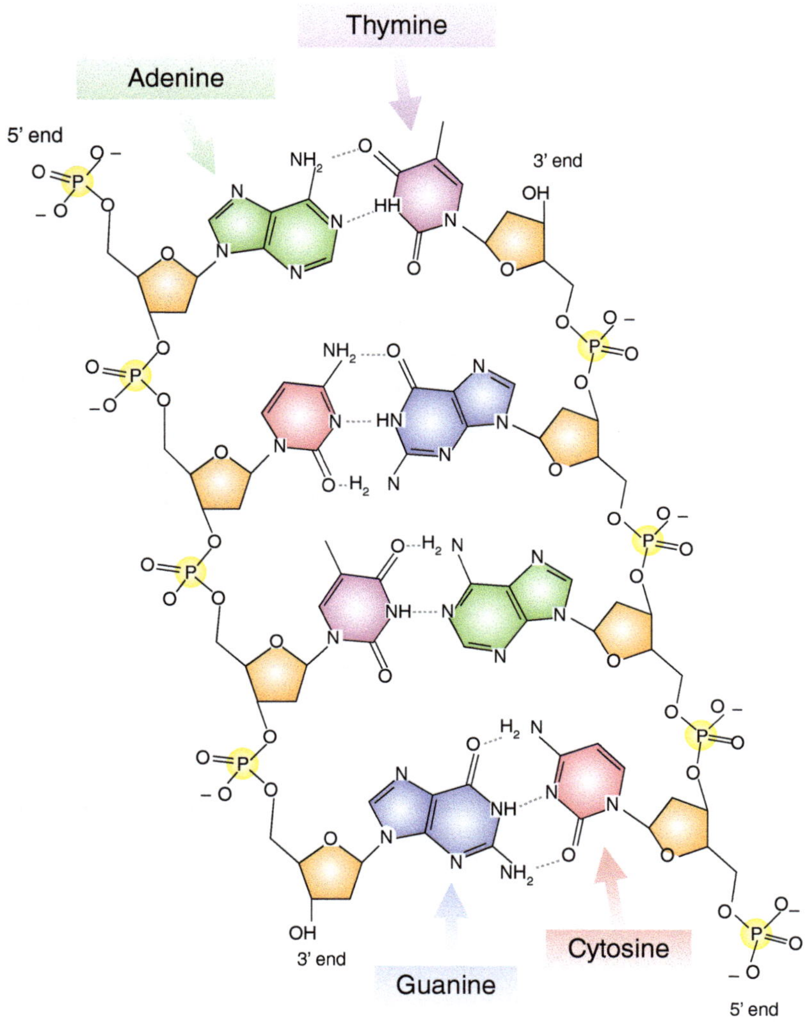

Figure 10.13 *Chemical structure of a DNA segment. Wikipedia, 2014 [33]. Reproduced with thanks to Madeleine Price Ball under the terms of the CC0 Creative Commons license.*

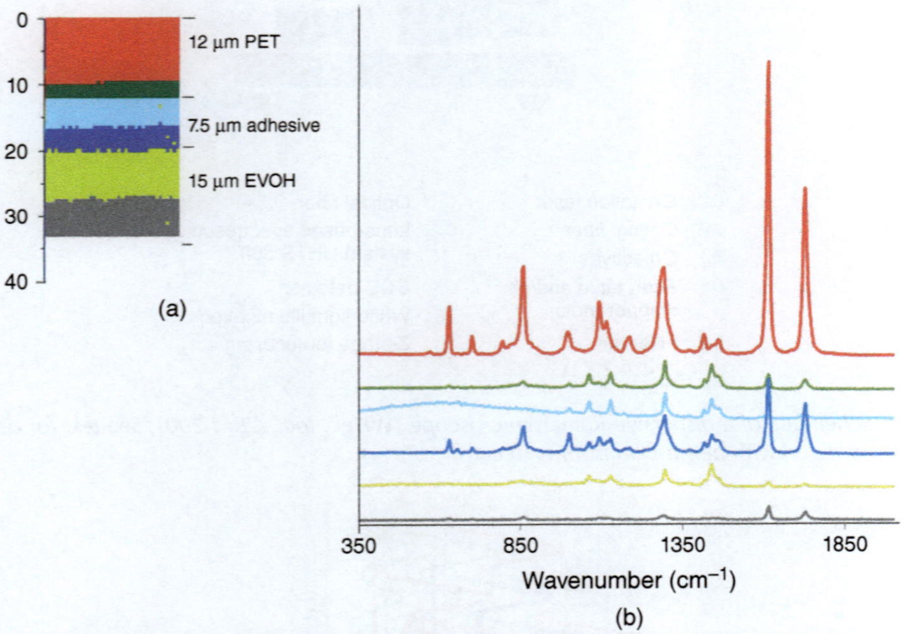

Figure 11.4 *(a) Cluster analysis results of a depth scan into a film laminate. (b) Mean cluster spectra for the distinct polymer and adhesive layers.*

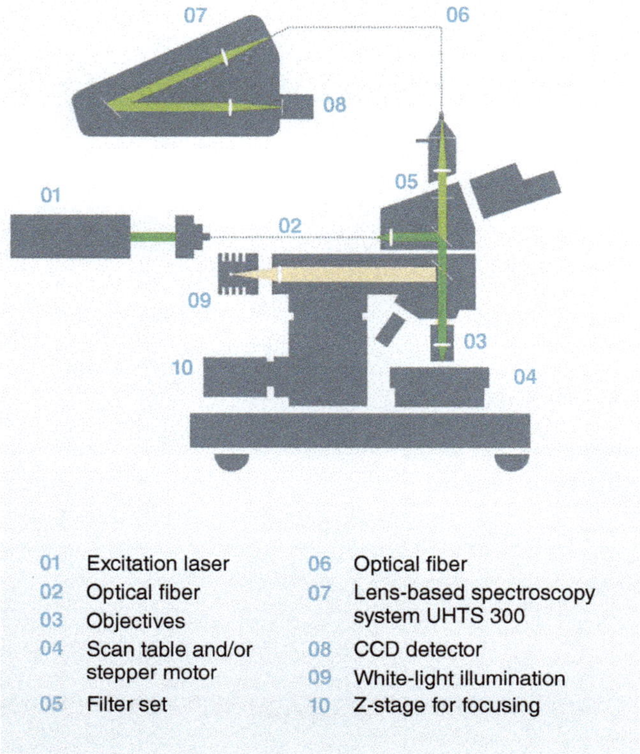

Figure 11.5 *Schematic of a dispersive Raman microscope (WITec, Inc., CRM 200). See text for details. WITec, 2014, [5]. Reproduced with permission from WITec.*

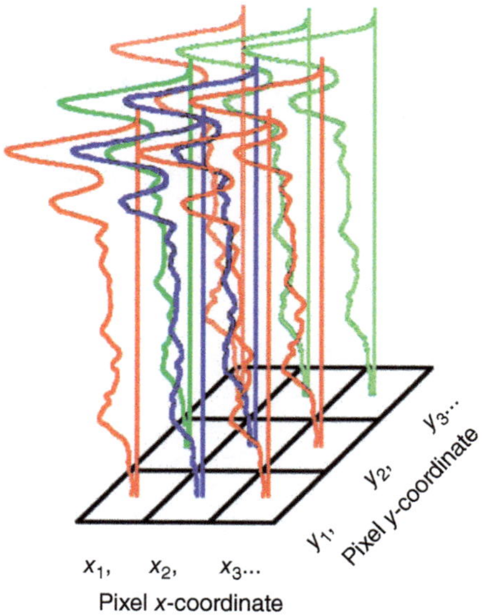

Figure 12.1 *Schematic representation of a hyperspectral data cube. Diem et al., 2013, [4]. Reproduced with permission from John Wiley & Sons.*

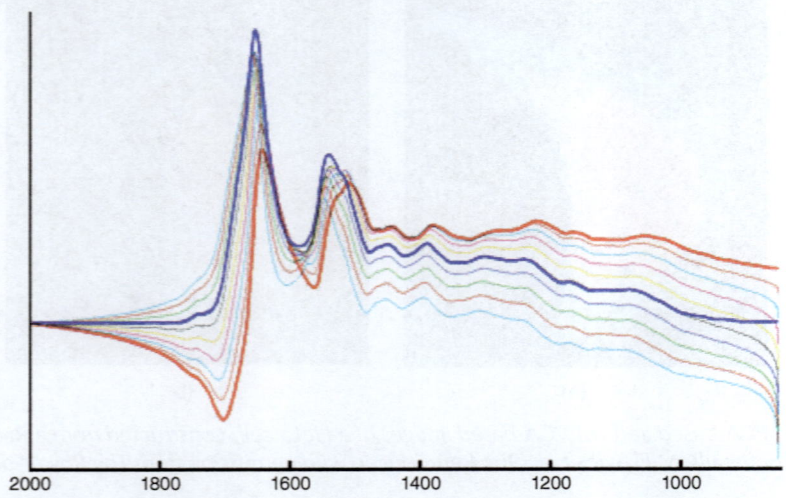

Figure 12.7 Effect of the phase correction algorithm on a distorted spectrum (red) to the corrected spectrum (blue) using different trial phase angles.

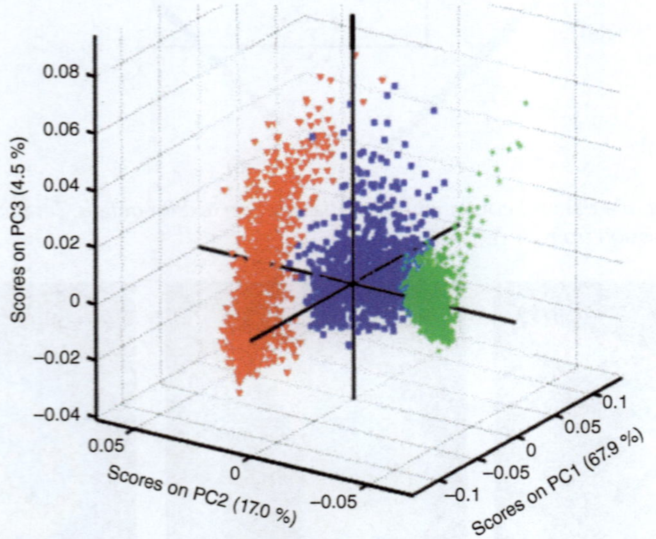

Figure 12.9 Example of a 3D "scores plot" of spectral data, demonstrating the separability of a data set in a space of vastly reduced dimensionality (3 dimensions vs. the original 450 dimensions).

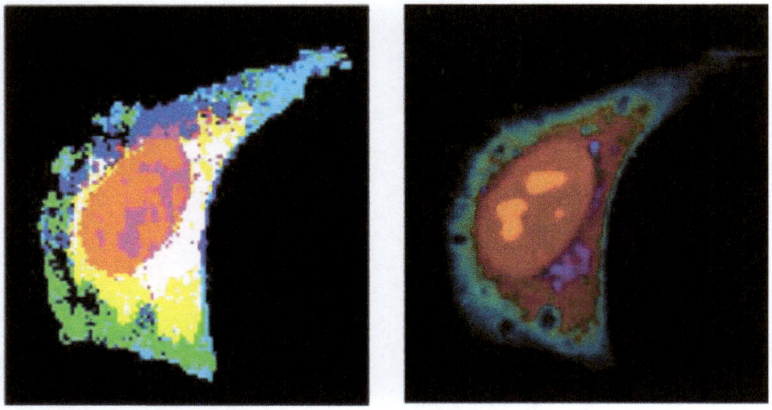

(a) (b)

Figure 12.10 *(a) PCA-based and (b) VCA-based images of a HeLa cell, constructed from a Raman hyperspectral data set. See text for details. Miljkovič et al. [30]. Reproduced with permission from The Royal Society of Chemistry.*

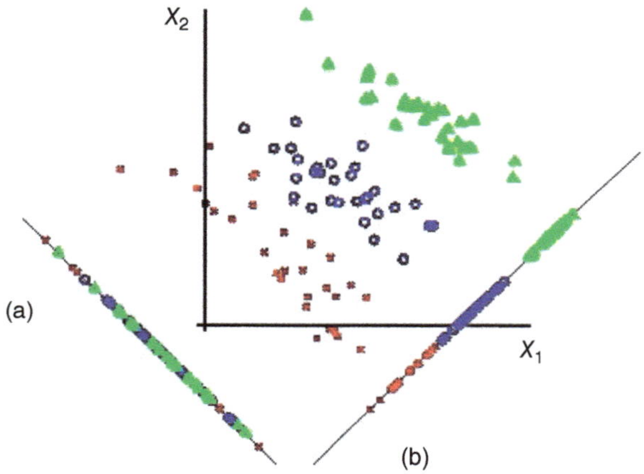

Figure 12.11 *LDA plot of a data set containing three classes. See text for details. Farag and Elhabian, 2008, [33]. Reproduced with permission from A. A. Farag.*

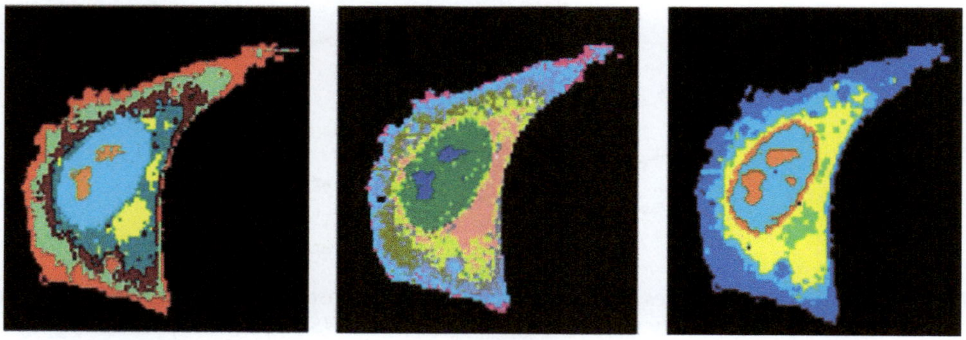

(a) (b) (c)

Figure 12.12 *(a) KMCA, (b) HCA, and (c) DCCA-based images of the same Raman data set shown before in Figure 2.10. See text for the different subcellular regions depicted. Miljkovič et al. [30]. Reproduced with permission from The Royal Society of Chemistry.*

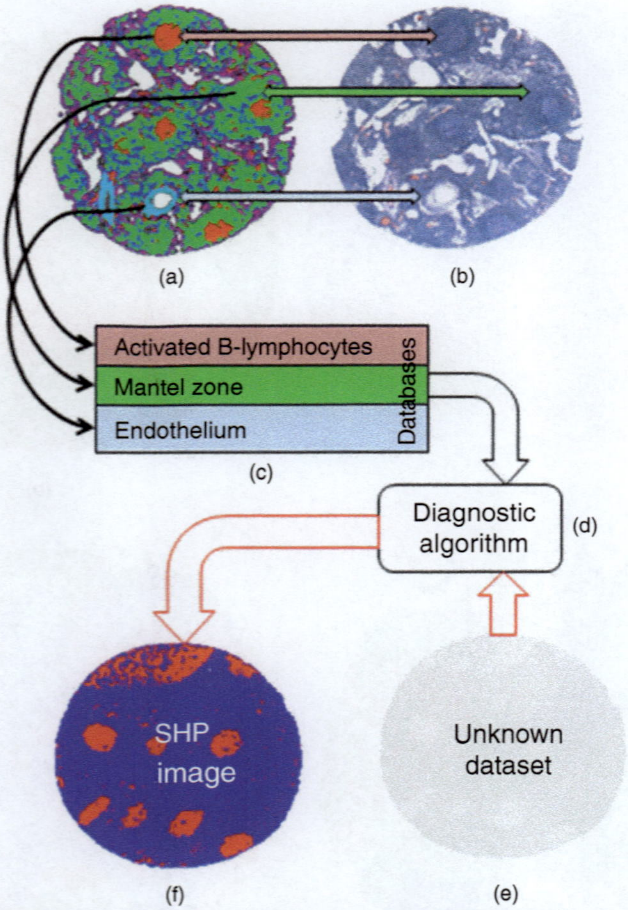

Figure 13.2 *Schematic of SHP process. (a) Pseudo-color image of tissue section obtained from hyperspectral data cube by HCA. (b) Photomicrograph of same tissue section, H&E-stained after IR data acquisition. (c) Databases extracted from HCA image during annotation process. (d) Diagnostic algorithm trained with database "C." (e) Hyperspectral data set from unknown sample. (f) Pseudo-color SHP image produced by diagnostic algorithm to distinguish activated B-lymphocytes from other tissue types. Diem et al., 2013 [10]. Reproduced with permission from Wiley-VCH.*

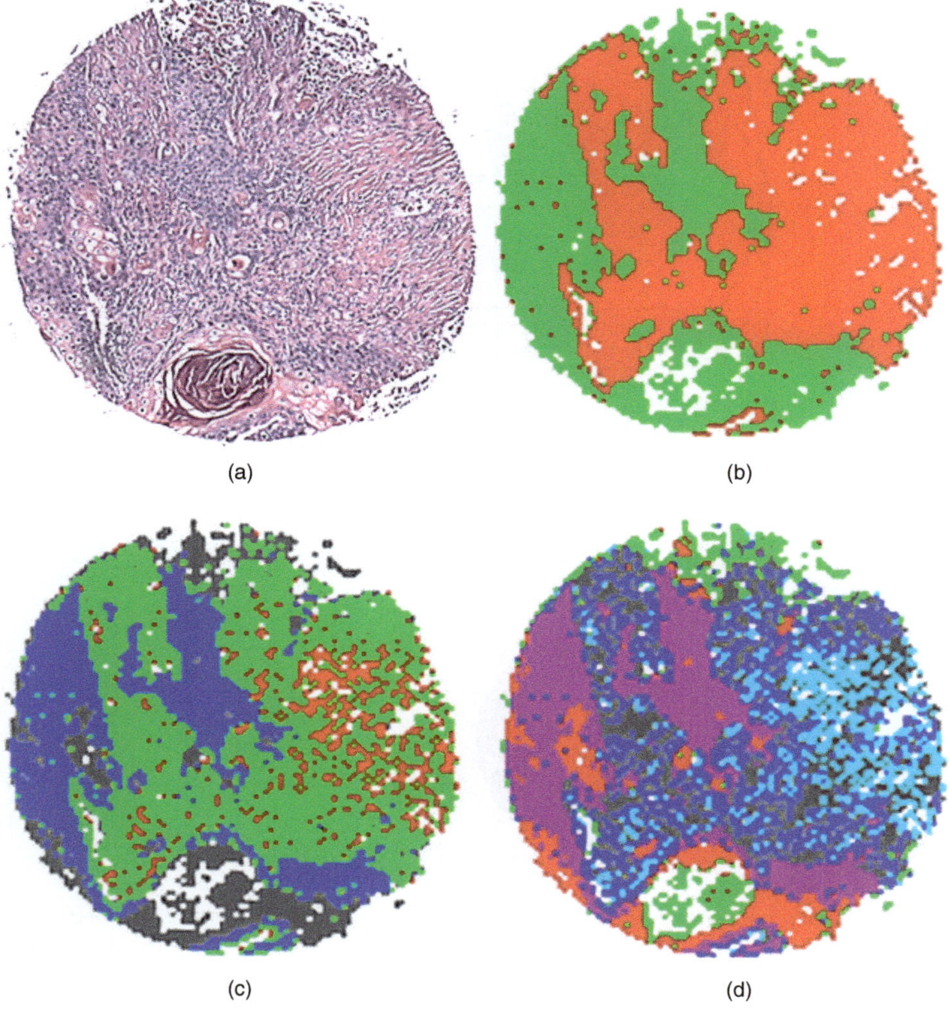

(a) (b) (c) (d)

Figure 13.3 *(a) Photomicrograph of an H&E-stained TMA spot. (b) 2-cluster, (c) 4-cluster, and (d) 6-cluster HCA pseudo-color images of same tissue spot. See text for details.*

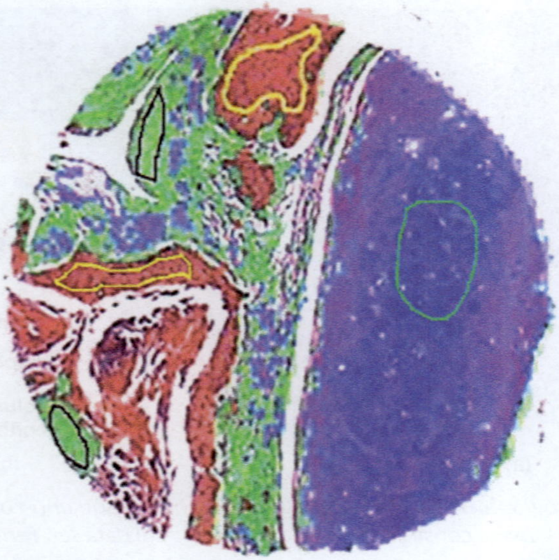

Figure 13.4 *Superposition of an H&E image and a HCA cluster image for annotation. The free-hand figures drawn by the pathologist identify regions of homogeneous tissue/disease state within one spectral cluster.*

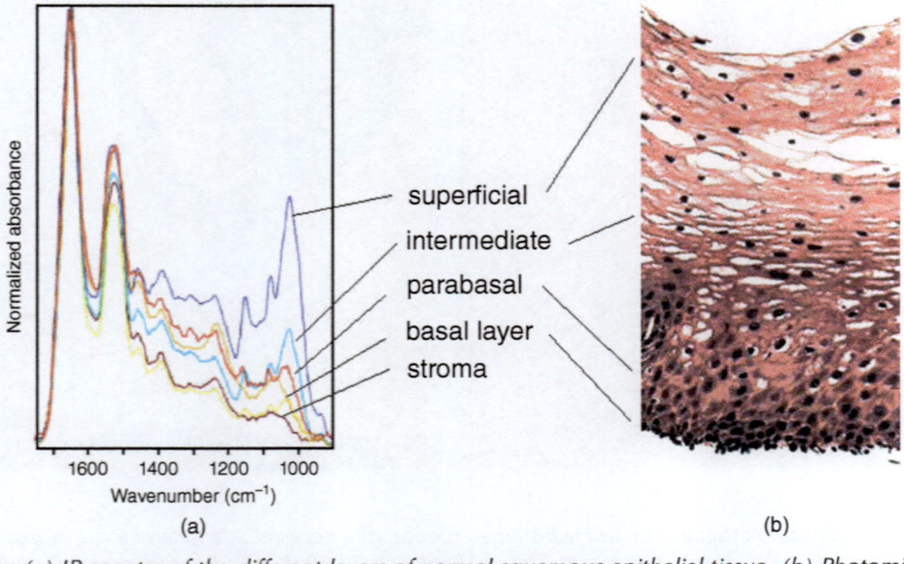

Figure 13.6 *(a) IR spectra of the different layers of normal squamous epithelial tissue. (b) Photomicrograph of H&E-stained epithelial tissue.*

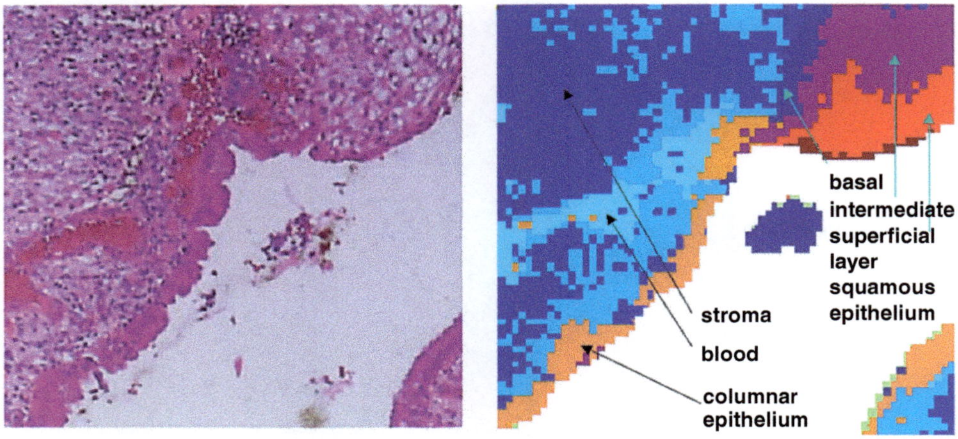

(a) (b)

Figure 13.7 *(a) Photomicrograph of an H&E-stained section of the squamo-columnar junction in a cervical biopsy. (b) Pseudo-color HCA image constructed from a hyperspectral data set. Fernandez et al. [29]. Reproduced with permission from Macmillan.*

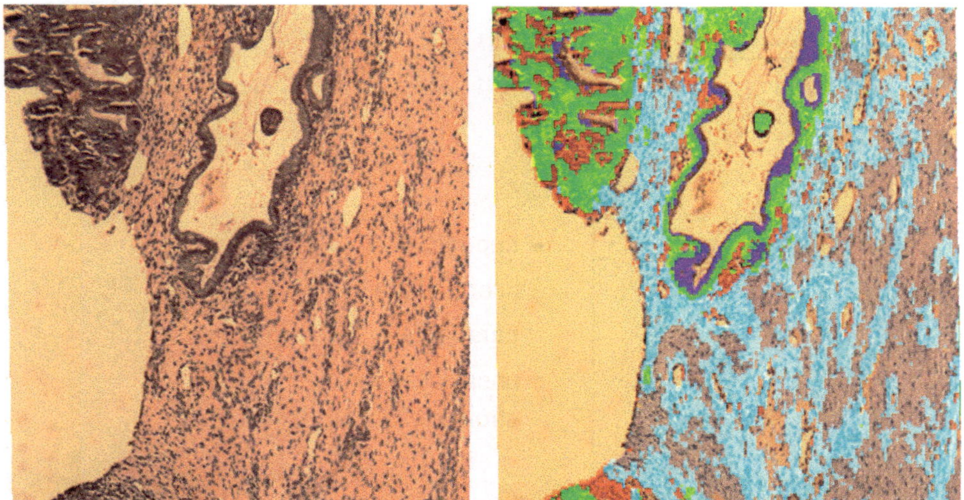

(a) (b)

Figure 13.8 *(a) Photomicrograph of an H&E-stained section of a cervical adenocarcinoma biopsy. (b) Overlay of HCA image on the H&E image. See text for details. Steller et al., 2006 [35]. Reproduced with permission from Springer Science and Business Media.*

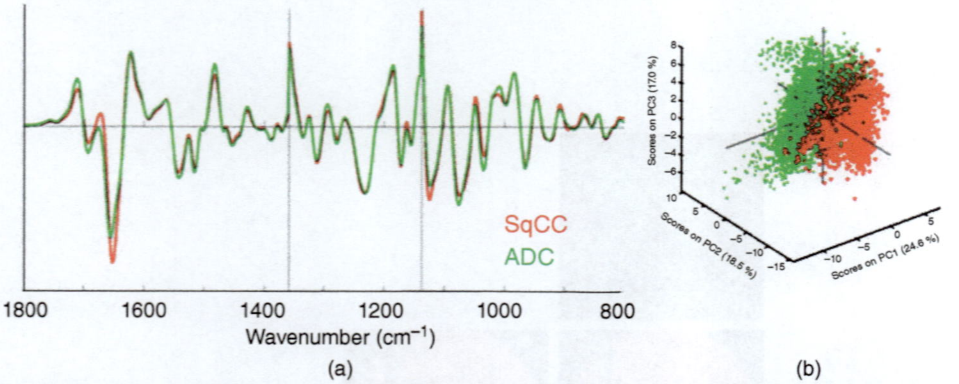

Figure 13.9 (a) Comparison of mean second derivative spectra of cervical adenocarcinoma (green) and cervical squamous cell carcinoma (red). Note the normalization of the spectra in three separate regions (see text). (b) PCA scores plot of the separation of ADC and SqCC.

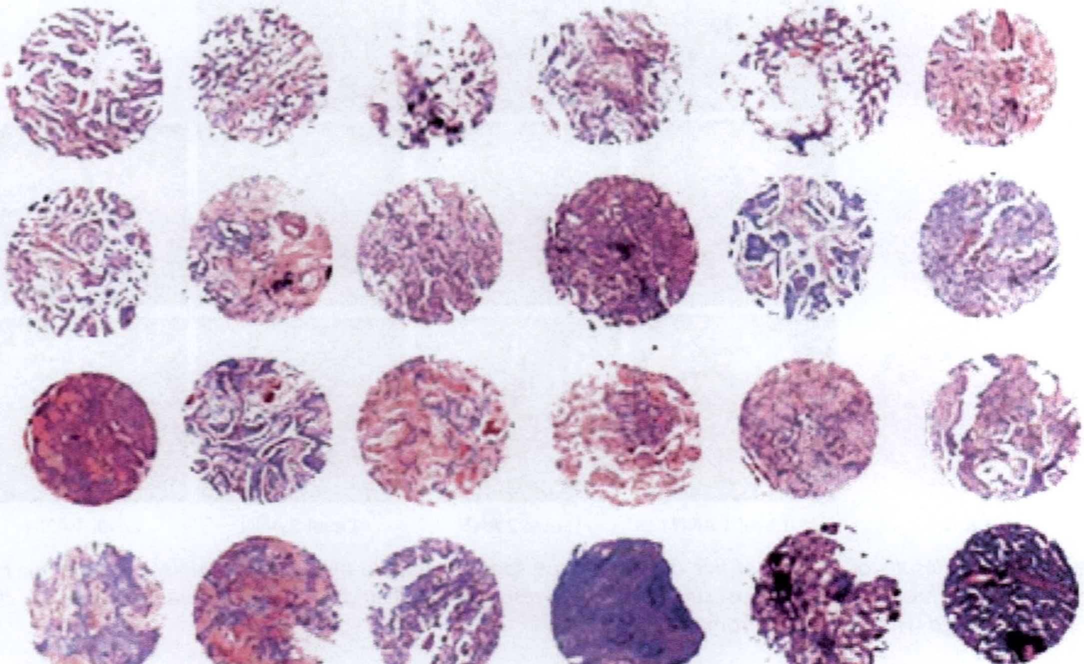

Figure 13.10 View of a section of a stained tissue microarray (TMA). Bird, et al., 2012 [15]. Reproduced with permission from Nature.

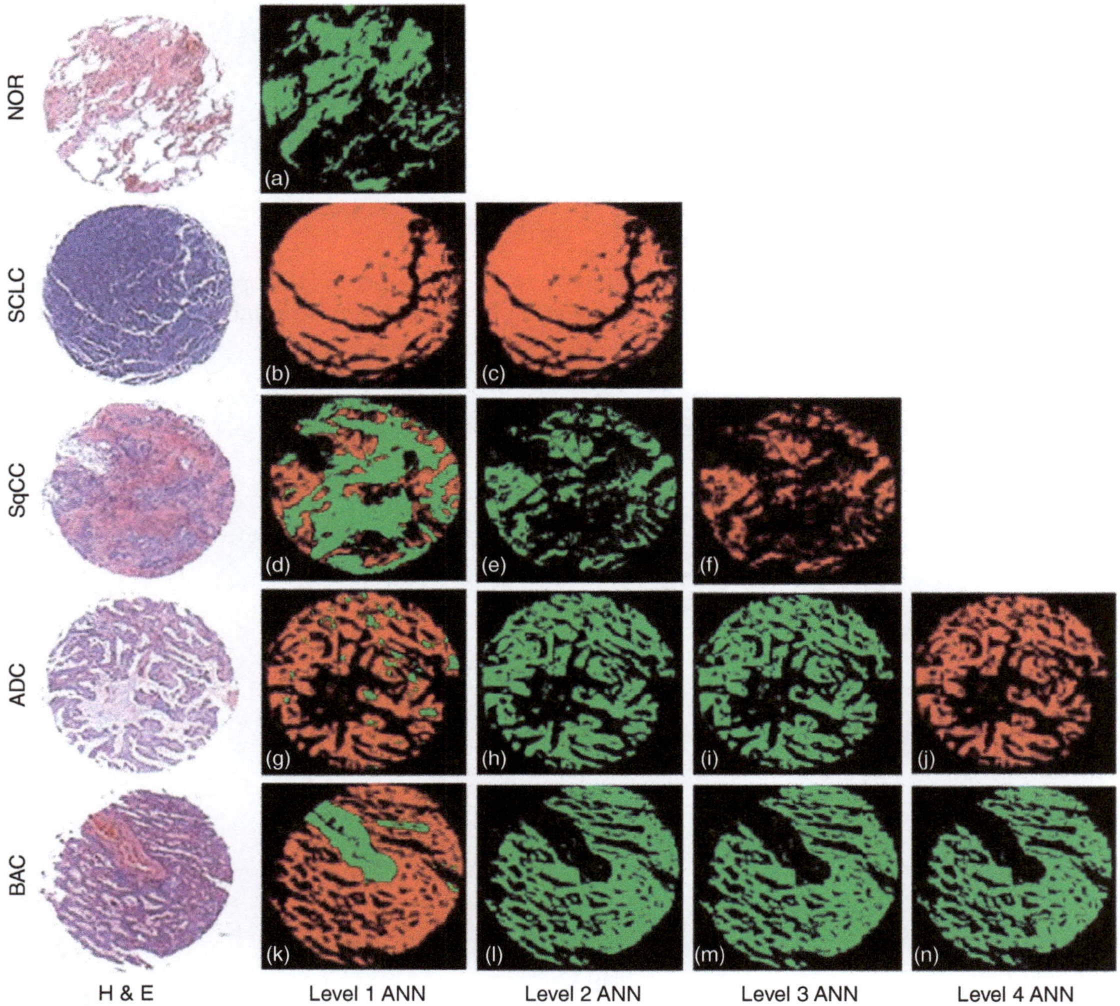

Figure 13.12 *Results of analysis of five different tissue spots by trained hierarchical ANNs. Left column: H&E images, second through fifth columns: successive ANN results, as indicated. See text for details. Bird, et al., 2012 [15]. Reproduced with permission from Nature.*

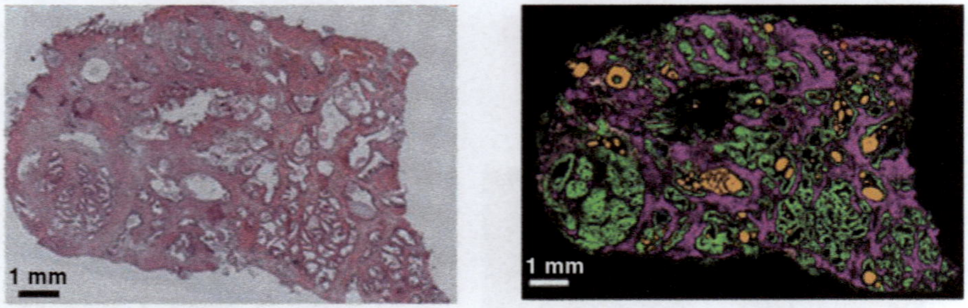

(a) (b)

Figure 13.13 *(a) Photomicrograph of an H&E-stained large prostate biopsy section and (b) SHP-based pseudocolor image of same section, embedded in paraffin. The color scheme is as follows: epithelium: green, smooth muscle: purple, blood: red, extracellular matrix: pale yellow, fibrous stroma: pink, concretion: orange, and lymphocytes: blue. Bassan, et al., 2014 [42]. Reproduced with permission of SPIE.*

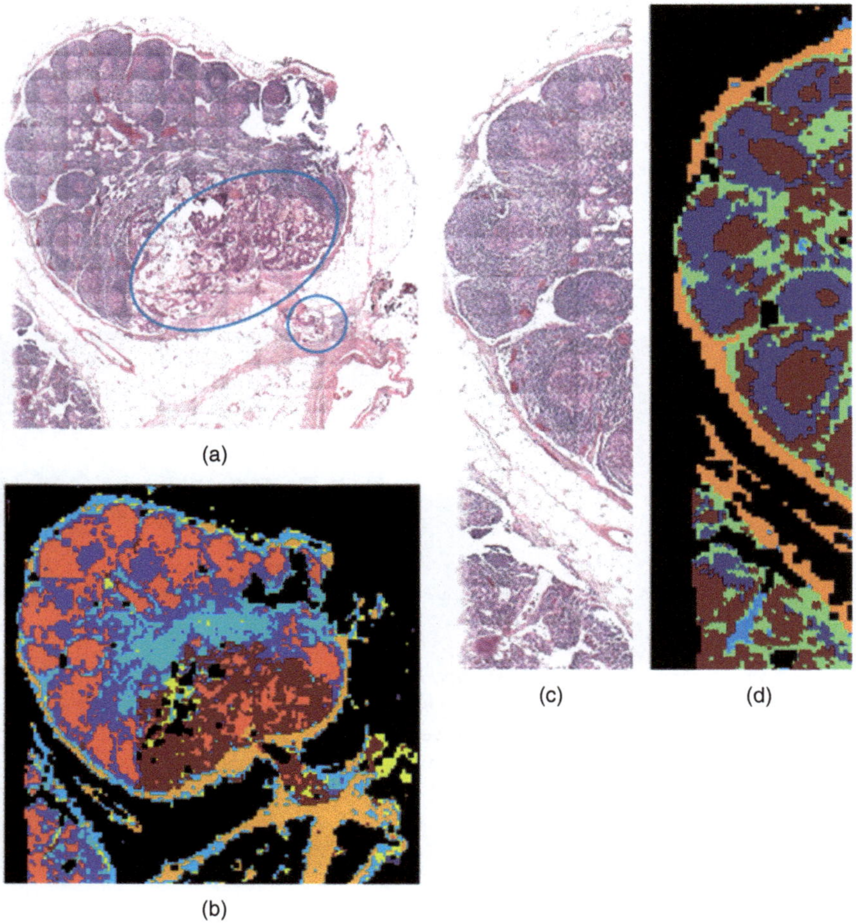

Figure 13.14 *(a) Photomicrograph of an 8 mm × 8 mm lymph node tissue section with a colon cancer metastasis (in blue ellipse and circle). (b) HCA-based pseudo-color image of section shown in (a) depicting different tissue types, including the adenocarcinoma (dark red). (c) Photomicrograph of partial section of same lymph node, showing reactive lymph node follicles. (d) HCA-based pseudo-color image of the section shown in (s); demonstrating the distinction between activated (dark brown) and nonactivated (dark blue) B-lymphocytes (dark blue) by SHP. Romeo, et al., 2005 [47].*

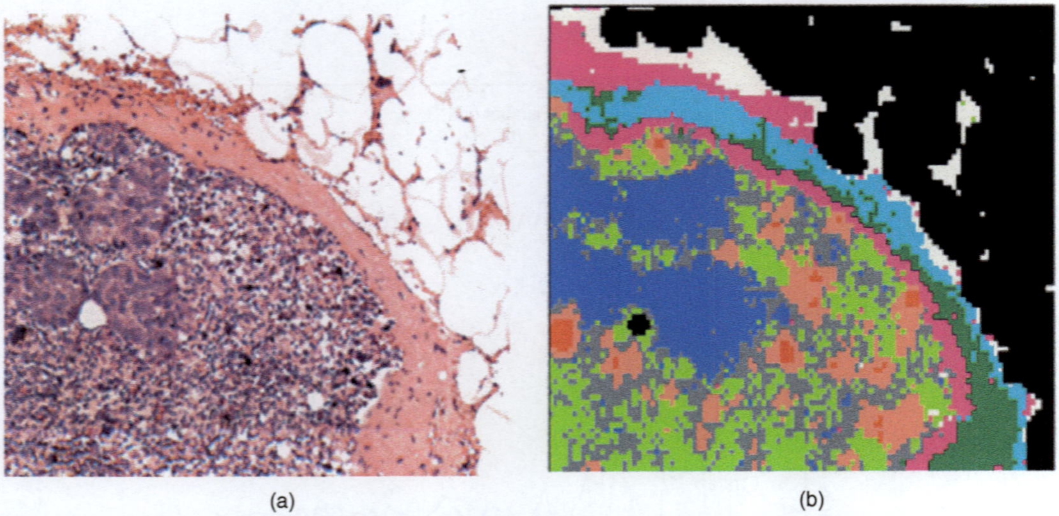

Figure 13.15 *(a) Photomicrograph of a lymph node tissue section, measuring about 0.8 mm on edge. (b) HCA-based pseudo-color image of section shown in (a). The capsule (green, blue, and purple), the micrometastasis (dark blue), and the lymphocytes (gray, green, and salmon) regions are clearly distinguished. Bird, et al., 2009 [50]. Reproduced with permission from Wiley-VCH.*

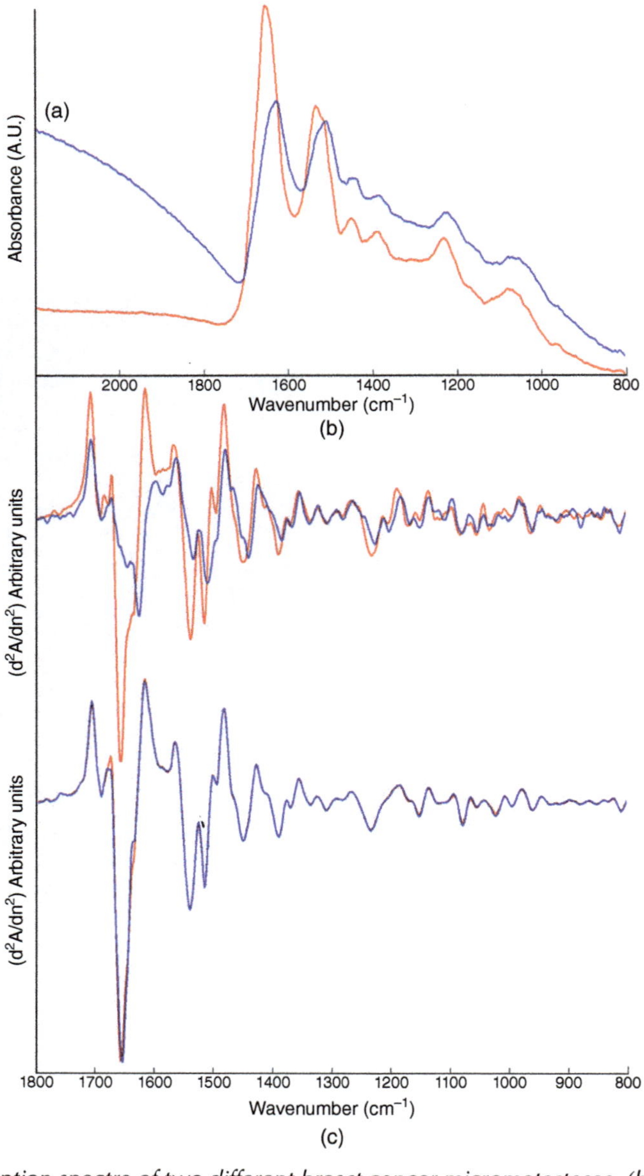

Figure 13.16 *(a) Absorption spectra of two different breast cancer micrometastases. (b) Corresponding second derivative spectra. (c) Spectra after R-Mie scattering correction. Bird et al., 2008 [49].*

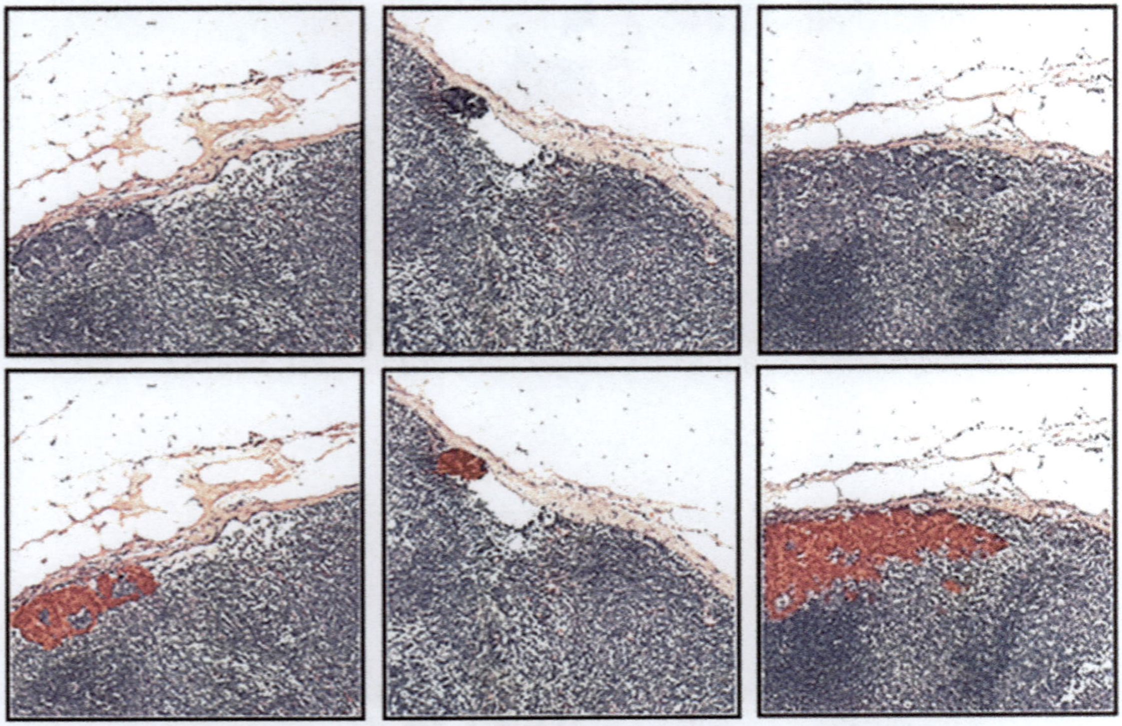

Figure 13.17 *Top row: photomicrographs of H&E-stained, 1 mm × 1 mm section of axillary lymph nodes harboring micrometastases. Bottom row: overlay of HCA-derived regions indicating tissue abnormalities. Bird et al., 2009 [50]. Reproduced with permission from Wiley-VCH.*

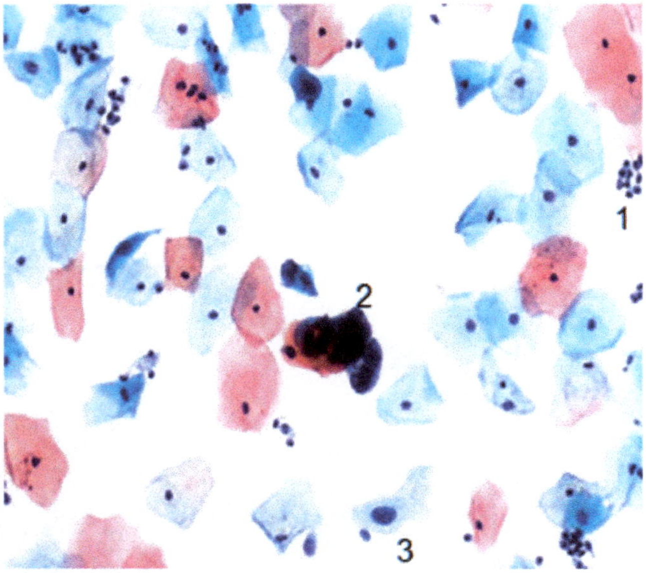

Figure 13.18 *Stained cytopathological sample of cervical epithelial cells. (1) Bacterial contamination, (2) clump of inflammatory cells, and (3) abnormal cell with enlarged nucleus. Schubert et al., 2010 [76].*

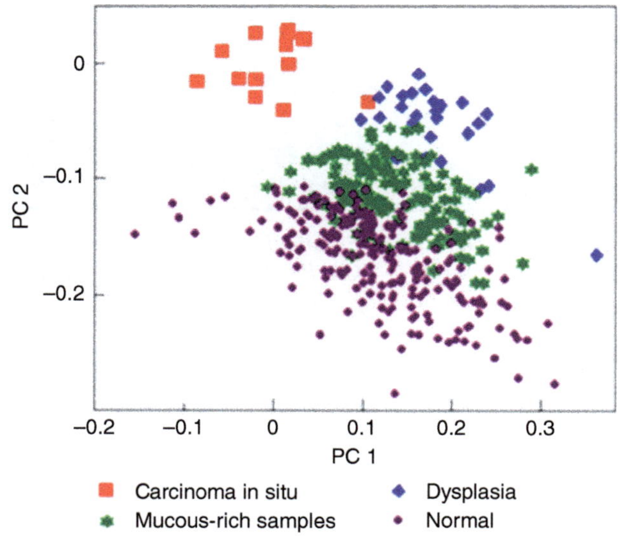

Figure 13.19 *PCA scores plot for macroscopically acquired cervical cell samples.*

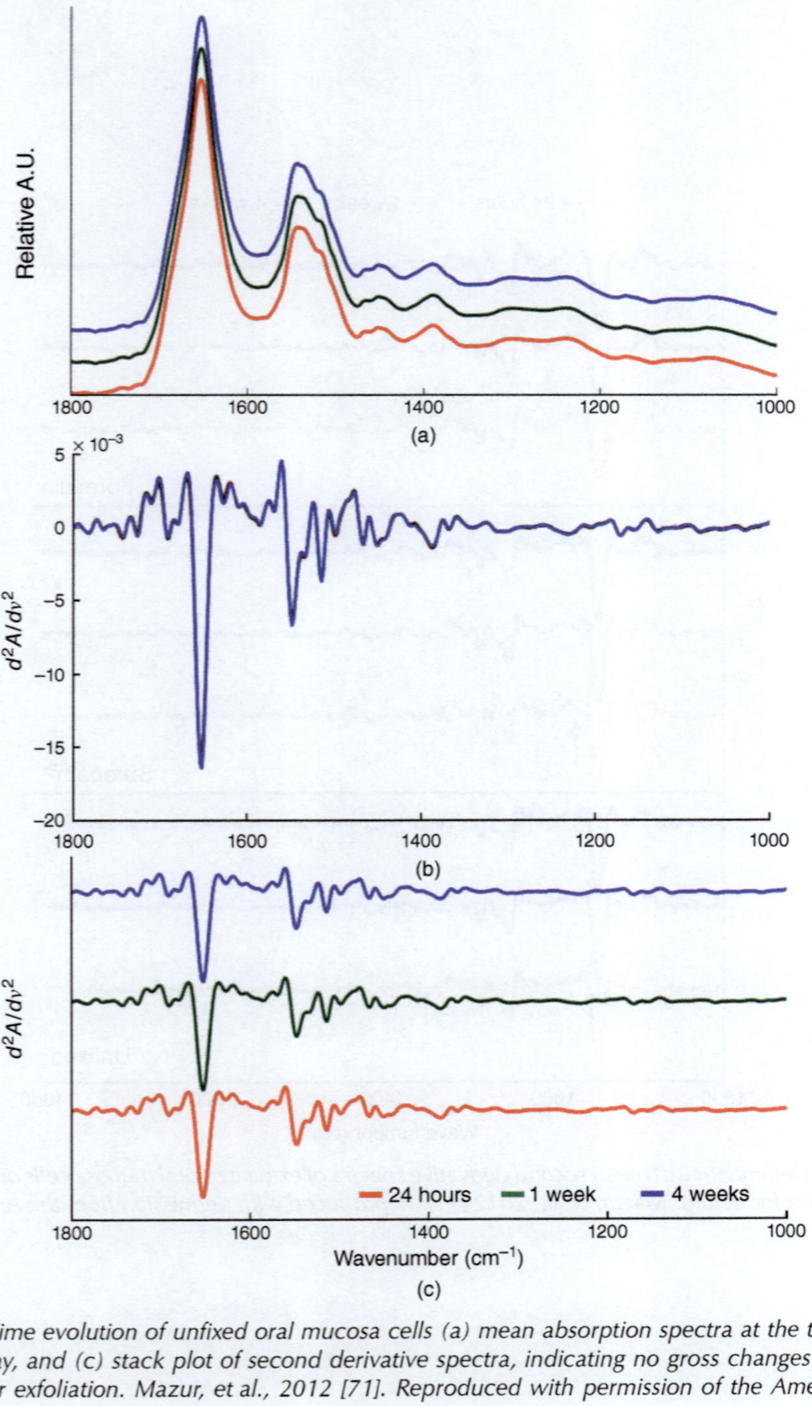

Figure 13.20 *Time evolution of unfixed oral mucosa cells (a) mean absorption spectra at the time points indicated, (b) overlay, and (c) stack plot of second derivative spectra, indicating no gross changes of unfixed cells for 4 weeks after exfoliation. Mazur, et al., 2012 [71]. Reproduced with permission of the American Chemical Society.*

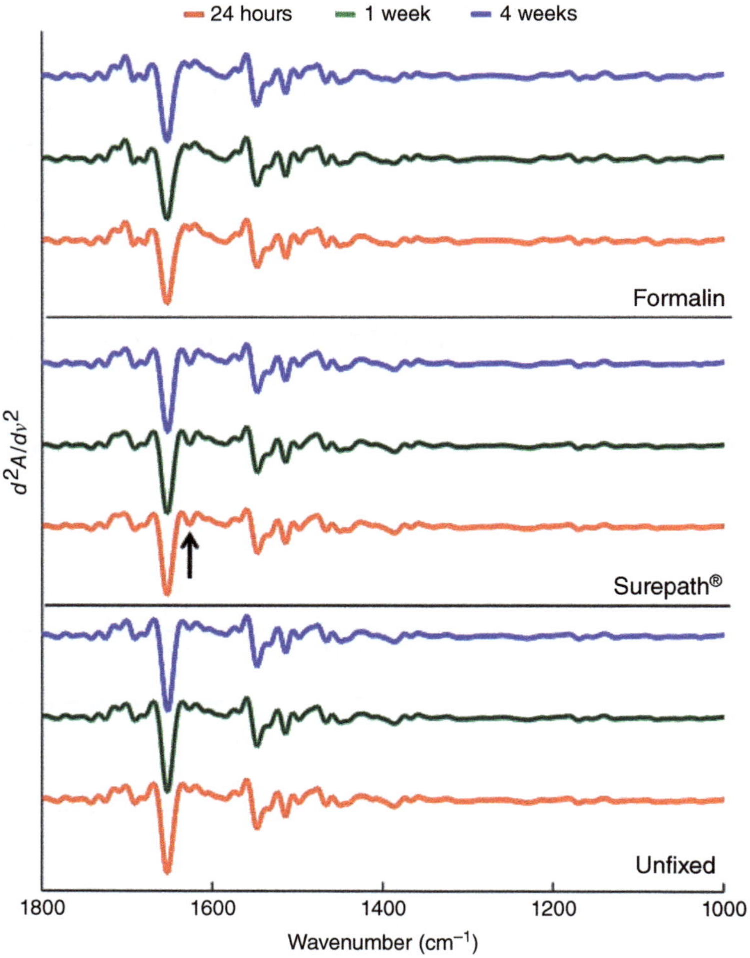

Figure 13.21 *Dependence of mean second derivative spectra of exfoliated oral mucosa cells on time and fixation protocol. See text for details. Mazur, et al., 2012 [71]. Reproduced with permission from the American Chemical Society.*

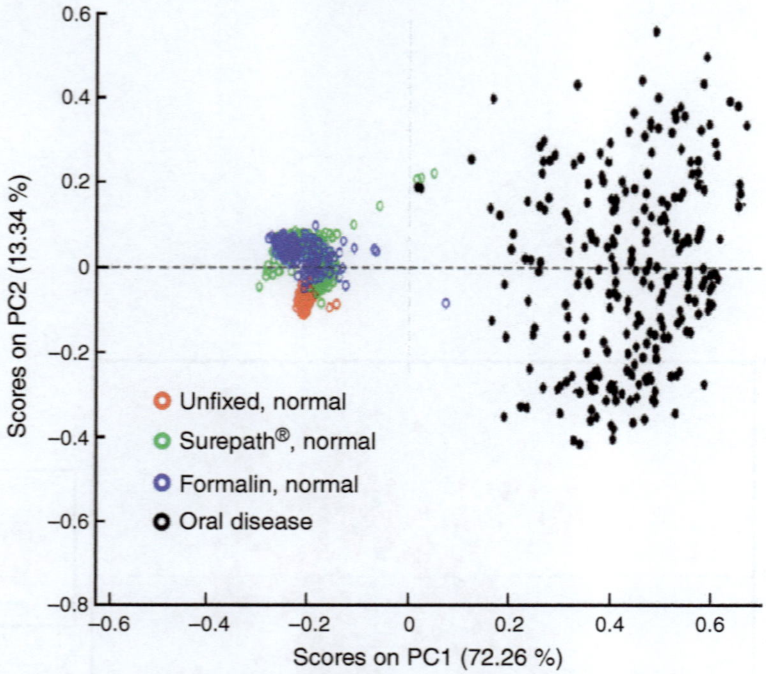

Figure 13.22 *PCA scores plot of spectral differences due to fixation (colored circles) and cellular abnormalities (black circles). Mazur, et al., 2012 [71]. Reproduced with permission from the American Chemical Society.*

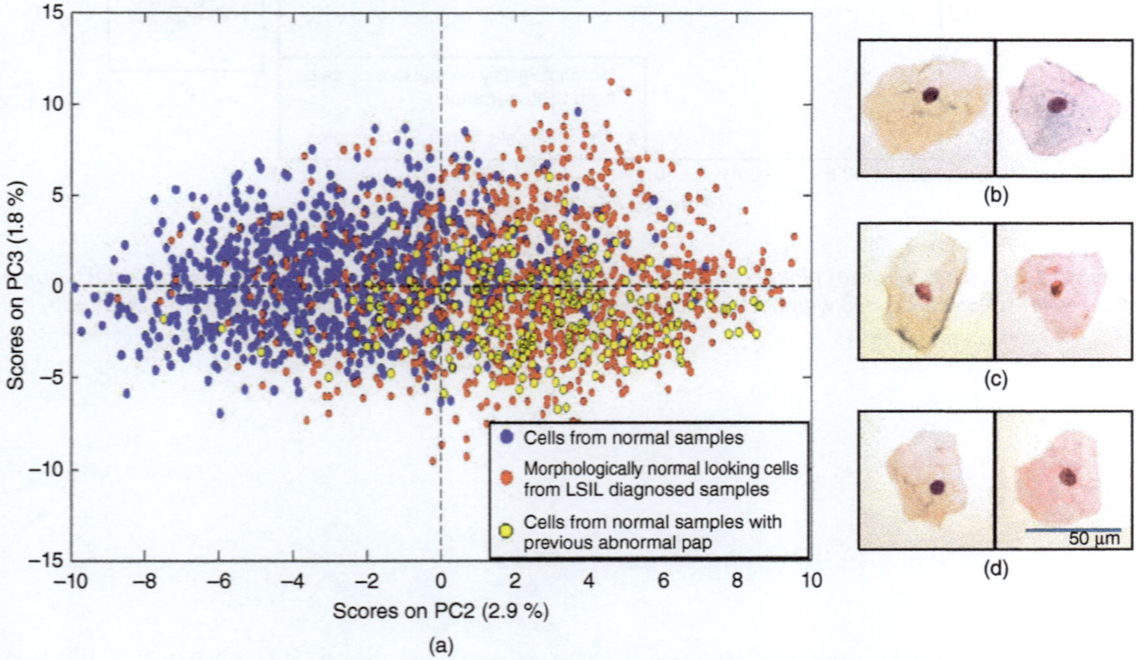

Figure 13.23 *(a) PCA scores plot of spectra of cervical squamous cells from five normal subjects (blue), from five patients diagnosed with LSIL (red) and two patients with history of HSIL (yellow). All cells exhibited normal morphology, as shown in panels (b), (c) and (d) for normal, LSIL and history of HSIL, respectively. (From Ref. [86]).*

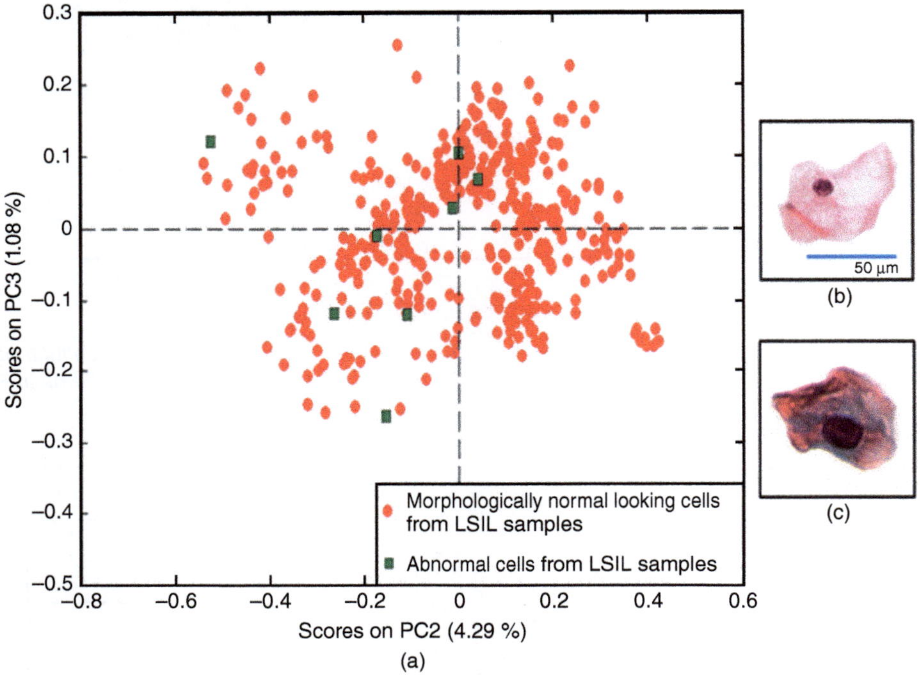

Figure 13.24 *(a) PCA scores plot of cells with normal morphology (red dots) and cells with abnormal morphology (green squares). (b) cell with normal morphology (c) cell with abnormal morphology (From Ref. [86]).*

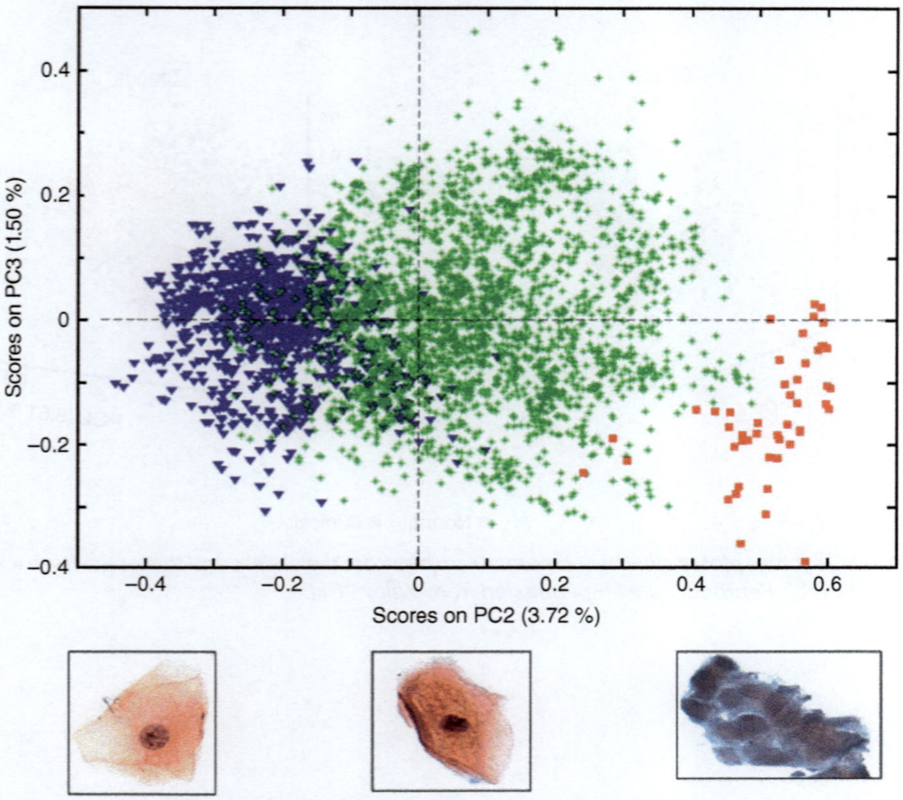

Figure 13.25 PCA scores plot of oral SCP. The cell images represent a normal cell (left), a morphologically normal cell from an abnormal sample (middle), and a clump of cancerous cells. All cells were harvested from the tongue. Papamarkakis, et al., 2010 [83]. Reproduced with permission from Nature.

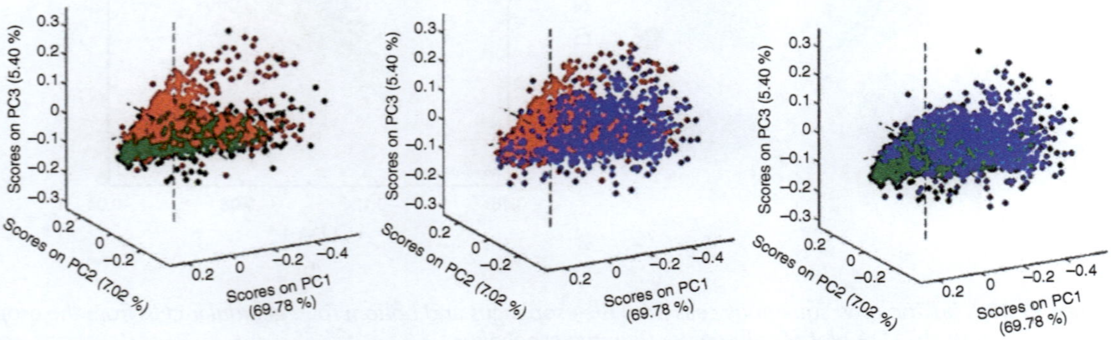

Figure 13.26 PCA scores plot of cells from different anatomical regions of the oral cavity (blue: tongue, red: mouth floor, and green: cheeks). Diem et al., 2014 [92]. Reproduced with permission from Wiley-VCH.

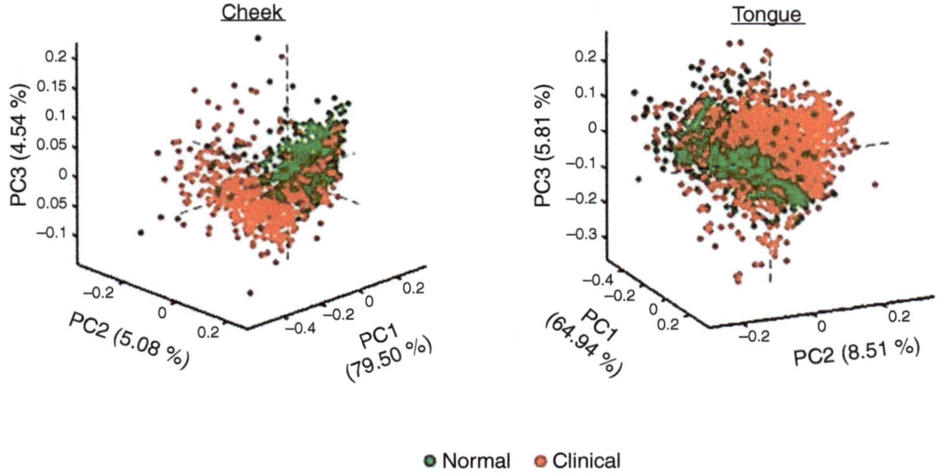

Figure 13.27 *PCA scores plot of clinical and normal samples from the cheeks and the tongue (see text for details). Diem et al., 2014 [92]. Reproduced with permission from Wiley-VCH.*

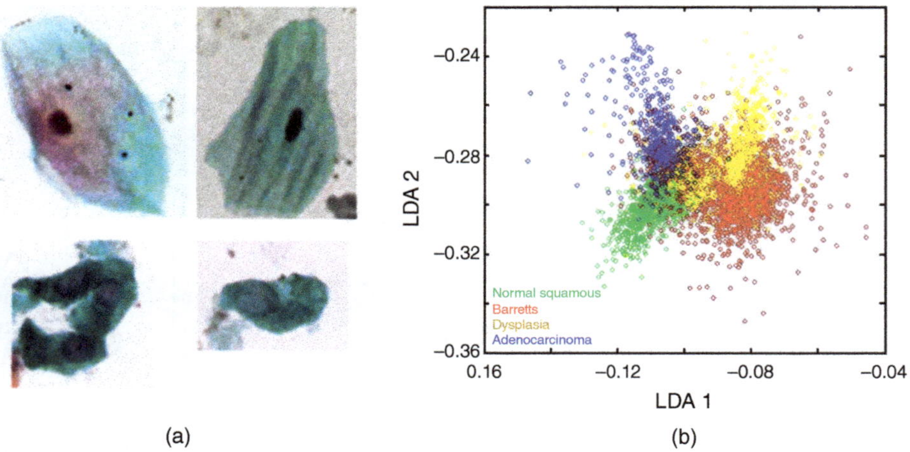

Figure 13.28 *(a) Top row: squamous cells from the esophagus and bottom row: glandular cells from the esophagus (Pap stain). (b) LDA plot of cell spectra from the esophagus.*

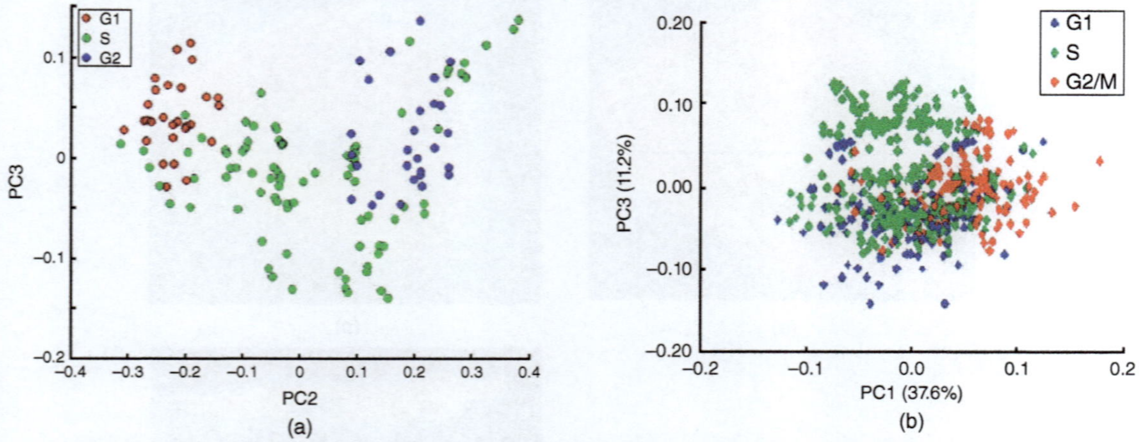

__Figure 13.30__ (a) PCA plot of cell spectra according to their stage in the cell division cycle. Spectra were not corrected for R-Mie scattering. <termDefinition xml:id="c13-tdef-0047">(From Ref. [98].)</termDefinition> (b) Partial least squares regression plot of R-Mie corrected spectra according to their stage in the cell division cycle. Jimenez-Hernandez, M., et al., 2013 [103]. Reproduced with permission from the Royal Society.

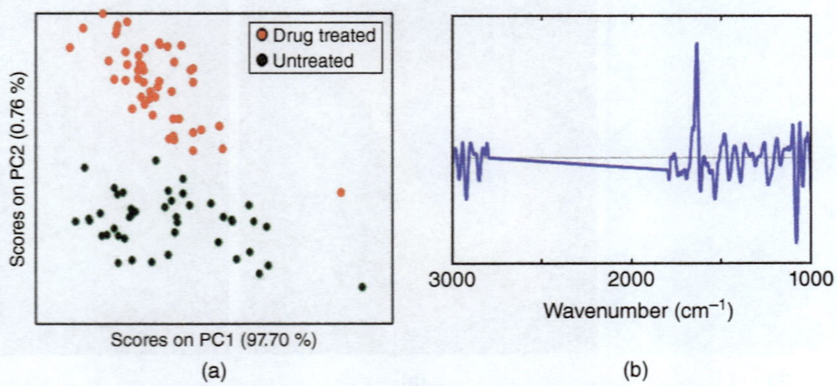

__Figure 13.31__ (a) Scores plot of untreated cells and cells treated with cyclophosphamide monohydrate. (b) Loading vector PC2 along which the PCA data are split. Marcsisin (2011) [107].

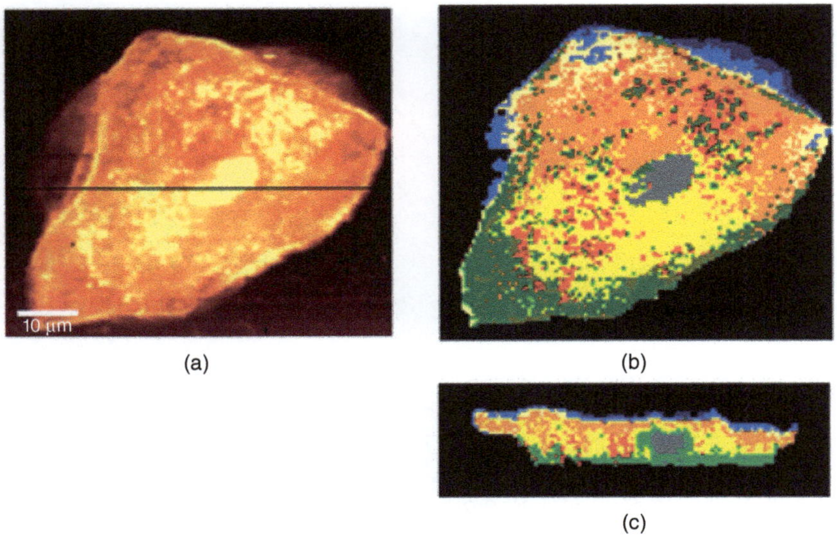

Figure 14.3 (a) Univariate image of integrated C–H Raman stretching intensities of an oral mucosa cell. Brighter yellow hues correspond to higher intensities. (b) Pseudo-color image created by HCA of the same data set. (c) Pseudo-color image of a depth scan along the black line in (a).

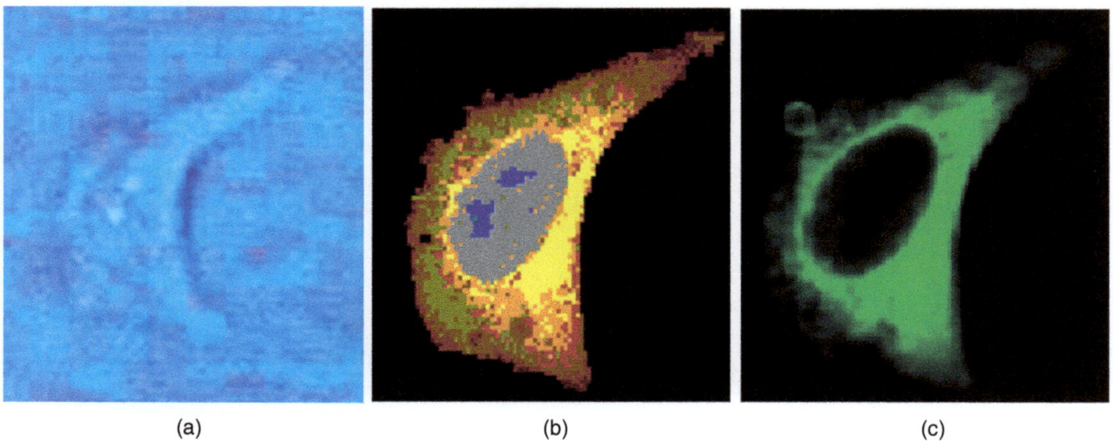

Figure 14.4 (a) Brightfield image of a HeLa cell in aqueous medium. (b) HCA-based Raman image of same cell. (c) Fluorescence image of same cell after staining with mitotracker stain. Matthäus et al. [1]. Reproduced with permission from Elsevier.

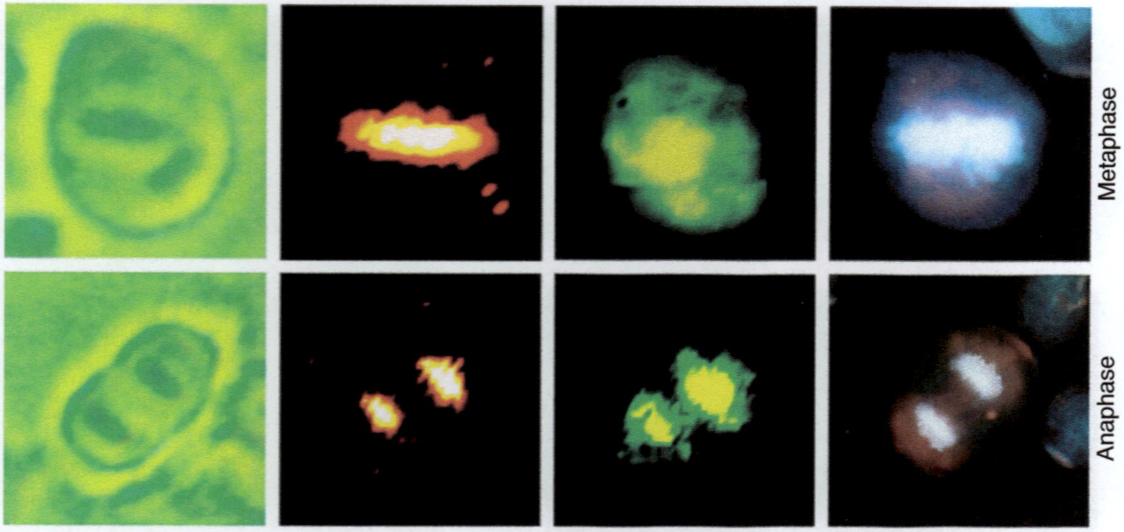

Figure 14.5 Raman intensity images of two different HeLa cells in the metaphase and anaphases of mitosis. See text for details. Matthäus et al., 2006 [18]. Reproduced with permission from Society for Applied Spectroscopy.

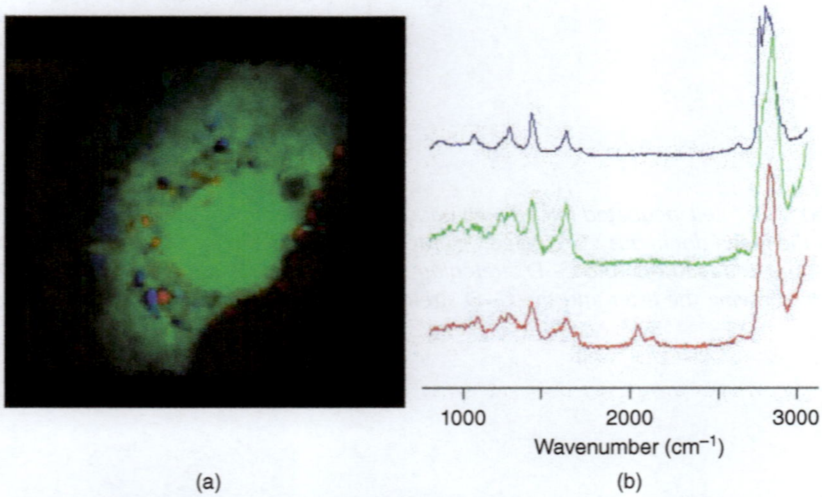

Figure 14.7 (a) VCA-based Raman image of the uptake of deuterated liposomes into an MCF7 cell. Bright green: nucleolus, green: nucleus, dark green: cytoplasm, blue: naturally occurring lipid inclusions in the cytoplasm, and red: deuterated liposomes. (b) Raman endmember spectra of (blue): naturally occurring phospholipids, (green): protein and nucleic acids, (red): deuterated liposomes.

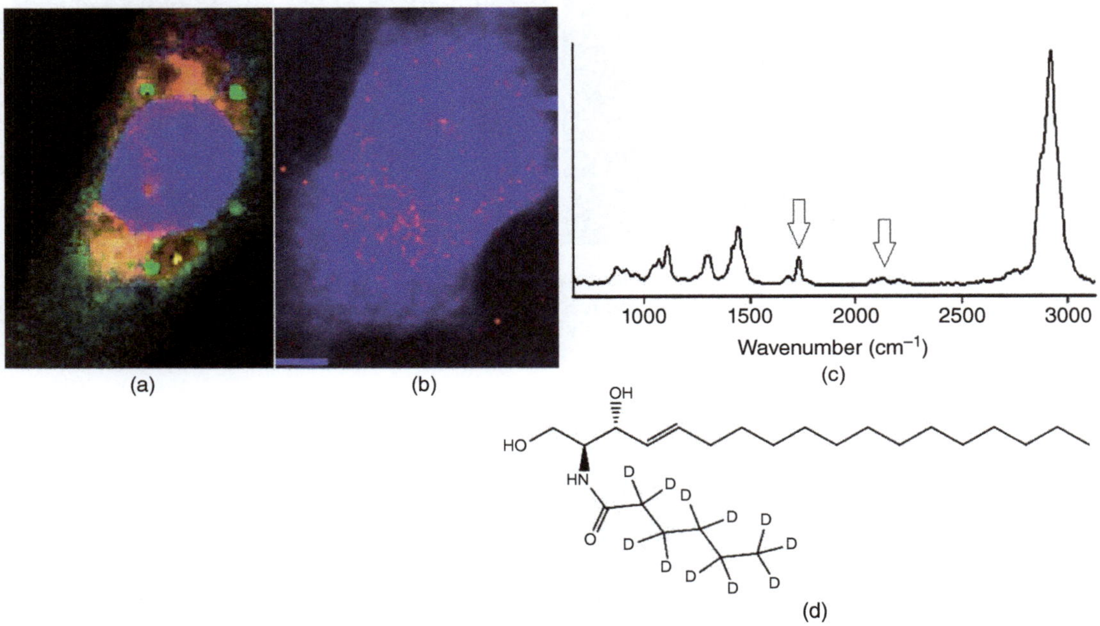

Figure 14.8 *(a) MCF7 cell incubated for 1 h with nanoparticles containing the drug shown in (d). The nanoparticles appear as the green dot in this VCA image. (b) MCF7 cell incubated for 6 h showing the release of the drug, identified by its spectral features (the C–D stretching modes) shown in (c). (c) Raman endmember spectrum of nanoparticles, showing the ester and the C–D stretching vibration (arrows). (d) Structure of the deuterated ceramide.*

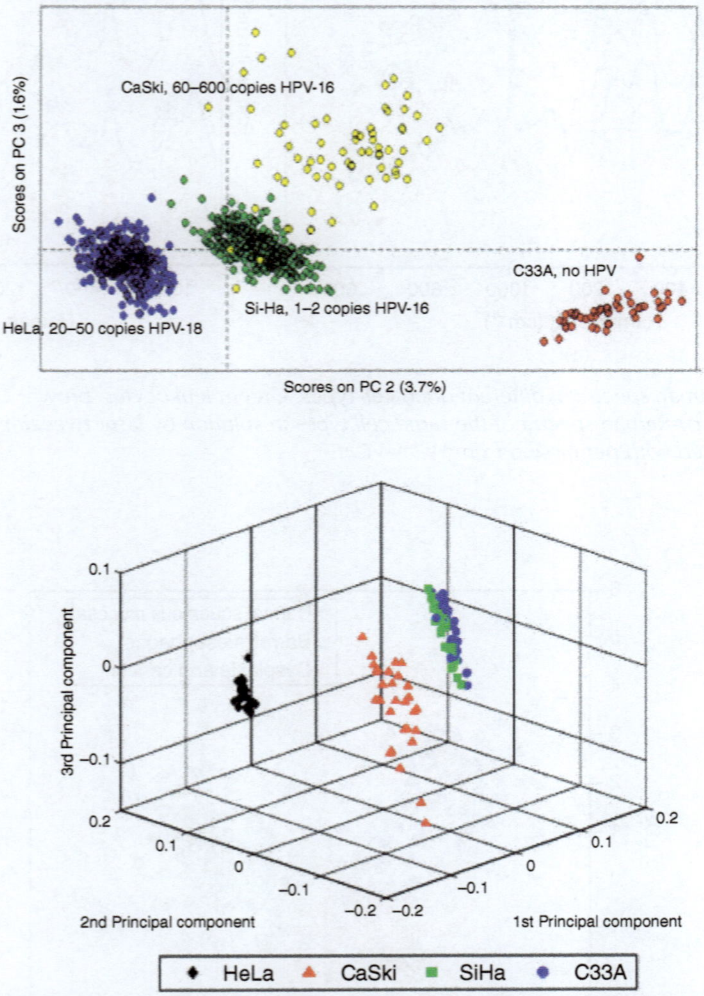

Figure 14.9 *(a) PCA scores plot of four different cervical cancer cell lines with different viral infection, obtained via infrared SCP. (b) PCA scores plot of Raman microscopic data of the same for cell lines.*

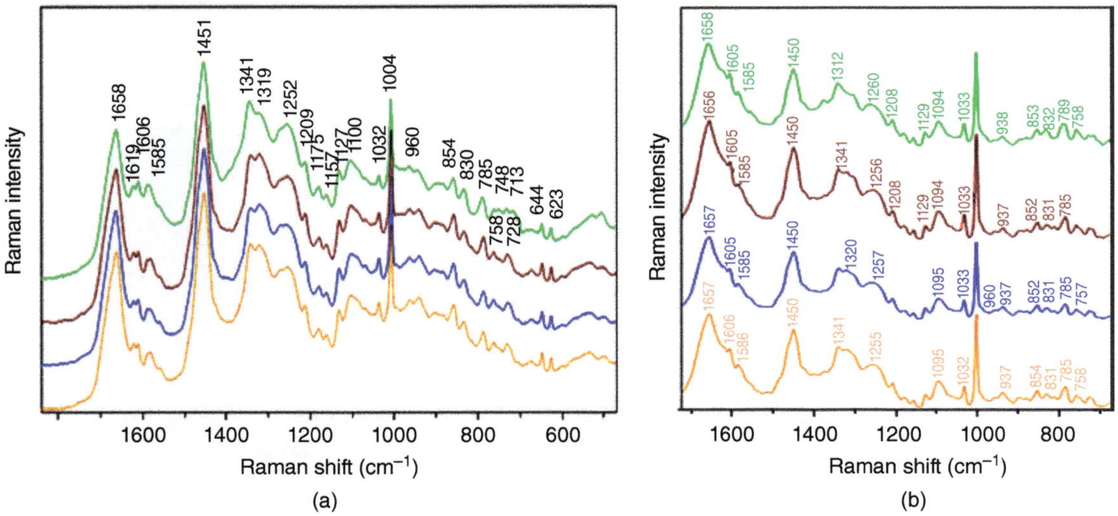

Figure 14.10 (a) Raman spectra of different dried cell types. Green: leukocytes, brown: OCI-AML3, blue: BT20, and orange: MCF7. (b) Raman spectra of the same cell types in solution by laser tweezing methods. Diem et al., 2013 [10]. Reproduced with permission from Wiley-VCH.

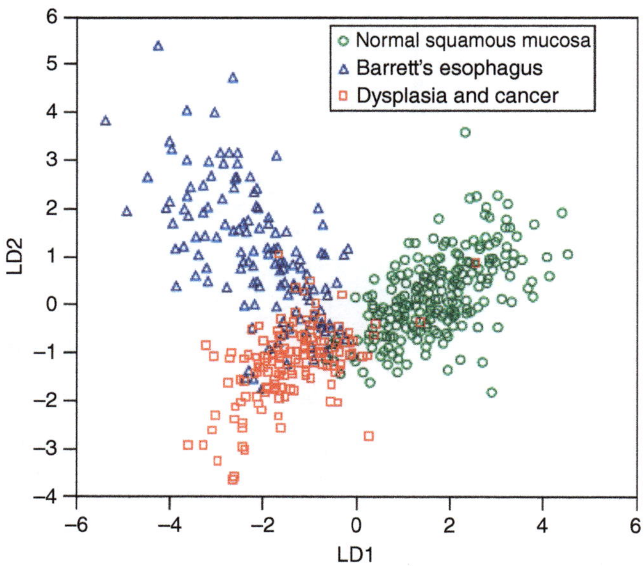

Figure 14.11 Linear discriminant analysis results for spectra collected from excised esophageal tissue. Barr et al., 2008 [29]. Reproduced with permission from John Wiley & Sons.

cell line and a related cell line that was rendered insensitive to a drug. The resulting scores plot demonstrated, for a relatively small but statistically significant, data set that macroscopic (ATR-based, attenuated total reflection) IR measurements can distinguish these data sets. The authors pointed out that careful preprocessing of the raw data is necessary for a successful separation. This study also pointed to the fact that cell cultures yield sufficiently homogeneous spectra and that macroscopic data acquisition is possible. The ATR approach taken avoided the pitfalls of R-Mie scattering. A similar study by the same group [100] demonstrated small differences in the spectral properties of cultured cells could be observed as a function of the confluence rate, but much larger changes were due to drug treatment. PCA revealed that the confluence- and the drug-induced changes caused separation of the cell spectra along different PCs, which indicates that the two sources of spectral variance are unrelated. The time evolution of spectral changes when the cell cultures were treated by the same drug used in the previously discussed study (ouabain) was reported by the same group [101]; here, the data were analyzed by 2D-IR spectroscopy (see Section 12.7). According to the discussion in this section, any strain can be imposed on a system for 2D-IR to be observable; here, the dimension along which 2D-IR was observed was exposure time to the drug. There were several very important facets to these studies. First, a plot of the raw spectral data in this study (as well as the previously discussed paper by the same authors) demonstrated that the spectral changes observed are so subtle that they cannot be perceived with the naked eye. In fact, it is highly laudable to display the similarity of such spectra, as in the past, artifacts based on enormous changes in spectral signatures appeared in the literature. The 2D-IR data suggested that only two wavenumber

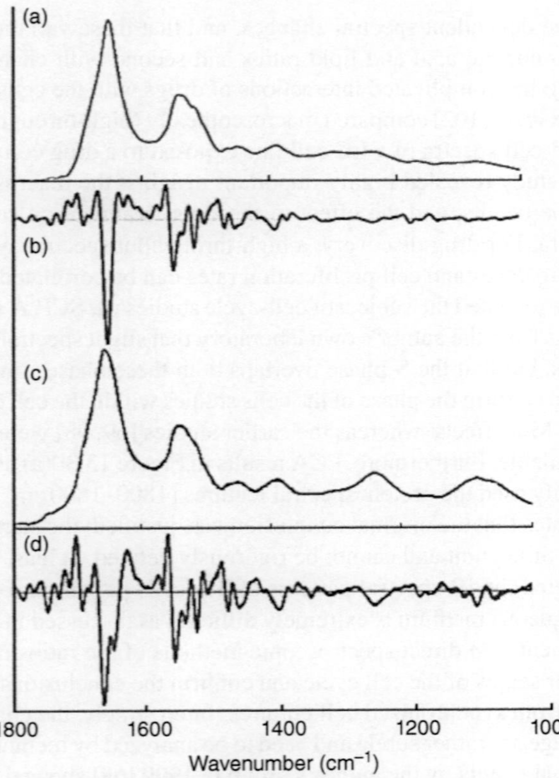

Figure 13.29 *(a, b) Averaged (about 50 cells) absorption and second derivative spectra of exfoliated oral mucosa cells, respectively. (c, d) Averaged (about 50 cells) absorption and second derivative spectra of cultured HeLa cells, respectively*

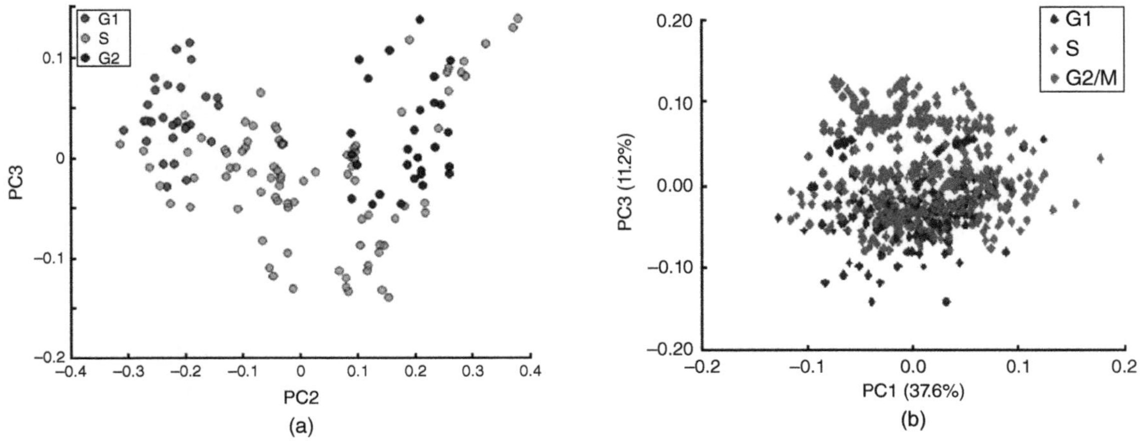

Figure 13.30 *(a) PCA plot of cell spectra according to their stage in the cell division cycle. Spectra were not corrected for R-Mie scattering. (From Ref. [98].) (b) Partial least squares regression plot of R-Mie corrected spectra according to their stage in the cell division cycle. Jimenez-Hernandez, M.,* et al.*, 2013 [103]. Reproduced with permission from the Royal Society. (*See insert for color representation of this figure.*)*

positions expressed the time-dependent spectral changes, and that these variations could be associated first with a change in protein to nucleic acid and lipid ratios and second with change in protein conformation. These studies shed light into the complicated interactions of drugs with the cells and their biology.

A study reported by Draux *et al.* [102] compared macroscopically (high-throughput system) acquired as well as synchrotron-based single-cell spectra of a LC cell line exposed to a drug commonly used in the treatment of LC (gemcitabine). This study revealed highly important insights: the macroscopic and microscopic data produced basically the same results, and the effect of the drug treatment on cell proliferation rates can be modeled from the IR spectra. For drug discovery, a high-throughput spectral system may prove invaluable, as exposure to sublethal drug doses and cell proliferation rates can be correlated.

Several recent papers have revisited the subject of cell-cycle studies *via* SCP. A study by Jimenez-Hernandez *et al.* [103] confirmed results from the author's own laboratory that slight spectral differences do exist for cells from the G1 and G2 phases, but that the S-phase overlaps both these phases. In both studies, immunohistochemical stains were used to confirm the phase of the cells studies within the cell cycle. The recent study [103] corrected cell spectra for R-Mie effects, whereas the earlier studies [69, 98] were carried out before such correction algorithms were available. Furthermore, PCA results in Figure 13.30(a) used the entire spectral vector, whereas the results in (b) only used the protein spectral features (1800–1600 cm^{-1}). The resemblance of these results is striking and indicates that the original conclusion was justified: the start and end of the S-phase represents a somewhat diffuse time point and cannot be rigorously defined, at least from the spectral viewpoint.

Whelan *et al.* [104] reported the IR spectral patterns of live cells passing through the cell cycle. Observation of such spectra in an aqueous medium is extremely difficult, as discussed in Section 13.10. These results present real-time measurements *via* direct spectroscopic methods of the ratios of protein, nucleic acids, and lipids during the three major stages of the cell cycle and confirm the conclusions reached by previous reports [98, 103] using dried cells from synchronized cell cultures. Interestingly, the changes of cellular spectra as a function of the cell-cycle stage are rather subtle and need to be analyzed by methods such as PCA to reveal significant changes. The original results by the author's group in 1999 [68] showed that much larger effect were contaminated by effects to cell morphology, which were not understood early in the development of SCP. Yet certain features observed in these very early studies certainly have been confirmed by the more recent efforts.

An interesting complementary study of the actual processes that occur in the final stage of the cell cycle, mitosis, was reported in 2006 [105]. This study employed Raman, IR, and fluorescence microspectral imaging techniques, and it will be discussed in Chapter 14 on Raman MSP.

13.10 Infrared spectroscopy of cells in aqueous media

The results reported so far have come from studies investigating individual, fixed, and dried cells. However, efforts also were undertaken to study cells in their native, aqueous environment. The aim of these studies was twofold: First, it was hoped that live cells could be observed in culture medium and to follow their progress through the cell division cycle. Second, the response of cells to treatment with certain anticancer drugs could be detected and used to predict cells' drug sensitivity or resistance. The overall aims of this research – keeping cells alive for hours in the spectrometer while recording their spectra and the effect of drug treatment on the spectra of the cells – were achieved. However, the observation of cells proceeding through the cell cycle was not achieved in the author's laboratory, because the cells refused to divide while confined to the IR sample chamber. This may have been, in part, because the cells were quite confined in this chamber, particularly, because cells need to round up before they go into mitosis.

Observation of live cells in aqueous media is a formidable task. This is because water possesses one of the largest IR transition dipole moments, producing very strong water absorption features even at short sample path lengths, which, in turn, make the observation of IR spectra in aqueous medium extremely difficult. Furthermore, cells have to be maintained at 37 °C and be continuously perfused with growth medium at a suitable pH for extended periods of time. To address the high background absorbance of the aqueous medium, thin path-length sample chambers need to be designed and methods to be developed to properly account for the liquid water contributions to the cells' spectra. However, a number of publications from the author's laboratory [106, 107] established that IR transmission and transflection measurements can be carried out microscopically in aqueous media and that cells can be kept alive for at least 24 h, but likely much longer, in suitable cell chambers.

To this end, a flow-through cell chamber was designed and constructed that utilized a CaF_2 window into which about a 10-μm-deep trough had been ion-etched and a flat CaF_2 top window with two small holes for the flow of cell culture medium [9]. Human cells were cultured directly onto the etched window; after a suitable culture time, the chamber was assembled and mounted in a thermostated aluminum block. Cell culture medium was gravity-fed from an IV-bag through the cell chamber. This flow system allowed the medium to be changed dynamically; for example, the cell culture medium could be substituted with pure buffer, starving the cells. Alternatively, a drug could be added to the medium flow to follow the drug-induced spectral changes in the cells. Software was developed to correct absorbed spectra for small, but unavoidable, liquid water contributions to the cell spectra. These contaminations occur because a spectrum measured through a cell contains less of a water contribution than the background spectrum collected through pure water (or medium); thus, the spectra of cells were overcompensated for by the water background and needed a correction step. As discussed before, second derivative spectra were used for data analysis that improved the situation enormously.

Although the high water background absorption degraded the signal-to-noise ratio of the measurement, the use of the PapMap^P method, including NA-PCA (Sections 13.7.3), allowed for the collection of excellent spectra of live cells and the effects of chemically perturbing these cells. Furthermore, it was demonstrated that cells could be kept alive for at least 24 h and that starvation of the cells by changing the medium to a buffer solution caused cell death in a relatively short time [9, 107].

In the course of these studies, a peculiar advantage of collecting cellular spectra in an aqueous environment was found. In the literature, only the disadvantages arising from water interferences are discussed, which include the need for short path lengths, the problem of maintaining constant path lengths between sample and

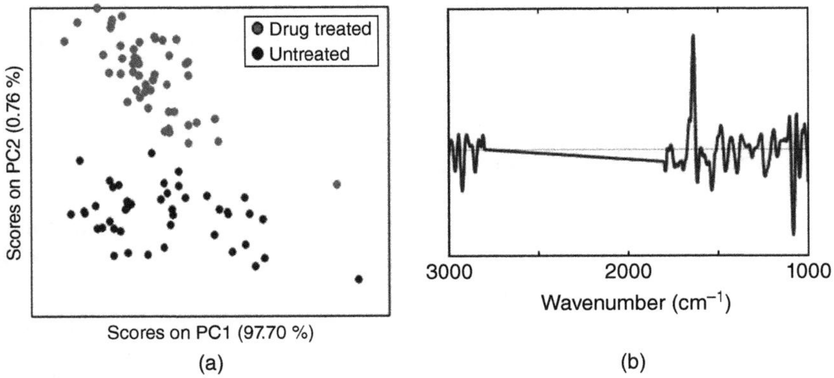

Figure 13.31 *(a) Scores plot of untreated cells and cells treated with cyclophosphamide monohydrate. (b) Loading vector PC2 along which the PCA data are split. Marcsisin (2011) [107]. (*See insert for color representation of this figure.*)*

background measurement, and the difficulties of compensating for the water contribution between sample and background spots. However, all these difficulties are accompanied by an enormous advantage of working in an aqueous environment: the absence of dispersive band shape contributions that often are observed in the IR spectra of cells and tissues. These effects result from R-Mie scattering (Section 12.3.2) and other effects that depend strongly on gradients of the reflective index. A significant advantage to live cell work in aqueous media is the absence of these effects. This can be attributed to the fact that the difference in refractive index between an individual hydrated cell and water is so small that scattering effects were not detected in the data presented here.

Two recent publications demonstrated that cells cultured directly onto the CaF_2 window of the live cell chamber did not exhibit any spectral nor morphological changes when kept in an environment suitable for cell survival, namely at 37°C and with fresh cell culture medium. When the flow of cell culture medium was stopped and the medium was replaced with pure buffer solution, morphological and spectral changes were observed [9]. The conclusion reached was that the cells in the presence of cell culture medium were alive, although cell division was never observed. This may have been because the thin space in which the cells were confined did not permit the morphological changes that precede mitosis.

Furthermore, experiments were conducted that demonstrated that SCP is capable of differentiating cells exposed to chemotherapeutical agents from untreated cells. Figure 13.31 shows the scores plot of live cells treated with an anticancer drug, cyclophosphamide monohydrate, and an untreated control group of cells. Cyclophosphamide monohydrate is an alkylating agent that acts on DNA at the 7-position nitrogen in guanine bases [108–110] and interrupts cell function by causing inter- and intra-strand cross-links as well as other DNA mutations [108–112]. It is a common chemotherapeutic in the treatment of various cancers [111, 112].

The cell spectra were collected *via* the PapMap^{P} image approach, coupled to NA-PCA image reconstruction as described earlier. To assure that the changes detected were a result of the drug acting on the cells and not a result of the detection of the drug in the surrounding culture medium, cells were only exposed to the drug during the incubation period. In the live cell sample chamber, cells were subjected to flow of normal culture medium. Thus, the observed changes reported here reflect overall changes occurring within the cells, not the drug itself.

Figure 13.31 demonstrates a sharp separation along PC2 of the drug-treated and the untreated cells. A plot of PC2 demonstrates that most of the spectral variations between the untreated and drug-exposed cells occur in the protein and low wavenumber regions. Specifically, there is a monosignate, positive band in the loading vector at about 1635 cm^{-1}; a monosignate band generally indicates the occurrence of a new spectral feature (band shifts in the PC vectors are indicated by bisignate bands). The band position coincides with amide I signatures observed previously due to necrosis and apoptosis [40] and could be an early indication of cell death upon drug treatment. The low-frequency negative bands in the loading vector at 1090 and 1240 cm^{-1} are typically associated the PO_2^- symmetric and antisymmetric stretching vibrations of DNA/RNA. Such changes could be directly due to drug-induced modifications of the DNA structure or spectral changes in DNA/RNA features due to cell death.

13.11 Future potential of SCP

Aside from the fixation studies, which were carried out for both exfoliated and cultured cells, and the viral load studies, which were carried out on cultured cells only, the majority of the work presented in the earlier sections has dealt with exfoliated cell; that is, it reported a truly new form of cytology, namely SCP. The size of the data sets at the author's exceeds all previously collected data sets combined [83, 85]. At the time of writing of this review, it appears that the SCP has matured to a level that allows for the detection of cellular abnormalities, such as dysplasia, cancer, and viral infection in exfoliated cells, and thus, it is poised to be applied to areas where classical cytology has very poor performance, in many cases below 50% accuracy.

The reason that the progress in SCP has been somewhat slower than that in other areas of spectral diagnostics is the fact that the correlation between classical cytopathology and SCP is difficult. In SCP, one has to rely on luck that within an ensemble of cells scrutinized by SCP, there is a diagnosable, abnormal cell. After a few thousands of cells from dysplastic patients, however, it will become very likely that some cells are found that display abnormal spectra and can, indeed, be diagnosed. Such a case is shown in Figure 13.11, which shows a clearly dysplastic cell whose spectrum clustered with other abnormal spectra [85].

For cultured cells, the efforts and research directions are more diversified and represent a number of other research groups as well as the LSpD. These efforts have demonstrated that IR MSP can detect the stages of a cells division cycle [69], the effect of drug treatment on cells [9, 101, 113], the degree of aggressiveness [114] of tumor cells, µRNA expression [115], cancer activation of fibroblasts [116], and a few others. In general, the results of these studies demonstrate that carefully carried out FTIR studies can reveal an extraordinary amount of information on the complex biochemical changes that occur when cells undergo natural or induced processes. A few general rules seem to apply for carefully planned and executed studies. The raw spectra, whether monitoring drug treatment or any other of the changes listed earlier, exhibit no or barely visible spectral changes and multivariate methods of analysis need to be employed to visualize spectral variations. A typical example is the study by the Brussels group [101], which demonstrated elegantly that the spectra of untreated cells and those treated with drugs are virtually identical to the naked eye, but that statistical (or, in this case, 2D) analysis of these data sets reveals changes that can be interpreted biochemically. Some of the spectral changes, for example, due to drug interactions are smaller than the changes due to cancerous disease; thus, the authors believe that the spectroscopy of cells can reveal much more information and can be used to reveal very subtle details. For bacterial cells, for example, IR spectroscopy, coupled to analysis by neural networks, could predict the mode of action of newly discovered drugs [117].

References

1. Blout, E.R. and Mellors, R.C. (1949) Infrared spectra of tissues. *Science*, **110**, 137–138.
2. Woernley, D.L. (1952) Infrared absorption curves for normal and neoplastic tissues and related biological substances. *Cancer Res.*, **12**, 516–523.
3. Mohlenhoff, B., Romeo, M.J., Diem, M. and Wood, B.R. (2005) Mie-type scattering and non-Beer–Lambert absorption behavior of human cells in infrared microspectroscopy. *Biophys. J.*, **88** (5), 3635–3640.

4. Bassan, P., Byrne, H.J., Bonnier, F. *et al.* (2009) Resonant Mie scattering in infrared spectroscopy of biological materials – understanding the 'dispersion artefact'. *Analyst*, **134**, 1586–1593.
5. Romeo, M.J., Boydston-White, S., Matthäus, C. *et al.* (2008) Infrared and Raman microspectroscopic studies of individual human cells, in *Vibrational Spectroscopy for Medical Diagnosis* (eds M. Diem, P.R. Griffiths and J.M. Chalmers), John Wiley & Sons, Ltd, Chichester, pp. 27–70.
6. Romeo, M.J. and Diem, M. (2005) Correction of dispersive line shape artifact observed in diffuse reflection infrared spectroscopy and absorption/reflection (transflection) infrared micro-spectroscopy. *Vib. Spectrosc.*, **38** (1-2), 129–132.
7. Bassan, P., Byrne, H.J., Lee, J. *et al.* (2009) Reflection contributions to the dispersion artefact in FTIR spectra of single biological cells. *Analyst*, **134**, 1171–1175.
8. Gulley-Stahl, H.J., Sommer, A.J. and Evan, A.P. (2010) Evanescent wave imaging, in *Vibrational Spectroscopic Imaging for Biomedical Applications* (ed G. Srinivasan), McGraw Hill, New York, pp. 99–121.
9. Marcsisin, E.J., Uttero, C.M., Miljkovic, M. and Diem, M. (2010) Infrared microspectroscopy of live cells in aqueous media. *Analyst*, **135**, 3227–3232.
10. Diem, M., Mazur, A., Lenau, K. *et al.* (2014) Molecular pathology via infrared and Raman spectral imaging, in *Next-Generation Spectroscopy-Based Diagnostics. In Vivo and Ex-Vivo Molecular Pathology* (ed P.J.), Wiley-VCH Verlag GmbH, Weinheim, pp. 45–102.
11. Diem, M., Mazur, A., Lenau, K. *et al.* (2013) Molecular pathology via infrared and Raman spectral imaging. *J. Biophotonics*, **6** (11-12), 855–886.
12. Matthäus, C., Bird, B., Miljković, M. *et al.* (2008) Infrared and Raman microscopy in cell biology, in *Biophysical Tools for Biologists*, Methods in Cell Biology, Vol. **89**, Vol. **2** (eds J.J. Correia, I. Detrich and H.W. Detrich), Elsevier, pp. 275–308.
13. Lasch, P., Pacifico, A. and Diem, M. (2002) Spatially resolved IR microspectroscopy of single cells. *Biopolymers*, **67** (4-5), 335–338.
14. Brauner, J.W., Flach, C.R. and Mendelsohn, R. (2005) A quantitative reconstruction of the amide I contour in the IR spectra of globular proteins: from structure to spectrum. *J. Am. Chem. Soc.*, **127**, 100–109.
15. Bird, B., Miljkovic, M., Remiszewski, S. *et al.* (2012) Infrared spectral histopathology (SHP): a novel diagnostic tool for the accurate classification of lung cancers. *Lab. Invest.*, **92**, 1358–1373.
16. Lasch, P. (1999) Computergestuetzte Bildrekonstuktion auf Basis FTIR-mikrospektrometrischer Daten humaner Tumoren, PhD Dissertation, *Humanmedizin*, Freie Universitaet Berlin, Berlin, p. 124.
17. Lasch, P., Haensch, W., Lewis, E.N. *et al.* (2002) Characterization of colorectal adenocarcinoma sections by spatially resolved FT-IR microspectroscopy. *Appl. Spectrosc.*, **56**, 1–9.
18. Lasch, P., Haensch, W., Naumann, D. and Diem, M. (2004) Imaging of colorectal adenocarcinoma using FT-IR microspectroscopy and cluster analysis. *Biochim. Biophys. Acta*, **1688** (2), 176–186.
19. Lasch, P., Diem, M., Hänsch, W. and Naumann, D. (2007) Artificial neural networks as supervised techniques for FT-IR microspectroscopic imaging. *J. Chemom.*, **20** (5), 209–220.
20. Pounder, F.N. and Bhargava, R. (2010) Toward automated breast histopathology using mid-IR spectroscopic imaging, in *Vibrational Spectroscopic Imaging for Biomedical Applications* (ed G. Srinivasan), McGraw-Hill, New York, pp. 1–26.
21. Romeo, M.J., Dukor, R.K. and Diem, M. (2008) Introduction to spectral imaging, and applications to diagnosis of lymph nodes, in *Vibrational Spectroscopy for Medical Diagnosis* (eds M. Diem, P.R. Griffiths and J.M. Chalmers), John Wiley & Sons, Ltd, Chichester, pp. 1–26.
22. Romeo, M.J., Boydston-White, S., Matthäus, C. *et al.* (2008) Vibrational microspectroscopy of cells and tissues, in *Biomedical Vibrational Spectroscopy* (eds P. Lasch and J. Kneipp), Wiley-Interscience, Hoboken, NJ, pp. 121–147.
23. Bird, B., Bedrossian, K., Laver, N. *et al.* (2009) Detection of breast micro-metastases in axillary lymph nodes by infrared micro-spectral imaging. *Analyst*, **134**, 1067–1076.
24. Ly, E., Piot, O., Wolthuis, R. *et al.* (2008) Combination of FTIR spectral imaging and chemometrics for tumour detection from paraffin-embedded biopsies. *Analyst*, **133**, 197–205.
25. O'Faolain, E., Hunter, M.B., Byrne, J.M. *et al.* (2005) A study examining the effects of tissue processing on human tissue sections using vibrational spectroscopy. *Vib. Spectrosc.*, **38**, 121–127.

26. Hackett, M.J., McQuillan, J.A., El-Assaad, F. *et al.* (2011) Chemical alterations to murine brain tissue induced by formalin fixation: implications for biospectroscopic imaging and mapping studies of disease pathogenesis. *Analyst*, **136**, 2941–2952.
27. Amharref, N., Beljebbar, A., Dukic, S. *et al.* (2006) Brain tissue characterization by infrared imaging in a rat glioma model. *Biochem. Biophys. Acta*, **1758** (7), 892–899.
28. Bambery, K.R., Schültke, E., Wood, B.R. *et al.* (2006) A Fourier transform infrared microscopic image investigation into an animal model exhibiting glioblastoma multiforme. *Biochem. Biophys. Acta*, **1758** (7), 900–907.
29. Fernandez, D.C., Bhargava, R., Hewitt, S.M. and Levin, I.W. (2005) Infrared spectroscopic imaging for histopathologic recognition. *Nat. Biotechnol.*, **23**, 469–474.
30. Schultz, C., Liu, K.Z., Kerr, P.D. and Mantsch, H.H. (1998) In situ infrared histopathology of keratinization in human oral/oropharyngeal squamous cell carcinoma. *Oncol. Res.*, **10**, 277–286.
31. Wood, B.R., Chiriboga, L., Yee, H. *et al.* (2004) Fourier transform infrared (FTIR) spectral mapping of the cervical transformation zone, and dysplastic squamous epithelium. *Gynecol. Oncol.*, **93** (1), 59–68.
32. Lasch, P. and Naumann, D. (1998) FT-IR microspectroscopic imaging of human carcinoma in thin sections based on pattern recognition techniques. *Cell. Mol. Biol.*, **44** (1), 189–202.
33. Diem, M., Chiriboga, L., Lasch, P. and Pacifico, A. (2002) IR spectra and IR spectral maps of individual normal and cancerous cells. *Biopolymers*, **67** (4-5), 349–353.
34. Chiriboga, L., Xie, P., Yee, H. *et al.* (1998) Infrared spectroscopy of human tissue. I. Differentiation and maturation of epithelial cells in the human cervix. *Biospectroscopy*, **4**, 47–53.
35. Steller, W., Einenkel, J., Horn, L.C. *et al.* (2006) Delimitation of a squamous cell cervical carcinoma using infrared microspectroscopic imaging. *Anal. Bioanal. Chem.*, **384**, 145–154.
36. Wood, B.R., Chiriboga, L., Yee, H. *et al.* (2004) FTIR mapping of the cervical transformation zone, squamous and glandular epithelium. *Gynecol. Oncol.*, **93** (4), 59–68.
37. Diem, M., Miljković, M., Bird, B. *et al.* (2012) Applications of infrared and Raman micro-spectroscopy of cells in medical diagnostics: present status and future promises. *Spectrosc. – Biomed. Appl.*, **27** (5-6), 463–496.
38. Papanicolaou, G.N. (1948) *The Epithelium of Women's Reproductive Organs*, The Commonwealth Fund, New York.
39. Miljković, M., Bird, B., Lenau, K. *et al.* (2013) Spectral cytopathology: new aspects of data collection, manipulation and confounding effects. *Analyst*, **138**, 3975–3982.
40. Jamin, N., Miller, L., Moncuit, J. *et al.* (2003) Chemical heterogeneity in cell death: combined synchrotron IR and fluorescence microscopy studies of single apoptotic and necrotic cells. *Biopolymers (Biospectroscopy)*, **72**, 366–373.
41. Bhargava, R. and Levin, I.W. (2008) Infrared spectroscopic imaging protocols for high-throughput histopathology, in *Vibrational Spectroscopy for Medical Diagnosis* (eds M. Diem, P.R. Griffiths and J.M. Chalmers), John Wiley & Sons, Ltd, Chichester, pp. 155–186.
42. Bassan, P., Sachdeva, A., Shanks, J.H. *et al.* (2014) Automated high-throughput assessment of prostate biopsy tissue using infrared spectroscopic chemical imaging. *Proc. SPIE*, 9041, **12** (4), (p.1 of 10).
43. Reddy, R.K. and Bhargava, R. (2010) Accurate histopathology from low signal-to-noise ratio spectroscopic imaging data. *Analyst*, **135**, 2818–2825.
44. Kumar, S., Desmedt, C., Larsimont, D. *et al.* (2013) Change in the microenvironment of breast cancer studied by FTIR imaging. *Analyst*, **138**, 4058–4065.
45. Kallenbach-Thieltges, A., Groβerüchkamp, F., Mosig, A. *et al.* (2013) Spectral histopathologic and immunohistochemical studies of colon adenocarcinomas. *J. Biophotonics*, **6** (1), 88–100.
46. Federman, S., Miller, L.M. and Sagi, I. (2002) Following matrix metalloproteinases activity near the cell boundary by infrared micro-spectroscopy. *Matrix Biol.*, **21** (7), 567–577.
47. Romeo, M.J. and Diem, M. (2005) Infrared spectral imaging of lymph nodes: strategies for analysis and artifact reduction. *Vib. Spectrosc.*, **38**, 115–119.
48. Mazur, A.I., Bird, B., Miljković, M. *et al.* (2013) Detection of B-lymphocyte activation by various primary tumors. *J. Biophotonics*, **6** (1), 101–109.
49. Bird, B., Miljković, M., Romeo, M.J. *et al.* (2008) Infrared micro-spectral imaging: automatic distinction of tissue types in axillary lymph node histology. *BMC J. Clin. Pathol.*, **8** (8), 1–14.

50. Bird, B., Romeo, M., Laver, N. and Diem, M. (2009) Spectral detection of micro-metastases in lymph node histopathology. *J. Biophotonics*, **2** (1-2), 37–46.
51. Bird, B., Miljković, M. and Diem, M. (2010) Two step resonant Mie scattering correction of infrared micro-spectral data: human lymph node tissue. *J. Biophotonics*, **3** (8-9), 597–608.
52. Bird, B., Miljković, M., Laver, N. and Diem, M. (2011) Spectral detection of micro-metastases and individual metastatic cells in lymph node histology. *Technol. Cancer Res. Treat.*, **10** (2), 135–144.
53. Wong, P., Wong, R., Caputo, T. *et al.* (1991) Infrared spectroscopy of exfoliated human cervical cells: evidence of extensive structural changes during carcinogenesis. *Proc. Natl. Acad. Sci. U.S.A.*, **88**, 10988–10992.
54. Wong, P., Lacelle, S., Fung, M. *et al.* (1995) Characterization of exfoliated cells and tissues from human endocervix and ectocervix by FTIR and ATR/FTIR spectroscopy. *Biospectroscopy*, **1**, 357–364.
55. Cohenford, M., Godwin, T.A., Cahn, F. *et al.* (1997) Infrared spectroscopy of normal and abnormal cervical smears: evaluation by principal component analysis. *Gynecol. Oncol.*, **66**, 59–65.
56. Papanicolaou, G.N. and Traunt, H.F. (1941) Smear diagnosis of carcinoma of the cervix. *Am. J. Obstet. Gynecol.*, **42** (2), 193–206.
57. Krafft, C. and Salzer, R. (2008) Neuro-oncological applications of infrared and Raman spectroscopy, in *Vibrational Spectroscopy for Medical Diagnosis* (eds M. Diem, P.R. Griffiths and J.M. Chalmers), John Wiley & Sons, Ltd, Chichester, pp. 231–259.
58. Krafft, C., Shapoval, L., Sobottka, S.B. *et al.* (2006) Identification of primary tumors of brain metastases by SIMCA classification of IR spectroscopic images. *Biochem. Biophys. Acta*, **1758** (7), 883–891.
59. Stone, N., Kendall, C. and Barr, H. (2008) Raman spectroscopy as a potential tool for early diagnosis of malignancies in esophageal and bladder tissues, in *Vibrational Spectroscopy for Medical Diagnosis* (eds M. Diem, P.R. Griffiths and J.M. Chalmers), John Wiley & Sons, Ltd, Chichester, PP. 203–230.
60. Naumann, D., Fijala, V., Labischinski, H. and Giesbrecht, P. (1988) The rapid differentiation of pathogenic bacteria using Fourier transform infrared spectroscopic and multivariate statistical analysis. *J. Mol. Struct.*, **174**, 165–170.
61. Helm, D., Labischinski, H., Schallen, G. and Naumann, D. (1991) Classification and identification of bacteria by FT-IR spectroscopy. *J. Gen. Microbiol.*, **137**, 69–79.
62. van der Mei, H.C., Naumann, D. and Busscher, H.J. (1993) Grouping oral streptococcal species using Fourier-transform infrared spectroscopy in comparison with classical microbial identification. *Arch. Oral Biol.*, **38** (11), 1993.
63. Helm, D. and Naumann, D. (1995) Identification of some bacterial components by FT-IR spectroscopy. *FEMS Microbiol. Lett.*, **126** (1), 75–79.
64. RiverDiagnostics (2013) http://www.riverd.com/. (accessed 2 January 2015).
65. Puppels, G.J., de Mul, F.F., Otto, C. *et al.* (1990) Studying single living cells and chromosomes by confocal Raman microspectroscopy. *Nature*, **347**, 301–303.
66. Wood, B.R., Quinn, M., Burden, F. *et al.* (1996) An investigation into FTIR spectroscopy as a biodiagnostic tool for cervical cancer. *Biospectroscopy*, **2**, 01–11.
67. Cohenford, M. and Rigas, B. (1998) Cytologically normal cells from neoplastic cervical samples display extensive structural abnormalities on IR spectroscopy: implications for tumor biology. *Proc. Natl. Acad. Sci. U.S.A.*, **95**, 15327–15332.
68. Boydston-White, S., Gopen, T., Houser, S. *et al.* (1999) Infrared spectroscopy of human tissue. V. Infrared spectroscopic studies of myeloid leukemia (ML-1) cells at different phases of the cell cycle. *Biospectroscopy*, **5**, 219–227.
69. Boydston-White, S., Romeo, M.J., Chernenko, T. *et al.* (2006) Cell-cycle-dependent variations in FTIR micro-spectra of single proliferating HeLa cells: principal component and artificial neural network analysis. *Biochim. Biophys. Acta*, **1758** (7), 908–914.
70. Papanicolaou, G.N. and Marshal, V.F. (1945) Urine sediment smears: a diagnostic procedure in cancers of urinary tract. *Science*, **101**, 519–520.
71. Mazur, A.I., Marcsisin, E.J., Bird, B. *et al.* (2012) Evaluating different fixation protocols for spectral cytopathology, part 1. *Anal. Chem.*, **84**, 1259–1266.

72. van Driel-Kulker, A.M.J., Ploem-Zaaijer, J.J., van der Zwan-van der Zwan, M. and Tanke, H.J. (1980) A preparation technique for exfoliated and aspired cells allowing different staining procedures. *Anal. Quant. Cytol.*, **2** (4), 243–246.
73. Stoler, M.H. and Schiffman, M. (2001) Interobserver reproducibility of cervical cytologic and histologic interpretations. *J. Am. Med. Assoc.*, **285** (11), 1500–1505.
74. Savitzky, A. and Golay, M.J.E. (1964) Smoothing and differentiation of data by simplified least-squares procedures. *Anal. Chem.*, **36** (8), 1627–1639.
75. Miljković, M., Bird, B. and Diem, M. (2012) Dispersive line shape effects in infrared spectroscopy. *Analyst*, **137**, 3954–3964.
76. Schubert, J.M., Mazur, A.I., Bird, B. *et al.* (2010) Single point vs. mapping approach for spectral cytopathology (SCP). *Biophotonics*, **3** (8-9), 588–596.
77. Davis, B.J., Carney, P.S. and Bhargava, R. (2010) Theory of midinfrared absorption microspectroscopy: I. Homogeneous samples. *Anal. Chem.*, **82**, 3474–3486.
78. Davis, B.J., Carney, P.S. and Bhargava, R. (2010) Theory of mid-infrared absorption microspectroscopy: II. Heterogeneous samples. *Anal. Chem.*, **82**, 3487–3499.
79. Science and Technology Council (2014) UK Central Laser Facility [Online].
80. Mantsch, H.H. and Jackson, M. (1995) Molecular spectroscopy in biodiagnostics. *J. Mol. Struct.*, **347**, 187.
81. Romeo, M., Mohlenhoff, B. and Diem, M. (2006) Infrared micro-spectroscopy of human cells: causes for the spectral variance of oral mucosa (buccal) cells. *Vib. Spectrosc.*, **42** (1), 9–14.
82. Schubert, J.M., Bird, B., Papamarkakis, K. *et al.* (2010) Spectral cytopathology of cervical samples: detecting cellular abnormalities in cytologically normal cells. *Lab. Invest.*, **90** (7), 1068–1077.
83. Papamarkakis, K., Bird, B., Schubert, J.M. *et al.* (2010) Cytopathology by optical methods: spectral cytopathology of the oral mucosa. *Lab. Invest.*, **90** (4), 589–598.
84. Romeo, M.J., Burden, F., Quinn, M. *et al.* (1998) Infrared microspectroscopy and artificial neural networks in the diagnosis of cervical cancer. *Cell. Mol. Biol.*, **44** (1), 179–187.
85. Schubert, J.M. (2011) Spectral cytology of human oral and cervical samples, PhD Dissertation, *Chemistry and Chemical Biology*, Northeastern University, Boston, MA.
86. Diem, M., Matthäus, C., Chernenko, T. *et al.* (2009) Infrared and Raman spectroscopy and spectral imaging of individual cells, in *Infrared and Raman Spectroscopic Imaging* (eds R. Salzer and H.W. Siesler), Wiley-VCH Verlag GmbH, Weinheim, pp. 173–202.
87. Ogden, G.R., Cowpe, J.G. and Green, M.W. (1990) The effect of distant malignancy upon quantitative cytologic assessment of normal oral mucosa. *Cancer*, **65**, 477–480.
88. van Oijen, M.G.C.T. and Slootweg, P.J. (2000) Oral field cancerization: carcinogen-induced independent events or micrometastatic deposits? *Cancer Epidemiol. Biomarkers Prev.*, **9**, 249–256.
89. zur Hausen, H. (2006) *Infections Causing Human Cancer*, Wiley-VCH Verlag GmbH, Weinheim.
90. Prevention, CfDCa (2007) Prevalence of High-Risk and Low-Risk Strains of HPV (2003–2004).
91. Ostrowska, K. (2011) Vibrational spectroscopy for cervical cytology, PhD Dissertation, *School of Physics*, Dublin Institute of Technology, Dublin.
92. Diem, M., Schubert, J., Miljković, M. *et al.* (2014) Infrared and Raman spectroscopy and spectral imaging of individual cells, in *Infrared and Raman Spectroscopic Imaging* (eds R. Salzer and H.W. Siesler), Wiley-VCH Verlag GmbH, Weinheim, pp. 181–223.
93. Bakker Schut, T.C., Van Dekken, H., Tilanus, H.W. *et al.* (1997) NIR Raman spectroscopy of healthy and diseased esophagus, in *Spectroscopy of Biological Molecules: Modern Trends*, Springer Verlag, Heidelberg, pp. 455–456.
94. Kendall, C., Day, J., Hutchings, J. *et al.* (2010) Evaluation of Raman probe for oesophageal cancer diagnostics. *Analyst*, **135**, 3038–3041.
95. Barr, H., Kendall, C. and Stone, N. (2004) The light solution for Barrett's esophagus: photodiagnosis and photo-dynamic therapy for columnar-lined esophagus. *Photodiagnosis Photodyn. Ther.*, **1** (1), 75–84.
96. Holman, H.N., Martin, M.C., Blakely, E.A. *et al.* (2000) IR spectroscopic characteristics of cell cycle and cell death probed by synchrotron radiation based Fourier transform IR spectromicroscopy. *Biopolymers (Biospectroscopy)*, **57**, 329–335.

97. Lasch, P., Boese, M., Pacifico, A. *et al.* (2002) FT-IR spectroscopic investigations of single cells on the subcellular level. *Vib. Spectrosc.*, **28** (1), 147–157.
98. Boydston-White, S., Chernenko, T., Regina, A. *et al.* (2005) Microspectroscopy of single proliferating HeLa cells. *Vib. Spectrosc.*, **38** (1-2), 169–177.
99. Gaigneaux, A., Ruysschaert, J.M. and Goormaghtigh, E. (2006) Cell discrimination by attenuated total reflection-Fourier transform infrared spectroscopy: the impact of preprocessing of spectra. *Appl. Spectrosc.*, **60** (9), 1022–1028.
100. Gasper, R. and Goormaghtigh, E. (2010) Effects of the confluence rate on the FTIR spectrum of PC-3 prostate cancer cells in culture. *Analyst*, **135**, 3048–3051.
101. Gaspar, R., Mijatovic, T., Kiss, R. and Goormaghtigh, E. (2011) Time dependence of cellular chemical changes induced in prostate PC-3 cancer cells by two structurally related cardenolides monitored by Fourier transform infrared (FT-IR) spectroscopy. *Appl. Spectrosc.*, **65** (6), 584–594.
102. Draux, F., Jeannesson, P., Gobinet, C. *et al.* (2009) IR spectroscopy evidences effect of non-cytotoxic doses of antitumor drug on cancer cells. *Anal. Bioanal. Chem*, in press, **395**(7), 2293–2301.
103. Jimenez-Hernandez, M., Hughes, C., Bassan, P. *et al.* (2013) Exploring the spectroscopic differences of Caki-2 cells progressing through the cell cycle while proliferating *in vitro*. *Analyst*, **138**, 3957–3966.
104. Whelan, D.R., Bambery, K.R., Puskar, L. *et al.* (2013) Synchrotron Fourier transform infrared (FTIR) analysis of single living cells progressing through the cell cycle. *Analyst*, **138**, 3891–3899.
105. Matthäus, C., Boydston-White, S., Miljković, M. *et al.* (2006) Raman and infrared microspectral imaging of mitotic cells. *Appl. Spectrosc.*, **60** (1), 1–8.
106. Miljković, M., Romeo, M., Matthäus, C. and Diem, M. (2004) Infrared microspectroscopy of individual human cervical cancer (HeLa) cells suspended in growth medium. *Biopolymers*, **74** (1-2), 172–175.
107. Marcsisin, E.J. (2011) Infrared spectroscopy to monitor drug response of individual live cells, PhD Dissertation, *Department of Chemistry*, Northeastern University, Boston, MA.
108. Ralhan, R. and Kaur, J. (2007) Alkylating agents and cancer therapy. *Expert Opin. Ther. Pat.*, **9** (17), 1061–1075.
109. Malayappan, B., Johnson, L., Nie, B. *et al.* (2010) Quantitative high-performance liquid chromatography-electrospray ionization tandem mass spectrometry analysis of bis-N7-guanine DNA-DNA cross-links in white blood cells of cancer patients receiving cyclophosphamide therapy. *Anal. Chem.*, **82** (9), 3650–3658.
110. Wang, J., Prorok, G. and Vaughan, W.P. (1992) Cytotoxicity, DNA cross-linking, and DNA single-strand breaks induced by cyclophophamide in a rat leukemia *in vivo*. *Cancer Chemother. Pharmacol.*, **31**, 381–386.
111. Souliotis, V.L., Dimopoulos, M.A. and Sfikakis, P.P. (2003) Gene-specific formation and repair of DNA monoadducts and interstrand cross-links after therapeutic exposure to nitrogen mustards. *Clin. Cancer Res.*, **9** (12), 4465–4474.
112. Hemminki, K. (1985) Binding of metabolites of cyclophosphamide to DNA in a rat liver microsomal system and *in vivo* in mice. *Cancer Res.*, **45** (9), 4237–4243.
113. Flower, K.R., Khalifa, I., Bassan, P. *et al.* (2011) Synchrotron FTIR analysis of drug treated ovarian A2780 cells: an ability to differentiate cell response to different drugs? *Analyst*, **136**, 498–507.
114. Gazi, E., Baker, M., Dwyer, J. *et al.* (2006) A correlation of FTIR spectra derived from prostate cancer tissue with Gleason grade, PSA and tumour stage. *Eur. Urol.*, **50**, 750–761.
115. Ozek, N.S., Tuna, S., Erson-Bensan, A.E. and Severcan, F. (2010) Characterization of microRNA-125b expression in MCF7 breast cancer cells by ATR-FTIR spectroscopy. *Analyst*, **135**, 3094–3102.
116. Holton, S.E., Walsh, M.J. and Bhargava, R. (2011) Subcellular localization of early biochemical transformations in cancer-activated fibroblasts using infrared spectroscopic imaging. *Analyst*, **136**, 2953–2958.
117. Schiffer, G., Udelhoven, T., Labischininski, H. *et al.* (2002) Application of FTIR spectroscopy in antibacterial drug research, in *Workshop on FTIR Spectroscopy in Microbiological and Medical Diagnostic*, Robert Koch Institute, Berlin.
118. Robichaux-Viehoefer, A., Kanter, E., Shappell, H. *et al.* (2007) Characterization of Raman Spectra Measured *in Vivo* for the Detection of Cervical Dysplasia. *Appl. Spectrosc.*, **61**(9), 986–993.

14

Raman Microspectroscopy of Cells and Tissue in Medical Diagnostics

14.1 Introduction

In analogy to the discussion in Chapter 13, Raman spectroscopy can be used to acquire spectral information from cells and tissues. The resulting techniques could be referred to as Raman Spectral Cytopathology (Ra-SCP) and Raman Spectral Histopathology (Ra-SHP), although this nomenclature has not found wide acceptance. As Raman and infrared (IR) spectroscopies are complementary techniques, both Ra-SHP and infrared spectral histopathology (IR-SHP) give similar information. However, because of the smaller excitation spot size in Ra-SHP, which in turn is due to the much smaller diffraction limit of visible light (see Chapter 11), the acquisition of Ra-SHP data from large tissue sections is impractical. However, Ra-SHP affords much higher spatial resolution that allows individual cellular components, such as the nucleus or even mitochondria [1, 2] to be detected in the spectral images, and the Raman spectra of intact chromosomes have been reported [3]. Furthermore, Raman spectroscopy has the enormous advantage of relative immunity from water interference and effects such as R-Mie scattering, which were introduced before as major confounding factors for IR-based techniques. The former of these points make Raman spectroscopy readily amenable for data collection of live tissue and cells and *in vivo* applications. In fact, several research groups have published numerous reports on Raman histopathology of hollow organs [4] or spectra acquired through needles fitted with fiber optics. Although the spatial resolution of these methods is less than that of microscopic Ra-SHP, the fact that an entire tissue fingerprint can be collected in a few seconds of data acquisition time bears an enormous advantage for *in vivo* tissue diagnosis.

In addition, Ra-SCP, the study of individual cells by Raman microscopy, is discussed. Ra-SCP can be carried out in two ways: first, one can utilize the inherent high spatial resolution of Ra-SCP to collect imaging data sets from one cell, consisting of 10,000 or more individual spectra, and reconstruct a spectral image of the cell. In such images, it is possible to differentiate individual cellular components such as the membrane, the cytoplasm, the nucleus and nucleoli, membrane-rich structures in the cytoplasm, and mitochondria [1]. This method can be viewed as a label-free analog of confocal fluorescence microspectroscopy and can produce images with similarly high spatial resolution. For example, the dynamics, subcellular origin, and chemical

Modern Vibrational Spectroscopy and Micro-Spectroscopy: Theory, Instrumentation and Biomedical Applications, First Edition.
Max Diem.

composition of lipid droplets were monitored in living hepatocytes. However, Ra-SCP imaging does not offer the same contrast that is afforded by antibody-bound fluorophores.

Second, Ra-SCP can also be carried out at low spatial resolution by collecting spectra from individual cells by utilizing longer wavelength excitation and lower numerical aperture (NA) objectives, and aiming the exciting laser beam at specific targets in the cell, typically the nucleus. This method is more labor intensive, but yields, in relatively short time, hundreds of spectra of selected cellular features that show enormous sensitivity toward many biochemical processes, such as the metabolic activity in live yeast cells [2]. This approach has been used very successfully to follow stem cell differentiation [5–7], or for the detection of circulating cancer cells in flow cytometry [8, 9].

Finally, owing to the very high spatial resolution of Ra-SCP, spectral information from individual bacterial cells can be collected. This aspect may be of tremendous importance for tracking the origin of bacterial contaminations or outbreaks of infectious diseases, because Raman spectroscopy can, in principle, identify one bacterial particle at the strain level.

In analogy to the discussion in Chapter 13, Figure 14.1 presents Raman spectra [10] of the same biochemical constituents shown in Figure 13.1. A casual inspection reveals that the Raman spectra are similarly rich in

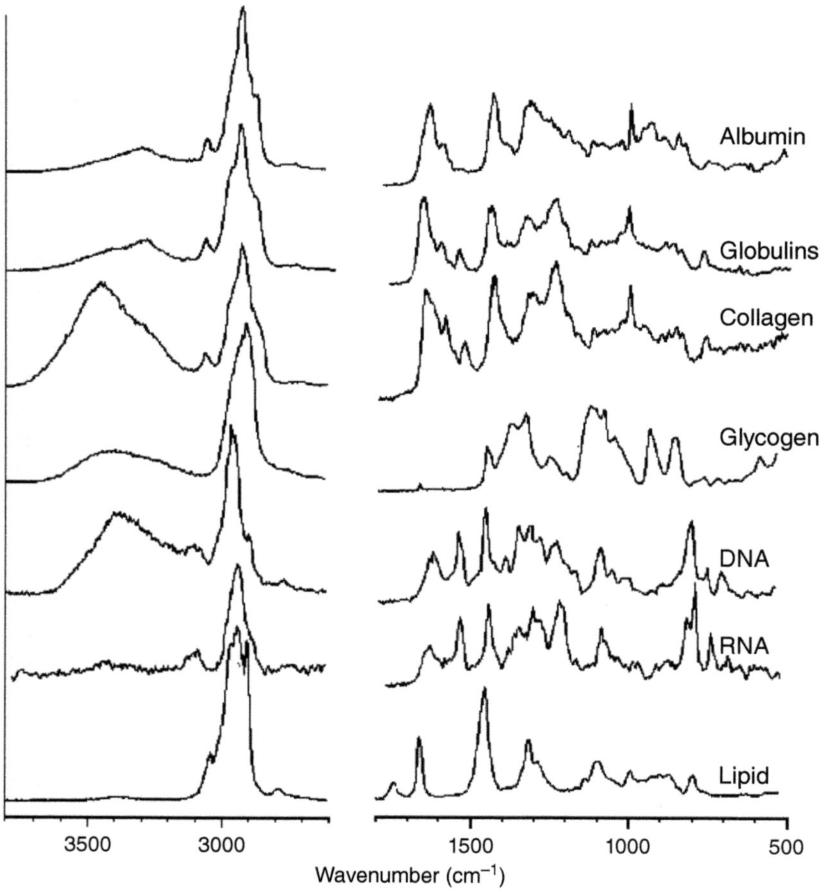

Figure 14.1 *Raman spectra of selected cellular components. Diem* et al.*, 2013 [10]. Reproduced with permission from Wiley-VCH*

information; in fact, Raman and IR spectroscopies are complementary in the sense that many nonpolar groups exhibit vibrations that are weak in the IR, but show strong Raman signals. Therefore, Raman spectroscopy yields more detail on C—H vibrations, which are particularly strong in this technique. However, Raman spectra are generally weaker, and more difficult to acquire; hence, Raman spectra are generally noisier than IR spectra. IR spectra contain more conformational information of the protein backbone and nucleic acids owing to the dipolar coupling mechanism that gives rise to the amide I manifold and the phosphate and base pair vibrations, respectively.

For Raman microscopy, the instrumentation is very similar to confocal fluorescence microscopy, and utilizes a basic visible microscope to which a laser source and a monochromator for the analysis of the scattered light have been added. The spatial resolution achievable (*cf.* Chapter 11) in Raman microscopy is similar to that of confocal fluorescence microscopy.

14.2 Experimental Consideration for Raman Microspectroscopy

The following discussion holds both for work on tissues or cells. Raman backscattering geometry (*cf.* Chapter 4) is used nearly exclusive in Raman microscopy, where the exciting laser light is focused through the same microscope objective that also collects the scattered light. As the depth resolution or confocal depth (see Chapter 11) in a Raman microscope is very small (on the order of 0.6–1 μm for mid-visible excitation and a high-NA objective), the spectral contribution of the substrate to the signal of the sample generally is relatively minor, and has been analyzed in detail by Everall [11]. Thick samples can be analyzed in a depth-resolved manner, up to a depth of about 100 μm. For biological and biomedical applications, the use of fluorescence-free quartz or CaF_2 slides or disks for sample substrates are perfectly adequate, although one should examine these materials upon receipt because some of them my contain fluorescent or chemoluminescent impurities. CaF_2 slides may contain minute amounts of lanthanide ions that co-crystallize with CaF_2 and produce strong, narrow emission lines that completely dominate the Raman spectra of cells and tissue.

For biological samples, the laser power threshold and photo-damage by the laser are major issues. A dried cultured cell that was well washed before data acquisition may withstand exposure to 5 mW of visible laser power focused to a spot size of 500 nm diameter for a few seconds for each pixel (voxel). Similarly, a piece of tissue that was cleaned of tissue processing media and paraffin will withstand similar laser power. For both sample systems, it is advantageous to work with a water-immersion objective for several reasons. The amount of collected light is significantly higher owing to, first, the index matching between sample and objective lens and, second, the better dissipation of the energy incident on a sample pixel. For work on a single cell, for example, the travel of the microscope stage is sufficiently small to be undisturbed by the presence of a drop of liquid on the sample. At the LSpD, cells were generally grown directly on, or spin-deposited onto a substrate, and then covered with a drop of water, buffer, or cell culture medium. The water-immersion objective was subsequently lowered to make contact with the water drop and to serve as a cover of the sample during data acquisition (see Figure 14.2).

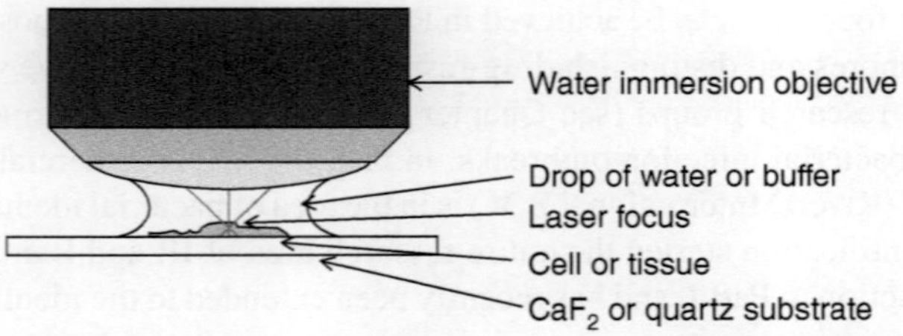

Figure 14.2 *Sampling arrangement for confocal Raman imaging using a water-immersion objective*

A major source of discussion among Raman microscopists is the excitation wavelength to be used. General wisdom has it that excitation at 632.8 nm (He—Ne) or 785 nm from a diode-pumped solid-state laser is optimal for biological samples. However, experiments carried out at the LSpD using 488-nm excitation resulted in fluorescence-free spectral data sets with the additional advantage of improved spatial resolution and enhanced signal intensities. These are due to the two effects: the ν^4 dependence of Raman scattering introduced earlier (see Eq. 4.9), which accounts for a factor of 6.7 scattering enhancement between 488- and 785-nm excitations, and the much higher sensitivity of the CCD detectors in the mid-visible rather than the near-IR wavelengths. Although one may expect higher fluorescence levels with mid-visible excitation, this often is not the case if the samples are well washed before data acquisition. Actually, early results from the LSpD were criticized of being erroneous because they did not contain any sample fluorescence.

Most of the other arguments presented in the section on infrared spectral cytopathology (IR-SCP) of cells and tissues discussed in Chapter 13 hold for Raman microscopy as well. These topics included the acquisition of the spectral datacubes, data preprocessing, correlation with visual images, and data analysis. For the work on individual cells, fluorescent stains were used frequently during the studies at the LSpD to correlate observed spectral features with biological organelles or structures that are commonly detected in cellular work. An example of this is the staining of a cell with a mitotracker stain that allows detection of mitochondria. The fluorescence of this stain could be detected without moving the cell, after Raman data acquisition and subsequent addition of the stain, using the same excitation wavelength and the Raman instrument operating as a broadband fluorescence microscope.

The preprocessing (see Section 12.2) should include cosmic spike removal, and conversion to linear wavenumber increments, as discussed earlier. As Raman data sets usually are noisier, and exhibit narrower peaks than IR spectra, conversion to derivative spectra often is not advantageous. For quantitative work, and comparison between different instruments, the intensity response of the Raman spectrometers should be established and used for intensity calibration of all acquired spectra [12]. This step is not necessary in IR-SCP because all measurements are ratioed against a background.

Several studies cited in Ref. [13] have concluded that Raman and IR microspectral methods coupled to similar multivariate analysis methods produce similar sensitivities and specificities and can detect comparable spectral changes. An example of such a comparison is briefly introduced in Section 14.4: the Raman studies of the Dublin group of Byrne on different cervical cell lines infected with different numbers of human papilloma viral copies and the corresponding IR studies at the LSpD revealed nearly identical results [14].

As pointed out earlier, Raman microspectroscopy allows label-free images of cells and tissue sections to be acquired with diffraction-limited spatial resolution or better than 0.5 μm. If this high resolution is not desirable, for example, when interrogating a large ensemble of cells for differences in cellular biology – it is advantageous to use a lower magnification and lower numeric aperture lens, and/or defocus the exciting laser to illuminate most of the cell, or at least most of the cellular nucleus. Although all these steps come with the penalty of reduced scattered intensity, they permit the collection of an "average" spectrum of a cell or nucleus. Such average spectra can then be used to carry out spectral cytology similar to the methods as described in Sections 13.8 and 13.9. Thus, Raman microspectroscopy affords the advantage of very high or low spatial resolution data acquisition, depending on the goal of the study.

Owing to the tight laser focus that can be achieved in Raman microscopy, it is possible to identify individual bacterial cells and even spores and distinguish drug-resistant from drug-sensitive species. These efforts have been pursued by several research groups (see Chapter 15) and may have enormous importance to rapidly determine the source of bacterial infection outbreaks. In fact, the first commercial application of vibrational spectroscopy diagnostics (RiverD International B.V.) is in the area of bacterial identification and classification. The field of bacterial identification started the entire research area of IR and Raman medical diagnostics, as pointed out in the Introduction to Part I, and has recently been extended to the identification of single bacterial particles without any culturing steps.

Neither the IR nor the Raman efforts in the area of bacterial classification and identification are covered in Chapters 13 and 14 of this book, which strictly emphasizes eukaryotic cells and tissue analysis, and also skips the analysis of body fluids and films that can be obtained by drying body fluids. For a review of both IR and Raman studies for bacterial classification and identification, the reader is referred to several recent review articles.

14.3 High-Resolution Raman Spectral Cytopathology

14.3.1 Subcellular organization

As discussed earlier, the best spatial resolution that can be achieved in far-field optical microscopy is determined by the wavelength of the exciting laser and the NA of the microscope objective. Using a blue laser for excitation, and a water-immersion object with an NA of 1, the spot from which the Raman spectra are collected measures about 300 nm in diameter. By raster scanning a cell immersed in buffer or cell culture medium through the laser spot in steps of between 300 and 500 nm, a hyperspectral Raman data set can be acquired that can be analyzed by the same multivariate methods discussed earlier (see Chapter 12). However, it was found advantageous to move the sample continuously at very low speeds through the laser focus, rather than employing a "stop-and-sample" mode.

Collection of a hyperspectral data set can be quite time consuming, because for a large cell, 100×100 voxel spectra may need to be collected. Typical data acquisition times for one spectrum range from 200 to 500 ms, such that the collection of an entire data set might require nearly an hour. This imaging modality produces confocal images of cells with similar spatial resolution than that afforded by confocal fluorescence microscopy, but does not require incubation of the cells with dyes or labels. Thus, the images produced are due to the natural spectral contrast of the different cellular compartments.

Figure 14.3 shows pseudo-color images of an oral squamous epithelial cell in an aqueous environment. Panel (a) represents a univariate image constructed from a Raman hyperspectral data set consisting of 10,000 individual spectra, collected on a 500-nm grid. The image in Figure 14.3(a) was obtained by integrating the entire C—H stretching region (2800–3050 cm^{-1}) and representing the integrated intensities by a color scale in which yellow indicates the highest value and black the lowest. Obviously, this image conveys nothing but the overall shape of the cell, and the location of the nucleus.

However, when hierarchical cluster analysis (HCA) is carried out on the data set, and the results are converted into a pseudo-color image (see Figure 14.3(b)), chemical information, encoded in the Raman spectra, provides the imaging contrast [15]. Here, the pyknotic nucleus is shown in gray, and the chemical composition encoded in these voxel spectra is sufficiently different from that of the cytoplasm to be detectable. Within the cytoplasm, an interesting feature is observed, represented by the occurrence of the red spots. The mean cluster spectra of these spots indicate a superposition of cytoplasmic protein and phospholipids, indicated by strong aliphatic CH_2 deformation and stretching modes at about 1445 and 2950 cm^{-1}, respectively. Natural phospholipid spots can be due to intracellular lipid droplets or due to structures such as the Golgi apparatus, vacuoles, or multilamellar vesicles. When exposing a cell to deuterated phospholipids [16] (e.g., liposomes produced entirely from deuterated phospholipids), it was found that the deuterated lipids equilibrate with the naturally occurring lipids within the cytoplasm. Thus, Raman imaging can be used to study the transport and exchange phenomena that are difficult to perform by other imaging methods. Figure 14.3(c) shows a confocal cross section of the cell, taken along the black line in Figure 14.3(a). Again, the nucleus is readily discernible in this image.

Figure 14.4 presents the results of a study on a cultured cell. The size of the nucleus is much larger than that of the previously analyzed cells and shows distinct nucleoli. This a common feature found when examining mature, exfoliated cells that may contain a pyknotic nucleus and actively growing cells like those found in

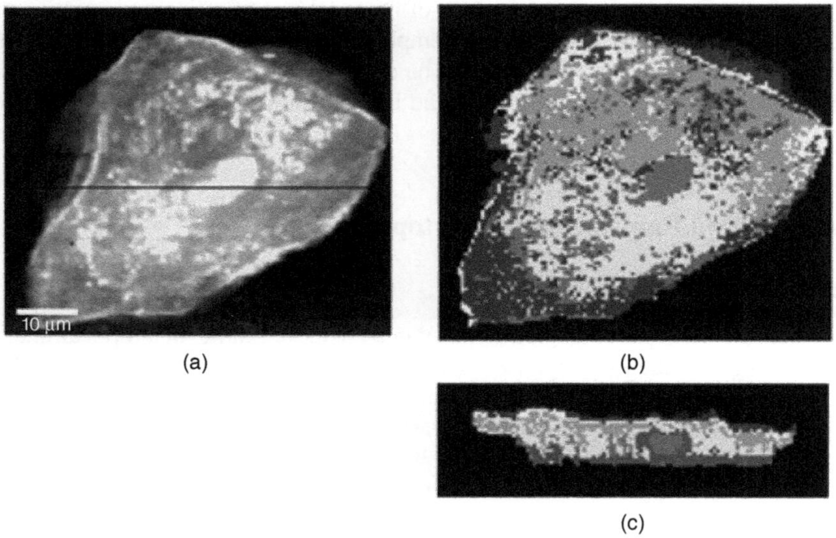

Figure 14.3 *(a) Univariate image of integrated C—H Raman stretching intensities of an oral mucosa cell. Brighter yellow hues correspond to higher intensities. (b) Pseudo-color image created by HCA of the same data set. (c) Pseudo-color image of a depth scan along the black line in (a). (*See insert for color representation of this figure.*)*

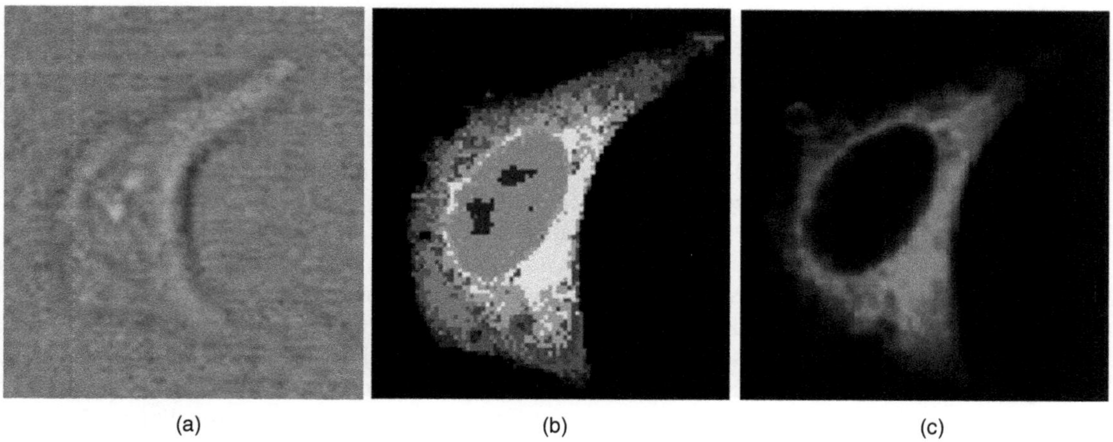

Figure 14.4 *(a) Brightfield image of a HeLa cell in aqueous medium. (b) HCA-based Raman image of same cell. (c) Fluorescence image of same cell after staining with mitotracker stain. Matthäus* et al. *[1]. Reproduced with permission from Elsevier. (*See insert for color representation of this figure.*)*

cell culture. In this study, the same methodologies described earlier for data collection and analysis were used and could distinguish the nucleus, the nucleoli, and mitochondria- and membrane-rich structures inside cells without the use of any contrast agents or dyes [17]. Panel (a) shows a bright field visual microscopic image of the HeLa cell in aqueous medium, taken through the Raman microscope. The contrast is poor, because the unstained cell is in aqueous medium with a refractive index nearly equal to that of the cell. Nevertheless, the outline of the cell, the nucleus and the nucleoli are discernible. Panel (b) shows the Raman image of the same cell, constructed *via* HCA (see Chapter 12) from a data set containing about 10,000 spectra. The cellular details available in this image are astounding: the large nucleus shown in gray is easily distinguished from the cytoplasm. Furthermore, two nucleoli shown in dark blue are detected within the nucleus. The spectra of the nucleus (mostly protein and DNA) and nucleolus (mostly protein and RNA) differ very minutely; in a single spectrum, such differences would not be significant. However, HCA detects correlated spectral differences that have physical significance; thus, the mean cluster spectra between nuclei and nucleoli show distinct spectral changes that could be interpreted biochemically [1], and were reproducible between different cells studied. Panel (c) shows a fluorescence image of the same cell, after a drop of mitotracker stain was added to the cell while in the microscope. Subsequently, the fluorescence image was collected, without reregistering the sample, using the Raman microscope as a confocal fluorescence microscope and integrating over the entire fluorescence emission spectrum collected at 488-nm excitation wavelength. The bright green areas in the fluorescence image colocalize well with the yellow and orange clusters in the HCA image; thus, these clusters were associated with mitochondria-rich areas of the cytoplasm. Incidentally, differently processed images of the same Raman data set are shown in Figures 12.10 and 12.12; some of these panels show even better details than the HCA image shown in Figure 14.4(b).

The next set of results to be discussed was reported by Otto and coworkers (Enschede, The Netherlands) and deals with earlier efforts to monitor the mitotic stages in actively dividing cells at the chromosome level and map the different chromatin regions in a chromosome [3]. In this study, the Raman spectra of a single chromosome of a cell in the metaphase of mitosis could be detected. Furthermore, in a polytene chromosome, the spectra collected in a chromosome band, the interband, and telomere region were noticeably different and indicate both the spatial resolution and the fingerprint sensitivity of Raman microspectral methods.

This work set the stage for subsequent high-resolution Raman studies and proved that visual excitation (660 nm) nonresonant Raman microspectroscopy provided an excellent tool to follow many biological processes. Excellent signal-to-noise spectra could be recorded in short acquisition times of the different "bands" of chromatin, providing the first label-free spectral identification of DNA–histone complexes in live chromosomes.

Figure 14.5 from the author's laboratory demonstrates the ability of Raman imaging to follow biological processes. This figure shows Raman images of two HeLa cells in the metaphase (top row) and anaphase (bottom row) of mitosis [18]. In Figure 14.5, the left-column images represent cells as seen in cell culture under an inverted phase contrast microscope. The second column shows intensity distribution maps of the nucleic acid band intensity at 785 cm^{-1}, normalized with respect to the amide I protein peak at 1655 cm^{-1}, whereas the third column depicts the protein amide I intensity (1655 cm^{-1}). The protein intensity delineates the microtubules and the dense histone-packed chromatin. The far right column shows, for comparison, the fluorescence images of the DAPI-stained cells. This image confirms the density and distribution of the condensed chromatin within the cell at each of the mitotic phases. The DNA-based Raman and DAPI fluorescence images show a high degree of agreement, indicating that it is possible to use Raman spectral imaging instead of fluorescence staining methods to visualize the DNA condensation during mitosis.

14.3.2 Subcellular transport phenomena

Another important application of this technology is to monitor transport processes into cells [16, 19], for example, the uptake of drug-loaded nanoparticles. Here, the inherent chemical differences between the cytoplasm and the nanoparticle provide the imaging contrast. In previous experiments, using fluorescence microscopy to follow the uptake of nanoparticles into cells, dyes, or labels needed to be incorporated in the nanoparticles for their detection.

Interestingly, HCA-based data analysis methods introduced earlier did not work as well as some of the spectral decomposition methods discussed in Section 12.4.1. The reason for this is not completely understood

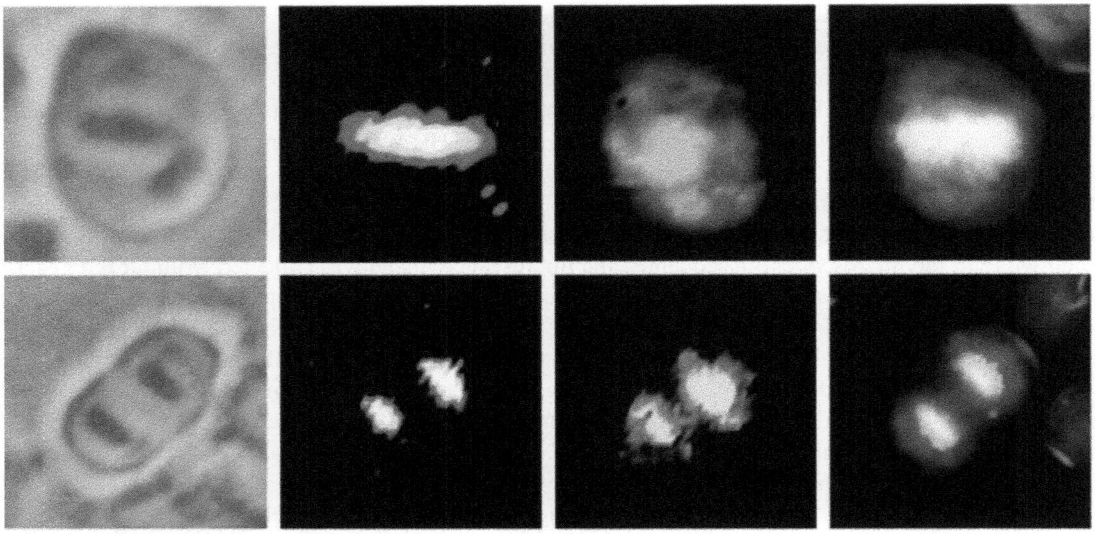

Figure 14.5 *Raman intensity images of two different HeLa cells in the metaphase and anaphases of mitosis. See text for details. Matthäus* et al.*, 2006 [18]. Reproduced with permission from Society for Applied Spectroscopy.* (See insert for color representation of this figure.)

at this point, but seems to be a function of the spectral variance to be detected in the particular experiment, as compared to the overall noise in the data set. Thus, it turned out that methods of spectral decomposition (vertex component analysis (VCA) and principal component analysis (PCA), see Chapter 12) gave better images in which small spectral changes due to transported nanoparticles into cells could be detected.

14.3.2.1 Uptake of Deuterated Liposomes

The first of the transport and drug-uptake phenomena studied at the LSpD was the uptake of liposomes into cells [16]. Liposomes are nanoparticles (100–200 nm in size) composed of a spherical phospholipid bilayer as shown in Figure 14.6; their interior can be used to transport chemotherapeutical drugs into cells. In order to distinguish the phospholipids of the liposomes from the naturally occurring phospholipids in a cell, the former incorporated fatty acids side chains that were perdeuterated, as shown by the structure in Figure 14.6(b), which is that of 1,2-dipalmitoyl-d_{62}-sn-glycero-3-phosphocholine (DPPC).

Figure 14.7 shows results demonstrating how Raman microspectroscopy can be used to study the uptake of DPPC liposomes into cells. Panel (a) shows a Raman-based image of a cell, reconstructed from VCA (see Chapter 12). Here, all green hues show the cytoplasm (darkest green), the nucleus, and the nucleolus (brightest green). As the spectra of these three regions are similar, they were grouped into one endmember spectrum, which is shown in (b), green trace. It is indicative of protein and nucleic acids in a variety of proportions. The blue spots in (a) denote naturally occurring phospholipid inclusions in the cell; the number of such inclusions varies enormously from cell type to cell type. The corresponding endmember spectrum (blue in b) exhibits typical phospholipid spectral features, devoid of protein features, and exhibiting a C—H stretching band profile typical of lipids. The red spots in (a) are due to spectral patterns (red trace in b) that contain distinct C—D stretching features between 2000 and 2200 cm^{-1}. The cell shown in Figure 14.7 was incubated with a medium that contained deuterated liposomes for 12 h. At this point of time, one observes C—D stretching features appearing in regions of phospholipid structures (blue) in the cell, indicating that a certain amount of

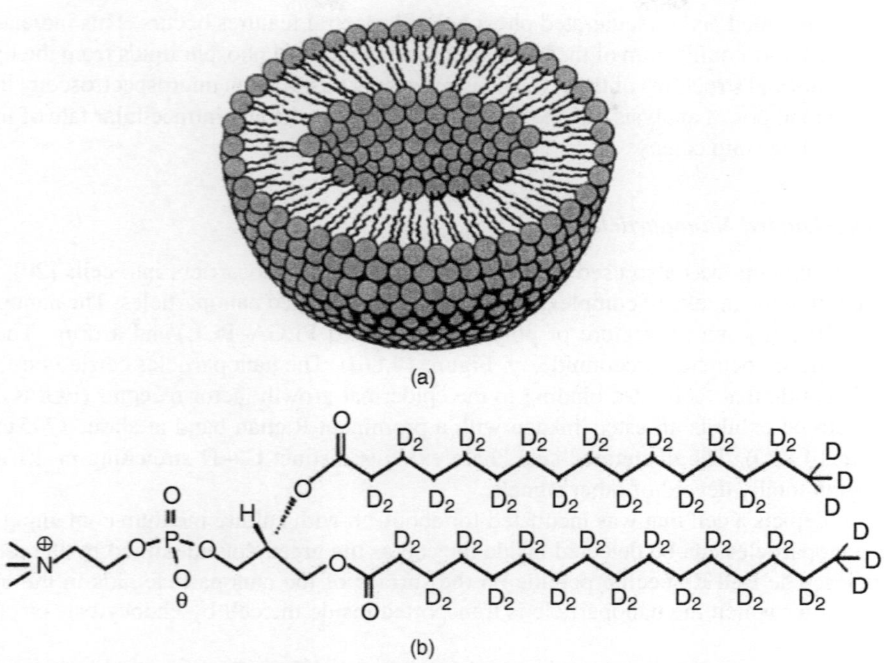

Figure 14.6 *(a) Schematic drawing of a liposome. (b) Example of a phospholipid with perdeuterated fatty acid side chains*

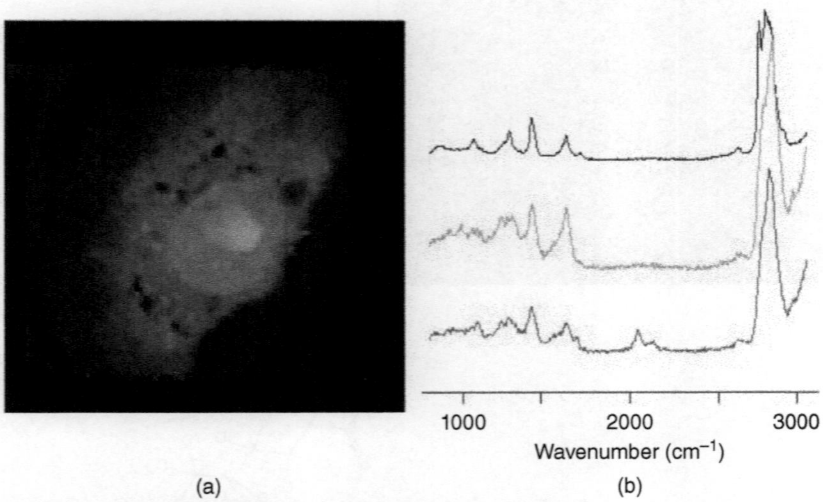

Figure 14.7 *(a) VCA-based Raman image of the uptake of deuterated liposomes into an MCF7 cell. Bright green: nucleolus, green: nucleus, dark green: cytoplasm, blue: naturally occurring lipid inclusions in the cytoplasm, and red: deuterated liposomes. (b) Raman endmember spectra of (blue): naturally occurring phospholipids, (green): protein and nucleic acids, (red): deuterated liposomes. (*See insert for color representation of this figure.*)*

equilibration of deuterated and undeuterated phospholipid spectral features occurs. This indicates that inside a cell, there is a dynamic equilibrium of the lipid structures: deuterated phospholipids from the liposomes will be found in phospholipid structures of the cell, and *vice versa.* Thus, Raman microspectroscopy in conjunction with multivariate methods of analysis may be used to study the uptake and intracellular fate of liposomes and other transport vehicles into cells.

14.3.2.2 Drug-Loaded Nanoparticles

Confocal Raman imaging was also used to monitor the uptake of nanocarriers into cells [20]. Here, results are reported detailing the uptake of complex, targeted, and drug-loaded nanoparticles. The nanoparticles were composed of a 10:1 (by mass) mixture of polymer (Pegylated PLGA–PCL)/and a drug. The drug was a pro-apoptotic, partially deuterated ceramide, *cf.* Figure 14.8(d). The nanoparticles carried on their surface a 17-amino acid peptide that facilitates binding to the epidermal growth factor receptor (EGFR) of cells. The nanoparticle material exhibits an ester linkage with a prominent Raman band at about 1745 cm^{-1} (see left arrow in Figure 14.8(c)). The deuterated side chain exhibits distinct C—D stretching peaks in the Raman spectra in a region totally devoid of other signals.

Figure 14.8(a) depicts a cell that was incubated for about 1 h with culture medium containing the nanoparticles. These nanoparticles can be detected inside the cell as the green dots identified by the ester linkage of the nanoparticles. The EGFR-specific peptide on the surface of the nanoparticle aids in the initial binding to the cell wall, after which the nanoparticle is transported inside the cell by endocytosis or other transport

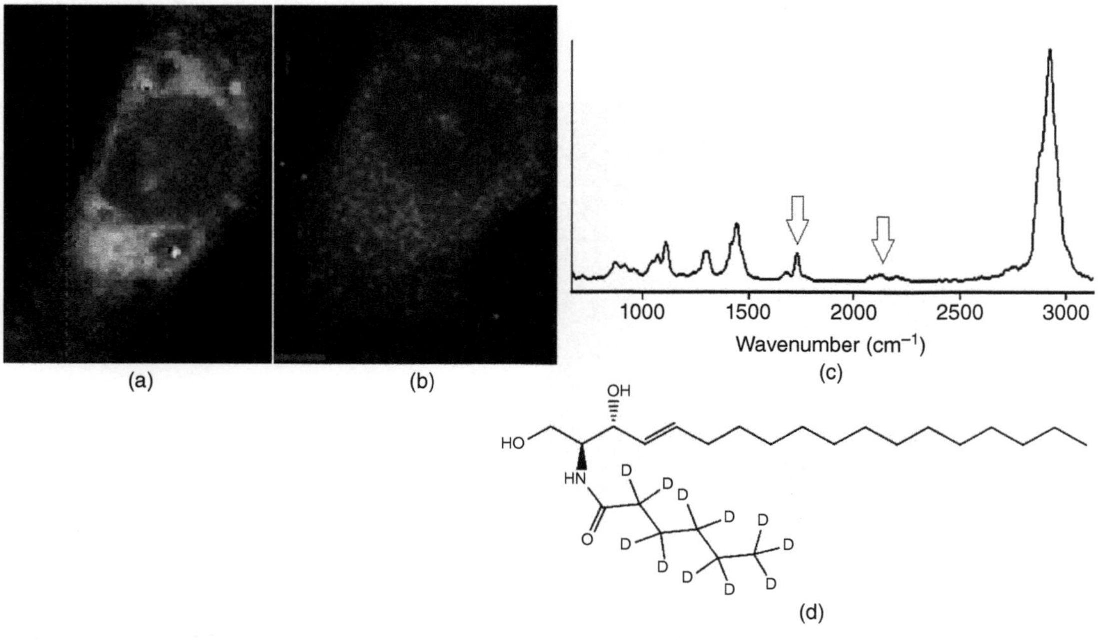

Figure 14.8 *(a) MCF7 cell incubated for 1 h with nanoparticles containing the drug shown in (d). The nanoparticles appear as the green dot in this VCA image. (b) MCF7 cell incubated for 6 h showing the release of the drug, identified by its spectral features (the C—D stretching modes) shown in (c). (c) Raman endmember spectrum of nanoparticles, showing the ester and the C—D stretching vibration (arrows). (d) Structure of the deuterated ceramide.* (See insert for color representation of this figure.)

phenomena. At some later time (typically 3–6 h), the nanoparticles disintegrate, and the drug load is released into the cytoplasm; see the small red dots in the cytoplasms shown in blue in (b). Owing to the specificity of the drug spectral signature, the released drug can be identified by the C—D stretching features marked by the right arrow in (c) at 2100–2200 cm^{-1}. As few as few thousand drug molecules in the laser focal volume can be detected by this methodology.

14.3.2.3 Monitoring Drug Uptake and Drug Metabolism

Drug uptake into cells was also reported by El-Mashtoly *et al.* [21]. Here, a drug with a distinct carbon–carbon triple bond, erlotinib, was incubated with an SW480 colon cancer cell line. The triple bond stretching vibration exhibits a strong, distinct band at 2110 cm^{-1}, in a spectral region that is devoid of any peaks (except, as pointed out earlier, the C—D stretching vibrations in deuterated compounds). Analysis of the resulting data set, using both univariate and multivariate approaches, demonstrated the distribution of the drug in the cytoplasmic periphery of cells. The drug concentration in the cell culture medium was kept at a very low level (100 μM), so that the cells still were viable, but the drug accumulation in the cells could be measured. Erlotinib is a known EGFR inhibitor. These receptors are normally associated with the outer cell membrane, as demonstrated by fluorescence microscopy. In cells incubated with erlotinib, internalization of the EGFR was found, and the accumulation of the drug in the cytoplasm, shown by the Raman microscopic results, could be verified in the fluorescence images. This study further demonstrated that structural changes (demethylation) of the drug molecule caused by its metabolic degradation can be monitored *via* Raman microscopy.

14.4 Low-Resolution Raman SCP of Cultured Cells *In Vitro*

In addition, a number of studies have been published in the literature that used Raman microspectroscopy (at lower spatial resolution, as pointed out earlier) for Ra-SCP in analogy to the IR-SCP studies reported in Chapter 13. For cultured cells, the accuracy of discrimination between cell types is similar to that reported for IR studies; however, all Raman data sets reported to date are much smaller than the corresponding IR data sets, owing to the much slower data acquisition in Raman microspectroscopy.

Several groups have used Raman SCP to study and monitor the differentiation of pluripotent embryonic stem cells, and the distinction between stem cells and cardiomyocytes (CMs) [7, 22–24]. The first of these studies, reported by Notingher *et al.* [6, 7], used murine embryonic stem cells (mEB) to follow their spontaneous differentiation from undifferentiated cells to embryoid bodies. In each of these stages, the cells exhibited distinct spectral patterns that allowed classification *via* PCA and HCA (see Chapter 12) with high accuracy. Raman data were collected from a laser spot measuring about 5 μm × 10 μm.

The authors interpreted the spectral changes upon differentiation as being due mostly to the increased mRNA concentration upon differentiation; they arrived at this conclusion by the comparison of mRNA Raman spectra with the principal component along which the Raman spectra of undifferentiated and differentiated mEB cells separated. Later studies by Chan *et al.* [22] used a similar approach to follow the differentiation of pluripotent human embryonic stem cells (hESCs) into CMs, the precursor cells for the human heart. The CMs are being considered for transplantation therapies in cardiac diseases. However, CMs lack the surface markers for identification; thus, methods such as immunohistochemistry (IHC) cannot be applied. Yet, the separation of CMs from undifferentiated hESCs is of prime importance, because the latter can form tumors after transplantation. Thus, Raman microspectroscopy provided an ideal tool to follow this differentiation.

In the study reported by Chan *et al.* [22], three cell lines, undifferentiated hESCs, precursors for cardiomyocytes (hESC-CMs), and human fetal left ventricular cardiomyocytes (FLV-CMs), were used. The laser spot utilized was much smaller, about 0.8 μm in diameter, and was centered in the cells to be studied. To account for the possibility of different cellular compartments being probed by the laser spot, average spectra collected over many collection sites for a given cell were used. However, the authors found that the spatial variations over a cell were relatively small.

As discussed in the case of IR-SCP, the spectral differences among the three cell types (hESCs, hESC-CMs, and FLV-CMs) were quite small, and the spectral data sets were noisy. Thus, multivariate methods of analysis (linear discriminant analysis, LDA) were employed, and the authors reported good separation between hESCs and FLV-CMs. The classification of the hESC-CMs from the other two cell types by LDA was not as good, and leave-one-out-cross-validation (LOOCV) (see Chapter 12) gave good results for the identification of hESCs and FLV-CMs (98% and 96%) but lower accuracy (66%) for the identification of hESC-CMs.

The two studies reported here suggest that sorting of certain cell types by Raman spectral methods – the analog of fluorescence-activated cell sorting (FACS) – may be practical in the future. This aspect is discussed in Section 14.5.

The studies introduced earlier probed cells that were plated onto substrates and formed colonies that are initially one-cell-layer thick. Thus, individual cells can be detected and analyzed by Raman microscopy. In another approach, cell pellets were formed and analyzed by collecting spectra at random positions on the pellet, rather than at the positions of cells. The advantage of this approach is that the confocal depth of the probing laser will sample cells at and below the surface and thus create an average measurement of more than one cell. The aim of this study was to establish whether or not Raman microspectroscopy can distinguish cell types that are infected with viruses, and contain different numbers of the viral genome. The cells studied were human cervical cancer cells, infected with the high-risk human papilloma virus (HPV) that is suspected to be the primary cause of cervical cancer. Four cell lines are commercially available: a cervical cancer cell line known as C33A that is not infected, the Si-Ha cell line that contains one to two copies of the viral genome of HPV 16, the HeLa cell line that contains between 20 and 50 copies of the HPV 18 genome, and the CaSki cell line that contains between 600 and 600 copies of HPV 16. Here, HPV 16 and HPV 18 are two different high-risk strains of the HPV virus. In a rare case of scientific coincidence, these four cell lines were studied independently, and nearly simultaneously, by the group of Byrne at the Dublin Institute of technology, and at the LSpD (see Section 13.8.3). These studies, for which the results are shown in Figure 14.9, used different techniques – Raman microscopy at Dublin [14] and IR-SCP at the LSpD [25] – and arrived at nearly identical conclusion: the different cell lines, although treated exactly the same, resulted in spectral data sets that showed distinct cluster patterns when analyzed by PCA. Even some of the details – for example, a much more diffuse cluster for the CaSki cell lines for which the number of viral copies varied the most – were reproduced by the two approaches. Owing to the ease with which to collect IR-SCP data, these data sets are much larger than the corresponding Raman data sets and offer much better statistics. The IR studies further suggested that the detected differences were due to the expression of different (amounts of) proteins by the viral infection and not due to the presence of the viral DNA *per se.*

Another application of low-resolution Ra-SCP is the typing and differentiation of cell lines. This work may be perceived as a precursor for work in the liquid phase, which has enormous potential for the detection of circulating tumor cells (CTCs) and for Raman-activated cell sorting (RACS), to be discussed in Section 14.5. In order to demonstrate that Ra-SCP is, indeed, capable to differentiate different cell types, the Jena group of Popp undertook a study [8–10] in which several cell types were studied to mimic what may be found in a sample of peripheral blood. The cell types included leukocytes extracted from blood of tumor-free patients, and cultured tumor cells from acute myeloid leukemia (OCI-AML3) and breast cancer (MCF7 and BT20). The cells were fixed in paraformaldehyde, transferred onto calcium fluoride slides and dried. Near-IR laser excitation at 785 nm was used to collect the Raman spectra of individual cells.

Figure 14.10(a) shows mean spectra collected for the different cell types. This figure reveals several highly important results. First, the spectra of the different cell types are rather similar. Any results where large spectral differences are observed are likely due to spurious effects. Second, multivariate methods such as support

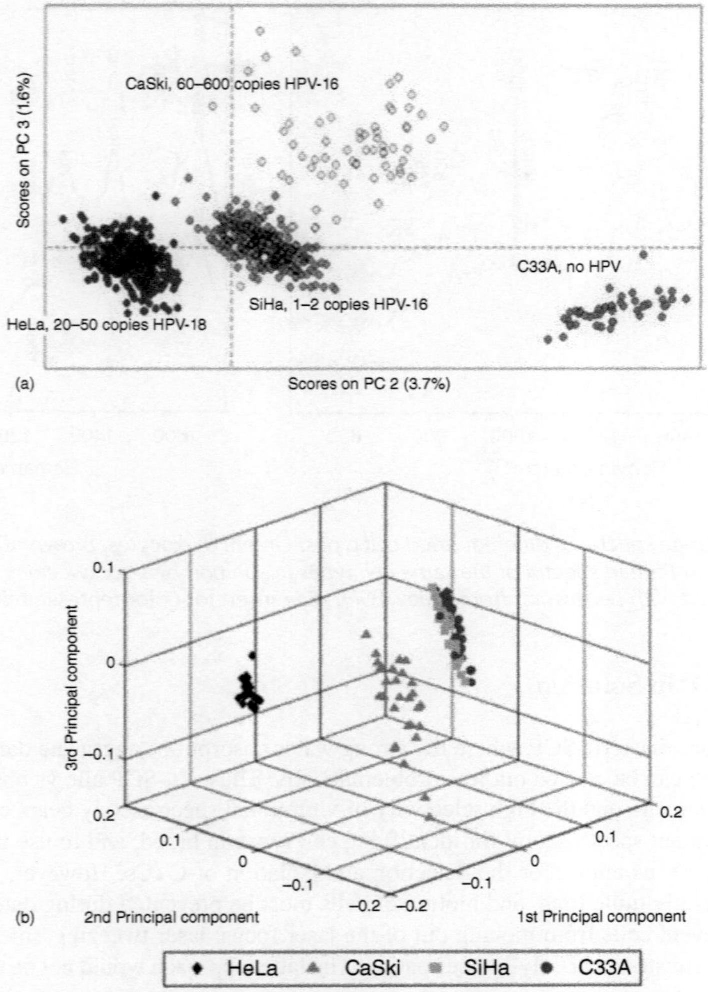

Figure 14.9 (a) PCA scores plot of four different cervical cancer cell lines with different viral infection, obtained via *infrared SCP. (b) PCA scores plot of Raman microscopic data of the same for cell lines. (*See insert for color representation of this figure.*)*

vector machine (SVM) (see Chapter 12) can differentiate the small spectral differences between cell types that are barely visible by visual inspection with high accuracy. Third, spectral quality is of prime importance. The spectra in Figure 14.10 exhibit excellent signal-to-noise ratios, facilitating the spectral differentiation of the cell types. A data set consisting of 3919 spectra was used to train an SVM that was able to classify, *via* LOOCV (see Chapter 12) the spectral data with prediction accuracy of better than 99%.

This model was applied to samples of mixed cell populations consisting of leukocytes, MCF7, BT20, and OCI-AML3 cells mixed at an equal ratio. The Raman data from the mixed sample were processed in the same way and subjected to the SVM model. Subsequently, the cells were identified by immunohistochemical stains. The fluorescence images were compared with the Raman-based classification, and it was found that the Raman images could classify the cell types better than fluorescence [10].

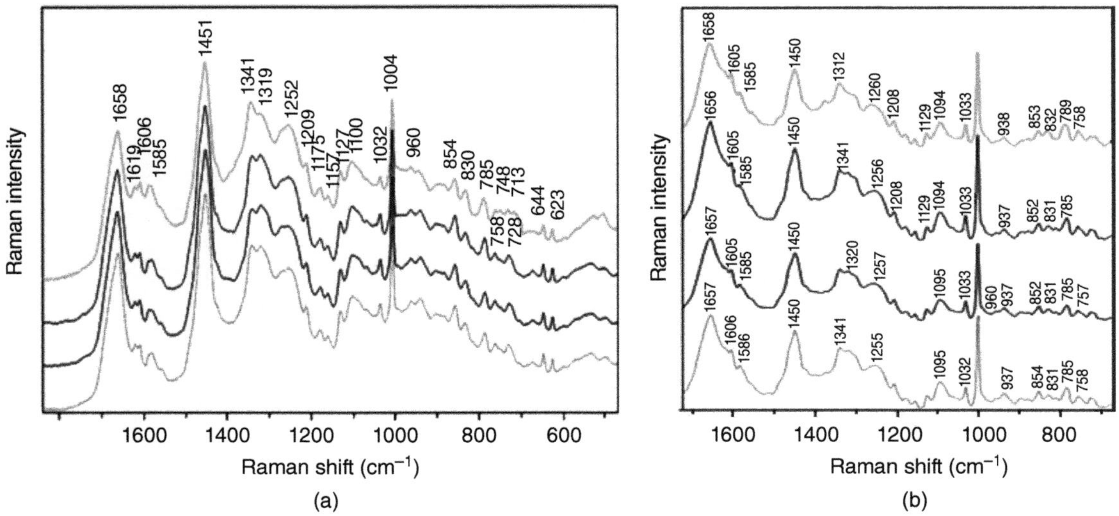

Figure 14.10 *(a) Raman spectra of different dried cell types. Green: leukocytes, brown: OCI-AML3, blue: BT20, and orange: MCF7. (b) Raman spectra of the same cell types in solution by laser tweezing methods. Diem* et al.*, 2013 [10]. Reproduced with permission from Wiley-VCH. (*See insert for color representation of this figure.*)*

14.5 Raman SCP in Solution

In contrast to solution-phase IR-SCP, where the strong water absorptions aggravate data collection, Ra-SCP on cells in suspension can be carried out less problematically. Thus, Ra-SCP allows observation of live cells in their native environment, and the high selectivity of vibrational spectroscopy bears enormous promise for the application of Raman spectroscopy for identifying cell types in blood, and to use this technique for cell sorting applications, for example, for the detection and isolation of CTCs. However, the acquisition times for Ra-SCP are generally quite long, and motion of cells must be prevented during data acquisition. Several methods exist to prevent cells from moving out of the laser focus: laser tweezing, microfluidic trapping, or attaching cells to the substrate *via* poly-lysine coating. The latter approach would not be useful for cell sorting.

14.5.1 Optical tweezing of cells in aqueous media[1]

Raman spectroscopy and optical tweezers technology are two complementary techniques because the laser used for Raman excitation and the high-NA microscope objective used for laser focusing and signal collection simultaneously form an optical tweezers. It can be assumed that the nucleus, as the densest portion of the cell, is forced into the focus and contributes significantly to the Raman spectrum. Laser tweezers Raman spectroscopy was recently reviewed [10].

The same four cell types described in the previous section were studied in phosphate-buffered saline (PBS) buffer after fixation with paraformaldehyde: 306 leukocytes from three different donors, 261 OCI-AML3, 301 MCF7, and 216 BT20 cells [18]. Cells were prepared from three independent cultivation batches and Raman spectra were collected on five consecutive days. Typical experimental conditions included excitation with a 785-nm single-mode laser (Toptica, Germany), and an upright Raman microscope (Microprobe, Kaiser

[1]This section is taken, in part, from Ref. [10].

Optical Systems, USA) with a 60×/NA 1.0 water-immersion objective (Nikon, Japan). Laser power at the sample was 75 mW, and the integration time was 10 s. Average Raman spectra are shown in Figure 14.10(b).

Apart from the spectral contributions of water near 1630 cm^{-1}, the Raman spectra of cells were very similar to the spectra of dried cells in (a). When SVMs were trained with 1210 spectra, high sensitivity (>99.7%) and specificity (>99.5%) were reached to distinguish normal and cancer cells.

14.5.2 Cells trapped in microfluidic chips

In contrast to the optical tweezers setup described earlier, where the cells are trapped in the laser field used for excitation of the Raman spectra, an optical trap was formed by two counter-propagating laser beams at a different wavelength than the Raman excitation. A major advantage of optical traps over laser tweezers is the drastically reduced power required to hold the cells in place that reduces the risk of damage to the cells. In the context of Raman spectroscopy, optical traps offer the possibility to hold and maneuver the cell relative to the excitation laser. An early study demonstrated this capability to collect solution-phase Raman spectra of the cell nucleus, the cytoplasm, and the membrane. After nondestructive Raman-based identification, cells can be sorted and subjected to further experiments.

A dedicated microfluidic quartz chip was developed [16] that incorporates three main components: a unit that focuses the flow for injecting single cells from a reservoir, the optical trap structure, and a Y-shaped branching unit for flow switching and cell sorting between the two chip outlets. Four channels accommodate the laser fibers for trapping. Additional channels were integrated for rinsing with cleaning solvents and filling with immersion fluid. The fiber channels had a diameter of 130 μm and the fluidic channels had a width of 70 μm and a height of 50 μm. The chip was mounted in a holder that provides simple, accurate, and stable adjustment of chip, microfluidic connections, and the trapping laser fibers.

The same four cell types discussed earlier (leukocytes, OCI-AML3, MCF7, and BT20) were used in the following study. The Raman microfluidic chip was installed in the same Raman instrument described earlier. The acquisition time was 10 s per spectrum using 100 mW laser intensity at 785 nm. Four hundred and five cell spectra were collected (100 leukocytes, 104 BT20, 100 MCF7, and 101 OCI-AML3). Twenty-one spectra were collected without cells and used for background compensation. The radiation of the 1070 nm trapping laser produces a very intense band in the Raman spectrum at about 3390 cm^{-1}, which must be removed from the spectra with a cut-off filter. In addition, a line from the laser that pumps the 1070 nm laser was observed as a very sharp line at 2364 cm^{-1}. To eliminate effects due to these intense contributions, the spectral range was cut to 650–1800 cm^{-1} for further processing.

Cells were identified based on Raman spectra by an LDA classifier. Mean sensitivity and specificity determined by iterated 10-fold cross validation were 96% and 99% for the distinction of cells. If the classification is synchronized with the micro flow control, an automated Raman-activated single-cell sorting will become possible. Complete automation involves cell injection, detection, trapping, spectra acquisition, data processing, classification according to a model, and cell sorting. Although the current approach with 10-s exposure time per spectrum constituted a significant progress in terms of cell number and sensitivity compared to earlier work with 120 s per spectrum, the throughput is still too low for clinical cell sorting applications. Further innovations are required to achieve higher cell sorting rates and complement the widely applied FACS. These innovations might include shorter exposure times owing to Raman signal enhancement and parallel detection channels. Slight spectral changes indicate the requirement to train individual model for each setup.

14.5.3 Resonance Raman spectroscopy of erythrocytes

Another application to be discussed here is the work on live erythrocytes (red blood cells) reported by Wood and McNaughton [26]. In this review, results of resonance Raman studies on fully functional erythrocytes and model studies using the main constituents of erythrocytes, hemoglobin, were discussed in detail. For these studies, live erythrocytes were attached to the bottom of a Petri dish *via* a poly-L-lysine layer to render the cells immobile. The Petri dish contained PBS solution and could be perfused with nitrogen or oxygen. The Petri dish was kept at physiological temperature, and a water-immersion objective was lowered into the buffer solution to excite and collect Raman microspectral information of the erythrocytes under different conditions.

The resonance Raman results of live erythrocytes mirror those observed in model studies of hemoglobin in solution. In general, with visible laser excitation, the "core size" and "spin state" marker bands are observable and provide similar insights that may be derived from the solution studies of hemoglobin and the spectra of oxygenated and deoxygenated erythrocytes can be separated very well by PCA.

Of particular interest are the findings that indicate the differences between the spectral characteristics of hemoglobin in solution and the spectral characteristics of hemoglobin in erythrocytes where the concentration of hemoglobin is orders of magnitude higher than what can be achieved in solution. In fact, in erythrocytes, one may assume that hemoglobin is in a semicrystalline state with extensive stacking interactions. These interactions can be probed by resonance Raman spectroscopy in the red end of the visible spectrum. The results from these studies show spectral features that were not observed previously for solution-phase resonance Raman studies and may be explained by an exciton state, formed by monomer stacking, from which the resonance Raman excitation originates. The stacked hemoglobin structure would not be likely to occur in the much lower concentration ranges achievable in solution studies.

14.6 Raman Spectral Histopathology (Ra SHP)

Raman spectroscopy has been used for *in vivo* and *ex vivo* diagnostic applications. In the former category, the efforts for *in vivo* diagnosis of cervical [27, 28] cancer and the work on *ex vivo* detection of esophageal and bladder cancer [29–31] have defined the enormous potential of fiber-optic-based Raman diagnostics. Stone's group, then at the Gloucestershire Hospital (UK), was instrumental in these efforts; they also have explored the possibility of intraoperative assessment of tissue using Raman spectroscopy [32, 33]. *Ex vivo* Raman images of tissue have been collected, using mostly near-IR excitation, for liver [13], colon [34], and lymph nodes [17, 35], and they have revealed information similar to images described in the section on IR-SHP, albeit with higher spatial resolution.

The first Raman tissue maps at high spatial resolution were reported by Krafft *et al.* [13] for liver tissue. The aim of this particular study was the detection and distinction of nuclei in tissue sections that are clearly discernible in stained tissue but are invisible in unstained section. This study used 785-nm excitation, and excellent, fluorescent-free spectra were obtained. Spectra were acquired at a data point spacing of 2.5 μm and 10-s exposure time per data point. As the nuclei of these liver cells are fairly large, the relatively coarse data grid nevertheless permitted the nuclei to be distinguished from the cytoplasm and extracellular matrix when the data sets were subject to multivariate curve resolution such as PCA or VCA algorithms (see Chapter 12). The principal components/endmember spectra for different tissue regions were quite different, presumably owing to the presence of hemoglobin in erythrocytes. The spectra of nuclei and cytoplasm/extracellular matrix are much more similar, yet differ by the nucleic acid signatures in the former. The corresponding IR-SHP data could not distinguish the nuclei, because the instrument used (a PerkinElmer Spotlight 400) has a spatial resolution of 10–12 μm, as established at the LSpD in the mid-IR spectral region.

In a more recent study that also encompasses IR- and Ra-SHP results [36, 37], Raman images of colon tissue were collected using 532-nm excitation. As pointed out earlier in the discussion of high-resolution Ra-SCP, the fluorescence contributions, in general, were similar to those observed with near-IR excitation. In the colon tissue study [37], data sets of about 10,000 spectral points were collected, covering about 90 μm × 110 μm. Owing to the shorter excitation wavelength and the tighter data grid, a spatial resolution of about 1 μm was achieved; however, this increased resolution came at the price of 19 h of data acquisition time for one tissue section. As in the previous paper where FTIR data could be related to IHC, the Raman images also picked up distinct spectral patterns in region where IHC detected mutated p53. The colonic crypts in normal and pre-cancerous tissue areas showed distinct spectral differences, and different distribution of the spectral patterns observed in the colonic crypts.

Point-by-point Ra-SHP was also carried out for unfixed, excised esophageal tissue. Here, the distinction of the relevant tissue types (see Section 13.8.5) could be carried out by Ra-SHP. The results of this study are shown in Figure 14.11. These spectra, reported in 2008, were taken from the surface of esophageal tissue [29], and were confirmed by similar discriminatory power of IR-SCP of exfoliated esophageal cell (see Section 13.8.5).

Thus, it appears that Ra-SHP offers a similar ability to discriminate between normal and diseased tissue states as does IR-SHP. The 1000-fold-long data acquisition times in Raman over IR-SHP presently make the former of these techniques impractical, although the higher spatial resolution and concomitant higher information content make Ra-SHP an interesting alternative. Here, one can hope that nonlinear Raman phenomena, such as coherent anti-Stokes Raman scattering (CARS) or femtosecond stimulated Raman scattering microscopy, will enable increased spatial resolution images to be collected in acquisition times equal or better than the corresponding FTIR SHP data. Efforts in this direction have recently been reported for high-resolution cellular studies that correlate Raman, CARS, and fluorescence techniques [38].

Ra-SHP proved to be particularly valuable for the study of brain tissue [39, 40]. The corresponding IR-SHP aspects of the efforts to determine the primary tumors for brain metastases were discussed in Section 13.5.3. Similar results hold for Ra-SHP of brain tissue as well, where different primary tumors produced metastases that could be readily classified. Ra-SHP detects the components of the tissue high in lipids and cholesterol with high sensitivity, owing to the high intensity of the apolar vibrational modes of lipids.

In vivo Raman imaging of the tumor margins during brain surgery is another area of high-impact research that is reviewed briefly with other *in vivo* techniques in Section 14.7.

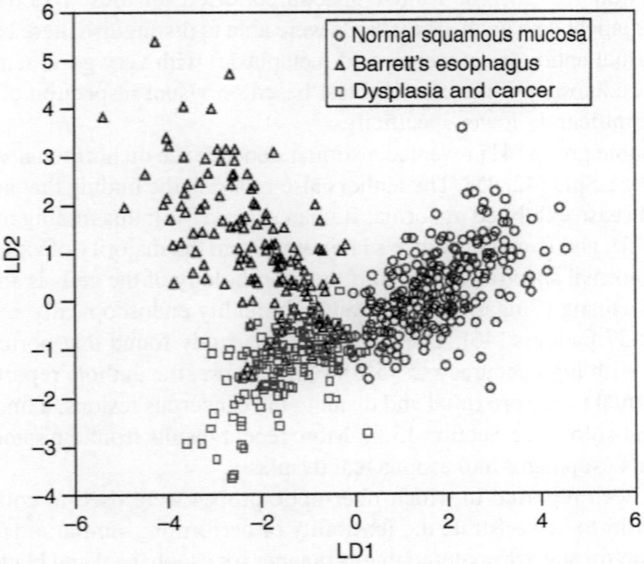

Figure 14.11 *Linear discriminant analysis results for spectra collected from excised esophageal tissue.* Barr et al., *2008 [29]. Reproduced with permission from John Wiley & Sons.* (See insert for color representation of this figure.)

14.7 *In Vivo* Raman SHP

As Raman spectroscopy is much less sensitive to interference from water, *in vivo* Raman spectra of tissue can be collected much more easily than the corresponding IR (attenuated total reflection) data. Measurement of Ra-SHP from hollow organs *via* specialized endoscopes is discussed in this section. These endoscopes contain fiber-optic Raman probes that permit the collection of Raman spectra from the surface of tissue in contact with the probe tip, or from a layer of tissue close to the surface.

In vivo Raman spectral diagnostics in hollow organs has been reported for the cervix and the esophagus, and preliminary studies for bladder tissues have been carried out. These studies promise enormous diagnostic potential, because the fingerprint sensitivity of Raman spectroscopy has much higher discriminatory power than the competing techniques, such as white light or fluorescence endoscopy. In the remainder of this discussion, the *in vivo* Raman diagnostic approach will be referred to Raman endoscopy.

Raman endoscopy is carried out using fiber-optic probes that are constructed such that they are compatible with standard endoscopes and consist of one excitation fiber surrounded by several optical fibers that collect the Raman scattering from the surface, or very close to the surface. The design and construction of such fibers is rather difficult: the end of the excitation fiber should be equipped with a band-pass filter that cuts out the Raman scattering of the fiber material (usually quartz). The collection fibers are usually beveled at the end to collect as much of the Raman scattered light while rejecting the reflected light. At the spectrometer end of the fiber bundle, the circularly arranged collection fibers usually are shaped to match the entrance slit shape of the spectrograph. For use in patients, these probes must withstand the harsh conditions of stringent cleaning methods, such as autoclaving after each use.

A study was reported by Mahadevan-Jansen and coworkers [28], which aimed at detecting cervical disease by Raman endoscopy. Rather than using exfoliated cells (see Section 13.8.3) or tissue specimens from biopsies, these measurements were carried out *in vivo* – by touching the Raman probe directly to the surface of the cervix. The following diagnostic classes were examined: normal ectocervix, normal endocervix, squamous metaplasia, and low-grade squamous intraepithelial lesion (LSIL) and high-grade squamous intraepithelial lesion (HSIL) (see Section 13.7.1). The Raman spectra collected for these five tissue classes appear quite similar, yet standard statistical data analysis methods were able to distinguish these classes, in particular HSIL from normal types (normal endo- or ectocervic and metaplasia) with very good sensitivity (89%) and specificity (~85%), whereas colcoscopy (the classical test based on visual inspection of the cervix after staining with acetic acid) had significantly lower specificity.

A later study by the same group [41] revealed a similar dependence on hormonal status that was reported in the IR-SCP studies at the LSpD [42, 43]. The authors also reported the finding that normal appearing tissue in patients with cervical disease exhibited abnormal Raman spectra [44]; this finding mirrors the SCP results by Cohenford and Rigas [45], and from the author's laboratory, that the majority of exfoliated cells from patients with disease exhibit abnormal spectra [25], even if the morphology of the cells is still normal.

In 2011, the group of Huang (Singapore) reported high-quality endoscopically acquired Raman spectra of esophageal tissue from 27 patients [46]. The authors of this study found that normal and cancerous tissues could be distinguished with high accuracy (>95%). Furthermore, the authors report small differences in the spectral patterns for normal tissue proximal and distal to the cancerous lesions, a finding that mirrors IR-SHP results from the lung and colon (see Section 13.4). More recent results from the same group [47] reported the classification of Barrett's esophagus and esophageal dysplasia.

Several studies have been reported in which fiber-optic probes were used to collect Raman spectra from biopsied tissue in an effort to demonstrate the feasibility of performing similar *in vivo* studies at a later date. Very good discrimination for spectra acquired in this manner for esophageal and bladder tissues were reported by Stone and coworkers [29]. Promising results for tumor margin detection were reported for brain biopsies [48, 49] and efforts for applications during brain surgery are underway.

14.8 Deep Tissue Raman SHP

In these measurements, Raman spectra are sought that originate in tissue layers a few millimeters to a few centimeters under the tissue surface. The former case, also referred to as "subcutaneous" measurements, has had a special role in diagnosing hard tissue diseases, such as osteoporosis in bone tissue. Here, the regions of interest lie a few millimeters under the skin, and methods have been developed to collect subcutaneous Raman spectra.

Deep tissue spectroscopy, on the other hand, aims at detecting and diagnosing disease in tissues up to centimeters under the surface. Application of this methodology to diagnose cancer in deep soft tissue is introduced later.

Both approaches use a technique referred to as "surface-offset Raman spectroscopy" (SORS). The penetration of red or near-IR laser radiation into tissue is quite deep, and can exceed a few centimeters, depending on the wavelength of the light. Factors determining this penetration are the water and hemoglobin absorptions, and the scattering properties of the tissue. In general, incident photons are absorbed or scattered; the Raman scattered photons may undergo multiple scattering and absorption processes (the so-called "photon migration") that can be modeled by Monte Carlo methods. The path of some photons through the tissue may lead deeper into the tissue, at which the photons are no longer of interest. Some photons that are scattered elastically and inelastically may end up at the tissue surface as shown in Figure 14.12. The further away from the excitation source a photon is detected, the deeper inside the tissue the photon has penetrated. Therefore, the penetration depth of the collected photons can be determined from the linear distance "*d*" between the excitation and the sampling spot. This method is ideally suited to study tissue such as bone close to the skin surface to detect abnormalities such as osteoporosis.

14.8.1 Hard tissue diagnostics

Raman and IR spectral methods are sensitive tools to determine many bone properties (matrix to mineral ratio, crystallinity, and age) of bone that can be directly correlated to bone strength, bone density, and osteoporosis. Mendelsohn and coworkers have reported on IR spectral characterization of bone tissue that consists primarily of an organic matrix (collagen) and a mixture of minerals, including calcium phosphate, calcium hydroxyapatite, and analogs containing carbonate ions. The observed spectral features can be assigned to the amide vibrations as well as the vibrations of the inorganic phosphate groups [51].

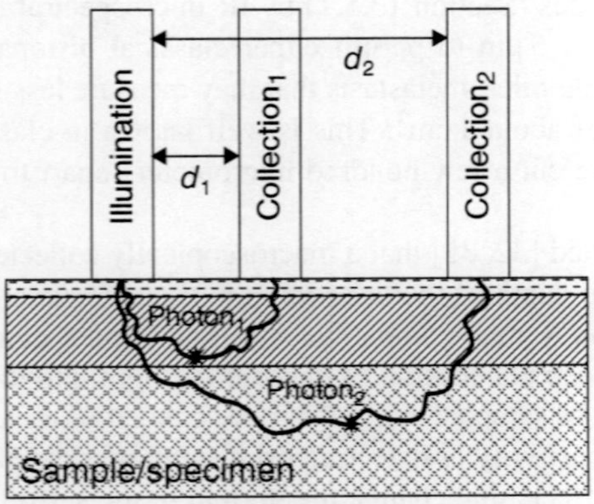

Figure 14.12 *Photon paths of scattered light in surface-offset Raman spectroscopy. Schulmerich* et al.*, 2009 [50]*

While IR spectroscopy was carried out on excised, thin section of bone, Raman spectroscopy offers the possibility of measuring exposed bone, or bone under a thin layer of skin. The first of these cases can be achieved in standard backscattering geometry. For the subcutaneous observation of Raman spectra of bone under a layer of skin, a SORS probe was developed that is described in detail in Ref. [50]. This arrangement of fibers permitted the subcutaneous collection of excellent bone spectra in a rat model in which the skin is quite thin.

By moving the exciting and collection optical fibers on the surface of the sample, a depth-resolved image of the bone structure was obtained; this techniques has been referred to as "Raman tomography" [52]. This exciting development, mostly from the laboratory of M. Morris (Michigan), opens the door for the development of spectroscopy-based tests for bone disease that are much less invasive than present medical procedures. The determination of bone density, and thereby, the progress of osteoporosis and brittle bone disease, is one possible application of SORS with enormous commercial potential. Progress in SORS application to hard tissue diagnostics was recently reviewed [52].

14.8.2 Deep tissue imaging of breast tissue and lymph nodes

SORS and several variants of this technique were recently applied to the analysis of tumors located deep in layers of surrounding tissue. The aim of these studies is to develop minimally invasive methods for breast cancer screening. While it may be impossible to actually localize a small cancerous lesion deep inside breast tissue by Raman tomography, it may be possible to distinguish the microcalcifications that occur around tumors from those that occur randomly in breast tissue. It was shown earlier [53] that chemical composition of disease-associated microcalcifications differs from those of normal calcifications. A later study, *via* FTIR imaging, from the group of Stone [54], demonstrated the origin of the spectral differences to be due to the substitution of carbonate ions in the calcium hydroxyapatite lattice by phosphate ions with progressively more severe disease. This finding may lead to a quantitative assessment of breast malignancies, because the mineral phase can be detected relatively deep inside breast tissue.

In phantom tissue section, the detection of calcifications up to 20 mm deep inside the phantom proved possible. For application for human breast cancer screening, the required penetration ranges into the tissue are about 19–50 mm; thus, the Raman-based methods have reached the lower limits of depth penetration necessary for successful screening.

Another application of deep tissue imaging, albeit not depth resolved, was reported by Stone and coworkers for the analysis of lymph nodes. This study may prove to be highly important for the detection of micrometastases in lymph nodes that may be missed by conventional methods. In the previous discussion of detection of micrometastases in lymph nodes (Section 13.5.1) by IR microspectral imaging, the lymph nodes needed to be sectioned to a thickness of 5 μm to permit either classical histopathology or SHP. However, these sections could easily miss a single micrometastasis that may measure less than 1 mm in diameter in a lymph node that may have a volume of about 1 cm^3. This is well known in classical histopathology; therefore, a series of microtome sections are cut a few hundred micrometers apart to sample most of the volume of a lymph node.

However, Stone's group reported [32, 35] that a macroscopically collected Raman spectra of lymph nodes provided a sufficient penetration depth of the laser into the tissue (several millimeters). Thus, lymph nodes were cut in half, and mounted on Raman-compatible substrates, and spectra were taken at random points of the lymph node. Owing to the deep penetration of the exciting light into the lymph node tissue, the majority of a lymph node's volume could be sampled, and subsurface cancerous areas could be detected with sufficient accuracy for clinical use. The sensitivity of detecting even small micrometastases may be helped by the previously discussed cancer field effects that change the normal tissue surrounding cancerous lesions, and that are detected by methods of vibrational spectroscopy.

References

1. Matthäus, C., van Vlerken, L., Devalapally, H. *et al.* (2007) Label-free detection of mitochondrial distribution in cells by nonresonant Raman micro-spectroscopy. *Biophys. J.*, **93**, 668–673.
2. Huang, Y.S., Karashima, T., Yamamoto, M. and Hamaguchi, H.O. (2005) Molecular-level investigation of the structure, transformation, and bioactivity of single living fission yeast cells by time- and space-resolved Raman spectroscopy. *Biochemistry*, **44**, 10009–10019.
3. Puppels, G.J., de Mul, F.F.M., Otto, C. *et al.* (1990) Studying single living cells and chromosomes by confocal Raman microspectroscopy. *Nature*, **347**, 301–303.
4. Bakker Schut, T.C., Van Dekken, H., Tilanus, H.W. *et al.* (1997) NIR Raman spectroscopy of healthy and diseased esophagus, in *Spectroscopy of Biological Molecules: Modern Trends*, Springer Verlag, Heidelberg, Germany pp. 455–456.
5. Notingher, I. (2011) Label-free imaging of phenotypic spectral markers in live cells derived from human stem cells. 14th European Conference on the Spectroscopy of Biological Molecules 2011, Coimbra University, Portugal.
6. Notingher, I., Bisson, I., Bishop, A.E. *et al.* (2004) In situ spectral monitoring of mRNA translation in embryonic stem cells during differentiation *in vitro*. *Anal. Chem.*, **76**, 3185–3193.
7. Notingher, I., Bisson, I., Polak, J.M. *et al.* (2004) In situ spectroscopic study of nucleic acids in differentiating embryonic stem cells. *Vib. Spectrosc.*, **35**, 199–203.
8. Neugebauer, U., Bocklitz, T., Clement, J.H. *et al.* (2010) Toward detection and identification of circulating tumor cells using Raman spectroscopy. *Analyst*, **135**, 3178–3182.
9. Neugebauer, U., Clement, J.H. Bocklitz, T. *et al.* (2010) Identification and differentiation of single cells from peripheral blood by Raman spectroscopic imaging. *J. Biophotonics*, **3** (8–9), 579–587.
10. Diem, M., Mazur, A., Lenau, K. *et al.* (2013) Molecular pathology via infrared and Raman spectral imaging. *J. Biophotonics*, **6** (11–12), 855–886.
11. Everall, N.J. (2004) Depth profiling with confocal Raman microscopy, part I. *Spectroscopy*, **19** (10), 22–27.
12. Stone, N., Kendall, C. and Barr, H. (2008) Raman spectroscopy as a potential tool for early diagnosis of malignancies in esophageal and bladder tissues, in *Vibrational Spectroscopy for Medical Diagnosis* (eds M. Diem, P.R. Griffiths and J.M. Chalmers), John Wiley & Sons, Ltd, Chichester, pp. 203–230.
13. Krafft, C., Diderhoshan, M.A., Recknagel, P. *et al.* (2011) Crisp and soft multivariate methods visualize individual cell nuclei in Raman images of liver tissue sections. *Vib. Spectrosc.*, **55** (1), 90–100.
14. Ostrowska, K.M., Malkin, A., Meade, A. *et al.* (2010) Investigation of the influence of high-risk human papilloma virus on the biochemical composition of cervical cancer cells using vibrational spectroscopy. *Analyst*, **135**, 3087–3093.
15. Matthäus, C., Bird, B., Miljković, M. *et al.*, Infrared and Raman microscopy in cell biology, in *Methods in Cell Biology*, Vol. **89**: Biophysical Tools for Biologists, Vol. 2, J.J. Correia and Detrich, H.W., Eds. 2008, Elsevier, pp. 275–308.
16. Matthäus, C., Kale, A., Chernenko, T. *et al.* (2008) New ways of imaging uptake and intracellular fate of liposomal drug carrier systems inside individual cells, based on Raman microscopy. *Mol. Pharmaceutics*, **5** (2), 287–293.
17. Orr, L.E., Christie-Brown, J., Hutchings, J.C. *et al.* (2010) Raman spectroscopy as a tool for the identification and differentiation of neoplasias contained within lymph nodes of the head and neck. *Proc. SPIE*, **7548**, 75481W–1.
18. Matthäus, C., Boydston-White, S., Miljković, M. *et al.* (2006) Raman and infrared microspectral imaging of mitotic cells. *Appl. Spectrosc.*, **60** (1), 1–8.
19. Matthäus, C., Boydston-White, S., Miljković, M. *et al.* (2008) Raman microscopic imaging of cells and applications monitoring the uptake of drug delivery systems. *Proc. SPIE Biophotonics: Photonic Solutions Better Health Care*, **6991**, 101–108.
20. Matthäus, C., Chernenko, T., Miljković, M. *et al.* (2011) Raman micro-spectral imaging of cells, and applications monitoring the uptake of drug delivery systems, in *Confocal Raman Imaging* (eds O. Hollricher and T. Dieing), Springer-Verlag, Heidelberg, pp. 137–162.
21. El-Mashtoly, S.F., Petersen, D., Yosef, H.K. *et al.* (2014) Label-free imaging of drug distribution and metabolism in colon cancer cells by Raman microscopy. *Analyst*, **139**, 1155–1161.
22. Chan, J.W., Lieu, D.K., Huser, T. and Li, R.A. (2009) Label-free separation of human embryonic stem cells and their cardiac derivatives using Raman spectroscopy. *Anal. Chem.*, **81**, 1324–1331.

23. Pascut, F.C., Goh, H.T., Welch, N. *et al.* (2011) Noninvasive detection and imaging of molecular markers in live cardiomyocytes derived from human embryonic stem cells. *Biophys. J.*, **100**, 251–259.
24. Pascut, F.C., Kalra, S., George, V. *et al.* (2013) Non-invasive label-free monitoring the cardiac differentiation of human embryonic stem cells *in vitro* by Raman spectroscopy. *Biochim. Biophys. Acta*, **1830**, 3517–3524.
25. Schubert, J.M. (2011) Spectral cytology of human oral and cervical samples, PhD Dissertation *Chemistry & Chemical Biology*, Northeastern University, Boston, MA.
26. Wood, B.R. and McNaughton, D. (2008) Resonance Raman spectroscopy of erythrocytes, in *Vibrational Spectroscopy for Medical Diagnosis* (eds M. Diem, P.R. Griffiths and J.M. Chalmers), John Wiley & Sons, Ltd, Chichester, pp. 261–309.
27. Utzinger, U., Heintzelman, D.L., Mahadevan-Jansen, A. *et al.* (2001) Near IR Raman spectroscopy for *in vivo* detection of cervical precancers. *Appl. Spectrosc.*, **55** (8), 955–959.
28. Robichaux-Viehoefer, A., Kanter, E., Shappell, H. *et al.* (2007) Characterization of Raman spectra measured *in vivo* for the detection of cervical dysplasia. *Appl. Spectrosc.*, **61** (9), 986–993.
29. Barr, H., Kendall, C. and Stone, N. (2008) Raman spectroscopy as a potential tool for early diagnosis of malignancies in esophageal and bladder tissues, in *Vibrational Spectroscopy for Medical Diagnosis* (eds M. Diem, P.R. Griffiths and J.M. Chalmers), John Wiley & Sons, Ltd, Chichester, pp. 203–230.
30. Barr, H., Kendall, C. and Stone, N. (2004) The light solution for Barrett's esophagus: photodiagnosis and photo-dynamic therapy for columnar-lined esophagus. *Photodiagn. Photodyn. Ther.*, **1** (1), 75–84.
31. Kendall, C., Day, J., Hutchings, J. *et al.* (2010) Evaluation of Raman probe for oesophageal cancer diagnostics. *Analyst*, **135**, 3038–3041.
32. Horsnell, J., Stonelake, P., Christie-Brown, J. *et al.* (2010) Raman spectroscopy - a new method for the intra-operative assessment of axillary lymph nodes. *Analyst*, **135**, 3042–3047.
33. Orr, L.E., Christie-Brown, J., Hutchings, J.C. *et al.* (2010) Raman spectroscopy as a tool for the identification and differentiation of neoplasias contained within lymph nodes of the head and neck. *Proc. SPIE*, **7548**, 7548W1–7548W11.
34. Krafft, C., Codrich, D., Pelizzo, G. *et al.* (2008) Raman and FTIR imaging of lung tissue: bronchopulmonary sequestration. *J. Raman Spectrosc.*, **40**, 595–603.
35. Lloyd, G.R., Orr, L.E., Christie-Brown, J. *et al.* (2013) Discrimination between benign, primary and secondary malignancies in lymph nodes from the head and neck utilising Raman spectroscopy and multivariate analysis. *Analyst*, **138**, 3900–3908.
36. Kallenbach, A., Großerüschkamp, F., Mosig, A. *et al.* (2012) Spectral histopathologic and immunohistochemical studies of colon adenocarcinomas. *J. Biophotonics*, **6** (1), 88–100.
37. Mavarani, L., Petersen, D., El-Mashtoly, S.F. *et al.* (2013) Spectral histopathology of colon cancer tissue sections by Raman imaging with 532 nm excitation provides label free annotation of lymphocytes, erythrocytes and proliferating nuclei of cancer cells. *Analyst*, **138**, 4035–4039.
38. El-Mashtoly, S.F., Niedieker, D., Petersen, D. *et al.* (2014) Automated identification of subcellular organelles by coherent anti-Stokes Raman scattering. *Biophys. J.*, **106**, 1910–1920.
39. Diem, M., Mazur, A.I., Lenau, K. *et al.* (2014) Molecular pathology via infrared and Raman spectral imaging, in *Next-Generation Spectroscopy-Based Diagnostics. In Vivo and Ex-Vivo Molecular Pathology* (ed P.J.), Wiley-VCH Verlag GmbH, Weinheim, pp. 45–102.
40. Koljenovic, S., Choo-Smith, L.P., Bakker Schut, T.C. *et al.* (2002) Discriminating vital tumor from necrotic tissue in human glioblastoma tissue samples by Raman spectroscopy. *Lab. Invest.*, **82** (10), 1265–1277.
41. Kanter, E.M., Vargis, E., Majumder, S. *et al.* (2009) Application of Raman spectroscopy for cervical dysplasia diagnosis. *J. Biophotonics*, **2** (1–2), 81–90.
42. Papamarkakis, K., Bird, B., Schubert, J.M. *et al.* (2010) Cytopathology by optical methods: spectral cytopathology of the oral mucosa. *Lab. Invest.*, **90**, 589–598.
43. Schubert, J.M., Bird, B., Papamarkakis, K. *et al.* (2010) Spectral cytopathology of cervical samples: detecting cellular abnormalities in cytologically normal cells. *Lab. Invest.*, **90**, 1068–1077.
44. Vargis, E., Byrd, T., Logan, Q. *et al.* (2011) Sensitivity of Raman spectroscopy to normal patient variability. *J. Biomed. Opt.*, **16** (11), 117004.

45. Cohenford, M. and Rigas, B. (1998) Cytologically normal cells from neoplastic cervical samples display extensive structural abnormalities on IR spectroscopy: implications for tumor biology. *Proc. Natl. Acad. Sci. U.S.A.*, **95**, 15327–15332.
46. Bergholt, M.S., Zheng, W., Lin, K. *et al.* (2011) Vivo diagnosis of esophageal cancer using image-guided Raman endoscopy and biomolecular modeling. *Tech. Cancer Res. Treat.*, **10** (2), 103–112.
47. Bergholt, M.S., Zheng, W., Ho, K.Y. *et al.* (2013) Fiberoptic confocal Raman spectroscopy for real-time *in vivo* diagnosis of dysplasia in Barrett's esophagus. *Gastroenterology*, **146** (1), 27–32.
48. Kirsch, M., Schackert, G., Salzer, R. and Krafft, C. (2010) Raman spectroscopic imaging for *in vivo* detection of cerebral brain metastases. *Anal. Bioanal. Chem.*, **398**, 1707–1713.
49. Krafft, C., Kirsch, M., Beleites, C. *et al.* (2007) Methodology for fiber-optic Raman mapping and FTIR imaging of metastases in mouse brains. *Anal. Bioanal. Chem.*, **389**, 1133–1142.
50. Schulmerich, M.V., Cole, J.H., Kreider, M. *et al.* (2009) Transcutaneous Raman spectroscopy of murine bone *in vivo*. *Appl. Spectrosc.*, **63** (3), 286–295.
51. Boskey, A. and Mendelsohn, R. (2005) Infrared analysis of bone in health and disease. *J. Biomed. Opt.*, **10** (3), 031102-1–031102-9.
52. Matousek, P. and Stone, N. (2013) Recent advances in the development of Raman spectroscopy for deep non-invasive medical diagnosis. *J. Biophotonics*, **6** (1), 7–19.
53. Haka, A.S., Shafer-Peltier, K.E., Fitzmaurice, M. *et al.* (2002) Identifying microcalcifications in benign and malignant breast lesions by probing differences in their chemical composition using Raman spectroscopy. *Cancer Res.*, **62**, 5375–5380.
54. Baker, R., Matousek, P., Ronayne, K.L. *et al.* (2007) Depth profiling of calcifications in breast tissue using picosecond Kerr gated Raman spectroscopy. *Analyst*, **132**, 48–53.

15

Summary and Epilogue

The discussions in Chapters 11, 13, and 14 have introduced Raman and infrared-based microspectroscopic techniques, and their applications toward medical diagnostics. Partially because of the author's own interest and partially because of page limits and time constraints in finishing the manuscript, the emphasis of Part II has been placed upon microspectroscopic investigations of disease, mostly for the detection and diagnosis of cancer. Hopefully, the review of the work from the author's laboratory, and of materials from other laboratories, will be adequate to introduce the reader to the state of the art in spectral diagnostics of cells and tissues.

However, both Raman and infrared techniques can be applied to other medical problems as well. Of these, bone diseases were mentioned briefly in Section 14.8 in the discussion of deep tissue imaging. Investigations of skin and cartilage have been published as well, and excellent sensitivity toward detection abnormalities has been reported. Other applications include the analysis of body fluids such as serum and plasma, saliva, urine, and synovial fluid, all of which have been published in the literature. Of particular interest were the studies to detect prion diseases, such as bovine spongiform encephalopathy in blood serum. Most of the measurements required for these efforts can be carried macroscopically, and in a high-throughput format. These efforts are highly promising in the diagnostic field, but were not covered here mostly for the reason that the emphasis of Part II was on microscopic imaging, data processing, and data analysis of data sets measuring in the millions of spectra.

One other broad subject of medical applications of vibrational spectroscopy that was – regretfully – omitted from this book is the entire field of classification of microorganisms by both infrared and Raman microspectroscopy. However, this field has become so large and diverse that it would necessitate a third part to this book. Yet, one should keep in mind that this field started the applications of vibrational spectroscopy to the analysis of biological matter, and incidentally is also the first field that developed into a commercial enterprise (River Diagnostics). The future of diagnostic methods *via* Raman microspectroscopy is particularly bright, since the tight focus of the laser beam in a Raman microscope allows identification of a single bacterium, or even a bacterial spore.

Obviously, these results will be of major interest for a broad range of researchers and – hopefully – investors. As a summary of the growth this field has experienced, a list of journal issues dedicated to progress in all of the diagnostic efforts using vibrational spectroscopy follows. These journals present a broader view of all

Modern Vibrational Spectroscopy and Micro-Spectroscopy: Theory, Instrumentation and Biomedical Applications, First Edition.
Max Diem.

aspects of diagnostic efforts using vibrational spectroscopic techniques than the more focused approach to cancer diagnostic presented in this book.

Cellular and Molecular Biology (France), **44**(1), 1998
Vibrational Spectroscopy, **28**(1), (2002) (Proceedings of Spec2000)
Technology in Research Cancer and Treatment, **1**(1), 2002
Vibrational Spectroscopy, **32**, (2003) (Proceedings of Spec2002)
Vibrational Spectroscopy, **38** (2005) (Proceedings of Spec2004)
J. Biomed. Optics, **10**(3), 2005
Biochimica et Biophysica Acta, **1758** (2006)
Analyst, **135** (2010) (Proceedings Spec2010)
J. Biophotonics, **3**(8–9), 2010
Analyst, **137** (2012)
J. Biophotonics, **6**(1), 2013
Analyst, **138** (2013) (Proceedings Spec2012).

Finally, several books and monographs dealing with spectral diagnostics have appeared among them.

"Vibrational Spectroscopy for Medical Diagnosis," M. Diem, P. Griffiths and J. Chalmers, Editors, J. Wiley-Interscience, Chichester, UK 2008.
"Biomedical Vibrational Spectroscopy," P. Lasch and J. Kneipp, Editors, J. Wiley (Blackwell Publishing), 2008.
"Infrared and Raman Spectroscopy and Spectral Imaging of Individual Cells," in "Infrared and Raman Spectroscopic Imaging", Editors: R. Salzer, H.W. Siesler, Wiley-VCH Publishing, Weinheim, Germany, 2009 (1st Edition), 2014 (2nd Edition).
"Next-generation spectroscopy-based diagnostics. In vivo and ex-vivo molecular pathology," Popp, J., Ed. Wiley-VCH Verlag, Weinheim, Germany.

Among the many conferences that now have section dedicated to vibrational spectroscopic diagnostics (IEEE, FACCS, PitCon, etc.), the Spec20xx series of conferences should be mentioned that are held biannually and are dedicated to spectral imaging and diagnostics.

Appendix A

The Particle in a Box: A Demonstration of Quantum Mechanical Principles for a Simple, One-Dimensional, One-Electron Model System

Real quantum mechanical systems have the tendency to become mathematically quite complicated and may discourage a novice in the field from pursuing the detailed steps to understand how the mathematical principles apply to physical systems. Thus, a simple scenario is presented here to illustrate the principles of Quantum Mechanics introduced in Section 1.4. The model to be presented is the so-called particle-in-a-box (henceforth referred to as the "PiB") that is an artificial system, yet with wide-ranging analogies to real systems. This model is very instructive, because it shows in detail how the quantum mechanical formalism works in a situation that is sufficiently simple to carry out the calculations step by step. Furthermore, the symmetry (parity) of the PiB wavefunctions is very similar to that of vibrational wavefunctions discussed in Section 1.4. Finally, the concept of transition from one stationary state to another can be demonstrated using the principles of the transition moment introduced in Section 1.5.

A.1 Definition of the Model System

The PiB model assumes a particle, such as an electron, to be placed into a potential energy well, or confinement shown in Figure A.1. This confinement (the "box") has zero potential energy for $0 \leq x \leq L$, where L is the length of the box. Outside the box, that is, for $x < 0$ and for $x > L$, the potential energy is assumed to be infinite. Thus, once the electron is placed inside the box, it has no chance to escape, and one knows for certain that the electron is in the box.

Next, the kinetic and potential energy expression will be defined, which subsequently allows writing the Hamiltonian, or the total energy operator of the system.

For any quantum mechanical system, the total energy is written as the sum of the kinetic and potential energies, T and V, respectively:

$$E = T + V \tag{A.1}$$

Modern Vibrational Spectroscopy and Micro-Spectroscopy: Theory, Instrumentation and Biomedical Applications, First Edition.
Max Diem.

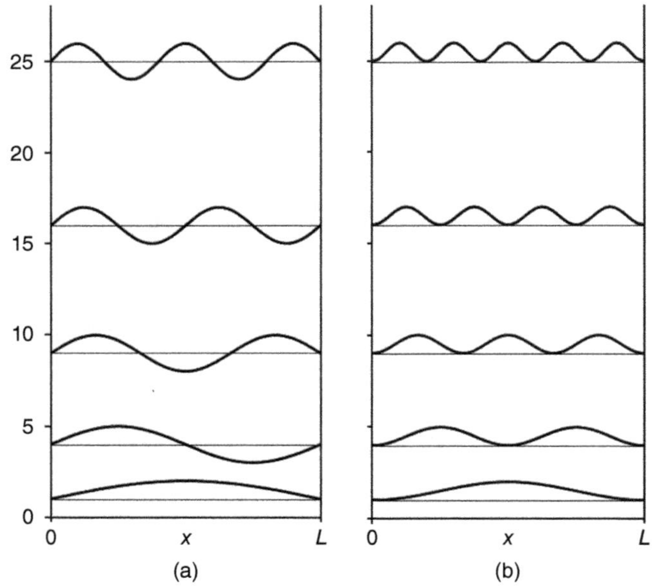

Figure A.1 *(a) Wavefunctions* $\psi_n(x) = \sqrt{\frac{2}{L}}\ \sin\left(\frac{n\pi}{L}\right)x$ *for n = 1, 2, 3, 4, and 5 drawn at their appropriate energy levels. Energy given in units of* $h^2/8mL^2$*. (b) Plot of the square of the wavefunctions shown in (a)*

As before, the kinetic energy of the particle is given

$$T = \frac{1}{2}mv^2 \quad \text{or} \quad T = \frac{p^2}{2m} \tag{A.2}$$

where m is the mass of the electron, and p is its linear momentum. Substituting, as before, the quantum mechanical momentum operator,

$$\hat{p} = \frac{\hbar}{i}\frac{d}{dx} \tag{A.3}$$

into Eq. A.2, the kinetic energy operator can be written as

$$\hat{T} = -\frac{\hbar^2}{2m}\frac{d^2}{dx^2} \tag{A.4}$$

As pointed out in Section 1.4, all the information one seeks about a quantum mechanical system is contained in a wavefunction ψ. In the case here, this wavefunction is a function of x only and is written as $\psi(x)$. Thus, Eq. A.4 is an instruction that prescribes: "to obtain the kinetic energy of a quantum mechanical system, take the second derivative of the wavefunction, and multiply the result by $-\hbar^2/2m$." The potential energy inside the box is zero; thus, the total energy of the particle inside the box is

$$\hat{T} = \hat{E} = -\frac{\hbar^2}{2m}\frac{d^2}{dx^2} \tag{A.5}$$

Since the potential energy outside the box is infinitely high, the particle cannot be there, because that case would correspond to infinite energy, and the discussion henceforth will deal with the inside of the box.

Thus, one may write the total Hamiltonian of the system as

$$\hat{H} = \hat{T} + 0 = -\frac{\hbar^2}{2m}\frac{d^2}{dx^2} \tag{A.6}$$

In the notation of linear algebra, an operator/eigenvector/eigenvalue equation is written as:

$$\hat{H}\psi = E\psi \tag{A.7}$$

Equation A.7 instructs to define an operator, such as the Hamiltonian shown in Eq. A.5, and operate with it on a set of yet unknown eigenfunctions to obtain the desired energy eigenvalues. The eigenfunctions typically form an n-dimensional vector space in which the eigenvalues are appearing on the diagonal. Thus, Eq. A.7 implies:

$$\hat{H}\begin{pmatrix}\psi_1\\ \psi_2\\ \psi_3\\ \psi_4\\ \vdots\end{pmatrix} = \begin{pmatrix}E_1 & 0 & 0 & & \\ 0 & E_2 & 0 & & \\ 0 & 0 & E_3 & \cdots & \\ 0 & 0 & 0 & & \\ & \vdots & & & \end{pmatrix}\begin{pmatrix}\psi_1\\ \psi_2\\ \psi_3\\ \psi_4\\ \vdots\end{pmatrix} \tag{A.8}$$

Each eigenvalue E_i is associated with one eigenfunction.

A.2 Solution of the Particle-in-a-Box Differential Equation

Next, one rewrites Eq. A.7 to explicitly include the operator

$$\left\{-\frac{\hbar^2}{2m}\frac{d^2}{dx^2}\right\}\psi(x) = E\psi(x) \tag{A.9}$$

which is a simple differential equation:

$$\frac{d^2}{dx^2}\psi(x) + \frac{2mE}{\hbar^2}\psi(x) = 0 \tag{A.10}$$

Next, the differential equation described by Eq. A.10 needs to be solved. The functions fulfilling this equation have to be of the form that their second derivative equals to the original function, multiplied by a constant. For example, the function

$$y = A\ \cos\ bx \tag{A.11}$$

could be solution of the differential Eq. A.10, since

$$\frac{d^2y}{dx^2} = -Ab^2\cos bx \tag{A.12}$$

Here, the term b^2 would correspond to $2mE/\hbar^2$, and A is a yet undefined amplitude factor. Similarly,

$$y = A\ \sin\ cx \tag{A.13}$$

or the sum of Eqs. A.11 and A.13 could be acceptable solutions. For reasons that will become obvious shortly, Eq. A.13 will be used as a trial function to fulfill Eq. A.10:

$$\psi(x) = A\sin\left(\frac{2mE}{\hbar^2}\right)^{\frac{1}{2}}x \tag{A.14}$$

and

$$\frac{d^2}{dx^2}\,\psi(x) = -\,A\,\frac{2mE}{\hbar^2}\sin\left(\frac{2mE}{\hbar^2}\right)^{\frac{1}{2}}x = -\,A\,\frac{2mE}{\hbar^2}\,\psi(x) \qquad \text{(A.15)}$$

At this point, it should be pointed out that the solutions of any differential equation depend to a large extent on the boundary conditions: the general solution of the differential equation may or may not describe the physical reality of the system, and it is the boundary conditions that force the solutions to be physically meaningful. In the case of the PiB, the boundary conditions are determined by one postulate of quantum mechanics that requires that wavefunctions are continuous. Thus, if the wavefunction outside the box is zero (since the potential energy outside the box is infinitely high and, therefore, the probability of finding the particle outside the box is zero), the wavefunction also must be zero at the inside boundaries of the box. Thus, one may write the boundary conditions for the PiB differential equation as

$$\psi(x) = 0 \quad \text{at} \quad x = 0 \quad \text{and} \quad \text{at} \quad x = L. \qquad \text{(A.16)}$$

Because of these conditions, the cosine function proposed as possible solutions (Eq. A.11) of Eq. A.10 was rejected, since the cosine function is nonzero at $x = 0$. Because of the required continuity at $x = L$, the function

$$\psi(x) = A\sin\left(\frac{2mE}{\hbar^2}\right)^{\frac{1}{2}}x$$

must be zero at $x = L$ as well. This can happen if the amplitude A is zero (this case is of no further interest, since a zero amplitude of the wavefunction implied that the particle is not inside the box), or if

$$\sin\left(\frac{2mE}{\hbar^2}\right)^{\frac{1}{2}}L = 0 \qquad \text{(A.17)}$$

The sine function is zero at multiples of π radians; that is,

$$\left(\frac{2mE}{\hbar^2}\right)^{\frac{1}{2}}L = n\pi, \qquad n = 1,\ 2,\ 3\ \ldots \qquad \text{(A.18)}$$

Solving Eq. A.18 for E yields

$$E = \frac{n^2\,\pi^2\hbar^2}{2\,m\,L^2} = \frac{n^2\,h^2}{8\,m\,L^2} \qquad \text{(A.19)}$$

Equation A.19 reveals that the energy levels of the particle in a box are quantized, that is, the energy can no longer assume any arbitrary value, but only values of $h^2/8\,m\,L^2, 4\,h^2/8\,m\,L^2, 9\,h^2/8\,m\,L^2$, and so on. This quantization is a direct consequence of the boundary conditions, which required wavefunctions to be zero at the edge of the box. Since the energy depends on this quantum number n, one usually writes Eq. A.19 as

$$E(n) = \frac{n^2\,h^2}{8\,m\,L^2} \qquad \text{(A.20)}$$

Substituting these energy eigenvalues back into Eq. A.14

$$\psi(x) = A\sin\left(\frac{2mE}{\hbar^2}\right)^{\frac{1}{2}}x \qquad \text{(A.14)}$$

one obtains

$$\psi_n(x) = A\sin\left(\frac{2m\,n^2\,h^2}{\hbar^2 8\,m\,L^2}\right)^{\frac{1}{2}}x = A\sin\frac{n\pi}{L}\,x \qquad \text{(A.21)}$$

"A" is a still undefined amplitude factor at this point. To determine "A," one argues as follows: since the square of the wavefunction is defined as the probability of finding the particle, the square of the wavefunction written in Eq. A.21, integrated over the length of the box, must be unity, since the particle is known to be in the box. This leads to the normalization condition

$$\int_0^L \psi_n^2(x)dx = 1 = A^2 \int_0^L \sin^2\left(\frac{n\pi}{L}x\right)\,dx \tag{A.22}$$

Using the integral relationship

$$\int \sin^2 ax\,dx = \frac{x}{2} - \frac{1}{4a}\sin 2ax \tag{A.23}$$

the amplitude A is obtained as follows:

$$A^2 \int_0^L \sin\left(\frac{n\pi x}{L}\right)^2 dx = A^2 \left[\frac{x}{2} - \frac{L}{4n\pi}\sin\left(\frac{2n\pi x}{L}\right)\right]_{x=0}^{x=L} = 1$$

$$A^2 \left[\frac{L}{2} - \frac{L}{4n\pi}\sin\left(\frac{2n\pi L}{L}\right) - 0 + \frac{L}{4n\pi}\sin 0\right] = A^2\left[\frac{L}{2}\right] = 1$$

$$A = \sqrt{\frac{2}{L}} \tag{A.24}$$

Thus, the stationary state wavefunctions for the particle in a box can be written in a final form as

$$\psi_n(x) = \sqrt{\frac{2}{L}}\sin\left(\frac{n\pi}{L}\right)x \tag{A.25}$$

The resulting stationary state (time-independent) wavefunctions and energies are depicted in Figure A.1(a). Although one refers to these wavefunctions as time-independent, they may be considered as standing waves in which the amplitudes oscillate between the extremes shown in Figure A.2 and resemble the motion of a plugged string. Time independency then refers to the fact that the system will stay in one of these standing wave patterns forever, or until perturbed by electromagnetic radiation.

The probability of finding the particle at any given position x is shown in Figure A.1(b). These traces are the squares of the wavefunctions, and depict that for higher levels of n, the probability of finding the particle

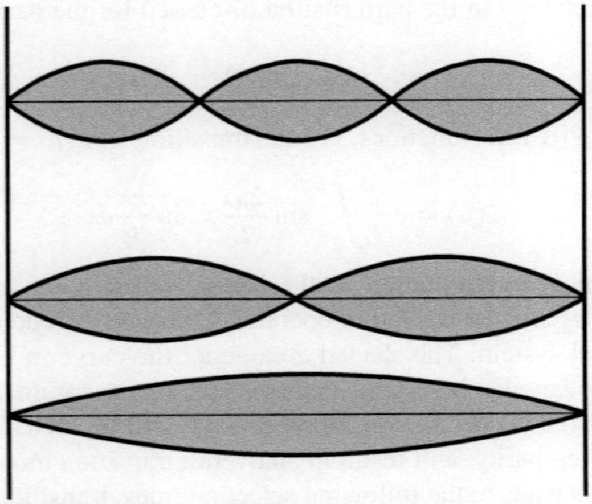

Figure A.2 *Representation of the particle-in-a-box wavefunctions shown in Figure A.1 as standing waves*

moves away from the center to the periphery of the box. If one draws a vertical line at the center of the box, one finds that the wavefunctions have the same odd/even parity as do the harmonic oscillator wavefunctions in their quadratic potential function (see Figure 1.5).

A.3 Orthonormality of the Particle-in-a-Box Wavefunctions

The PiB wavefunctions form an orthonormal vector space as defined before for the vibrational wavefunctions (Eq. 1.53), which implies that

$$\int_0^L \psi_n(x)\, \psi_m(x) dx = \delta_{mn} = \begin{cases} = 1 \; if\, n = m \\ = 0 \; if\, n \neq m \end{cases} \tag{A.26}$$

The wavefunctions' normality was established above by normalizing them (Eq. A.24); in order to demonstrate that they are orthogonal, the integral

$$\frac{2}{L}\int_0^L \sin\frac{n\pi x}{L}\, \sin\frac{m\pi x}{L} dx \tag{A.27}$$

needs to be evaluated. This can be established using the integral relationship

$$\int \sin ax\, \sin bx\, dx = -\frac{\sin(a+b)x}{2(a+b)} + \frac{\sin(a-b)x}{2(a-b)} \tag{A.28}$$

For any two adjacent wavefunction, say $m = 1$ and $n = 2$, or $m = 2$ and $n = 3$, the numerator of the first term in Eq. A.28 will contain the sine function of odd multiples of π, whereas the numerator of the second term will contain the sine function of even multiples of π. Since the sine function of odd and even multiples of π is zero, the total integral described by Eq. A.27 is zero. An identical argument can be presented for any $n \neq m$ case.

A.4 Dipole-Allowed Transitions for the Particle in a Box

The PiB model system may also be used to study transitions between stationary states, induced by electromagnetic radiation. In complete analogy to the perturbation discussed for the harmonic oscillator, the transition moment

$$\langle \psi_m | \mu | \psi_n \rangle \tag{A.29}$$

needs to be evaluated for the PiB wavefunctions. For the transition from $n = 1$ to $n = 2$, Eq. A.29 becomes

$$\langle \mu \rangle = e\,\frac{2}{L}\int_0^L \sin\frac{2\pi x}{L} x \sin\frac{\pi x}{L} dx \tag{A.30}$$

In analogy with the graphical interpretation of the harmonic oscillator transition moment, Figure A.3 demonstrates the plot of ψ_1, ψ_2 and the transition operator. The latter is represented by a straight line, since $\mu = ex$ for a one-dimensional system. The shaded area under the curve in Figure A.3(b), when integrated from 0 to L, is nonzero. This can also be established from the parity argument introduced earlier: since ψ_2 and the dipole operator $\mu = ex$ are odd functions, their product will be even. This product, multiplied by the ground state ψ_1, which has even parity, will result in an overall transition moment with even parity.

For the particle in a box, this leads to the following selection rules: transitions with $n = \pm 1, \pm 3, \pm 5$, and so on are allowed, whereas transitions with $n = \pm 2,\ \pm 4$, and so on are forbidden because the transition moment

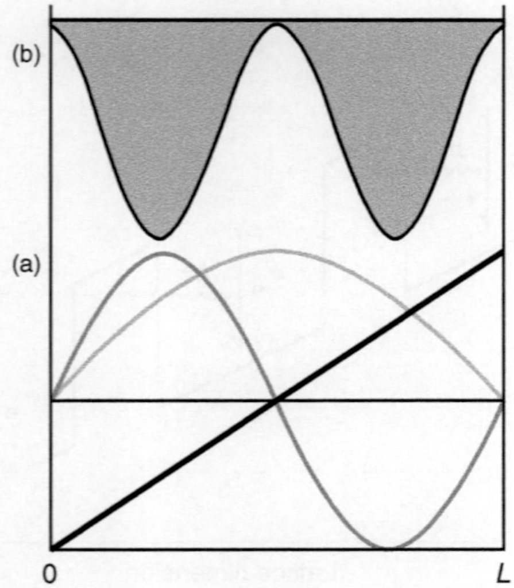

Figure A.3 *(a) Plot of particle-in-a-box wavefunction ψ_1 (light gray), ψ_2 (gray) and transition moment operator μ. (b): product of the three functions shown in (a)*

integrals are zero. Thus, one encounters here the transitions being allowed or forbidden depending on the symmetry of the wavefunctions. The transition moment is the quantity that needs to be evaluated in order to determine whether or not a transition may occur.

A.5 Real-World PiBs

Although the PiB was introduced here as a model to demonstrate quantum mechanical principles in a mathematically manageable system, there are physical examples that can be treated adequately using the PiB formalism. One of these is frequently incorporated as an experiment in physical chemistry laboratories, and involves a conjugated dye such as 1,6-diphenyl-1,3,5-hexatriene. In this molecule, the Lewis structure suggests three double and four single bonds. The UV-vis absorption spectrum shows one broad absorption peak that is, in this experiment, commonly assigned as a PiB transition of one electron from the $n = 3$ to $n = 4$ PiB energy level. Here, one assumes that the six conjugated π-electrons occupy the energy levels 1, 2, and 3. Modeling this system with a conjugated length of about 1 nm, corresponding to seven alternating single and double bonds, reproduces the observed transition frequency satisfactorily.

Certain quantum dot structures can also be modeled by a two-dimensional particle in a box. Quantum dots are typically manufactured by creating n-type semiconductors whose electrons are free to move over the entire size of the dot. In these quantum dots, the energy levels of the free electrons follow a 2-D PiB model; consequently, the color of electronic transitions can be tuned by changing the size of the quantum dot.

Finally, the super lattice found in quantum cascade lasers (QCLs) crystal gives rise to energy levels that can be explained using a PiB model. In these super-lattice structures, layers of high, and low potential energy alternate, produced by different doping levels in the layers. These potential wells mimic a Pib with finite potential energy barriers (see Figure A.4). Furthermore, these energy wells are superimposed on a sloping potential energy background. The slope of the background has two consequences: first, the wavefunctions are distorted to have higher amplitudes at the left side of each potential well, and second, the electrons in the ground state of each well can tunnel through the barrier between the wells to end up in an excited state of the next lower

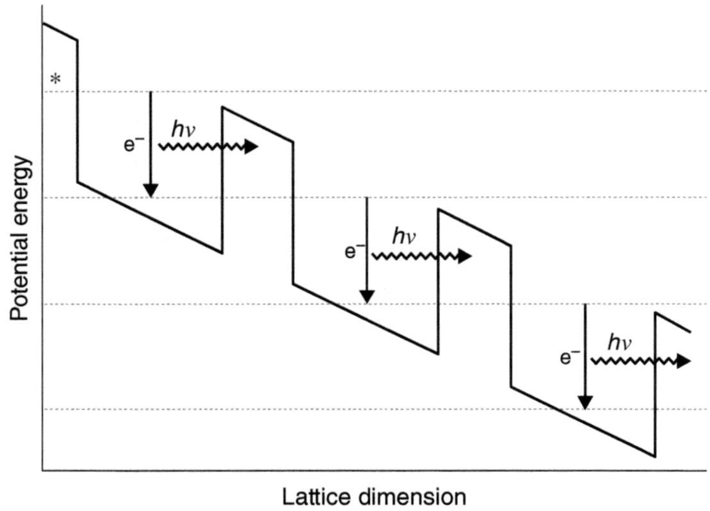

Figure A.4 *The super-lattice structure in a quantum cascade laser modeled by successive PiB energy states*

energy well. Electrons are injected at a potential energy marked by the * -symbol, and undergo a transition as indicated by the left-most down arrow. During this transition, an (infrared) photon is emitted and subsequently, the electron may tunnel through the finite-heights barrier and arrives in the next quantum well, and undergo another transition. The emission and tunneling processes are repeated as many times as there are quantum wells in the superstructure. The term "cascade" in QCL is due to the fact that one electron can undergo many consecutive emission process in the super-lattice structure. By placing the super-lattice crystal into an optic cavity, stimulated emission from the excited states into the ground states of each well can be achieved.

Appendix B

A summary of the Solution of the Harmonic Oscillator (Hermite) Differential Equation

In Section 1.4, the one-dimensional vibrational Schrödinger equation for the harmonic oscillator was derived as

$$\left\{-\frac{\hbar^2}{2m}\frac{d^2}{dx^2} + \frac{1}{2}kx^2\right\}\psi(x) = E\psi(x) \tag{1.46}$$

This differential equation, also known as "Hermite's" differential equation, is quite difficult to solve, and an outline for this procedure will be presented here.[1]

First, Eq. 1.46 is reformatted to read

$$\frac{d^2\psi}{dx^2} + \left(\frac{2mE}{\hbar^2} - \alpha^2x^2\right)\psi = 0 \tag{B.1}$$

where

$$\alpha = \frac{2\pi m}{\hbar} \tag{1.49}$$

Equation B.1 is solved for the case of large values of x, for which

$$\alpha^2x^2 \gg \frac{2mE}{\hbar^2} \tag{B.2}$$

that is, the potential energy part dominates the expression in the parentheses in Eq. B.1. This results in a simplified equation,

$$\frac{d^2\psi}{dx^2} - \alpha^2x^2\psi = 0 \tag{B.3}$$

which has, for large values of x, approximate solutions

$$\psi(x) = A\, e^{-\frac{\alpha x^2}{2}} \tag{B.4}$$

[1]The author is indebted to Prof. David Budil from the Department of Chemistry and Chemical Biology of Northeastern University for writing up this procedure from several references as part of the lecture notes of the undergraduate Physical Chemistry course; these lecture notes have been shared and expanded by several faculty members teaching this class.

Modern Vibrational Spectroscopy and Micro-Spectroscopy: Theory, Instrumentation and Biomedical Applications, First Edition.
Max Diem.

that are known as the "asymptotic solution." Next, one assumes that the final results are given by a trial solution of the form

$$\psi(x) = f(x)e^{-\frac{\alpha x^2}{2}} \tag{B.5}$$

that is, Eq. B.4, multiplied by a yet unknown function $f(x)$. The second derivative needed in Eq. B.1 is then

$$\frac{d^2\psi}{dx^2} = e^{-\frac{\alpha x^2}{2}} (f'' - 2\alpha f' - \alpha f + \alpha^2 x^2 f) \tag{B.6}$$

where the primed quantities denote derivative with respect to x. When these trial solutions are substituted into the original differential Eq. B.1, one obtains a new differential equation

$$\left[\frac{2mE}{\hbar^2} - \alpha\right] f(x) - 2\alpha f'(x) + f''(x) = 0 \tag{B.7}$$

which is solved by a series expansion for $f(x)$:

$$f(x) = \sum_{n=0}^{\infty} a_n x^n \tag{B.8}$$

The necessary derivatives of $f(x)$ are

$$f'(x) = \sum_{n=1}^{\infty} n a_n x^{n-1} \quad \text{and} \quad f''(x) = \sum_{n=1}^{\infty} n(n-1) a_n x^{n-2} \tag{B.9}$$

Substituting these derivatives back into Eq. B.7 and rearranging yields

$$\sum_{n=0}^{\infty} \left[(n+2)(n+1)a_{n+2} - 2\alpha n\, a_n + \left(\frac{2mE}{\hbar^2} - \alpha\right) a_n\right] x^n = 0 \tag{B.10}$$

The sum of this series is zero for all x; therefore, each of the coefficients in the series must separately equal zero as well:

$$(n+2)(n+1)a_{n+2} - 2\alpha n\, a_n + \left(\frac{2mE}{\hbar^2} - \alpha\right) a_n = 0 \tag{B.11}$$

from which one obtains a recursion formula for the coefficients of $f(x)$:

$$a_{n+2} = \frac{2\alpha n + \alpha - 2mE/\hbar^2}{(n+2)(n+1)} a_n \tag{B.12}$$

Thus, the functions

$$\psi(x) = e^{-\frac{\alpha x^2}{2}} \sum_{n=0}^{\infty} a_n x^n \tag{B.13}$$

in principle, fulfill the original differential equation, Eq. B.1, where the coefficients a_n are given by Eq. B.12. Next, one proceeds by defining the boundary conditions for these functions. This step involves rather tricky arguments. It can be shown that the function $f(x)$, given by Eq. B.13, goes to infinity for large values of x. This is by no means self-evident, since the Gaussian part of the function approaches zero, whereas the term x^n approaches infinity. For the wavefunctions $\psi(x)$ to be finite (which is a requirement for valid wavefunctions, as defined by one of the postulates of quantum mechanics), the summation in Eq. B.13 must be terminated at some finite value of n. This will occur if the numerator of Eq. B.12 becomes zero:

$$2\alpha n + \alpha - \frac{2mE}{\hbar^2} = 0 \tag{B.14}$$

In this case, the coefficients for $(n+2)$ and higher will be zero. This leads to

$$E = \alpha(2n+1)\frac{\hbar^2}{2m} \tag{B.15}$$

with

$$\alpha = \frac{2\pi v m}{\hbar} \tag{1.49}$$

the energy eigenvalues

$$E_n = \left(n + \frac{1}{2}\right) hv \tag{B.16}$$

are obtained. The energy eigenvalues result by applying the boundary conditions to the trial solution, Eq. B.13, quite similar to the situation of the particle in a box, discussed in Appendix 1, where the energy eigenvalues,

$$E_n = \frac{n^2 h^2}{8\ mL^2} \tag{B.19}$$

were obtained after imposing the boundary conditions on the trial solutions.

Next, the final form of the vibrational wavefunctions will be derived. According to Eq. B.13, these wavefunctions contain a Gaussian factor and a series expansion in x with the expansion coefficients given by Eq. B.12. As pointed out earlier, the expansion given in Eq. B.13 stops at some value of n; the polynomials up to this value are referred to as $H_n(x)$. Depending on whether n is odd or even, two equations are obtained to define these functions:

$$H_n(x) = \sum_{j=0}^{n/2} a_{2j} x^{2j} \quad \text{for } n = \ 0, 2, 4, \ldots \tag{B.20}$$

$$H_n(x) = \sum_{j=1}^{(n+1)/2} a_{2j-1} x^{2j-1} \quad \text{for } n = 1, 3, 5, \ldots \tag{B.21}$$

The form of the polynomials $H_n(x)$ will be established next. For this, one uses the recursion formula given by Eq. B.12 and starts, most conveniently, at the value of n at which the series expansion given in Eq. B.13 was terminated. At this value of n,

$$a_n = (2\sqrt{\alpha})^n \tag{B.22}$$

Using the recursion formula (B.12), one determines the previous terms a_{n-2} and a_{n-1} repeatedly until $n = 0$ or 1. This, in turn, leads to the Hermite polynomials introduced in Section 1.4, Eq. 1.50, and the recursion formula Eq. 1.51. It is, indeed, very interesting to realize that the solution of a differential equation that appears quite simple requires many steps and quite a few approximations, as discussed earlier. The author, as a spectroscopist and not a mathematician, always marvels at the insight and ingenuity of the mathematical researchers who first proposed solutions to the equations encountered in physical chemistry. In this case, the differential equation was solved by a 19th-century French mathematician Charles Hermite.

Appendix C

Character Tables for Chemically Important Symmetry Groups

C.1 The nonaxial groups

C_1	E
A	1

C_s	E	σ_h		
A'	1	1	$x,\ y,\ R_z$	$x^2,\ y^2$ $z^2,\ xy$
A''	1	−1	$z,\ R_x,\ R_y$	$yz,\ xz$

C_i	E	i		
A_g	1	1	$R_x,\ R_y,\ R_z$	$x^2,\ y^2,\ z^2$ $xy,\ xz,\ yz$
A_u	1	−1	$x,\ y,\ z$	

C.2 The C_n groups

C_2	E	C_2		
A	1	1	$z,\ R_z$	$x^2,\ y^2,\ z^2,\ xy$
B	1	−1	$x,\ y,\ R_x,\ R_y$	$yz,\ xz$

C_3	E	C_3	C_3^2		$\varepsilon = \exp(2\pi i/3)$
A	1	1	1	$z,\ R_z$	$x^2 + y^2,\ z^2$
E	$\begin{Bmatrix} 1 \\ 1 \end{Bmatrix}$	$\begin{matrix} \varepsilon \\ \varepsilon^* \end{matrix}$	$\begin{matrix} e^* \\ \varepsilon \end{matrix}$	$(x,\ y)(R_x,\ R_y)$	$(x^2 - y^2,\ xy)(yz,\ xz)$

Modern Vibrational Spectroscopy and Micro-Spectroscopy: Theory, Instrumentation and Biomedical Applications, First Edition.
Max Diem.

C_4	E	C_4	C_2	C_4^3		
A	1	1	1	1	$z,\ R_z$	$x^2+y^2,\ z^2$
B	1	−1	1	−1		$x^2-y^2,\ xy$
E	1	i	−1	$-i$	$(x,\ y)(R_x,\ R_y)$	$(yz,\ xz)$
	1	−1	−1	i		

C_5	E	C_5	C_5^2	C_5^3	C_5^4		$\varepsilon = \exp(2\pi i/5)$
A	1	1	1	1	1	$z,\ R_z$	$x^2-y^2,\ z^2$
E_1	1	ε	ε^2	ε^{2*}	ε^*	$(x,\ y)(R_x,\ R_y)$	$(yz,\ xz)$
	1	ε^*	ε^{2*}	ε^2	ε		
E_2	1	ε^2	ε^*	ε	ε^{2*}		$(x^2-y^2,\ xy)$
	1	ε^{2*}	ε	ε^*	ε^2		

C_6	E	C_6	C_3	C_2	C_3^2	C_6^5		$\varepsilon = \exp(2\pi i/6)$
A	1	1	1	1	1	1	$z,\ R_z$	$x^2+y^2,\ z^2$
B	1	−1	1	−1	1	−1		
E_1	1	ε	$-\varepsilon^*$	−1	$-\varepsilon$	ε^*	$(x,\ y)$ $(R_x,\ R_y)$	$(xz,\ yz)$
	1	ε^*	ε	−1	$-\varepsilon^*$	ε		
E_2	1	$-\varepsilon^*$	$-\varepsilon$	1	$-\varepsilon^*$	$-\varepsilon$		$(x^2-y^2,\ xy)$
	1	$-\varepsilon$	$-\varepsilon^*$	1	$-\varepsilon$	$-\varepsilon^*$		

C_7	E	C_7	C_7^2	C_7^3	C_7^4	C_7^5	C_7^6		$\varepsilon = \exp(2\pi i/7)$
A	1	1	1	1	1	1	1	$z,\ R_z$	$x^2-y^2,\ z^2$
E_1	1	ε	ε^2	ε^3	ε^{3*}	ε^{2*}	ε^*	$(x,\ y)$ $(R_x,\ R_y)$	$(xz,\ yz)$
	1	ε^*	ε^{2*}	ε^{3*}	ε^3	ε^2	ε		
E_2	1	ε^2	ε^{3*}	ε^*	ε	ε^3	ε^{2*}		$(x^2-y^2,\ xy)$
	1	ε^{2*}	ε^3	ε	ε^*	ε^{3*}	ε^2		
E_3	1	ε^3	ε^*	ε^2	ε^{2*}	ε	ε^{3*}		
	1	ε^{3*}	ε	ε^{2*}	ε^2	ε^*	ε^3		

C_8	E	C_3	C_4	C_2	C_7^3	C_8^3	C_8^5	C_8^7		$\varepsilon = \exp(2\pi i/8)$
A	1	1	1	1	1	1	1	1	$z,\ R_z$	$x^2-y^2,\ z^2$
B	1	−1	1	1	1	−1	−1	−1		
E_1	1	ε	i	−1	$-i$	$-\varepsilon^*$	$-\varepsilon$	ε^*	$(x,\ y)$ $(R_x,\ R_y)$	$(xz,\ yz)$
	1	ε^*	$-i$	−1	i	$-\varepsilon$	ε^*	ε		
E_2	1	i	−1	1	−1	$-i$	i	$-i$		$(x^2-y^2,\ xy)$
	1	$-i$	−1	1	−1	i	$-i$	i		
E_3	1	$-\varepsilon$	i	−1	$-i$	ε^*	ε	$-\varepsilon^*$		
	1	$-\varepsilon^*$	$-i$	−1	i	ε	ε^*	$-\varepsilon$		

C.3 The D_n groups

D_2	E	$C_2(z)$	$C_2(y)$	$C_2(x)$		
A	1	1	1	1		x^2, y^2, z^2
B_1	1	1	−1	−1	z, R_z	xy
B_2	1	−1	1	−1	y, R_y	xz
B_3	1	−1	−1	1	x, R_x	yz

D_3	E	$2C_3$	$3C_2$		
A_1	1	1	1		x^2-y^2, z^2
A_2	1	1	−1	z, R_z	
E	2	−1	0	$(x, y)(R_x, R_y)$	$(x^2-y^2, xy)(xz, yz)$

D_4	E	$2C_4$	$C_2(=C_4^2)$	$2C_2'$	$2C_2''$		
A_1	1	1	1	1	1		x^2-y^2, z^2
A_2	1	1	1	−1	−1	z, R_z	
B_1	1	−1	1	1	−1		x^2-y^2
B_2	1	−1	1	−1	1		xy
E	2	0	−2	0	0	$(x, y)(R_x, R_y)$	(xz, yz)

D_5	E	$2C_5$	$2C_5^2$	$5C_2$		
A_1	1	1	1	1		x^2-y^2, z^2
A_2	1	1	1	−1	z, R_z	
E_1	2	2 cos 72°	2 cos 144°	0	$(x, y)(R_x, R_y)$	(xz, yz)
E_2	2	2 cos 144°	2 cos 72°	0		(x^2-y^2, xy)

D_6	E	$2C_6$	$2C_3$	C_2	$3C_2'$	$3C_2'$		
A_1	1	1	1	1	1	1		x^2-y^2, z^2
A_2	1	1	1	1	−1	−1	z, R_z	
B_1	1	−1	1	−1	1	−1		
B_2	1	−1	1	−1	−1	1		
E_1	2	1	−1	−2	0	0	$(x, y)(R_x, R_y)$	(xz, yz)
E_2	2	−1	−1	2	0	0		(x^2-y^2, xy)

C.4 The C_{nv} groups

C_{2v}	E	C_2	$\sigma_v(xz)$	$\sigma_v'(yz)$		
A_1	1	1	1	1	z	x^2, y^2, z^2
A_2	1	1	−1	−1	R_z	xy
B_1	1	−1	1	−1	x, R_y	xz
B_2	1	−1	−1	1	y, R_x	yz

C_{3v}	E	$2C_3$	$3\sigma_v$		
A_1	1	1	1	z	x^2-y^2, z^2
A_2	1	1	−1	R_z	
E	2	−1	0	$(x, y)(R_x, R_y)$	$(x^2-y^2, xy)(xz, yz)$

C_{4v}	E	$2C_4$	C_2	$2\sigma_v$	$2\sigma_d$		
A_1	1	1	1	1	1	z	$x^2-y^2,\ z^2$
A_2	1	1	1	−1	−1	R_z	
B_1	1	−1	1	1	−1		x^2-y^2
B_2	1	−1	1	−1	1		xy
E	2	0	−2	0	0	$(x,\ y)(R_x,\ R_y)$	$(xz,\ yz)$

C_{5v}	E	$2C_5$	$2C_5^2$	$5\sigma_v$		
A_1	1	1	1	1	z	$x^2+y^2,\ z^2$
A_2	1	1	1	−1	R_z	
E_1	2	2 cos 72°	2 cos 144°	0	$(x,\ y)(R_x,\ R_y)$	$(xz,\ yz)$
E_2	2	2 cos 144°	2 cos 72°	0		$(x^2-y^2,\ xy)$

C_{6v}	E	$2C_6$	$2C_3$	C_2	$3\sigma_v$	$3\sigma_d$		
A_1	1	1	1	1	1	1	z	$x^2+y^2,\ z^2$
A_2	1	1	1	1	−1	−1	R_z	
B_1	1	−1	1	−1	1	−1		
B_2	1	−1	1	−1	−1	1		
E_1	2	1	−1	−2	0	0	$(x,\ y)(R_x,\ R_y)$	$(xz,\ yz)$
E_2	2	−1	−1	2	0	0		$(x^2-y^2,\ xy)$

C_{2h}	E	C_2	i	σ_h		
A_g	1	1	1	1	R_z	$x^2,\ y^2,\ z^2,\ xy$
B_g	1	−1	1	−1	$R_x,\ R_y$	$xz,\ yz$
A_u	1	1	−1	−1	z	
B_u	1	−1	−1	1	$x,\ y$	

C_{3h}	E	C_3	C_3^2	σ_h	S_3	S_3^3		$\varepsilon = \exp(2\pi i/3)$
A'	1	1	1	1	1	1	R_z	x^2+y^2, z^2
E'	$\left\{\begin{matrix}1\\1\end{matrix}\right.$	$\begin{matrix}\varepsilon\\\varepsilon^*\end{matrix}$	$\begin{matrix}\varepsilon^*\\\varepsilon\end{matrix}$	$\begin{matrix}1\\1\end{matrix}$	$\begin{matrix}\varepsilon\\\varepsilon^*\end{matrix}$	$\left.\begin{matrix}\varepsilon^*\\\varepsilon\end{matrix}\right\}$	(x, y)	(x^2-y^2, xy)
A''	1	1	1	−1	−1	−1	z	
E''	$\left\{\begin{matrix}1\\1\end{matrix}\right.$	$\begin{matrix}\varepsilon\\\varepsilon^*\end{matrix}$	$\begin{matrix}\varepsilon^*\\\varepsilon\end{matrix}$	$\begin{matrix}-1\\-1\end{matrix}$	$\begin{matrix}-\varepsilon\\-\varepsilon^*\end{matrix}$	$\left.\begin{matrix}-\varepsilon^*\\-\varepsilon\end{matrix}\right\}$	(R_x, R_y)	(xz, yz)

C_{4h}	E	C_4	C_2	C_4^3	i	S_4^3	σ_h	S_4		
A_g	1	1	1	1	1	1	1	1	R_z	x^2+y^2, z^2
B_g	1	-1	1	-1	1	-1	1	-1		x^2-y^2, xy
E_g	$\{1$	i	-1	$-i$	1	i	-1	$-i\}$	(R_x, R_y)	(xz, yz)
	$\{1$	$-i$	-1	i	1	$-i$	-1	$i\}$		
A_u	1	1	1	1	-1	-1	-1	-1	z	
B_u	1	-1	1	-1	-1	1	-1	1		
E_u	$\{1$	i	-1	$-i$	-1	$-i$	1	$i\}$	(x, y)	
	$\{1$	$-i$	-1	i	-1	i	1	$-i\}$		

C_{3h}	E	C_5	C_5^2	C_5^3	C_5^4	σ_h	S_5	S_5^7	S_5^3	S_5^9		$\varepsilon = \exp(2\pi i/5)$
A'	1	1	1	1	1	1	1	1	1	1	R_z	$x^2+y^2,\ z^2$
E_1'	$\{1$	ε	ε^2	ε^{2*}	ε^*	1	ε	ε^2	ε^{2*}	$\varepsilon^*\}$	$(x,\ y)$	
	$\{1$	ε^*	ε^{2*}	ε^2	ε	1	ε^*	ε^{2*}	ε^2	$\varepsilon\}$		
E_2'	$\{1$	ε^2	ε^*	ε	ε^{2*}	1	ε^2	ε^*	ε	ε^{2*}		$(x^2-y^2,\ xy)$
	$\{1$	ε^{2*}	ε	ε^*	ε^2	1	ε^{2*}	ε	ε^*	ε^2		
A''	1	1	1	1	1	-1	-1	-1	-1	-1	z	
E_1''	$\{1$	ε	ε^2	ε^{2*}	ε^*	-1	$-\varepsilon$	$-\varepsilon^2$	$-\varepsilon^{2*}$	$-\varepsilon^*\}$	$(R_x,\ R_y)$	$(xz,\ yz)$
	$\{1$	ε^*	ε^{2*}	ε^2	ε	-1	$-\varepsilon^*$	$-\varepsilon^{2*}$	$-\varepsilon^*$	$-\varepsilon\}$		
E_2''	$\{1$	ε^2	ε^*	ε	ε^{2*}	-1	$-\varepsilon^2$	$-\varepsilon^*$	$-\varepsilon$	$-\varepsilon^{2*}$		
	$\{1$	ε^{2*}	ε	ε^*	ε^2	-1	$-\varepsilon^{2*}$	$-\varepsilon$	$-\varepsilon^*$	$-\varepsilon^2$		

C_{6h}	E	C_6	C_3	C_2	C_3^2	C_6^3	i	S_3^3	S_6^5	σ_h	S_6	S_3		$\varepsilon = \exp(2\pi i/6)$
A_g	1	1	1	1	1	1	1	1	1	1	1	1	R_z	$x^2+y^2,\ z^2$
B_g	1	-1	1	-1	1	-1	1	-1	1	-1	1	-1		
E_{1g}	$\{1$	ε	$-\varepsilon^*$	-1	$-\varepsilon$	ε^*	1	ε	$-\varepsilon^*$	-1	$-\varepsilon$	$\varepsilon^*\}$	$(R_x,\ R_y)$	$(xz,\ yz)$
	$\{1$	ε^*	$-\varepsilon$	-1	$-\varepsilon^*$	ε	1	ε^*	$-\varepsilon$	-1	$-\varepsilon^*$	$\varepsilon\}$		
E_{2g}	$\{1$	$-\varepsilon^*$	$-\varepsilon$	1	$-\varepsilon^*$	$-\varepsilon$	1	$-\varepsilon^*$	$-\varepsilon$	1	$-\varepsilon^*$	$-\varepsilon\}$		$(x^2-y^2,\ xy)$
	$\{1$	$-\varepsilon$	$-\varepsilon^*$	1	$-\varepsilon$	$-\varepsilon^*$	1	$-\varepsilon$	$-\varepsilon^*$	1	$-\varepsilon$	$-\varepsilon^*\}$		
A_u	1	1	1	1	1	1	-1	-1	-1	-1	-1	-1	z	
B_u	1	-1	1	-1	1	-1	-1	1	-1	1	-1	1		
E_{1u}	$\{1$	ε	$-\varepsilon^*$	-1	$-\varepsilon$	ε^*	-1	$-\varepsilon$	ε^*	1	ε	$-\varepsilon^*\}$	$(x,\ y)$	
	$\{1$	ε^*	$-\varepsilon$	-1	$-\varepsilon^*$	ε	-1	$-\varepsilon^*$	ε	1	ε^*	$-\varepsilon\}$		
E_{2u}	$\{1$	$-\varepsilon^*$	$-\varepsilon$	1	$-\varepsilon^*$	$-\varepsilon$	-1	ε^*	ε	-1	ε^*	$\varepsilon\}$		
	$\{1$	$-\varepsilon$	$-\varepsilon^*$	1	$-\varepsilon$	$-\varepsilon^*$	-1	ε	ε^*	-1	ε	$\varepsilon^*\}$		

C.5 The D_{nh} groups

D_{2h}	E	$C_2(z)$	$C_2(y)$	$C_2(x)$	i	$\sigma(xy)$	$\sigma(xz)$	$\sigma(yz)$		
A_g	1	1	1	1	1	1	1	1		x^2, y^2, z^2
B_{1g}	1	1	−1	−1	1	1	−1	−1	R_z	xy
B_{2g}	1	−1	1	−1	1	−1	1	−1	R_y	xz
B_{3g}	1	−1	−1	1	1	−1	−1	1	R_x	yz
A_u	1	1	1	1	−1	−1	−1	−1		
B_{1u}	1	1	−1	−1	−1	−1	1	1	z	
B_{2u}	1	−1	1	−1	−1	1	−1	1	y	
B_{3u}	1	−1	−1	1	−1	1	1	−1	x	

D_{3h}	E	$2C_3$	$3C_2$	σ_h	$2S_3$	$3\sigma_v$		
A_1'	1	1	1	1	1	1		x^2+y^2, z^2
A_2'	1	1	−1	1	1	−1	R_z	
E'	2	−1	0	2	−1	0	(x, y)	(x^2-y^2, xy)
A_1''	1	1	1	−1	−1	−1		
A_2''	1	1	−1	−1	−1	1	z	
E''	2	−1	0	−2	1	0	(R_x, R_y)	(xz, yz)

D_{4h}	E	$2C_4$	C_2	$2C_2'$	$2C_2''$	i	$2S_4$	σ_h	$2\sigma_u$	$2\sigma_d$		
$A^1{}_g$	1	1	1	1	1	1	1	1	1	1		x^2+y^2, z^2
A_{2g}	1	1	1	−1	−1	1	1	1	−1	−1	R_z	
B_{1g}	1	−1	1	1	−1	1	−1	1	1	−1		x^2-y^2
B_{2g}	1	−1	1	−1	1	1	−1	1	−1	1		xy
E_g	2	0	−2	0	0	2	0	−2	0	0	(R_x, R_y)	(xz, yz)
A_{1u}	1	1	1	1	1	−1	−1	−1	−1	−1		
A_{2u}	1	1	1	−1	−1	−1	−1	−1	1	1	z	
B_{1u}	1	−1	1	1	−1	−1	1	−1	−1	1		
B_{2u}	1	−1	1	−1	1	−1	1	−1	1	−1		
E_u	2	0	−2	0	0	−2	0	2	0	0	(x, y)	

D_{5h}	E	$2C_5$	$2C_5^2$	$5C_2$	σ_h	$2S_5$	$2S_5^3$	$5\sigma_v$		
A_1'	1	1	1	1	1	1	1	1		$x^2+y^2,\ z^2$
A_2'	1	1	1	−1	1	1	1	−1	R_z	
E_1'	2	2 cos 72°	2 cos 144°	0	2	2 cos 72°	2 cos 144°	0	$(x,\ y)$	
E_2'	2	2 cos 144°	2 cos 72°	0	2	2 cos 144°	2 cos 72°	0		$(x^2-y^2,\ xy)$
A_1''	1	1	1	1	−1	−1	−1	−1		
A_2''	1	1	1	−1	−1	−1	−1	1	z	
E_1''	2	2 cos 72°	2 cos 144°	0	−2	−2 cos 72°	−2 cos 144°	0	$(R_x,\ R_y)$	$(xz,\ yz)$
E_2''	2	2 cos 144°	2 cos 72°	0	−2	−2 cos 144°	−2 cos 72°	0		

D_{6h}	E	$2C_6$	$2C_3$	C_2	$3C_2'$	$3C_2''$	i	$2S_3$	$2S_6$	σ_h	$3\sigma_d$	$3\sigma_v$		
A_{1g}	1	1	1	1	1	1	1	1	1	1	1	1		$x^2+y^2,\ z^2$
A_{2g}	1	1	1	1	−1	−1	−1	1	1	1	1	−1	R_z	
B_{1g}	1	−1	1	−1	1	−1	1	−1	1	−1	1	−1		
B_{2g}	1	−1	1	−1	−1	1	1	−1	1	−1	−1	1		
E_{1g}	2	1	−1	−2	0	0	2	1	−1	−2	0	0	$(R_x,\ R_y)$	$(xz,\ yz)$
E_{2g}	2	−1	−1	2	0	0	2	−1	−1	2	0	0		$(x^2-y^2,\ xy)$
A_{1u}	1	1	1	1	1	1	−1	−1	−1	−1	−1	−1		
A_{2u}	1	1	1	1	−1	−1	−1	−1	−1	−1	1	1	z	
B_{1u}	1	−1	1	−1	1	−1	−1	1	−1	1	−1	1		
B_{2u}	1	−1	1	−1	−1	1	−1	1	−1	1	1	−1		
E_{1u}	2	1	−1	−2	0	0	−2	−1	1	2	0	0	$(x,\ y)$	
E_{2u}	2	−1	−1	2	0	0	−2	1	1	−2	0	0		

D_{8h}	E	$2C_8$	$2C_8^3$	$2C_4$	C_2	$4C_2'$	$4C_2''$	i	$2S_8$	$2S_8^3$	$2S_4$	σ_h	$4\sigma_d$	$4\sigma_v$		
A_{1g}	1	1	1	1	1	1	1	1	1	1	1	1	1	1		x^2+y^2, z^2
A_{2g}	1	1	1	1	1	−1	−1	1	1	1	1	1	−1	−1	R_z	
B_{1g}	1	−1	−1	1	1	1	−1	1	−1	−1	1	1	1	−1		
B_{2g}	1	−1	−1	1	1	−1	1	1	−1	−1	1	1	−1	1		
E_{1g}	2	$\sqrt{2}$	$-\sqrt{2}$	0	−2	0	0	2	$\sqrt{2}$	$-\sqrt{2}$	0	−2	0	0	(R_x, R_y)	(xz, yz)
E_{2g}	2	0	0	−2	2	0	0	2	0	0	−2	2	0	0		(x^2-y^2, xy)
E_{3g}	2	$-\sqrt{2}$	$\sqrt{2}$	0	−2	0	0	2	$-\sqrt{2}$	$\sqrt{2}$	0	−2	0	0		
A_{1u}	1	1	1	1	1	1	1	−1	−1	−1	−1	−1	−1	−1		
A_{2u}	1	1	1	1	1	−1	−1	−1	−1	−1	−1	−1	−1	1	z	
B_{1u}	1	−1	−1	1	1	1	−1	−1	1	1	−1	−1	−1	1		
B_{2u}	1	−1	−1	1	1	−1	1	−1	1	1	−1	−1	1	−1		
E_{1u}	2	$\sqrt{2}$	$-\sqrt{2}$	0	−2	0	0	−2	$-\sqrt{2}$	$\sqrt{2}$	0	2	0	0		
E_{2u}	2	0	0	−2	2	0	0	−2	0	0	2	−2	0	0	(x, y)	
E_{3u}	2	$-\sqrt{2}$	$\sqrt{2}$	0	−2	0	0	−2	$\sqrt{2}$	$-\sqrt{2}$	0	2	0	0		

C.6 The D_{nd} groups

D_{2d}	E	$2S_4$	C_2	$2C_2'$	$2\sigma_d$		
A_1	1	1	1	1	1		$x^2+y^2,\ z^2$
A_2	1	1	1	−1	−1	R_z	
B_1	1	−1	1	1	−1		x^2-y^2
B_2	1	−1	1	−1	1	z	xy
E	2	0	−2	0	0	$(x,\ y)$; $(R_x,\ R_y)$	(xz, yz)

D_{3d}	E	$2C$	$3C_2$	i	$2S_6$	$3\sigma_d$		
A_{1g}	1	1	1	1	1	1		$x^2+y^2,\ z^2$
A_{2g}	1	1	−1	1	1	−1	R_z	
E_g	2	−1	0	2	−1	0	$(R_x,\ R_y)$	$(x^2-y^2,\ xy)$, $(xz,\ yz)$
A_{1u}	1	1	1	−1	−1	−1		
A_{2u}	1	1	−1	−1	−1	1	z	
E_u	2	−1	0	−2	1	0	$(x,\ y)$	

D_{4d}	E	$2S_8$	$2C_4$	$2S_8^3$	C_2	$4C_2'$	$4\sigma_d$		
A_1	1	1	1	1	1	1	1		$x^2+y^2,\ z^2$
A_2	1	1	1	1	1	−1	−1	R_z	
B_1	1	−1	1	−1	1	1	−1		
B_2	1	−1	1	−1	1	−1	1	z	
E_1	2	$\sqrt{2}$	0	$-\sqrt{2}$	−2	0	0	$(x,\ y)$	
E_2	2	0	−2	0	2	0	0		$(x^2-y^2,\ xy)$
E_3	2	$-\sqrt{2}$	0	$\sqrt{2}$	−2	0	0	$(R_x,\ R_y)$	$(xz,\ yz)$

D_{5d}	E	$2C_5$	$2C_3^2$	$5C_2$	i	$2S_{10}^3$	$2S_{10}$	$5\sigma_d$		
A_{1g}	1	1	1	1	1	1	1	1		x^2+y^2, z^2
A_{2g}	1	1	1	−1	1	1	1	−1	R_z	
E_{1g}	2	2 cos 72°	2 cos 144°	0	2	cos 72°	2 cos 144°	0	(R_x, R_y)	(xz, yz)
E_{2g}	2	2 cos 144°	2 cos 72°	0	2	2 cos 144°	2 cos 72°	0		(x^2-y^2, xy)
A_{1u}	1	1	1	1	−1	−1	−1	−1		
A_{2u}	1	1	1	−1	−1	−1	−1	1	z	
E_{1u}	2	2 cos 72°	2 cos 44°	−2	−2	−2 cos 72°	−2 cos 14°	0	(x, y)	
E_{2u}	2	2 cos 14°	2 cos 72°	0	−2	−2 cos 144°	−2 cos 72°	0		

D_{6d}	E	$2S_{12}$	$2C_6$	$2S_4$	$2C_3$	$2S_{12}^5$	C_2	$6C_2'$	$6\sigma_d$		
A_1	1	1	1	1	1	1	1	1	1		x^2+y^2, z^2
A_2	1	1	1	1	1	1	1	−1	−1	R_z	
B_1	1	−1	1	−1	1	−1	1	1	−1		
B_2	1	−1	1	−1	1	−1	1	−1	1	z	
E_1	2	$\sqrt{3}$	1	0	−1	$-\sqrt{3}$	−2	0	0	(x, y)	
E_2	2	1	−1	−2	−1	1	2	0	0		(x^2-y^2, xy)
E_3	2	0	−2	0	2	0	−2	0	0		
E_4	2	−1	−1	2	−1	−1	2	0	0		
E_5	2	$-\sqrt{3}$	1	0	−1	$\sqrt{3}$	−2	0	0	(R_x, R_y)	(xz, yz)

C.7 The S_n groups

S_4	E	S_4	C_2	S_4^3		
A	1	1	1	1	R_z	x^2+y^2, z^2
B	1	−1	1	−1	z	x^2-y^2, xy
E	$\left\{\begin{matrix}1\\1\end{matrix}\right.$	$\begin{matrix}i\\-i\end{matrix}$	$\begin{matrix}-1\\-1\end{matrix}$	$\left.\begin{matrix}-i\\i\end{matrix}\right\}$	$(x, y); (R_x, R_y)$	(xz, yz)

S_6	E	C_3	C_3^2	i	S_6^5	S_6		$\varepsilon = \exp(2\pi i/3)$
A_g	1	1	1	1	1	1	R_z	$x^2 + y^2, z^2$
E_g	1	ε	ε^*	1	ε	ε^*	(R_x, R_y)	$(x^2 - y^2, xy)$;
	1	ε^*	ε	1	ε^*	ε		(xz, yz)
A_u	1	1	1	−1	−1	−1	z	
E_u	1	ε	ε^*	−1	$-\varepsilon$	$-\varepsilon^*$	(x, y)	
	1	ε^*	ε	−1	$-\varepsilon^*$	$-\varepsilon$		

S_8	E	S_8	C_4	S_8^3	C_2	S_8^5	C_4^3	S_8^7		$\varepsilon = \exp(2\pi i/8)$
A	1	1	1	1	1	1	1	1	R_z	$x^2 + y^2, z^2$
B	1	−1	1	−1	1	−1	1	−1	z	
E_1	1	ε	i	$-\varepsilon^*$	−1	$-\varepsilon$	$-i$	ε^*	(x, y);	
	1	ε^*	$-i$	$-\varepsilon$	−1	$-\varepsilon^*$	i	ε	(R_x, R_y)	
E_2	1	i	−1	$-i$	1	i	−1	$-i$		$(x^2 - y^2, xy)$
	1	$-i$	−1	i	1	$-i$	−1	i		
E_3	1	$-\varepsilon^*$	$-i$	ε	−1	ε^*	i	$-\varepsilon$		(xz, yz)
	1	$-\varepsilon$	i	ε^*	−1	ε	$-i$	$-\varepsilon^*$		

C.8 The cubic groups

T	E	$4C_3$	$4C_3^2$	$3C_2$		$\varepsilon = \exp(2\pi i/3)$
A	1	1	1	1		$x^2 + y^2 + z^2$
E	1	ε	ε^*	1		$(2z^2 - x^2 - y^2,$
	1	ε^*	ε	1		$x^2 - y^2)$
T	3	0	0	−1	(R_x, R_y, R_z); (x, y, z)	(xy, xz, yz)

T_h	E	$4C_3$	$4C_3^2$	$3C_2$	i	$4S_6$	$4S_6^5$	$3\sigma_h$		$\varepsilon = \exp(2\pi i/3)$
A_g	1	1	1	1	1	1	1	1		$x^2 + y^2 + z^2$
A_u	1	1	1	1	−1	−1	−1	−1		
E_g	1	ε	ε^*	1	1	ε	ε^*	1		$(2z^2 - x^2 - y^2,$
	1	ε^*	ε	1	1	ε^*	ε	1		$x^2 - y^2)$
E_u	1	ε	ε^*	1	−1	$-\varepsilon$	$-\varepsilon^*$	−1		
	1	ε^*	ε	1	−1	$-\varepsilon^*$	$-\varepsilon$	−1		
T_g	3	0	0	−1	−3	0	0	−1	(R_x, R_y, R_z)	(xz, yz, xy)
T_u	3	0	0	−1	−3	0	0	1	(x, y, z)	

T_d	E	$8C_3$	$3C_2$	$6S_4$	$6\sigma_d$		
A_1	1	1	1	1	1		$x^2+y^2+z^2$
A_2	1	1	1	−1	−1		
E	2	−1	2	0	0		$(2z^2-x^2-y^2,$ $x^2-y^2)$
T_1	3	0	−1	1	−1	(R_x, R_y, R_z)	
T_2	3	0	−1	−1	1	(x, y, z)	(xy, xz, yz)

O	E	$6C_4$	$3C_2(=C_4^2)$	$8C_3$	$6C_2$		
A_1	1	1	1	1	1		$x^2+y^2+z^2$
A_2	1	−1	1	1	−1		
E	2	0	2	−1	0		$(2z^2-x^2-y^2,$ $x^2-y^2)$
T_1	3	1	−1	0	−1	$(R_x, R_y, R_z); (x, y, z)$	
T_2	3	−1	−1	0	−1		(xy, xz, yz)

O_h	E	$8C_3$	$6C_2$	$6C_4$	$3C_2(=C_4^2)$	i	$6S_4$	$8S_6$	$3\sigma_h$	$6\sigma_d$		
A_{1g}	1	1	1	1	1	1	1	1	1	1		$x^2+y^2+z^2$
A_{2g}	1	1	−1	−1	1	1	−1	1	1	−1		
E_g	2	−1	0	0	2	2	0	−1	2	0		$(2z^2-x^2-y^2,$ $x^2-y^2)$
T_{1g}	3	0	−1	1	−1	3	1	0	−1	−1	(R_x, R_y, R_z)	
T_{2g}	3	0	1	−1	−1	3	−1	0	−1	1		(xz, yz, xy)
A_{1u}	1	1	1	1	1	−1	−1	−1	−1	−1		
A_{2u}	1	1	−1	−1	1	−1	1	−1	−1	1		
E_u	2	−1	0	0	2	−2	0	1	−2	0		
T_{1u}	3	0	−1	1	−1	−3	−1	0	1	1	(x, y, z)	
T_{2u}	3	0	1	−1	−1	−3	1	0	1	−1		

C.9 The groups C_∞, and $D_{\infty h}$ for linear molecules

$C_{\infty v}$	E	$2C_\infty^\Phi$	$\cdots$	$\infty\sigma_v$		
$A_1 \equiv \Sigma^+$	1	1	$\cdots$	1	z	x^2+y^2, z^2
$A_2 \equiv \Sigma^-$	1	1	$\cdots$	−1	R_z	
$E_1 \equiv \Pi$	2	$2\cos\Phi$	$\cdots$	0	$(x, y); (R_x, R_y)$	(xz, yz)
$E_2 \equiv \Delta$	2	$2\cos 2\Phi$	$\cdots$	0		(x^2-y^2, xy)
$E_3 \equiv \Phi$	2	$2\cos 3\Phi$	$\cdots$	0		
$\cdots$	$\cdots$	$\cdots$	$\cdots$	$\cdots$		

$D_{\infty h}$	E	$2C_\infty^\Phi$	$\cdots$	$\infty\sigma_v$	i	$2S_\infty^\Phi$	$\cdots$	∞C_2		
Σ_g^+	1	1	$\cdots$	1	1	1	$\cdots$	1		x^2+y^2, z^2
Σ_g^-	1	1	$\cdots$	−1	1	1	$\cdots$	−1	R_z	
Π_g	2	$2\cos\Phi$	$\cdots$	0	2	$-2\cos\Phi$	$\cdots$	0	(R_x, R_y)	(xz, yz)
Δ_g	2	$2\cos 2\Phi$	$\cdots$	0	2	$2\cos 2\Phi$	$\cdots$	0		(x^2-y^2, xy)
$\cdots$	$\cdots$	$\cdots$	$\cdots$	$\cdots$	$\cdots$	$\cdots$	$\cdots$	$\cdots$		
Σ_u^+	1	1	$\cdots$	1	−1	−1	$\cdots$	−1	z	
Σ_u^-	1	1	$\cdots$	−1	−1	−1	$\cdots$	1		
Π_u	2	$2\cos\Phi$	$\cdots$	0	−2	$2\cos\Phi$	$\cdots$	0	(x, y)	
Δ_u	2	$2\cos 2\Phi$	$\cdots$	0	−2	$-2\cos 2\Phi$	$\cdots$	0		
$\cdots$	$\cdots$	$\cdots$	$\cdots$	$\cdots$	$\cdots$	$\cdots$	$\cdots$	$\cdots$		

C.10 The icosahedral groups[1]

I_h	E	$12C_5$	$12C_5^2$	$20C_3$	$15C_3$	i	$12S_{10}$	$12S_{10}^3$	$20S_6$	15σ		
A_g	1	1	1	1	1	1	1	1	1	1		$x^2+y^2+z^2$
T_{1g}	3	$\frac{1}{2}(1+\sqrt{5})$	$\frac{1}{2}(1-\sqrt{5})$	0	−1	3	$\frac{1}{2}(1-\sqrt{5})$	$\frac{1}{2}(1+\sqrt{5})$	0	−1	(R_x, R_y, R_z)	
T_{2g}	3	$\frac{1}{2}(1-\sqrt{5})$	$\frac{1}{2}(1+\sqrt{5})$	0	−1	3	$\frac{1}{2}(1+\sqrt{5})$	$\frac{1}{2}(1-\sqrt{5})$	0	−1		
G_g	4	−1	−1	1	0	4	−1	−1	1	0		
H_g	5	0	0	−1	1	5	0	0	−1	1		$(2z^2-x^2-y^2,$ $x^2-y^2,$ $xy, yz, zx)$
A_u	1	1	1	1	1	−1	−1	−1	−1	−1		
T_{1u}	3	$\frac{1}{2}(1+\sqrt{5})$	$\frac{1}{2}(1-\sqrt{5})$	0	−1	−3	$-\frac{1}{2}(1-\sqrt{5})$	$-\frac{1}{2}(1+\sqrt{5})$	0	1	(x, y, z)	
T_{2u}	3	$\frac{1}{2}(1-\sqrt{5})$	$\frac{1}{2}(1+\sqrt{5})$	0	−1	−3	$-\frac{1}{2}(1+\sqrt{5})$	$-\frac{1}{2}(1-\sqrt{5})$	0	1		
G_u	4	−1	−1	1	0	−4	1	1	−1	0		
H_u	5	0	0	−1	1	−5	0	0	1	−1		

[1]For the pure rotation group 1, the outlined section in the upper left is the character table; the g subscripts should of course, be dropped and (x, y, z) assigned to the T_1 representation.

Appendix D

Introduction to Fourier Series, the Fourier Transform, and the Fast Fourier Transform Algorithm

In this appendix, a short review of Fourier series, the Fourier transforms, and methods to compute Fourier transforms is presented. In view that a large portion of this monograph deals with "Fourier Transform Infrared Spectroscopy" (FTIR), the review of this basic mathematical process is more than appropriate.

D.1 Data Domains

For researchers without engineering or physics background, the concept of Fourier transforms may at first appear foreign. Thus, an effort has been made to present this field on a conceptual basis. This is attempted using an example that is easy to absorb. Consider, for example, the musical note "A," the 440 Hz sound produced by a violin's "A" string. A plot of such a sound wave shown in Figure D.1(a) shows the amplitude of the sound wave as a function of time; since this signal varies periodically with time, it can be represented as a sine or cosine wave with frequency $\nu = 440$ Hz and an intensity (amplitude) S. However, one could depict the same information by plotting the intensity S in a graph that has frequency on the abscissa. This results in a graph shown in Figure D.1(b), which indicates that all frequencies, except 440 Hz, have no amplitude, or are not represented, while the frequency at 440 Hz appears with amplitude S. The "frequency domain" representations shown in (b) and (d) have the advantage that a sound signal that contains more than one frequency becomes more obvious to interpret. This is shown in Figure D.1(c) and (d). Here, a signal consisting of two sine waves, at 440 and 880 Hz, are shown superimposed in (c). This more complex and harder to interpret amplitude *vs.* time representation is much clearer when depicted in the amplitudes *vs.* frequency plot shown in (d).

The two representations depicted above are mathematically related by a Fourier transform. To introduce the concept of Fourier transforms, it is best to first discuss the concept of Fourier series, and transition to Fourier transforms from there.

Modern Vibrational Spectroscopy and Micro-Spectroscopy: Theory, Instrumentation and Biomedical Applications, First Edition.
Max Diem.

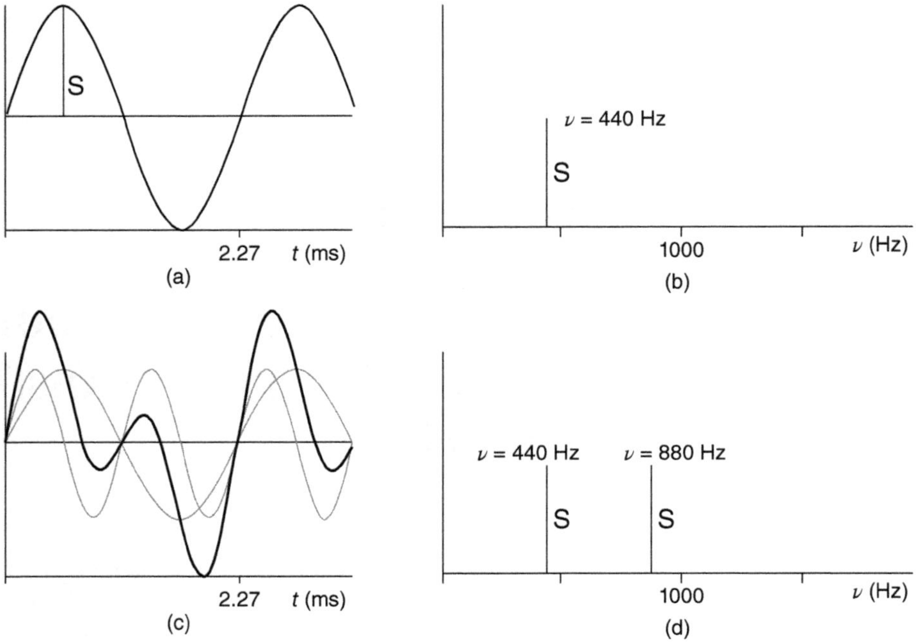

Figure D.1 *Example of representing data in the time and frequency domain*

D.2 Fourier Series

Fourier series expansion or harmonic analysis extracts appropriately weighted harmonic components from a general periodic waveform. Any function $f(x)$, which is periodic between $-\pi$ and $+\pi$(or L to $+L$) can be expanded in this interval by a Fourier series. The Fourier series expansion of the function $f(x)$ is defined by

$$f(x) = \sum_{n=0}^{\infty} c(n)\, e^{in\pi x/L} \tag{D.1}$$

where the expansion coefficients c_n are given by

$$c_n = \frac{1}{2L} \int_{-L}^{+L} f(x) e^{in\pi x/L}\, dx \tag{D.2}$$

In the case of real functions, the expansion given in Eq. D.1 takes the form

$$f(x) = a_0 + \sum_{n=0}^{\infty} a_n \cos\left(\frac{n\pi x}{L}\right) + b_n \sin\left(\frac{n\pi x}{L}\right) \tag{D.3}$$

with the real expansion coefficients given by

$$a_n = \frac{1}{2L}\int_{-L}^{+L} f(x) \cos\left(\frac{n\pi x}{L}\right) dx \quad \text{and}$$

$$b_n = \frac{1}{2L}\int_{-L}^{+L} f(x) \sin\left(\frac{n\pi x}{L}\right) dx \tag{D.4}$$

An example of Fourier series expansion is presented next. Consider a square wave, given by Eq. D.5:

$$f(x) = \begin{cases} +1 & \text{for} \quad 0 \leq x \leq \pi \\ -1 & \text{for} \quad \pi \leq x \leq 2\pi \end{cases} \tag{D.5}$$

This function is shown as the heavy black trace in Figure D.2, and is assumed to repeat periodically. Substituting Eq. D.5 into the equations for the expansion coefficients a_n and b_n (Eq. D.4) and integrating from 0 to 2π, one finds that all the terms a_n are zero:

$$a_0 = \left(\frac{1}{\pi}\right) \int_0^{2\pi} f(x)\,dx = \left(\frac{1}{\pi}\right) \left\{ \int_0^{\pi} (+1)\,dx + \int_{\pi}^{2\pi} (-1)dx \right\} = 0 \tag{D.6}$$

$$a_n = \left(\frac{1}{\pi}\right) \int_0^{2\pi} f(x)\cos(nx)dx$$

$$= \left(\frac{1}{\pi}\right) \left\{ \int_0^{\pi} \cos(nx)\,dx + \int_{\pi}^{2\pi} -\cos(nx)dx \right\} = 0 \tag{D.7}$$

whereas the terms b_n assume the values:

$$b_n = \left(\frac{1}{\pi}\right) \int_0^{2\pi} f(x)\sin(nx)dx$$

$$= \left(\frac{1}{\pi}\right) \left\{ \int_0^{\pi} \sin(nx)\,dx + \int_{\pi}^{2\pi} -\sin(nx)dx \right\} = 0 \tag{D.8}$$

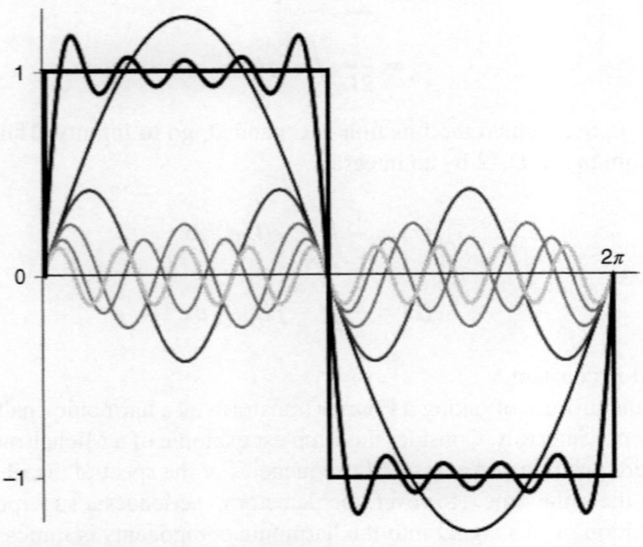

Figure D.2 *Approximation of a square wave function (heavy black line) by a scaled sum of harmonic frequencies (n = 1–9). See text for details*

for even values of *n*, and

$$= \left(\frac{1}{\pi}\right)\left\{\int_0^{\pi} \sin(nx)\,dx + \int_{\pi}^{2\pi} -\sin(nx)dx\right\} = \left(\frac{1}{\pi}\right)\left\{\frac{2}{n}+\frac{2}{n}\right\}$$

$$= \frac{4}{\pi n} \tag{D.9}$$

for odd values of n. Thus, a square wave can be expanded into an infinite series of all odd harmonics, scaled by $1/n$:

$$f(x) = \left(\frac{4}{\pi}\right)\sum_{n=1,3,5\,\ldots}^{\infty} \frac{1}{n}\sin nx \tag{D.10}$$

This is shown in Figure D.2 for $n = 1 - 9$. It is obvious from this graph that an increasing number of higher harmonics improves the fit between the square wave and the sum of all the co-added harmonics. Similar expansions can be carried out to approximate any periodic function (such as a sawtooth function and other, more complex functions) by a sum of harmonics.

D.3 Fourier Transform

The concept of Fourier series is quite useful for introducing the concept of harmonic analysis, and the concept of the discrete Fourier transform to be discussed later. Next, the standard concepts of the Fourier transform are introduced. A logical connection between the principles of Fourier expansions and Fourier transforms can be made by substituting

$$k = n\pi/L \tag{D.11}$$

into Eqs. D.1 and D.2 to obtain

$$f(x) = \sum_{n=-\infty}^{\infty} c(n)e^{ikx} \tag{D.12}$$

$$c_n = \frac{1}{2L}\int_{L}^{+L} f(x)e^{ikx}dx \tag{D.13}$$

and letting the interval, *L*, over which the function is expanded, go to infinity. Thus, *k* gets very small, and one can substitute the sum in Eq. D.12 by an integral:

$$f(x) = \frac{1}{c}\int_{-\infty}^{+\infty} c(k)e^{ikx}dx \tag{D.14}$$

$$c(k) = \frac{1}{c}\int_{-\infty}^{+\infty} f(x)e^{ikx}dk \tag{D.15}$$

where *c* is a normalization constant.

Thus, one may view the process of taking a Fourier transform as a harmonic analysis with infinitely small increments in the frequency intervals. Consider the simplest example of a Michelson interferometer, with the movable mirror at its zero position and at rest. All frequencies of the spectral distribution $S(\tilde{\upsilon})$ (see Eq. 3.25) arrive at the detector at the same time. However, the detector experiences a superposition of all amplitudes, and a unique decomposition of this signal into the harmonic components is impossible. Moving the mirror by an amount *dx* gives a different superposition of all amplitudes, since now a different set of wavelengths undergo destructive or constructive interference. If one collects the interferogram at infinitely small-spaced mirror displacements, the spectral distribution $S(\tilde{\upsilon})$ and the interferogram $J(x)$ are just the Fourier transforms of each other (cf. Eqs. 3.28 and (3.29)).

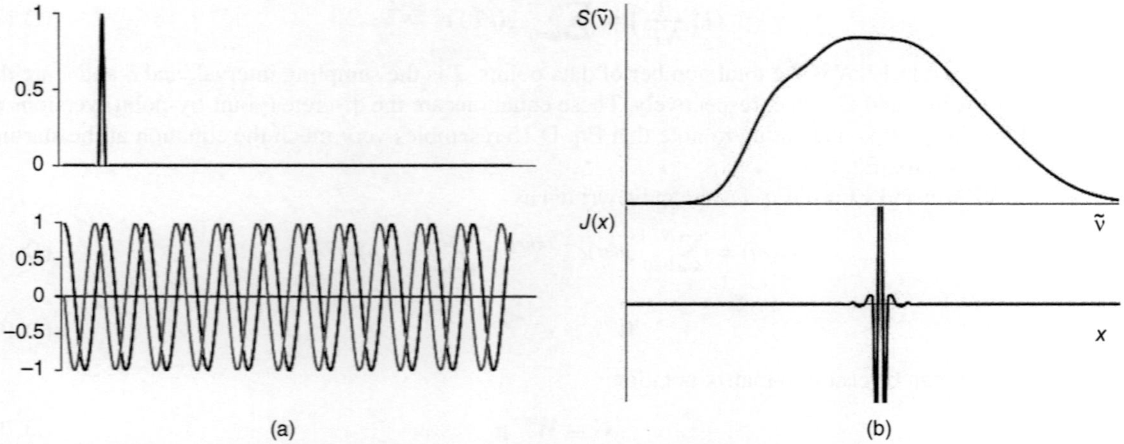

Figure D.3 *Examples of Fourier transforms (FTs). (a) The FT of a delta function is a cosine function (black: real part, gray: imaginary part). Note that the real and imaginary parts are 90° out of phase. (b) The FT of a broad spectral distribution is a sharp, symmetric "interferogram" J(x). The FT's shown here were calculated via the Microsoft Excel fast FT implementation. The interferogram shown in (b) was unfolded*

Some very important properties of Fourier transforms are discussed next. The Fourier transform of a shifted δ-function, $\delta(\tilde{\nu} - \tilde{\nu}_0)$, for example, an infinitely narrow-band (monochromatic) light source shown in Figure D.3(a), is a periodic (sine or cosine) function with frequency $\tilde{\nu}_0$. This was also demonstrated in Figure D.1. This fact is used, as discussed in Section 3.3 to calibrate the path difference between the moving and fixed mirror in an interferometer, by using a reference laser and counting laser fringes.

The Fourier transform of a broad spectral distribution, as shown in Figure D.3(b) gives a very sharp interferogram with a strong "center burst" at the zero path difference (ZPD) with only a few additional frequency components aside from the center burst. If narrow spectral features are superimposed on this broad intensity distribution, an interferogram similar to the one shown in Figure 3.7 is obtained. This interferogram (which is presented inverted, that is, with the ZPD peak pointing down) shows the typical sharp features of a broad intensity distribution, but also many fringes away from the ZPD peak, indicating sharp spectral components. It is also noteworthy to realize that the interferogram in Figure 3.7 is not completely symmetric. The aspects of dealing with "chirped" (asymmetric) interferograms is discussed in Section 3.3.

D.4 Discrete and Fast Fourier Transform Algorithms

The equations given so far for Fourier transform pairs describe the process and concepts but not a practical method to carry out Fourier transforms of a set of data. The transformation from interferogram to frequency (wavenumber) domain is presently performed by an algorithm known as fast Fourier transform (FFT).

FFT is based on the principles of the discrete Fourier transform: since the interferogram is not sampled continuously, but at discrete times, the interferogram is obtained as a one-dimensional vector of digital values. For such a situation, the process of taking the Fourier transform can be written as

$$g(kT) = \frac{1}{N}\sum\nolimits_{n=0}^{N-1} G\left(\frac{n}{NT}\right) e^{\frac{2\pi ink}{N}} \tag{D.16}$$

$$G\left(\frac{n}{NT}\right) = \sum_{n=0}^{N-1} g(kT)\, e^{-\frac{2\pi ink}{N}} \tag{D.17}$$

In Eqs. D.16 and D.17, N is the total number of data points, T is the sampling interval, and n and k are the running indexes in g and G space, respectively. These equations are the discrete (point-by-point) versions of Eqs. D.14 and D.15. It is interesting to note that Eq. D.16 resembles very much the equation at the starting point of this appendix, Eq. D.1.

Setting $n/NT = m$ and $kT = p$, Eq. D.16 can be written as

$$G(m) = \sum_{k=0}^{N-1} g(p)e^{-\,2\pi imp} = \sum_{k=0}^{N-1} g(p)W^{mp} \tag{D.18}$$

where

$$W^{mp} = e^{-\,2\pi imp} \tag{D.19}$$

Equation D.19 can be cast into matrix notation:

$$\mathbf{G} = \mathbf{W}^{mp}\mathbf{g} \tag{D.20}$$

Thus, the computation of a discrete Fourier transform from g (interferogram) to G (spectrum) space is reduced to computing a (complex) transformation matrix $\mathbf{W}^{mp}$ and multiplying the vector of discrete points with this matrix. For typical infrared spectroscopic applications, a sample set may consist of 8 K data points (i.e., 8192 points in the $g(p)$ vector). The Fourier transform operation, according to Eq. D.18, requires for each of the 8192 points in G space an 8 K × 8 K matrix to be multiplied by an 8 K vector. Such matrix manipulations are slow, since for each data point, 8 K multiplications and 8 K additions are required.

This problem was alleviated by the FFT algorithm, developed by Cooley and Tukey [1], which avoids the problem of the large number of multiplications and additions by factoring the W matrix into sparse matrices that have many zero elements. It can be shown that such a factoring is always possible, but the factoring will require reordering the entries in the G and g vectors. Furthermore, the FFT algorithm only works for data vectors that have integer powers of 2 (256, 512, 1024, etc.) entries. A detailed discussion of the FFT algorithm is beyond the scope of this chapter, and the reader is referred to Ref. [2]. With increased computational power of modern desktop machines, and implementation of FFT routines in the MATLAB environment, FFT computations can be carried out for a 1024-point data vector in a few milliseconds.

A final comment is appropriate about the presentation of the FFT results, for both MATLAB and EXCEL implementations. As pointed out in the caption of Figure D.3, all Fourier transform examples shown in this appendix as well as in Chapter 3 were carried out using Microsoft EXCEL. As a spectroscopist, one may visualize an interferogram to have the appearance as shown in Figure 3.7 or D.3(b), that is, with a distinct center burst at the ZPD, and decreasing fringe intensities away from the center burst. However, if the opposite process is performed, namely a spectrum is transformed into interferogram domain *via* MATLAB or EXCEL-based FFT routines, both the real and imaginary portion obtained look notably different, as shown in Figure D.4(a) and (b), for a 512-point FFT of a real spectrum of a human cell. Here, the "center burst" is shown on the far left of the graph, with the previous data point at the far right of the graph. When folding the interferogram that points 1–256 are moved to positions 257–512, and the points 257–512 are moved to positions 1–256, the folded interferogram shown in (c) is obtained. This form of the interferogram is what one would expect by scanning the movable mirror from point −256 to 256 with respect to the ZPD. Note that the real part of the interferogram is symmetric about the ZPD. Panels (b) and (d) show the corresponding situations for the imaginary part of the Fourier transform. Note that the imaginary part is antisymmetric about the ZPD.

The fact that the interferograms are folded as shown in (a) and (b) requires that zero filling (see Section 3.3) is carried out by filling the zeros in the middle of the interferogram. That is, in the case of the 512-point interferogram shown in Figure D.4, 512 data are filled in starting at point 256.

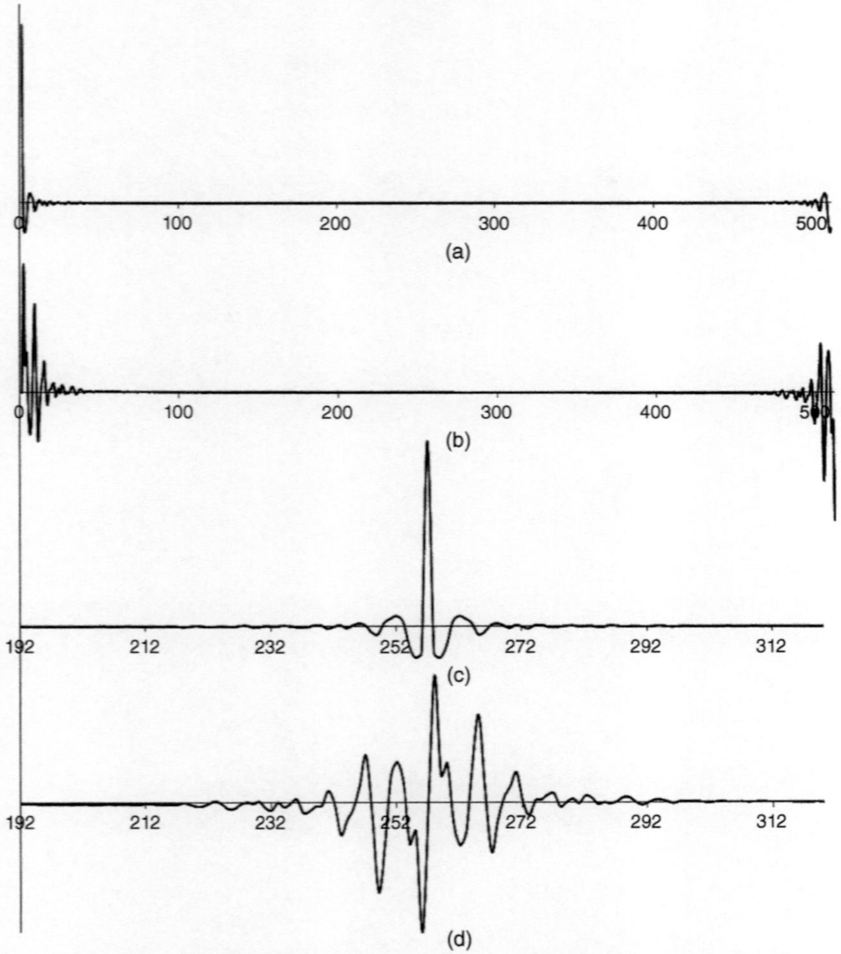

Figure D.4 *(a) Real part of a reverse transform of a spectrum back to interferogram domain. Note the unfolded representation of the interferogram. (b) Imaginary part of interferogram shown in (a). (c) Interferogram shown in (a) folded to demonstrate the symmetry of the real part of the interferogram. (d) Interferogram shown in (b) folded to demonstrate the antisymmetric nature of the imaginary part of the interferogram*

References

1. Cooley, J.W. and Tukey, J.W. (1965) An algorithm for the machine calculation of complex Fourier series. *Math. Comput.*, **19**, 297–310.
2. Brigham, E.O. (1974) *The Fast Fourier Transform*, Prentice Hall, Englewood, NJ.

Appendix E

List of Common Vibrational Group Frequencies (cm^{-1})

Wavenumber Range	Group	Description
3400	ROH	OH stretch
3330–3400	$—NH_2$	NH antisymmetric stretch
3250–3300	$—NH_2$	NH symmetric stretch
3250–3300	—NH	Amide A
3300–3330	C ≡ C—H	Acetylenic CH stretch
3050–3065	═CH	Aromatic CH stretch
2960–3020	$—CH_3$	Antisymmetric CH stretch
2910–2930	$—CH_2$	Antisymmetric CH stretch
2880–2970	$—CH_3$	Symmetric CH stretch
2850–2860	$—CH_2$	Symmetric CH stretch
3000–3030	—CH	CH stretch
2560–2590	RSH	Thiole SH stretch
2350–2600	R_2BH	BH stretch
2275–2450	R_2PH	PH stretch
2230–2300	C ≡ C	CC triple bond stretch
2230–2260	C ≡ N	CN triple bond stretch
2100–2250	R_3SiH	SiH stretch
2230–2250	$—CD_3$	Antisymmetric CD stretch
2250–2300	$—CD_2$	Antisymmetric CD stretch
2120–2130	$—CD_3$	Symmetric CD stretch
2200–2010	$—CD_2$	Symmetric CD stretch
2250	C—D	C—D stretch
1760	C═O	Organic acids

Modern Vibrational Spectroscopy and Micro-Spectroscopy: Theory, Instrumentation and Biomedical Applications, First Edition.
Max Diem.

Wavenumber Range	Group	Description
1700–1725	C═O	Ketones
1690–1710	C═O	C═O in DNA bases
1640–1690	C═O	Amide I in —NH—CO—
1650–1660	C═N	—
1640–1660	C═C	Alkenes, aromatics such as DNA bases
1630–1690	C═O	Amide I′ in —ND—CO—
1620–1630	C═O	Amide I in tertiary amides
1610–1630	C═O	Amide I in —NH—^{13}CO—
1595–1600	—CO_z—	Antisymmetric carboxylate stretch
1450–1600	N═O	NO stretch in organic nitrates
1440–1470	—CH_3	Antisymmetric deformation
1440–1470	—CH_z	Symmetric deformation
1420–1430	C—OH	Carbon–oxygen stretch in organic acids
1390–1410	—CO_z—	Symmetric carboxylate stretch
1370–1390	—CH_3	Symmetric deformation
1350	—CF_3	—
1250–1340	(see text)	Amide III
1250–1330	C ≡ C—H	Methine deformation
1250–1265	—COH	COH deformation, organic acids
1235	PO_2^{-}	Antisymmetric PO stretch in DNA, RNA, phospholipids
1090	PO_2^{-}	Symmetric PO stretch in DNA, RNA, phospholipids
1100–1200	S═O	Sulfoxides, sulfonic acids
1060–1120	C—F	—
1100–1300	C—O	—
950–1150	C—C	—
1040–1070	S═O	—
1060	C═S	—
1000–1030	CCH	Aromatic in-plane C—H deformation
992	—	Benzene ring breathing
830–930	C—O—C	Symmetric stretch, ethers
800–830	C—O—P—O—C	Symmetric and antisymmetric phosphodiester stretch
720–730	C—Cl	—
705	C—Cl_2	Symmetric stretch
665	C—Cl_3	Symmetric stretch
700–750	—	Amide V N—H out-of-plane deformation
640–650	C—Br	—
630–750		Amide IV O—C—N deformation
610	C—Br_2	Symmetric stretch
~600		Amide VI C—O out-of-plane deformation
540	C—Br_3	Symmetric stretch
510–525	—S—S—	S—S stretch
520	C—I	—

Appendix F

Infrared and Raman Spectra of Selected Cellular Components

In this appendix, the infrared and Raman spectra of the following biochemical components of cells and tissues are shown:

1. Nucleic acids: DNA (Figure F.1) RNA (Figure F.2)
2. Proteins: Albumin (mostly α-helical) (Figure F.3) γ-Globulin (mostly β-sheet) (Figure F.4) Collagen (triple helical) (Figure F.5)
3. Lipids (Figure F.6) Phospholipids (Figure F.7)
4. Carbohydrate: Glycogen (Figure F.8).

In each panel, the top trace represents the infrared spectrum, which is represented as absorption spectra with amplitudes between 0 and 0.5 AU. The middle trace represents the second derivative of the corresponding infrared spectrum, after multiplication by (−1). The bottom trace represents the Raman spectra, which are presented in arbitrary intensity units.

All spectra were acquired microscopically from thin films cast on CaF_2 windows after dissolving the samples in appropriate solvents, and air drying. Infrared data were collected at 4 cm^{-1} spectral resolution from sample spots measuring about 40 μm on edge, using a PerkinElmer (Shelton, CT) Spotlight 300 microspectrometer. 64 or 128 interferograms were co-added to improve the signal-to-noise ratio. Raman data were collected using a WiTEC (Ulm, Germany) confocal Raman microspectrometer CRM 200 using either 514.5 or 632.8 nm excitation (about 15 mW power at the sample). The spectral resolution of the Raman spectra, dispersed by a 30-cm focal length monochromator incorporating a 600 mm^{-1} grating, varies between about 3 and 5 cm^{-1}. Data acquisition time for each spectrum was between 3 and 10 seconds.

The author wishes to thank Dr Tatyana Chernenko and Dr Christian Matthäus from the Laboratory for Spectral Diagnosis, Department of Chemistry and Chemical Biology, Northeastern University in Boston for collecting and presenting these reference data.

Modern Vibrational Spectroscopy and Micro-Spectroscopy: Theory, Instrumentation and Biomedical Applications, First Edition.
Max Diem.

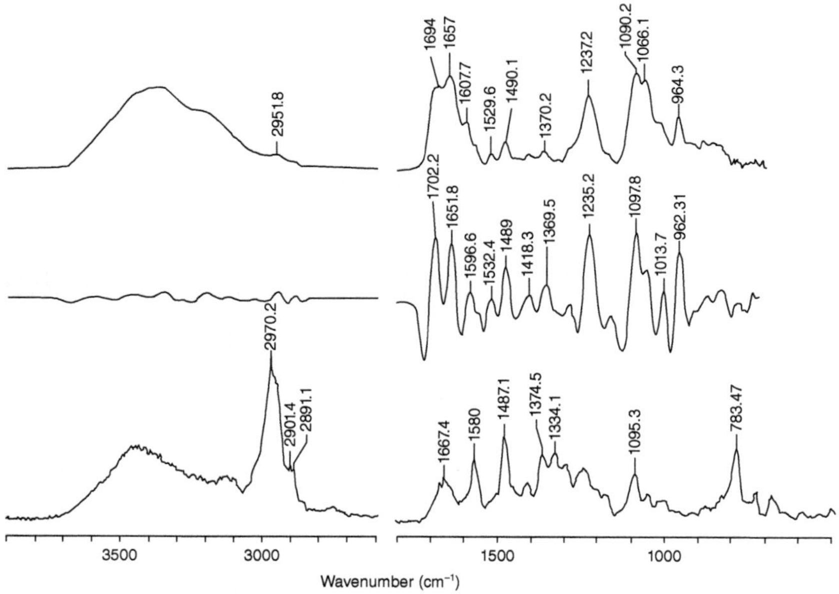

Figure F.1 *Infrared (top), second derivative infrared (middle), and Raman spectra of DNA (bottom)*

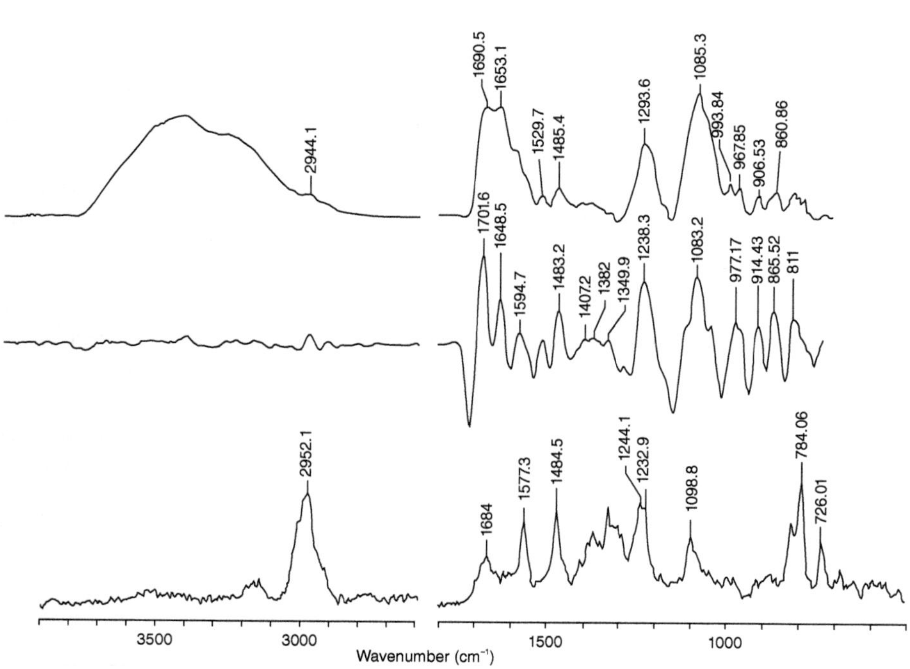

Figure F.2 *Infrared (top), second derivative infrared (middle), and Raman spectra of RNA (bottom)*

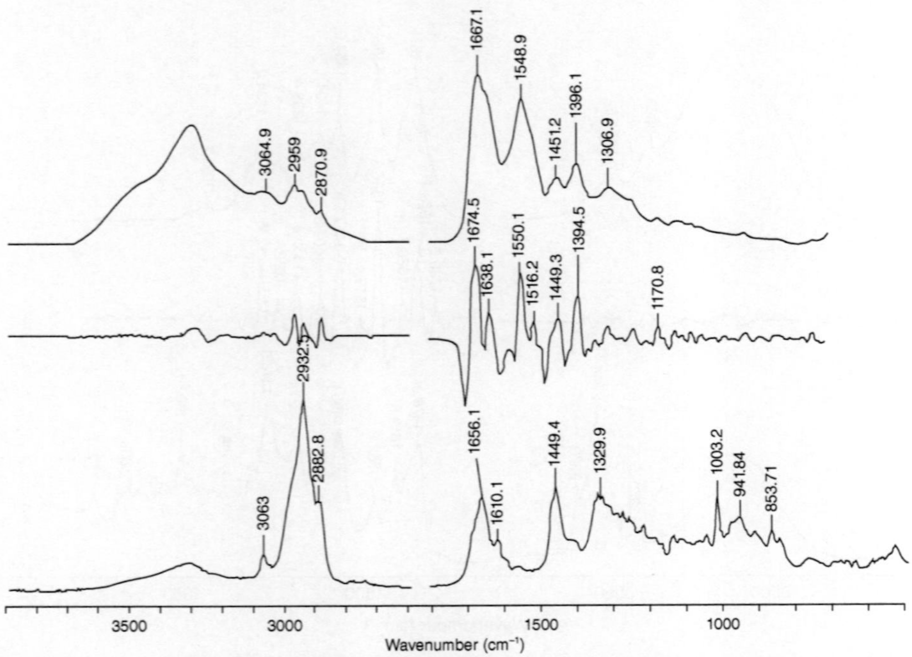

Figure F.3 *Infrared (top), second derivative infrared (middle), and Raman spectra (bottom) of albumin (mostly α-helical protein)*

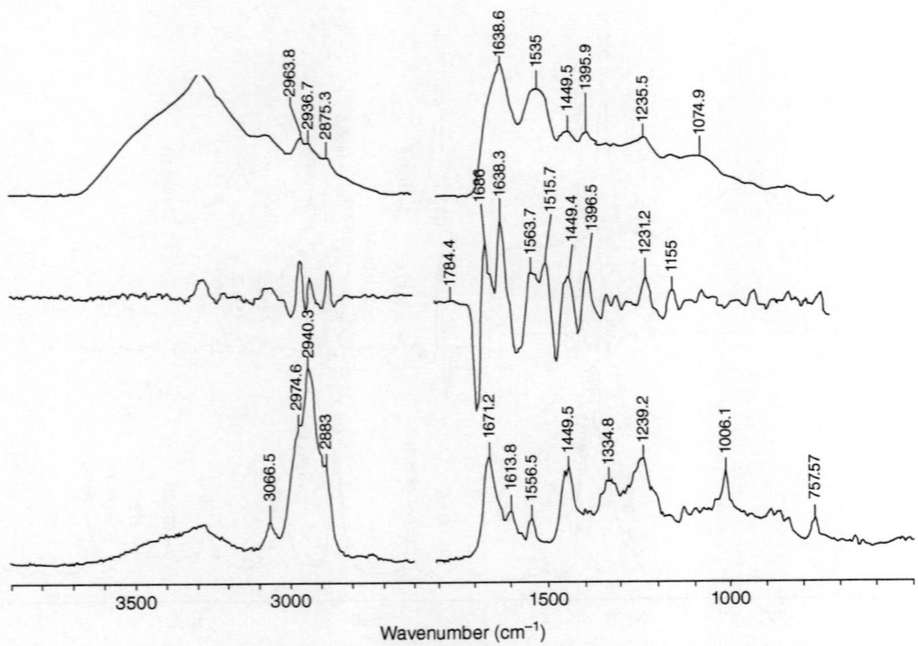

Figure F.4 *Infrared (top), second derivative infrared (middle), and Raman spectra (bottom) of γ-globulin (mostly β-sheet protein)*

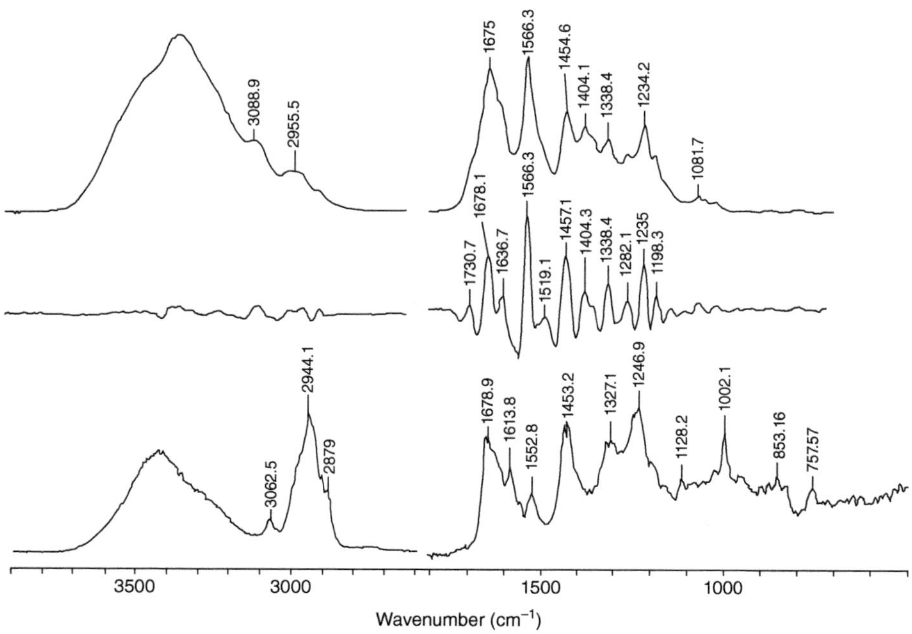

Figure F.5 *Infrared (top), second derivative infrared (middle), and Raman spectra (bottom) of collagen (triple helical protein)*

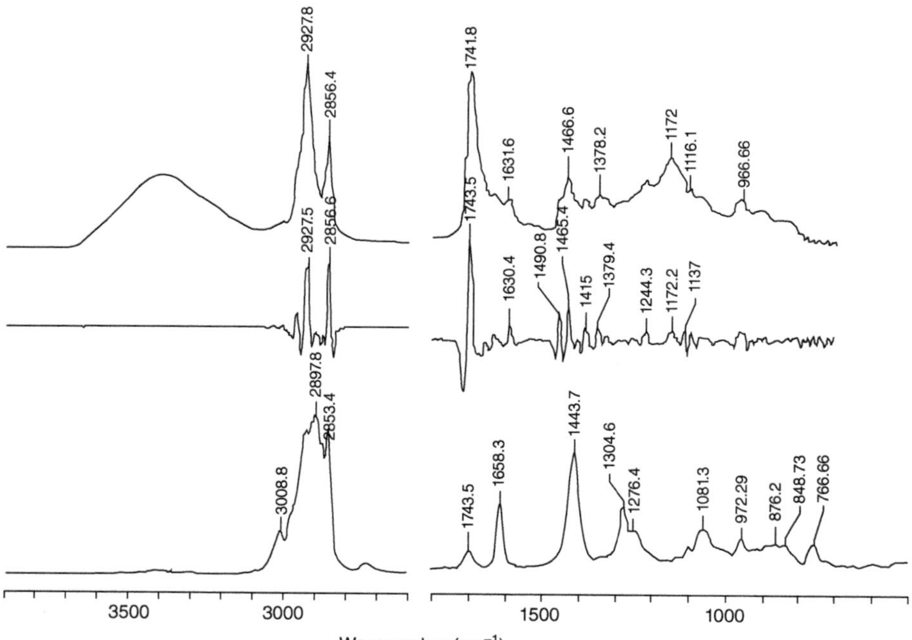

Figure F.6 *Infrared (top), second derivative infrared (middle), and Raman spectra (bottom) of a lipid*

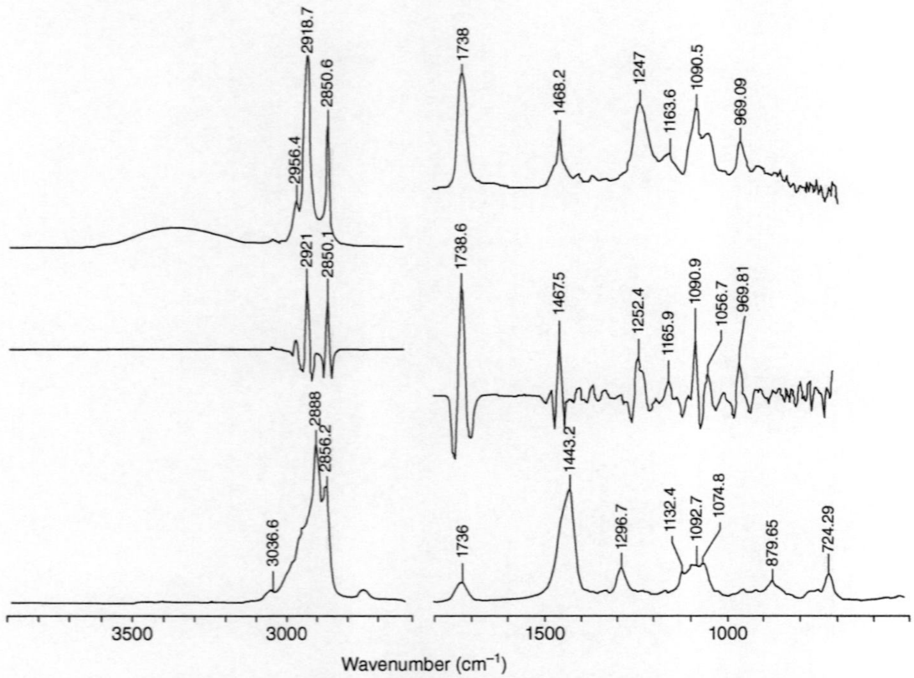

Figure F.7 *Infrared (top), second derivative infrared (middle), and Raman spectra (bottom) of a phospholipid*

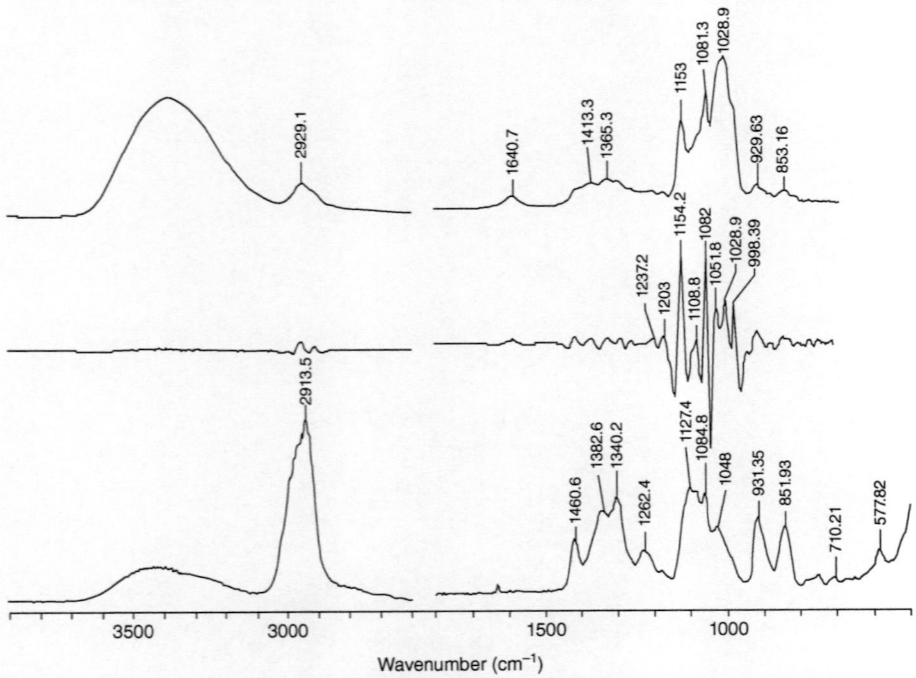

Figure F.8 *Infrared (top), second derivative infrared (middle), and Raman spectra (bottom) of a glycogen*

Index

Modern Vibrational Spectroscopy and Micro-Spectroscopy: Theory, Instrumentation and Biomedical Applications, First Edition.
Max Diem.

www.ingramcontent.com/pod-product-compliance
Ingram Content Group UK Ltd.
Pitfield, Milton Keynes, MK11 3LW, UK
UKHW051039091026
15624UKWH00011B/575